Immunology of Type 1 Diabetes
Second Edition

Recent Volumes in this Series

Volume 506
LACRIMAL GLAND, TEAR FILM, AND DRY EYE SYNDROME 3: BASIC SCIENCE AND CLINICAL RELEVANCE
Edited by David A. Sullivan, Michael E. Stern, Kazuo Tsubota, Darlene A. Dartt, Rose M. Sullivan, and B. Britt Bromberg

Volume 507
EICOSANOIDS AND OTHER BIOACTIVE LIPIDS IN CANCER, INFLAMMATION, AND RADIATION INJURY 5
Edited by Kenneth V. Honn, Lawrence J. Marnett, Santosh Nigam, and Charles Serhan

Volume 508
SENSORIMOTOR CONTROL OF MOVEMENT AND POSTURE
Edited by Simon C. Gandevia, Uwe Proske and Douglas G. Stuart

Volume 509
IRON CHELATION THERAPY
Edited by Chiam Hershko

Volume 510
OXYGEN TRANSPORT TO TISSUE XXIII: OXYGEN MEASUREMENTS IN THE 21ST CENTURY: BASIC TECHNIQUES AND CLINICAL RELEVANCE
Edited by David F. Wilson, John Biaglow and Anna Pastuszko

Volume 511
PEDIATRIC GENDER ASSIGNMENT: A CRITICAL REAPPRAISAL
Edited by Stephen A. Zderic, Douglas A. Canning, Michael C. Carr and Howard McC. Snyder III

Volume 512
LYMPHOCYTE ACTIVATION AND IMMUNE REGULATION IX: LYMPHOCYTE TRAFFIC AND HOMEOSTASIS
Edited by Sudhir Gupta, Eugene Butcher and William Paul

Volume 513
MOLECULAR AND CELLULAR BIOLOGY OF NEUROPROTECTION IN THE CNS
Edited by Christian Alzheimer

Volume 514
PHOTORECEPTORS AND CALCIUM
Edited by Wolfgang Baehr and Krzysztof Palczewski

Volume 515
NEUROPILIN: FROM NERVOUS SYSTEM TO VASCULAR AND TUMOR BIOLOGY
Edited by Dominique Bagnard

Volume 516
TRIPLE REPEAT DISORDERS OF THE NERVOUS SYSTEM
Edited by Lubov T. Timchenko

Volume 517
DOPAMINERGIC NEURON TRANSPLANTATION IN THE WEAVER MOUSE MODEL OF PARKINSON'S DISEASE
Edited by Lazaros C. Triarhou

Volume 518
ADVANCES IN MALE MEDIATED DEVELOPMENTAL TOXICITY
Edited by Bernard Robaire and Barbara F. Hales

Volume 519
POLYMER DRUGS IN THE CLINICAL STAGE
Edited by Maeda, et al.

Volume 520
CYTOKINES AND CHEMOKINES IN AUTOIMMUNE DISEASE
Edited by Pere Santamaria

Volume 521
IMMUNE MECHANISMS OF PAIN AND ANALGESIA
Edited by Halina Machelska and Christoph Stein

Immunology of Type 1 Diabetes Second Edition

Edited by
George S. Eisenbarth, M.D., Ph.D.
Executive Director, Barbara Davis Center for Childhood Diseases
Professor of Pediatrics, University of Colorado Health Sciences Center
Denver, Colorado, U.S.A.

Kluwer Academic / Plenum Publishers
New York, Boston, Dordrecht, London, Moscow
Landes Bioscience / Eurekah.com
Georgetown, Texas U.S.A.

Kluwer Academic / Plenum Publishers
Landes Bioscience / Eurekah.com

Printed in the U.S.A.

Kluwer Academic / Plenum Publishers, 233 Spring Street, New York, New York, U.S.A. 10013
http://www.wkap.nl/

Please address all inquiries to the Publishers:
Eurekah.com / Landes Bioscience, 810 South Church Street, Georgetown, Texas, U.S.A. 78626
Phone: 512.863.7762; Fax: 512.863.0081
http://www.eurekah.com
http://www.landesbioscience.com

Immunology of Type 1 Diabetes, Second Edition edited by George S. Eisenbarth, Landes / Kluwer dual imprint / Landes series: Medical Intelligence Unit

ISBN: 0-306-47871-4

Library of Congress Cataloging-in-Publication Data

Immunology of type 1 diabetes / edited by George S. Eisenbarth.-- 2nd ed.
p. ; cm.
Rev ed. of: Type I diabetes. Oxford University Press, 1996.
Includes bibliographical references and index.
ISBN 0-306-47871-4
1. Diabetes--Immunological aspects. I. Title: Immunology of type one diabetes. II. Eisenbarth, George S. III. Type I diabetes.
[DNLM: 1. Diabetes Mellitus, Insulin-Dependent--immunology. WK 810 I3385 2004]
RC660.T97 2004
616.4'62079--dc22

2003021860

INTRODUCTION

Type 1 diabetes has become the most intensively studied autoimmune disorder with contributions from basic and physician scientists throughout the world. Many of the authors for the current volume are at the Barbara Davis Center for Childhood Diabetes, a clinical care and research institute devoted to type 1 diabetes. The first edition of this book, edited by Dr. Kevin Lafferty (Research Director of the Barbara Davis Center) and George Eisenbarth was published in 1996. The first edition formed the basis for the creation of a web based book of the same title that allows multiple timely revisions (see www.barbaradaviscenter.org). Unfortunately Dr. Lafferty, one of the pioneers of the immunology of diabetes, passed away in 2001. His humor, enthusiasm, and logical mind will be greatly missed by those who worked directly with him but also by the larger scientific community attempting to solve the puzzle that is type 1 diabetes. It is hoped that the solution to this puzzle will be the cure and prevention of this devastating illness.

PARTICIPANTS

Rodabe Amaria
Barbara Davis Center
for Childhood Diabetes
University of Colorado
Health Sciences Center
Denver, Colorado
U.S.A.

Donald Bellgrau
Department of Immunology
University of Colorado
Health Sciences Center
Denver, Colorado
U.S.A.

Regine Bergholdt
Steno Diabetes Center
Gentofte
Denmark

Nicholas Bright
Cambridge Institute for Medical
Research
Cambridge
United Kingdom

H. Peter Chase
Barbara Davis Center
for Childhood Diabetes
University of Colorado
Health Sciences Center
Denver, Colorado
U.S.A.

Marilyne Coulombe
Barbara Davis Center
for Childhood Diabetes
University of Colorado
Health Sciences Center
Denver, Colorado
U.S.A.

John Creemers
Department for Human Genetics
University of Leuven and Flanders
Interuniversity Institute
for Biotechnology
Herestraat, Leuven
Belgium

Dana Dabelea
University of Colorado
Health Sciences Center
Denver, Colorado
U.S.A.

George S. Eisenbarth
Barbara Davis Center
for Childhood Diabetes
University of Colorado
Health Sciences Center
Denver, Colorado
U.S.A.

Stephanie C. Eisenbarth
Immunology Department
Yale University
New Haven, Connecticut
U.S.A.

Ronald G. Gill
Barbara Davis Center
for Childhood Diabetes
University of Colorado
Health Sciences Center
Denver, Colorado
U.S.A.

Anthony R. Hayward
National Institutes of Health
Bethesda, Maryland
U.S.A.

Peter Heding
Steno Diabetes Center
Gentofte
Denmark

John Hutton
Barbara Davis Center
for Childhood Diabetes
University of Colorado
Health Sciences Center
Denver, Colorado
U.S.A.

Jan Jensen
University of Colorado
Health Sciences Center
Barbara Davis Center
for Childhood Diabetes
Denver, Colorado
U.S.A.

Kirstine Juhl
Barbara Davis Center
for Childhood Diabetes
University of Colorado
Health Sciences Center
Denver, Colorado
U.S.A.

Katalin Kelemen
Barbara Davis Center
for Childhood Diabetes
University of Colorado
Health Sciences Center
Denver, Colorado
U.S.A.

Julie Lang
Department of Immunology
University of Colorado
Health Sciences Center
Denver, Colorado
U.S.A.

Thomas Mandrup-Poulsen
Steno Diabetes Center
Gentofte
Denmark

Jørn Nerup
Steno Diabetes Center
Gentofte
Denmark

Karin Nielsen
Steno Diabetes Center
Gentofte
Denmark

Runa Nolsøe
Steno Diabetes Center
Gentofte
Denmark

Jill Norris
University of Colorado
Health Sciences Center
Denver, Colorado
U.S.A.

Jan Nygaard Jensen
Barbara Davis Center
for Childhood Diabetes
University of Colorado
Health Sciences Center
Denver, Colorado
U.S.A.

Flemming Pociot
Steno Diabetes Center
Gentofte
Denmark

Alberto Pugliese
Diabetes Research Institute
University of Miami
Miami, Florida
U.S.A.

Marian Rewers
Barbara Davis Center
for Childhood Diabetes
University of Colorado
Health Sciences Center
Denver, Colorado
U.S.A.

Derry C. Roopenian
The Jackson Laboratory
Bar Harbor, Maine
U.S.A.

David V. Serreze
The Jackson Laboratory
Bar Harbor, Maine
U.S.A.

Thomas Sparre
Steno Diabetes Center
Gentofte
Denmark

Joachim Størling
Steno Diabetes Center
Gentofte
Denmark

Matthias G. von Herrath
La Jolla Institute for Allergy
and Immunology
San Diego, California
U.S.A.

Christina Wasmeier
Barbara Davis Center
for Childhood Diabetes
University of Colorado
Health Sciences Center
Denver, Colorado
U.S.A.

Liping Yu
Barbara Davis Center
for Childhood Diabetes
University of Colorado
Health Sciences Center
Denver, Colorado
U.S.A.

CONTENTS

Chapter 1

Primer:
Immunology/Autoimmunity

Stephanie C. Eisenbarth and George S. Eisenbarth

Immunologists over the past two decades have solved a series of fundamental genetic and biochemical questions related to immune function. Discoveries include elucidation of the mechanism by which antibody and T cell receptor diversity is generated[1,2] the characterization of a series of lymphokines essential for lymphocyte activation[3,4] the characterization of cell surface lymphocyte molecules (and in many cases their function[5,6] including molecules of the "innate" immune system[7] and molecules mediating what are termed "second" signals of T lymphocyte activation,[8,9] the definition of the crystal structure and peptide-binding properties of human leukocyte antigen (HLA)[10] class II[11-13] and class I molecules, and characterization of CD4 T cell subsets with divergent functions.[14,15] The field of immunology is now focused on defining mechanisms by which the immune system discriminates between self and nonself antigens or alternately "danger," and therefore by necessity the determinants of autoimmunity.[16] For some diseases and for some animal models, answers to a subset of the basic questions concerning autoimmunity posed in Table 1 are now available. It is clear that a single genetic or triggering mechanism to explain all autoimmune diseases affecting even a single tissue will be inadequate. For example, myasthenia gravis that results from autoantibodies to the acetylcholine receptor can be triggered by the drug penicillamine or a thymoma. For most patients with myasthenia gravis no triggering factor has been defined.

Normal Immune Response

In many respects, the modern age of immunology began with the independent descriptions by Talmage and Burnet in 1957 of the clonal selection theory.[17] "The basic concept was of the immunologically competent cell...; that is a cell which is susceptible to specific stimulation by contact with the appropriate antigenic determinant. The main possibilities of reaction are: (a) destruction, especially if physiologically immature; (b) proliferation without essential change of character – to what we would now call memory cells; and (c) proliferation and antibody production as a clone of plasma cells." In 1957 the mechanism underlying the huge diversity (which was fundamental to the clonal selection theory) of immunological reactivity of single lymphocytes was unknown. It is now understood that B cells that produce antibodies and T cells responsible for cell-mediated immunity bear on their surface unique antigen receptors created by genetic recombinatorial processes.

The receptor of B cells is the immunoglobulin molecule. For each B cell its antibody receptor has the same specificity as the antibodies that will eventually be secreted by the mature plasma cell. B cells can recognize soluble antigen. Antibody genes are created by a combinatorial process in which any of a family of multiple variable-region gene segments are combined with short joining and diversity immunoglobulin gene segments and then with the constant-region gene segments of immunoglobulin. Heavy and light chains of immunoglobulin are then joined, and billions of different antibodies (differing by their amino acid sequence) are generated. Antibody genes (but not T cell receptor genes) also undergo a high degree of

Immunology of Type 1 Diabetes, Second Edition, edited by George S. Eisenbarth.

Table 1. Basic questions

What genes underlie susceptibility to autoimmunity?
What triggers autoimmunity?
What are the target autoantigens?
What effector systems lead to disease?

mutation that increases the affinity with which antigens are recognized by clones of B lymphocytes.

Although T cells have a similar genetic mechanism as B cells for the generation of T cell receptor diversity, they differ in that they react only with antigen bound to class I or class II molecules of the major histocompatibility complex (MHC) on antigen-presenting cells. In fact, for a T cell to respond to antigen, antigen must be presented on the surface of antigen-presenting cells (e.g., macrophages – Fig. 1) in the context of MHC molecules. A crucial element of antigen-presenting cells (e.g., macrophages, B cells, and dendritic cells) is the ability to present peptides of antigens in the groove of HLA molecules. These "transplantation" antigens (HLA), as they were initially termed (because differences between transplantation antigens of donor and hosts determined ability to accept grafts), function by binding short peptides for presentation to T cells. T cells expressing a T cell receptor complex that includes the coreceptor CD4 interact with class II HLA molecules (DR, DP, and DQ of man, I-A and I-E of the mouse). The CD8 coreceptor performs a similar function for T cells that interact with class I molecules (HLA A, B, and C of man; K and D molecules of the mouse). Figure 1 illustrates the uptake, cleavage, and presentation of an "external" antigen to a CD4-positive T cell.

An additional name applied to the genes coding for class II HLA antigens is that of immune response genes. These molecules are highly polymorphic, differing between individuals in their amino acid sequence.[18,19] Each different sequence is given a number. Such differences are inherited in a Mendelian fashion. The sequence differences determine which peptides can be bound and presented. Thus, such inherited differences in HLA molecules between individuals can determine which antigen an individual can respond to, and importantly, which autoimmune disorder they are likely to develop. As modeled in Figure 1, the developing immune system has the potential to respond to essentially every foreign antigen, but also to every self-antigen.

In addition to the requirement of peptide presentation in the context of MHC molecules, a T cell requires an activation signal from the antigen-presenting cell. This "second signal" can be provided by cell surface molecules CD80 or 86 (B7.1 and B7.2, respectively). Thus, recognizing its cognate peptide in the groove of the MHC molecule is not enough to activate a naive T cell and therefore is also not enough to break tolerance to a self-peptide. On the contrary, if a T cell recognizes its cognate peptide on the surface of an APC (Antigen Presenting Cell) without appropriate costimulation, tolerance or anergy is induced. This is a potent mechanism for preventing nonspecific or self-specific T cell inflammatory reactions. This, however, leads to the question of how the APC knows when to activate a T cell with a second signal in order to get an immune response.

One T cell can recognize up to a million different peptides with its receptor (please see section on "Target Antigens") but it would obviously be disadvantageous for the immune system to activate a T cell in the absence of a real threat (e.g., infection). This puzzle was recently elucidated with the discovery of pattern recognition receptors (PRR) and in particular, Toll-like receptors first discovered in Drosophilia and then in humans and mice.[20] PRRs are considered a part of the Innate Immune system (as opposed to the Adaptive Immune system); Innate Immunity reacts faster to infections and is comprised of phagocytic cells, the complement system and a number of soluble inflammatory mediators such as antibacterial peptides and

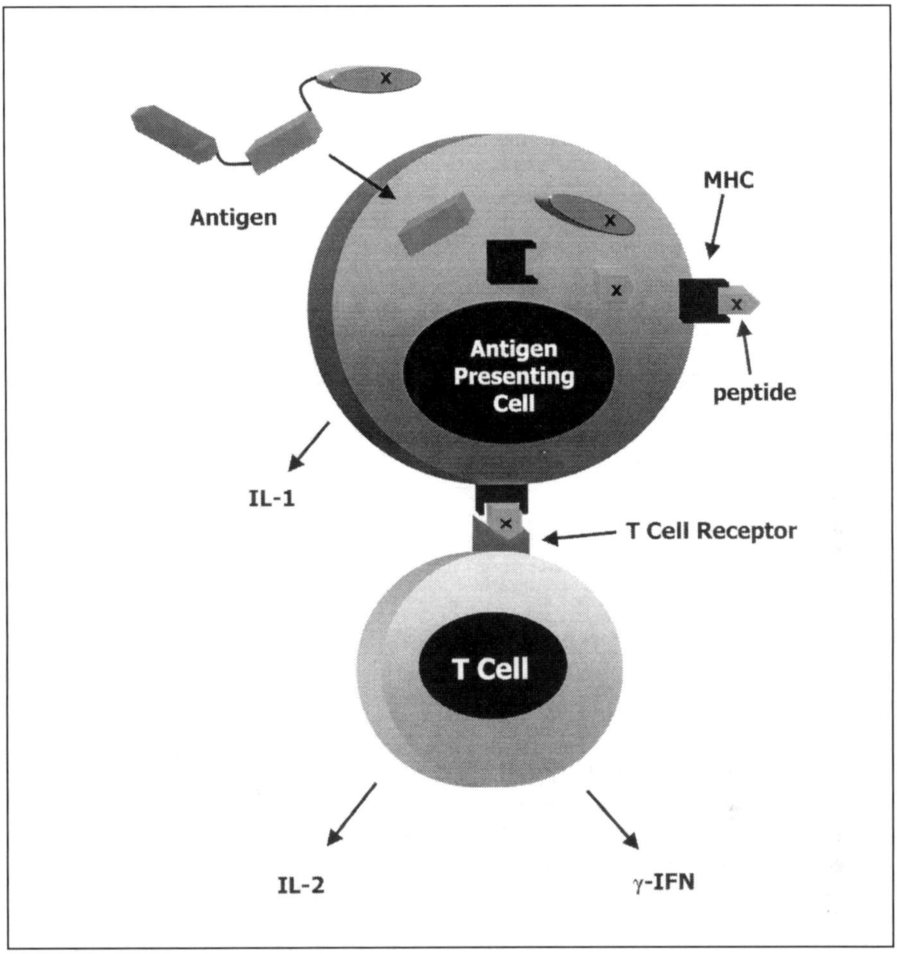

Figure 1. Antigen presentation by antigen-presenting cell with peptide in the groove of a major histocompatibility complex class II molecule (e.g., DR, DP or DQ).

cytokines. PRRs recognize highly conserved sequences within bacterial structures, for example, lipopolysaccharide (LPS) in the cell wall of Gram-negative bacteria.[21] These receptors are located on macrophages and dendritic cells—antigen presenting cells—and signals from receptor ligation induce many changes in presentation capabilities.[7] Therefore, when a dentritic cell (DC) encounters an invading pathogen, it not only engulfs it for antigen presentation, but also is stimulated to upregulate MHC class I and II, costimulatory molecules B7.1 and B7.2 and produces a host of inflammation-stimulating cytokines. The combination of these changes in the APC enables effective stimulation of naive T cells and is the link between the Innate and Adaptive Immune systems.

Once an antigen-specific T cell has been appropriately activated, it in turn, helps activate the humoral arm of Acquired Immunity – the B cell. The majority of humoral responses are initiated by Helper CD4+ T cells, although there are a few exceptions with certain classes of pathogens, but we will focus on the former here. T cell help is initiated within the lymph nodes or spleen when an antigen-specific, armed T cell recognizes its antigen in MHC II on the surface of the B cell (whose immunoglobulin receptor recognized the same pathogen, internal-

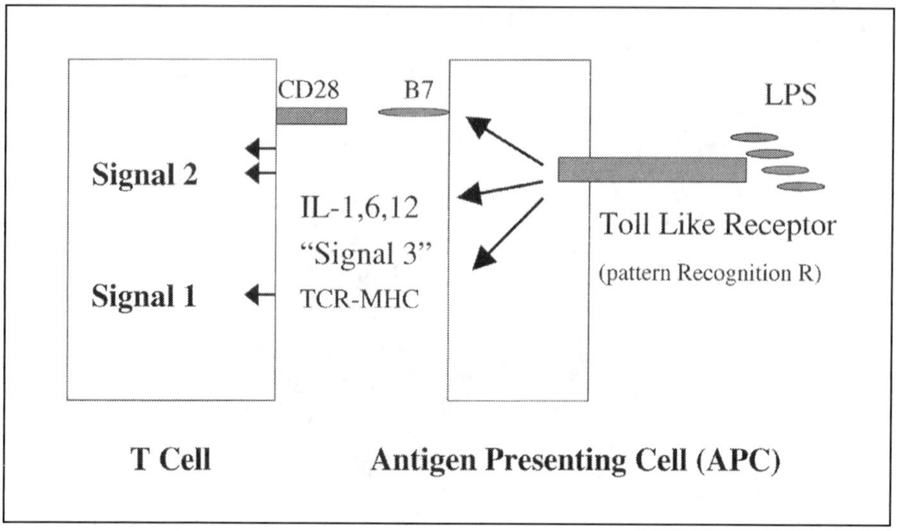

Figure 2. Innate immune system: Toll-like receptor activation of APC. A series of pattern recognition receptors (toll-like) are present on antigen presenting cells. They recognize common molecules on infectious agents and stimulate the antigen presenting cell signaling "danger" as described by Dr. Matzinger. The B7 molecule reacts with CD28 on T cells, MHC molecules (react with the T cell receptor(TCR)) are upregulated, and interleukins are produced such as IL-12 that stimulates a Th1 T cell response. The T cell requires multiple signals for activation with signal 1 from the T cell receptor.

ized it and presented fragments of the pathogen on MHC class II). This is termed Linked Recognition because it ensures that the T and B cells recognize the same pathogen (although not necessarily the same epitope). Engagement of the T cell receptor (TCR) induces upregulation of CD40 ligand on the surface of the T cell and production of secreted activators of B cells, such as IL-4, 5 and 6. The combination of these stimuli activates the B cell to proliferate and secrete antibody. Therefore, this is another stage at which tolerance to self-peptides is censured or can be broken. If an autoreactive B cell does not encounter an activated T cell specific for the same self-protein, then no autoantibodies can be produced (unless the antigen is a large repeating molecule and T cell independent). This step could be circumvented in autoimmunity, however, through the chance meeting of an armed T cell specific for nonself peptide recognizing a B cell which has phagocytosed and presented that T cell's antigen but has antibodies specific for another self-protein. Activation of that B cell would lead to the production of autoantibodies.

Because the immune effector mechanisms that have evolved are capable of rapidly killing cells, the potential for response to self is a consequence of having a potent immune system (analogous to having firearms in a house to protect from burglars—weapons that may nevertheless accidentally kill family members). As stated by Burnet, "Autoimmune disease can be defined as a condition in which structural or functional damage is produced by the reaction of immunocytes or antibodies with normal components of the body. The central theme is the emergence of 'forbidden clones' of pathogenic immunocytes and the various ways by which these can arise and find ways of escaping the normal controls."

Of note, such a pathogenic mechanism related to disease of the lymphoproliferative (lpr) mouse.[22,23] Lpr mice have either of two independent autosomal recessive mutations of the Fas gene, which influences "programmed" lymphocyte cell death (apoptosis). Lpr mice have marked lymphoproliferation and a disorder resembling lupus erythematosus and rheumatoid arthritis. The Fas gene is the same as a human gene termed Apo-I, which is involved in the control of apoptosis. Apoptosis is the fate of more than 90% of developing thymocytes that react with self-antigens in the thymus. "Central" tolerance to self antigens is determined by deletion of T cells whose T cell receptor have high affinity for self MHC antigens or self MHC antigens with bound self peptides. It is hypothesized that inhibition of the apoptotic pathway because of mutations or deletions of Fas leads to the failure to eliminate autoreactive T cells.

Tolerance to self antigens is thought to result from at least three different mechanisms: deletion of autoreactive cells; clonal inactivation (anergy) where T cells reactive with self are present but resistant to activation upon antigen (Ag) encounter; and potential silencing of self-reactive cells by other cells (immune regulation). Tolerance can also be subdivided by the site of acquisition (central tolerance (thymus) and peripheral tolerance). The thymus is the primary site of T cell deletion, with more than 90% of T cells destroyed within the thymus. At a functional level, the breaking of "tolerance" is fundamental to concepts of autoimmunity. Recent studies indicate that the thymus and other lymphoid organs contain "self-antigen-presenting cells".[24] For example, insulin message and protein are present in the thymus, lymph node, and spleen. Functional studies suggest that minute quantities of such self-antigens are important for the maintenance of tolerance.[25]

In addition to the concept of "forbidden" clones arising because of mutations that bypass normal regulatory events, it has become clear that, given appropriate context of antigen presentation, "autoimmune" responses can be generated to essentially all proteins. For example, subcutaneous injection of human insulin (the current therapy for type 1A diabetes) almost invariably leads to low levels of anti-insulin autoantibodies. For most destructive autoimmune disorders, a characteristic group of autoantibodies arise. These antibodies frequently react with a series of molecules from the same subcellular localization. For lupus erythematosus, it is the ribonucleoprotein complex; for type 1 diabetes, insulin secretory granules and islet synaptic-like microvesicles. This suggests that once tissue destruction commences, in the context of an inflammatory lesion (insulitis), tolerance is broken to a series of autoantigens.

Genetics of Autoimmunity

The general concept that autoimmunity develops in the setting of genetic susceptibility, and in particular in association with a series of specific HLA alleles, applies to most human autoimmune disorders. It is important to note, however, that the same HLA allele (e.g., MHC class II alleles) may protect from one disorder and may be associated with a different disease (e.g., HLA DR2, DQA1*0102, DQB1*0602 is rare in patients with type 1 diabetes but it is associated with multiple sclerosis).[26,27] It appears that susceptibility encoded by alleles within the HLA region do not globally influence the development of autoimmunity but rather influence the likelihood of specific disorders.

If HLA alleles determine disease susceptibility by their function as "immune" response genes, selective targeting is likely. An alternative manner by which HLA alleles may determine susceptibility for autoimmune disease is by altering the developing T cell repertoire. This may be particularly important for diseases associated with what have been termed "superantigens" such as Kawasaki's disease. Superantigens bind to whole classes of T cells bearing specific VbetaT cell receptor families. Superantigens thereby activate large numbers of T cells (many more than the usual antigen that interacts only with clones of T cells with a specific complementary binding site).

Recent experiments with transgenic mice have provided convincing evidence that the predilection of T cells for MHC is due to the way the T cell receptor repertoire is derived. Developing thymocytes express a multitude of T cell receptors, and only the infrequent thymocyte that expresses a receptor with low affinity for MHC antigens presented in the thymus matures further; the selection signal is supplied via the engaged T cell receptor complex. This selection on self MHC (termed positive selection) is potentially dangerous in that T cells with overt autoreactivity can be and most likely are selected. Those developing thymocytes with high affinity for self MHC, and which are therefore autoreactive, transduce a signal via their T cell receptor that leads to cell death rather than further differentiation.

Experiments in transgenic nonobese diabetic (NOD) mice suggest that simple skewing of the T cell repertoire can, but does not always alter diabetes susceptibility in this animal model.[28] For example, Santamaria and coworkers have produced a transgenic mouse with a T cell receptor that targets islets and causes diabetes. Multiple class II alleles when bred onto the mice delete the autoreactive T cells and prevent disease.[29]

In autoimmune diseases studied to date, concordance of identical twins is never 100% and is usually between 30% and 70%.[30,31] Concordance of nonidentical twins is often in the range of 5% and similar to the risk of siblings. Finally, for most disorders (multiple sclerosis, type 1 diabetes, etc.), the concordance is higher for monozygotic twins as contrasted with HLA identical siblings. The above information strongly suggests that alleles of genes outside the major histocompatibility complex contribute to disease susceptibility. In animal models, a number of alleles influencing disease susceptibility have been defined.[32] In humans, for example, autoimmunity is associated with complement deficiency (e.g., lupus erythematosus)[33] alleles of insulin (type 1 diabetes),[34] and nucleoside phosphorylase deficiency (hemolytic anemia). It is likely that complement deficiencies contribute to autoimmunity by altering processing of antigen-antibody complexes. Insulin gene polymorphisms influence insulin synthesis and therefore may influence "tolerance" to insulin.[25,35] T cells are particularly sensitive to nucleotide metabolites and it is thought that nucleoside phosphorylase deficiency may be associated with autoimmunity secondary to T cell toxicity. The genes underlying two remarkable syndromes of autoimmunity APS-I (Autoimmune Polyendocrine Syndrome Type I) and XPID (X-linked Polyendocrinopathy, Immune dysregulation and Diarrhea) are now cloned. For both these syndromes mutations of transcription factors lead to multiple autoimmune disorders including type 1 diabetes.

With refined techniques for genetic mapping and the knowledge provided by the genome project it is likely that in the next decade great progress will be made in the definition of alleles underlying genetic susceptibility for common autoimmune disorders. Even as genes causing autoimmunity are discovered their mechanism of action will not be immediately clear. Mutations of the AIRE (Autoimmune regulator) gene on chromosome 21 cause the APS-I (Autoimmune Polyendocrine Syndrome) with Addison's disease, hypoparathyroidism, mucocutaneous candidiasis, hepatitis, and for 18%, type 1 diabetes. This transcription factor that is expressed in the thymus leads to so many autoimmune disorders probably by regulating thymic expression of "peripheral" antigens.[35a]

Triggering of Autoimmunity

A large group of experimental autoimmune disorders can be induced by immunization with self-proteins. To induce disease, and even to generate an immune response, such molecules are usually injected in a depot form with adjuvants that activate antigen-presenting cells. A model of multiple sclerosis follows immunization with myelin basic protein.[36] There are conceptually similar models for experimental autoimmune thyroiditis (immunization with thyroglobulin), experimental myocarditis (immunization with myosin peptides), and experimental autoimmune oophoritis (immunization with the oocyte sperm cell receptor).[37] Once disease is induced in these animal models, T cell clones reacting with immunizing antigens are sufficient to transfer disease. In that disease can be induced by simple immunization with self-molecules, it is clear that T cells and B cells reactive to self exist in normal animals. Thus, the context of immunization determines whether pathogenic clones are expanded. In part, the ease of inducing autoimmunity raises the question as to what mechanism prevents expansion of autoimmune clones.

In animal models, both genetic (e.g., T cell immunodeficiency gene on chromosome 4 of the BB rat) and environmental manipulations of the immune system lead to autoimmunity (neonatal thymectomy, injection of anti-RT6 antibodies in nonlymphonic BB rats, and neonatal therapy with cyclosporine A, injection of poly-IC into RT1-U rats).[38] It appears that neonatal thymectomy induces autoimmunity by decreasing a specific subset of "regulatory" T lymphocytes that express CD25 (IL-2 receptor) and are CD4 positive without antigenic stimulation.[39,40]

Table 2. "Etiologic" classification of autoimmunity

Classification	Example
Oncogenic	Ovarian carcinoma and cerebellar degeneration
Dietary	Gliadin and celiac disease
Drug	Penicillamine and myasthenia gravis
Infectious	Streptococci and rheumatic fever
Idiopathic	Type 1A diabetes (Autoimmune DM)

Table 2 presents an "etiologic" classification of human autoimmunity based upon identified triggering factors. Infectious agents,[41] neoplasms, and drugs have all been found to induce specific autoimmune diseases. Tumors that induce autoimmunity are characterized by the expression of specific autoantigens. One of the best-characterized oncogenic autoimmune syndromes results in cerebellar degeneration. Oncogenic cerebellar degeneration is associated with specific anti-Purkinje cell antibodies. It is induced by ovarian carcinomas expressing what have been termed CDR (cerebellar degeneration related) antigens.[42] Additional oncogenic autoimmune disorders include pemphigus associated with lymphoma, myasthenia gravis and pure red cell aplasia associated with thymomas, and retinopathy associated with small-cell carcinoma. It is thought that, similar to the induced autoimmune disorders of animals described above, presentation to the immune system by tumor cells of self peptides induces "remote" autoimmunity.

The drug penicillamine is associated with a large number of different autoimmune disorders, including myasthenia gravis, bullous pemphigoid, lupus erythematosus, polymyositis, and dermatomyositis.[43] In addition to overt disease, penicillamine induces antibodies such as anti-insulin autoantibodies in the absence of overt pathology. It is hypothesized that penicillamine (dimethylcysteine) may induce autoimmunity by haptenation of multiple proteins.

One of the best examples of autoimmunity induced by food is celiac disease, or gluten sensitive enteropathy.[44,45] Celiac disease is characterized by marked lymphocytic infiltrates of the small bowel associated with villous atrophy. Ingestion of the wheat protein gliadin leads to the production of anti-transglutaminase autoantibodies and destruction of the villi. The disease is "cured" by elimination of wheat gluten from the diet. Like most autoimmune disorders, celiac disease occurs only in the presence of specific HLA alleles; in particular, with DQ alpha and beta sequences (DQA1*0501 and DQB1*0201) provided by HLA haplotypes DR5 (DQA1*0501) and DR7 (DQB1*0201) in trans or in individuals with DR3 haplotypes (DQA1*0501/DQB*0201).

Autoimmunity associated with mononucleosis is related to the unique ability of the virus to directly infect B cells.[46] The Epstein-Barr virus (EBV) stimulates B cell proliferation and polyclonal antibody production. These polyclonal antibodies reacting with self -antigens can lead to disease. The proliferation of B cells is self-limited in the presence of an effective T cell response. The atypical lymphocytes of EBV infection are activated T cells that are essential to suppress uncontrolled B cell growth. In addition to environmental factors triggering autoimmunity, infections can suppress autoimmunity. Almost any viral infection of NOD mice prevents diabetes, as does a single injection of complete Freund's adjuvant.

Target Antigens

Exogenous antigens are taken into a cell by specific receptors (e.g., binding of antigen to surface immunoglobulin of B cells, immunoglobin Fc receptor-mediated uptake of immune complexes) or nonantigen-specific mechanisms (e.g., pinocytosis). Such antigens are then cleaved to peptides and coupled to class II MHC molecules for presentation on the cell surface and recognition by specific T cells (CD4-positive T cells). Endogenous antigens, synthesized by the cell (e.g., viruses), are cleaved by peptidases in a proteosome complex, transported into vesicles by specific peptide transporters (TAP1, TAP2),[47] coupled to class I MHC molecules and then presented on the cell surface for recognition by T cells bearing CD8 accessory molecules. In contrast to the above usual antigen processing, superantigens are molecules that bind outside of the peptide binding "groove" of class I and class II molecules and trigger whole families of T cells by binding to common sequences of T cell receptors. Thus, for T cell responses to superantigens, it is predicted that responding cells will bear T cell receptors of limited families (e.g., Vbeta8.2), but the T cell receptor antigen-binding region ("complementarity determining region") will differ in amino acid sequence.

The antibody and T cell responses characteristic of autoimmune disorders appear to be "antigen driven." The amino acid sequences of autoantibodies are mutated relative to immunoglobulin germ-line sequences and autoantibodies are usually of high affinity. In addition, the large number of different autoantibodies to different autoantigens for even a single autoimmune disease suggests that whatever the initial lesion leading to autoimmunity, the immune response "spreads" to a series of antigenic epitopes and molecules of the involved tissue. For example, large families of antibodies react with islet autoantigens during the beta cell destruction associated with type 1 diabetes. To date, these autoantibodies target molecules within two islet secretory organelles, insulin secretory granules (e.g., insulin, proinsulin, ICA512(IA-2), carboxypeptidase H, GM2-1 ganglioside) and synaptic-like microvesicles where gamma amino butyric acid (GABA) is stored (e.g., glutamic acid decarboxylase).[48]

The detection of a family of autoantibodies for each autoimmune disease complicates the elucidation of which autoantigen (if any) is central to disease pathogenesis. In the past, the major criteria for defining such autoantigens rested primarily on the ability to immunize with the given antigen and create disease. This is probably an inadequate criterion to define pathogenic autoantigens of spontaneous autoimmune disorders of man and animal models.

It is known that for many tissues, immunization with several different autoantigens can induce disease (e.g., myelin basic protein and proteolipid protein for experimental autoimmune encephalitis). With the development of molecular genetic techniques, more definitive tests of an antigen's disease relevance are now possible. For example, genes for specific autoantigens can be introduced into ectopic tissues by creating transgenic mice with the gene coding for the autoantigen coupled to a promoter directing ectopic expression. If, for example, destruction of the ectopic tissue is induced, the single autoantigen is sufficient to target autoimmunity. In a similar manner, induction of thymic expression of an autoantigen may be sufficient to induce tolerance to the antigen. If all autoimmunity is blocked in such transgenic mice, then the given autoantigen is essential for disease pathogenesis. Introduction of autoantigen synthesis by transplanted tissues (e.g., adenovirus gene transfer systems) may allow a series of autoantigens to be rapidly tested to determine whether they are "sufficient" for tissue destruction. Just as important, gene knockouts can be used to directly test the relevance of specific antigens. Baekkeskov and coworkers have found that NOD mice lacking GAD65 develop diabetes.[49] Thus, GAD65 is not essential, though GAD67 may play a role.

Autoantibodies usually react with conformational epitopes of their target antigen and for many autoimmune disorders autoantibodies to multiple different epitopes of even single antigens are targets. In contrast T lymphocytes target relatively short peptides (8-12 amino acids in length). Thus "molecular" mimicry (cross-reaction between different molecules) of high affinity autoantibodies is likely to be limited for autoantibodies. In contrast, it is now estimated that each T cell receptor can react with approximately 1 million different peptides. Peptides that activate a single T lymphocyte clone have been engineered such that they bind to the appropriate MHC molecule, and then might only have one or two (occasionally none) similar amino acids, yet activate the same T cell receptor. Thus there is enormous potential for a given T cell receptor to react with peptides from multiple different molecules and some of these molecules will have no "obvious" sequence homology. This suggests that many molecules may be "molecular mimics" (peptide mimotopes) and activate autoimmunity. In that most individuals do not have autoimmune disease, the immune system usually deletes or appropriately regulates autoreactive T cells.

Effector Mechanisms

In most immune disorders, both humoral and cellular arms of the immune system contribute to disease pathogenesis. Autoantibody-mediated disorders include autoimmune hemolytic anemia, immune-mediated thrombocytopenia (ITP), myasthenia gravis, Lambert-Eaton myasthenic syndrome, pemphigus, and Graves' disease.[50] For each of these disorders, if the mother has the autoantibodies a transient neonatal disease follows transplacental passage of antibodies.

Autoantibodies can induce pathology by blocking the function of specific receptors (e.g., hypothyrodisim,[51] "geriatric" hypoparathyroidism),[52] stimulating receptors (Graves' disease, hypoglycemia associated with anti-insulin receptor autoantibodies), by direct cytotoxicity (hemolytic anemia), by opsonizing cells (hemolytic anemia and thrombocytopenic purpura), and by deposition of immune complexes (lupus nephritis).

Cell-mediated immunity is usually associated with tissue destruction (e.g., type 1 diabetes, thyroiditis), either through direct cellular cytotoxicity or by indirect cytotoxic mechanisms (e.g., delayed-type hypersensitivity). The classic cytotoxic pathway (that which underlies rejection of allogeneic tissue) is mediated by direct recognition of antigenic peptides complexed with MHC class I molecules by CD8-positive cytotoxic T cells; killing by cytotoxic T cells involves direct cellular contact between T cells and the target cell. The function and expansion of clones of such cytotoxic T cells are dependent upon "help" provided by CD4-positive T cells that respond to antigenic stimulation by releasing a series of lymphokines, including IL-2 (interleukin-2). Recently, CD4-positive T cells have been subdivided based on the lymphokines they produce[53,54] into three groups of "helper cells": Th0, Th1, and Th2. Th0 cells constitute a small subset postulated to be related in the thymus to reactivity with autoantigens and produce IL-2 but low levels of interferon gamma (IFN-γ).[55] Th1 cells determine delayed hypersensitivity and produce IL-2 and IFN-γ, Th2 cells produce IL-4, 5+13. Th2 cells predominately lead to humoral immunity and they may even be protective relative to destructive autoimmunity.

An additional mechanism of cell-mediated destruction of target cells involves indirect cytotoxicity mediated by CD4-positive (Th1) lymphocytes. Such CD4-positive lymphocytes, following recognition of their target antigen presented in the context of a class II MHC molecule, secrete a group of lymphokines (e.g., interferon gamma), which in turn activates cells such as macrophages to release IL-1, nitric oxide, and free radicals. Each of these molecules can directly kill selected target tissues.[56,57] This second mechanism may be primarily responsible for the autoimmune beta cell destruction of type I diabetes and for destruction of xenogeneic tissue transplants. For example, anti-islet CD4-positive clones (as will be discussed in later chapters) are sufficient to transfer diabetes.[58]

Therapy of Autoimmunity

A series of novel therapies for the prevention and therapy of autoimmunity are being studied. One can broadly divide therapies of autoimmunity into therapies which are antigen specific and those which are antigen nonspecific. Examples of antigen-specific therapies include prevention of diabetes in NOD mice by parenteral or oral administration of insulin[59-61] MHC-binding peptide therapies for prevention of experimental autoimmune encephalitis,[62] and T cell or peptide vaccination strategies.[63,64] Antigen nonspecific therapies include drugs such as rapamycin (sirolimus), cyclosporine A, glucocorticoids, anti-T cell monoclonal antibodies, IL-2-diptheria toxin conjagates[65-67] and FK506 (tacrolimus). The disadvantage of nonantigen specific therapies is the likely suppression of important immune function with increased risk of infection and malignancy. In addition, selective drugs have unique toxicity unrelated to immunosuppression, such as osteoporosis and diabetes with glucocorticoids and the renal toxicity associated with cyclosporine A. Nevertheless, it is clear that potent immunosuppressive drugs such as cyclosporine A, which act by blocking lymphokine production by T cells, can prevent certain forms of autoimmunity. When it is possible to use such drugs at low doses with marked therapeutic effect (e.g., cyclosporine A for psoriasis), they become important therapeutic agents. It is also hoped that certain strategies of short-term immunosuppression may lead to a state of long-term tolerance, thereby lessening the deleterious effects of immunosuppression. For example, treatment with specific monoclonals directed against the T cell coreceptor CD4 can lead to long-term engraftment of both heart transplants and islets and nonactivating anti-CD3 antibodies can cause long-term remission of diabetes in NOD mice[68] and blocking of "costimulation" can block autoimmunity.[69-72]

Antigen-specific therapies include the removal of a disease-inciting antigen (if it is identified), deviation of an immune response to nonpathogenic pathways (Th2 versus Th1, with oral "tolerance"[73,74] or parenteral immunization), and targeting specific autoreactive T cells or B cells through their receptors for targeting specific HLA alleles.

For example, celiac disease is successfully treated by the removal of the wheat protein gliadin from the diet. Studies of the effects of eliminating bovine milk products in neonates are planned for infants at high genetic risk or type 1 diabetes.

For endocrine glands it is possible to suppress cellular activity through feedback inhibitory circuits (e.g., administration of thyroxine or insulin inhibits thyroid or beta cells, respectively). Therapy with insulin not only prevents development of diabetes of BB rats[59] and NOD mice,[75] but also prevents lymphocytic infiltrates into the islets and beta cell destruction. Protection by insulin in the BB rat requires metabolically active insulin and induction of hypoglycemia with concomitant inhibition of insulin secretion by B-cells. In contrast, nonmetabolically active peptides of insulin prevent diabetes of NOD mice.

A single report suggests that glucocorticoid therapy in patients with anti-adrenal autoantibodies leads to loss of autoantibodies and prevention of Addison's disease.[76] The mechanism by which such "isohormonal" therapy is effective is unknown, but we hypothesize that a component of auto-antigen presentation occurs within the target gland.[77]

A single injection of Freund's adjuvant is sufficient to prevent the development of diabetes of the NOD mouse.[78,79] As reviewed in Chapter 5, such therapy may act by shifting islet-infiltrating T cells to Th2-like lymphocytes, producing IL-4 rather than interferon gamma though the mechanism is almost certainly more complex.[79]

Pilot studies in man and experimental studies of animal models have been completed of oral antigen therapy for diabetes, multiple sclerosis (myelin basic protein),[80] rheumatoid arthritis (collagen),[61] and uveitis (retinal S protein).[81] For the most part, studies of oral tolerance in man have failed to achieve clinical benefit. It is hypothesized that the feeding of autoantigens induces T cells, producing suppressive lymphokines TGF-β and IL-4 when such T cells encounter antigen.[82] Studies of the mechanism underlying oral tolerance are most advanced for experimental autoimmune encephalitis, where, in vivo antibodies to TGF-β block induction of 'oral tolerance'.[83]

The trimolecular complex (MHC molecule, peptide antigen, and T cell receptor) underlying the specificity of T cell responses is an obvious target for therapy of autoimmunity. In experimental disorders with a very limited T cell receptor response, such as experimental autoimmune encephalitis (almost exclusive use of T cells bearing Vbeta8.2 T cell receptor chain), targeting of a specific family of T cell receptors is efficacious in preventing disease. Another approach being actively pursued is the production of peptides that bind to MHC alleles associated with disease.[62] For example, a peptide binding to a high-risk allele for type I diabetes, DQA1*0301/DQB1*0302, might prevent diabetes by preventing the binding of "diabetogenic" peptides.

One of the most exciting pathways for the prevention of autoimmunity in animal models goes under the rubric of "immunologic vaccination".[84] It has been found that a series of spontaneous and experimental autoimmune disorders can be prevented by administering target autoantigens or peptides of given autoantigens. A single injection of the insulin B chain peptide, amino acids B:9-23, can prevent type 1 diabetes for the life of NOD mice.[85] Multiple routes of antigen administration have been studied including subcutaneous injection with or without adjuvant, oral or nasal mucosal administration, administration of DNA constructs coding for the autoantigen, and in specific models all can prevent disease. For a number of peptide autoantigens, the peptide is modified producing an "altered peptide ligand". Such

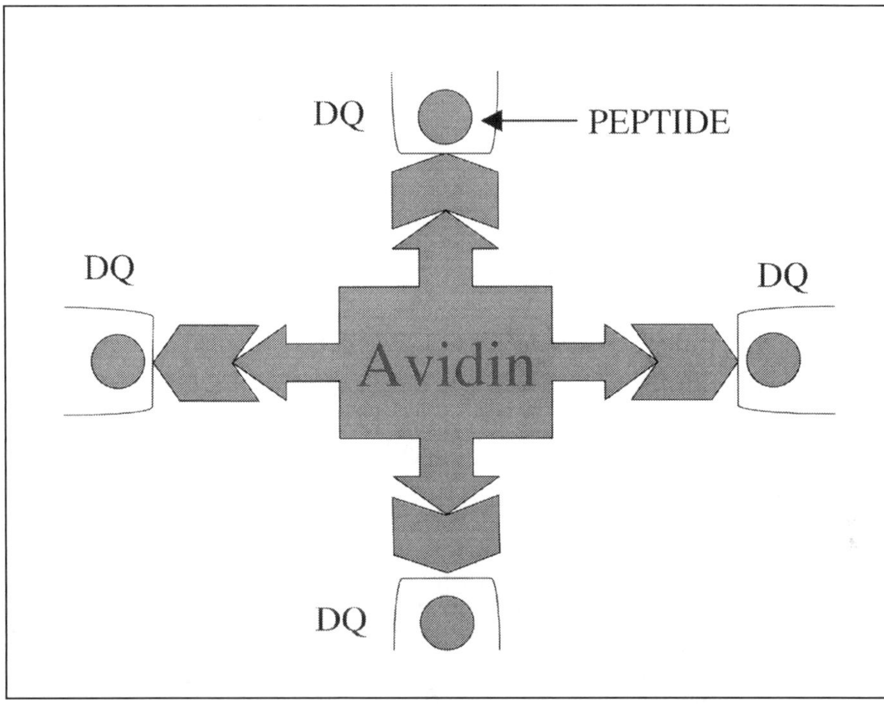

Figure 3. Tetramers are so-called because avidin binds four biotinylated HLA molecules (DQ in this example). The DQ molecule with peptide, especially as a multimer as indicated, then binds to T cell receptor of T cells specific for that DQ and specific peptide.

altered peptide ligands are engineered such that they bind to relevant class II HLA molecules and are presented to T lymphocytes. Because of changes in amino acid sequence of the residues interacting with the T cell receptor, altered peptides can either antagonize T cell activation, induce anergy in the responding T cell, or lead to the generation of antigen specific regulatory T cells. Such regulatory T cells are thought to suppress disease by production to Th2 and related cytokines at the site of tissue injury. "Immunologic vaccination" is now entering clinical trials, particularly for patients developing or with new onset type 1A diabetes. For example altered peptide ligands of insulin peptide B:9-23 and altered peptide ligand of myelin basic protein are in clinical trials. For both altered peptide ligands and native peptides used for "immunologic vaccination" there is a risk that the peptide will actually exacerbate the disease one is seeking to treat. Development of improved assays to quantitate T cell function and autoreactive T cell frequencies will be important for progress in improving the risk benefit ratio for such therapies. Two general assay formats, namely ELISPOT assays[86] and "tetramer" (also termed "multimer")[87] assays have already greatly impacted studies of T cell responses to infections and are likely to be important for autoimmunity. The ELISPOT assays rely upon culturing lymphocytes and then detecting individual cells and the cytokines they produce. Tetramer assays utilize avidin to couple together four biotin labeled MHC molecules containing a relevant peptide. By having four MHC molecules the affinity for reaction with the receptors of T cells is greatly increased. Such molecular complexes are labeled with fluorescent dyes and are able to bind to the T cell receptors specific for the restricting MHC element and the peptide. Using tetramers and a fluorescent cell sorter autoreactive T lymphocytes can be identified. It would be a major advance if one could rapidly evaluate therapies for autoimmune disorders in terms of effects upon the pathogenic autoimmune T cells with ELISPOT and/or tetramer analysis.

This chapter is just a brief primer. We strongly recommend "Immunobiology" by Charles A. Janeway that includes detailed illustrations of current concepts of immunology for those needing a more in-depth review of this rapidly expanding field of study.

References

1. Early P, Huang H, Davis M et al. An immunoglobulin heavy chain variable region is generated from three segments of DNA: Vh, D and Jh. Cell 1980; 19:981-992.
2. Davis MM, Bjorkman PJ. T cell antigen receptor genes and T cell recognition. Nature 1988; 334:395-402.
3. Or R, Renz H, Terada N et al. IL-4 and IL-2 promote human T-cell proliferation through symmetrical but independent pathways. Clin Immunol Immunopathol 1992; 64:210-217.
4. Romagnani S. Human TH1 and TH2 subsets: Regulation of differentiation and role in protection and immunopathology. Int Arch Allergy Immunol 1992; 98:279-285.
5. Marshall LS, Aruffo A, Ledbetter JA et al. The molecular basis for T cell help in humoral immunity: CD40 and its ligand, gp39. J Clin Immunol 1993; 13:165-174.
6. Larson RS, Springer TA. Structure and function of leukocyte integrins. Immunol Rev 1990; 114:181-217.
7. Medzhitov R, Janeway C. Innate immunity. N Engl J Med 2000; 343:338-344.
8. Kay AB. Review Articles: Advances in immunology: Allergy and allergic diseases (Second of two parts). N Engl J Med 2001; 344:109-113.
9. Blazar BR, Taylor PA, Noelle RJ et al. $CD4^+$ T cells tolerized ex vivo to host alloantigen by anti-CD40 ligand (CD40L;CD154) antibody lose their graft-versus-host disease lethality capacity but retain nominal antigen responses. J Clin Invest 1998; 102:473-482.
10. Beck S, Trowsdale J. The human major histocompatibility complex: Lessons from the DNA sequence. Annu Rev Genomics Hum Genet 2000; 1:117-137.
11. Chicz RM, Urban RG, Gorga JC et al. Specificity and promiscuity among naturally processed peptides bound to HLA-DR alleles. J Exp Med 1993; 178:27-47X.
12. Lee KH, Wucherpfennig KW, Wiley DC. Structure of a human insulin peptide/HLA-DQ8 complex and susceptibility to type 1 diabetes. Nat Immunol 2001; 2:501-507.
13. Brown JH, Jardetzky TS, Gorga JC et al. Three-dimensional structure of the human class II histocompatibility antigen HLA-DR1. Nature 1993; 364:33-39.
14. Bjorkman PJ, Saper MA, Samraoui B et al. Structure of the human class I histocompatibility antigen, HLA-A2. Nature 1987; 329:506-512.
15. Bjorkman PJ, Saper MA, Samraoui B et al. The foreign antigen binding site and T cell recognition regions of class I histocompatiblity antigens. Nature 1987; 329:512-518.
16. Kamradt T, Mitchison NA. Tolerance and autoimmunity. N Engl J Med 2001; 344:655-664.
17. Burnet M. Cellular Immunology 1969; 1.
18. Nepom GT, Erlich H. MHC class-II molecules and autoimmunity. Ann Rev Immunol 1991; 9:493-525.
19. Nepom GT. Immunogenetics and IDDM. Diabetes Rev 1993; 1:93-103.
20. Medzhitov R, Preston-Hurlburt P, Janeway CA. A human homologue of the Drosophila toll protein signals activation of adaptive immunity. Nature 1997; 388:394-397.
21. Takeuchi O, Hoshino K, Kawai T et al. Differential roles of TLR2 and TLR4 in recognition of gram-negative and gram-positive bacterial cell wall components. Immunity 1999; 11:443-451.
22. Watanabe-Fukunaga R, Brannan CI, Copeland NG et al. Lymphoproliferation disorder in mice explained by defects in fas antigen that mediate apoptosis. Nature 1992; 356:314-317.
23. Lichter P, Walczak H, Weitz S et al. The human APO-1(APT) antigen maps to 10q23, a region that is syntenic with mouse chromsome 19. Genomics 1992; 14:179-180.
24. Pugliese A, Brown D, Garza D et al. Self-Antigen presenting cells expressing IDDM-associated autoantigens exist in both thymus and peripheral lymphoid organs in humans. J Clin Invest 2001; 107(5):555-564.
25. Hanahan D. Peripheral-antigen-expressing cells in thymic medulla: Factors in self- tolerance and autoimmunity. Curr Opin Immunol 1998; 10:656-662.
26. Erlich HA, Griffith RL, Bugawan TL et al. Implication of specific DQB1 alleles in genetic susceptibility and resistance by identification of IDDM siblings with novel HLA-DQB1 allele and unusual DR2 and DR1 haplotypes. Diabetes 1991; 40:478-481.
27. Atkinson MA, Maclaren NK. What causes diabetes? Sci Am 1990; 263:62-63.
28. Lipes MA, Rosenzweig A, Tan KN et al. Progression to diabetes in nonobese diabetic (NOD) mice with transgenic T cell receptors. Science 1993; 259:1165-1169.
29. Schmidt D, Amrani A, Verdaguer J et al. Autoantigen-independent deletion of diabetogenic CD4+ thymocytes by protective MHC class II molecules. J Immunol 1999; 162:4627-4636.

30. Olmos P, A'Hern R, Heaton DA et al. The significance of the concordance rate of type I (insulin-dependent) diabetes mellitus in identical twins. Diabetologia 1988; 31:747-750.
31. Ebers GC, Bulman DE, Sadovnick AD et al. A population based study of multiple sclerosis in twins. N Engl J Med 1986; 315:1638-1642.
32. Rossini AA, Greiner DL, Friedman HP et al. Immunopathogenesis of diabetes mellitus. Diabetes Rev 1993; 1:43-75.
33. Hartung K, Baur MP, Coldewey R et al. Major histocompatibility complex haplotypes and complement C4 alleles in systemic lupus erythematosus. J Clin Invest 1992; 90:1346-1351.
34. Julier C, Hyer RN, Davies J et al. Insulin-IGF2 region on chromosome 11p encodes a gene implicated in HLA-DR4-dependent diabetes susceptibility. Nature 1991; 354:155-159.
35. Smith KM, Olson DC, Hirose R et al. Pancreatic gene expression in rare cells of thymic medulla: Evidence for functional contribution to T cell tolerance. Int Immunol 1997; 9:1355-1365.
35a. Anderson MS, Vernanzi ES, Klein L et al. Projection of an Immunological Self-Shadow Within the Thymus by the Aire Protein. Science 2002; 298:1395-1401.
36. Massacesi L, Joshi N, Lee-Parritz D et al. Experimental allergic encephalomyelitis in cynomolgus monkeys: Quantitation of T cell responses in peripheral blood. J Clin Invest 1992; 90:399-404.
37. Rhim SH, Millar SE, Robey F et al. Autoimmune disease of the ovary induced by a ZP3 peptide from the mouse zona pellucida. J Clin Invest 1992; 89:28-35.
38. Ellerman KE, Like AA. Susceptibility to diabetes is widely distributed in normal class IIu haplotype rats. Diabetologia 2000; 43(7):890-8.
39. Itoh M, Takahashi T, Sakaguchi N et al. Thymus and autoimmunity: Production of CD25+CD4+ naturally anergic and suppressive T cells as a key function of the thymus in maintaining immunologic self-tolerance. J Immunol 1999; 162:5317-5326.
40. Salomon B, Lenschow DJ, Rhee L et al. B7/CD28 costimulation is essential for the homeostasis of the CD4+CD25+ immunoregulatory T cells that control autoimmune diabetes. Immunity 2000; 12:431-440.
41. Stollerman GH. Variation in group A streptococci and the prevalence of rheumatic fever: A half-century vigil. Ann Intern Med 1993; 118:467-469.
42. Hetzel DJ, Stanhope R, O'Neill BP et al. Gynecologic cancer in patients with subacute cerebellar degeneration predicted by anti-purkinje cell antibodies and limited in metastatic volume. Mayo Clin Proc 1990; 65:1558-1563.
43. Carroll GJ, Will RK, Peter JB et al. Penicillamine induced polymyositis and dermatomyositis. J Rheumatol 1987; 14:995-1001.
44. Reunala T, Salmi J and Karvonen J. Dermatitis herpetiformis and celiac disease associated with Addison's disease. Arch Dermatol 1987; 123:930-932.
45. Savilahti E, Simell O, Koskimies S et al. Celiac disease in insulin-dependent diabetes mellitus. J Pediatr 1986; 108:690-693.
46. Ehrenstein M, Longhurst C, Isenberg DA. Production and analysis of IgG monoclonal anti-DNA antibodies from systemic lupus erythematosus (SLE) patients. Clin Exp Immunol 1993; 92:39-45.
47. Colonna M, Bresnahan M, Bahram S et al. Alleleic variants of the human putative peptide transporter involved in antigen processing. Proc Natl Acad Sci USA 1992; 89:3932-3936.
48. Thai A-C, Eisenbarth GS. Natural history of IDDM. Diabetes Rev 1993; 1:1-14.
49. Baekkeskov S, Kanaani J, Jaume JC et al. Does GAD have a unique role in triggering IDDM? J Autoimmun 2000; 15:279-286.
50. Costagliola S, Swillens S, Niccoli P et al. Binding assay for thyrotropin receptor autoantibodies using the recombinant receptor protein. J Clin Endocrinol Metab 1992; 75:1540-1544.
51. Takasu N, Yamada T, Takasu M et al. Disappearance of thyrotropin-blocking antibodies and spontaneous recovery from hypothyroidism in autoimmune thyroiditis. N Engl J Med 1992; 326:513-518.
52. Posillico JT, Wortsman J, Srikanta S et al. Parathyroid cell surface autoantibodies that inhibit parathyroid hormone secretion from dispersed human parathyroid cells. Bone Miner Res 1986; 1:475-483.
53. Wang R, Abrams SI, Loh DY et al. Separation of CD4+ functional responses by peptide dose in Th1 and Th2 subsets expressing the same transgenic antigen receptor. Cell Immunol 1993; 148:357-370.
54. Ebel F, Schmitt E, Peter-Katalinic J et al. Gangliosides: Differentiation markers for murine T helper lymphocyte subpopulations TH1 and TH2. Biochemistry 1992; 31:12190-12197.
55. Romagnani S. Th1 and Th2 subsets of CD4+ T lymphocytes. Science Med 1994; 1:68-77.
56. Kolb H, Kolb-Bachofen V. Nitric oxide: A pathogenetic factor in autoimmunity. Immunol Today 1992; 13:157-160.
57. Corbett JA, Wang JL, Sweetland MA et al. Interleukin 1β induces the formation of nitric oxide by β-cells purified from rodent islets of langerhans. J Clin Invest 1992; 90:2384-2391.

58. Bradley BJ, Haskins K, La Rosa FG et al. CD8 T cells are not required for islet destruction induced by a CD4+ islet-specific T-cell clone. Diabetes 1992; 41:1603-1608.
59. Gotfredsen CF, Buschard K, Frandsen EK. Reduction of diabetes incidence of BB Wistar rats by early prophylactic insulin treatment of diabetes-prone animals. Diabetologia 1985; 28:933-935.
60. Keller RJ, Eisenbarth GS, Jackson RA. Insulin prophylaxis in individuals at high risk of type I diabetes. Lancet 1993; 341:927-928.
61. Zhang JZ, Lee CS, Lider O et al. Suppression of adjuvant arthritis in lewis rats by oral administration of type II collagen. J Immunol 1990; 145:2489-2493.
62. Gautam AM, Pearson CI, Sinha AA et al. Inhibition of experimental autoimmune encephalomyelitis by a nonimmunogenic nonself peptide that binds to I-Au. J Immunol 1992; 148:3049-3054.
63. Gaur A, Wiers B, Liu A et al. Amelioration of autoimmune encephalomyelitis by myelin basic protein synthetic peptide-induced anergy. Science 1992; 258:1491-1494.
64. Haynes BF, Hemler ME, Mann DL et al. Characterization of a monoclonal antibody (4F2) that binds to human monocytes and to a subset of activated lymphocytes. J Immunol 1981; 126:1409-1414.
65. Kelley VE, Bacha P, Pankewycz O et al. Interleukin 2 diptheria toxin fusion protein can abolish cell-mediated immunity in vivo. Proc Natl Acad Sci USA 1988; 85:3980-3984.
66. Strom TB, Kelley VR, Murphy JR et al. Interleukin-2 receptor-directed therapies: Antibody-or cytokine- based targeting molecules. Ann Rev Med 1993; 44:343-353.
67. Strom TB, Kelley VR, Woodworth TG et al. Interleukin-2 receptor-directed immunosuppressive therapies: Antibody- or cytokine-based targeting molecules. Immunol Rev 1992; 129:131-163.
68. Shizuru JA, Taylor-Edwards C, Banks BA et al. Immunotherapy of the nonobese diabetic mouse: Treatment with an antibody to T-helper lymphocytes. Science 1988; 240:659-662.
69. Howard LM, Miga AJ, Vanderlugt CL et al. Mechanisms of immunotherapeutic intervention by anti-CD40L (CD154) antibody in an animal model of multiple sclerosis. J Clin Invest 1999; 103:281-290.
70. Rossini AA, Mordes JP, Markees TG et al. Induction of islet transplantation tolerance using donor specific transfusion and anti-CD154 monoclonal antibody. Transplant Proc 1999; 31:629-632.
71. Kirk AD, Burkly LC, Batty DS et al. Treatment with humanized monoclonal antibody against CD154 prevents acute renal allograft rejection in nonhuman primates. Nat Med 1999; 5:686-693.
72. Smith JA, Tso JY, Clark MR et al. Nonmitogenic anti-CD3 monoclonal antibodies deliver a partial T cell receptor signal and induce clonal anergy. J Exp Med 1997; 185:1413-1422.
73. Khoury SJ, Hancock WW, Weiner HL. Oral tolerance to myelin basic protein and natural recovery from experimental autoimmune encephalomyelitis are associated with downregulation of inflammatory cytokines and differential upregulation of transforming growth factor beta, interleukin 4 and prostaglandin E expression in the brain. J Exp Med 1992; 176:1355-1364.
74. Hancock WW, Sayegh MH, Kwok CA et al. Oral, but not intravenous, alloantigen prevents accelerated allograft rejection by selective intragraft Th2 cell activation. Transplantation 1993; 55:1112-1118.
75. Atkinson MA, Maclaren NK, Luchetta R. Insulitis and diabetes in NOD mice reduced by prophylactic insulin therapy. Diabetes 1990; 39:933-937.
76. De Bellis A, Bizzarro A, Rossi R et al. Remission of subclinical adrenocortical failure in subjects with adrenal autoantibodies. J Clin Endocrinol Metab 1993; 76:1002-1007.
77. Schloot N, Eisenbarth GS. Isohormonal therapy of endocrine autoimmunity. Immunol Today 1995; 16:289-294.
78. Sadelain MW, Qin H-Y, Lauzon J et al. Prevention of type I diabetes in NOD mice by adjuvant immunotherapy. Diabetes 1990; 39:583-589.
79. Shehadeh NN, LaRosa F, Lafferty KJ. Altered cytokine activity in adjuvant inhibition of autoimmune diabetes. J Autoimmun 1993; 6:291-300.
80. Weiner HL, Mackin GA, Matsui M et al. Double-blind pilot trial of oral tolerization with myelin antigens in multiple sclerosis. Science 1993; 1321-1324.
81. Nussenblatt RB, Caspi R, Mahdi R et al. Inhibition of S-antigen induced experimental autoimmune uveoretinitis by oral induction of tolerance with S-antigen. J Immunol 1990; 144:1689-1695.
82. Miller A, Lider O, Weiner HL. Antigen-driven bystander suppression after oral administration of antigens. J Exp Med 1991; 174:791-798.
83. Miller A, Lider O, Roberts AB et al. Suppressor T cells generated by oral tolerization to myelin basic protein suppress both in vitro and in vivo immune responses by the release of transforming growth factor B after antigen-specific triggering. Proc Natl Acad Sci USA 1992; 421:425.
84. Simone EA, Wegmann DR, Eisenbarth GS. Immunologic "vaccination" for the prevention of autoimmune diabetes mellitus (type 1A). Diabetes Care 1999; 22:B7-B15.

85. Daniel D, Wegmann DR. Protection of nonobese diabetic mice from diabetes by intranasal or subcutaneous administration of insulin peptide B-(9-23). Proc Natl Acad Sci USA 1996; 93:956-960.
86. Witherden D, van Oers N, Waltzinger C et al. Tetracycline-controllable selection of CD4(+) T cells: Half-life and survival signals in the absence of major histocompatibility complex class II molecules. J Exp Med 2000; 191:355-364.
87. Kwok WW, Liu AW, Novak EJ et al. HLA-DQ tetramers identify epitope-specific T cells in peripheral blood of herpes simplex virus type 2-infected individuals: Direct detection of immunodominant antigen-responsive cells. J Immunol 2000; 164:4244-4249.

Chapter 2A

Cell Therapy of Diabetes

Jan Nygaard Jensen and Jan Jensen

With the beginning of the new century we now experience greatly increased focus on the possible use of cell therapy in diabetes treatment, not only due to advances within the area of cell-based disease therapy itself, but due to the successful islet transplantation of 7 diabetic patients by Shapiro and coworkers, in Edmonton, Canada. The remarkable increase in success rate compared to previous allo-islet transplant procedures is apparently due to use of a glucocorticoid-free immuno-suppression regimen much less toxic to the transplanted islet cells themselves than those drugs previously used. Presently, this protocol is now evaluated in independent islet transplant centers around the world, and results are encouraging. The less encouraging fact is the immediate shortage of islets to be used for transplantation. For that reason, the research interest in reprogramming non β-cells into fully functional insulin-producing β-cells is almost exploding. Much ado is caused by the unexpected finding of plasticity in embryonic (ES) and adult (AS) stem cells.[1] These cells appear to be able to cross previously believed insurmountable barriers between distant phenotypic cell states. But to actually control such differentiation changes rigidly and reproducibly may well turn out to be the next overwhelming barrier hindering their use. However, ES-, or AS-cells are far from the only possibilities as starting material for the generation of β-cells for transplantation. With this review, we will go through the present possibilities that one-day may yield a sufficiently large transplantable cell mass, capable of restoring blood glucose homeostasis when transferred to a diabetic patient.

Islet Cell Transplantation – The Beginning of a New Era in Diabetes Treatment

Type 1A diabetes is caused by the autoimmune destruction of the pancreatic β-cells. What causes this initial destruction is not known, but the end result is clear: the cell type is lost, and an endocrine disorder is manifest. Ever since the initial description of diabetes by Egyptian physicians, Type 1 diabetes has been an incurable disease. Up until the beginning of the 20th century it was the cause of certain death shortly after diagnosis. With the discovery of insulin in 1921, death became preventable, but devastating complications caused by the disease, were not. Today, there is still no cure available, and the diabetic individual will, although being on intense glucose monitoring and insulin therapy, still face a future with disease related complications, and a somewhat diminished lifespan.[2] However, this might change within the near future, as recent successes in islet transplantation has lead to a remarkably increase in the success rate of this procedure.[3] Initially performed by the group of J. Shapiro, at Edmonton University, this procedure is now evaluated by over 40 islet transplantation centers around the world, and new centers are spawning (see the link to International Pancreas and Islet Transplantation Association (IPITA) at www.ipita.org for further information). Islet transplantation has been performed globally in the last 10 years, and although certain individuals were found to be insulin-independent for longer periods, only a minor fraction of all transplantations were successful. In that respect, the combined kidney and pancreas transplantation procedure has shown to have the highest degree of success. Further information of the combined data

Immunology of Type 1 Diabetes, Second Edition, edited by George S. Eisenbarth.

from these patients is available from the International islet Transplant Registry (www.med.uni-giessen.de/itr/), at U. Giessen, Germany. Unfortunately, as promising as the current transplantation procedure may be, as disheartening is the fact that human islets are simply not available to treat but a fraction of the current type I diabetics. An unlimited source, or a procedure for generation of β-cells is required. Investigators around the world have united to this case, strongly supported by the major interest organizations of the world's diabetics, such as the Juvenile Diabetes Foundation International, JDFI (www.jdf.org), American Diabetes Association, ADA (www.diabetes.org), European Association for the Study of Diabetes, EASD (www.easd.org), together with research programs funded by the NIH (www.nih.gov) and NIDDK (www.niddk.nih.gov).

The Cell Type in Demand: The Highly Specialized β-Cell

The β-cell is a highly specialized cell type that phylogenetically developed as the main regulator of glucose homeostasis in higher organisms. The hallmark of the β-cell is its exclusive insulin production, which can not be copied by any other cell type in the organism under normal circumstances. The specialized function of the β-cell, however, extends far beyond the production of insulin, as this cell type has developed elaborate mechanisms whereby it controls insulin production, storage, and release, via an integration of total energy homeostasis of the organism, including those sources as fat and protein in addition to carbohydrates.[4] Apparently, β-cell metabolism has gradually developed in a way that allows it to interpret metabolic rate, and secrete insulin accordingly. Importantly, the metabolic rate of the β-cell, in stark contrast to most cells of the organism, directly reflects the extra-cellular glucose levels, thus allowing it to secrete insulin as long as blood glucose levels are above median levels. It is now known that the main enzyme that causes the β-cell glucose metabolic rate to reflect circulating glucose, is the high-K_m glucose phosphorylating enzyme, glucokinase (hexokinase IV).[5] This enzyme has a K_m value (6-11 mM) that reflects physiological concentrations of glucose, and will therefore phosphorylate glucose taken up by the cell with a rate that is proportional to the extracellular glucose concentration. Thus, this enzyme is not as active as lower Km hexokinases (several hundred fold lower K_m values) which are present in most other cells, where these allow efficient phosphorylation of all glucose taken up by the cell. Phosphorylated glucose (glucose-6-phosphate) cannot escape the cell membrane through diffusion or retro-transport through glucose transporters, and represents the first metabolic step in the ATP-generating process of glycolysis. This process is the cornerstone of β-cell glucose dependent insulin release (Fig. 1).

In addition to the cell-autonomous glucose sensing mechanism of the β-cell, it also responds to endocrine signaling from primarily the gut. Normally, energy intake is initially sensed by the organism through its digestive system, as food is ingested, metabolic substrates become present in the intestine due to proteolysis, lipolysis and carbohydrate hydrolysis, and subsequently transported via the enterocytes to the blood. Specialized endocrine cells reside in the gut wall, capable of sensing energy supply from recent food intake. Two such cells are the GIP (K-cells), and GLP1 producing cells (L-cells). GIP cells are present in the more proximal region of the small intestine, where GLP1 cells reside further down the gut tract. Both cell types are via GIP and GLP1 secretion capable of potentiating insulin release from the β-cells. The β-cell expresses both GLP1 and GIP receptors, which signals through G-protein adaptors (GTP binding proteins), which in turn promote cAMP generation by regulating adenylate cyclase. These hormones can not elicit an insulin release effect by themselves. Only in presence of increased glucose will they increase the total amount of insulin released. Thus, these are also known as the incretin hormones. However, this endocrine system efficiently allows the β-cell to be "posed and ready" for the increase in circulating glucose after a meal. In addition to these mechanisms, it also appears that the β-cells themselves may function via a "group-release", as it has been observed that cells within isolated islets share small metabolites and second messengers via gap-junctions between individual cells. Thus, as the β-cell stimulus secretion pathway is triggered by membrane depolarization, causing a wave of free Ca^{2+}-ion secondary messenger

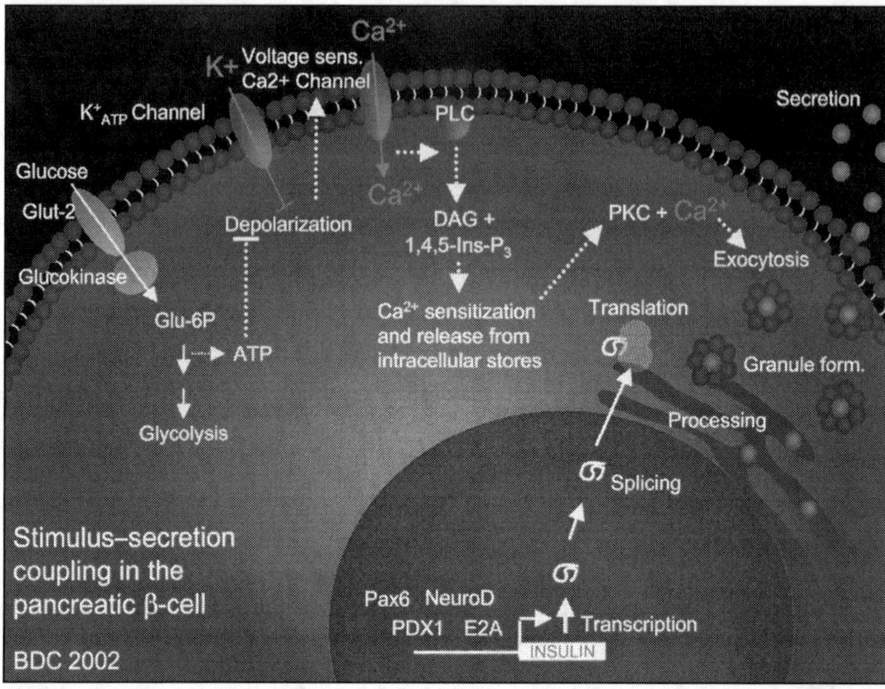

Figure 1. Glucose-dependent insulin release of the β-cell. The β-cell produces insulin via a high level mRNA production. The level of insulin transcription is determined by a set of critical transcription factors binding the insulin promoter (NeuroD, E2A, Pdx1 and Pax6). Recently, nuclear uptake of the Pdx1 factor was found to be regulated in a glucose-dependent fashion, partly explaining how insulin transcription is stimulated by glucose. The two introns in the insulin gene hnRNA are spliced out, and the mature mRNA is transported to the cytoplasm, where ribosomes translate the immature preproinsulin, and transport the product into the endoplasmatic reticulum. The prepeptide is cleaved off in the process. Pro-insulin gets transferred, and sorted to the Golgi due to the presence of its pro-peptide, which also is cleaved off during this process. Further processing occurs in the Golgi, where the intervening C-peptide is cleaved by prohormone convertase-1/3 and 2, generating mature insulin consisting of it's disulfide-bonded A and B chains. Insulin-containing secretory granules form from the Golgi apparatus, containing insulin/Zn^{2+} hexamers. Acidification of the granules occurs, and these are now ready for membrane fusion, and content release. Several factors associate to the membranous component of the granules, involved in the actual release mechanism (SNARE's, SNAP25). Two proteins that may be involved modulating the Ca^{2+}-regulated release are the protein tyrosine phosphatases IA-2 and phogrin (IA-2b) both of which associate to insulin granules. The major pathway for glucose-dependent insulin release is triggered by glycolysis, generating ATP. The ATP-production rate is corresponding to levels of extracellular glucose due to the High Km glucose transporter Glut 2 and High Km glucokinase. The increased ratio of ATP/ADP closes the ATP-sensitive K-channels, which in turn leads to a loss of the negative resting potential of across the membrane (depolarization). This opens voltage gated Ca-channels, dramatically increasing the intracellular Ca^{2+} concentration. As a second messenger, Ca^{2+} has diverse effects, including the activation of phospholipase C (PLC), which cleaves phosphatidylinositol bisphosphate into InsP3 and diacylglycerol (DAG) – both of which have secondary messenger effects. InsP3 mobilized further Ca^{2+} from intracellular stores, and DAG binds to, and activates protein kinase C. The activity of protein kinase C leads to granule exocytosis via a mechanism not yet completely understood.

release, such ions may travel via the gap-junctions through several cells and trigger insulin release, quite possibly in cells that had not met or exceeded the threshold for secretion as measured by that cell itself. This effect can account for the pulsatile insulin release observed in individual islets in culture. Also, as β-cells almost invariably are found as clustered into islets of

Langerhans, or similar types of aggregates, in all species where such a cell type is present this could well signify that this community-action is quite important for optimal function of the cell.

Thus, in a quite a number of ways the β-cell phenotype has specifically adapted to a very specialized work function. Consequently, we probably fare best by viewing this cell as specialized as any muscle, liver, or neuronal cell type. This also means that to generate this cell type from another non β-cell type several concerns arise and certain questions should be addressed as to the nature and requirements of artificially generated β-cells. The first demand is of course that no matter what, the resulting cell "has to be able to produce, store and release insulin in a glucose-stimulatory fashion, with a response window reflecting the normal β-cell". This does not impose a rigid demand that the cells are phenotypical identical to normal β-cells, but we have to ask how much we can deviate from the full phenotype of the normal β-cell and still fulfill the needed behavioral characteristics of the cells to be transplanted. One view is that it may not be possible to come close to fulfilling the demand without generating a cell type that is almost indistinguishable from a normal β-cell. Alternatively, we may use the phrase "If it walks like a duck, and talks like a duck, it is a duck", and believe that the few characteristics investigated will fulfill the requirement. At the moment, this is a very difficult and overlooked question to answer. For instance, the insulin-producing ES-cells by Soria et al (described later), obtained through selection have so far only been tested on their property of relieving streptozotocin (STZ) treated animals from overt diabetes. Whether these fully functions as a normal β-cell mass, and will continue to do so over time is not clear. Similarly, the "converted" hepatocyte cells observed by Ferber et al was also only tested for their capability of rescuing STZ-treated animals – in fact, these cells are most likely far from resembling normal β-cells in a whole range of aspects. Such reservations can be applied to all presently published data of de-novo generated β-cells, as sufficient insight of the phenotype and behavior of these artificially generated β-cells has not been obtained.

The Beta-Cell: A Specialized Team-Player Packing a Punch but with a Poor Constitution

A second special aspect of the β-cell behaviour that deserves mentioning here, is β-cell mass control (see Fig. 2). A more complete understanding of this may help us better in therapeutic matters. Most tissue masses in the adult are dynamic but tightly regulated according to need. Especially during development, total organ sizes have to be critically balanced, but in some cases the proper balancing of tissue is taken care of by excessive formation followed by reduction to need. Apparently, the β-cell mass is adjusted this way after birth. In rodents, apoptosis of endocrine pancreatic cells are observed during early post-natal stages,[6] and recent data have shown a similar reduction in humans. β-cell mass adjustment is not limited to post-natal adjustment. Following pregnancy, the β-cell mass of the mother undergoes a post-partum involution, to adjust to the lower demand after delivery. This is because during gestation only the mother's β cells were contributing to glucose homeostasis. This involution again occurs through apoptosis. In rat models of insulinoma-carrying hosts, the pancreatic β-cell mass is similarly reduced due to apoptosis, and a cease of insulin production in the remaining β-cells is observed. Thus, the β-cell is capable of sensing whether it is present in excess, and thereby adjust accordingly. The mechanism for this is unknown, but appears at least partly to be caused via a prolonged hypoglycaemic state, although this might occur indirectly. Conversely, the β-cell mass can also be upregulated according to need. Obesity increases the demand for additional β-cells, and the existing β-cells respond through an increase in mass, mainly through proliferation.[7] This is believed to be caused by a continuous hyperglycaemic state,[8] and such prolonged hyperglycaemia combined with hyperlipidemia with a progressive peripheral insulin resistance as observed in preType 2 diabetic patients, may cause deterioration of β-cell function, either by causing β-cell dysfunction,[7] or lead to β-cell apoptosis.[9] This, in turn, leads to further strain on residual β-cells, and a worsening condition, that eventually may lead to overt Type 2 diabetes.[10] Pathological upregulation of the β-cell mass is also observed in several cases. Persistent hyperinsulinemia and hypoglycaemia of infancy (PHHI),[11] previously known as nesidioblastosis,

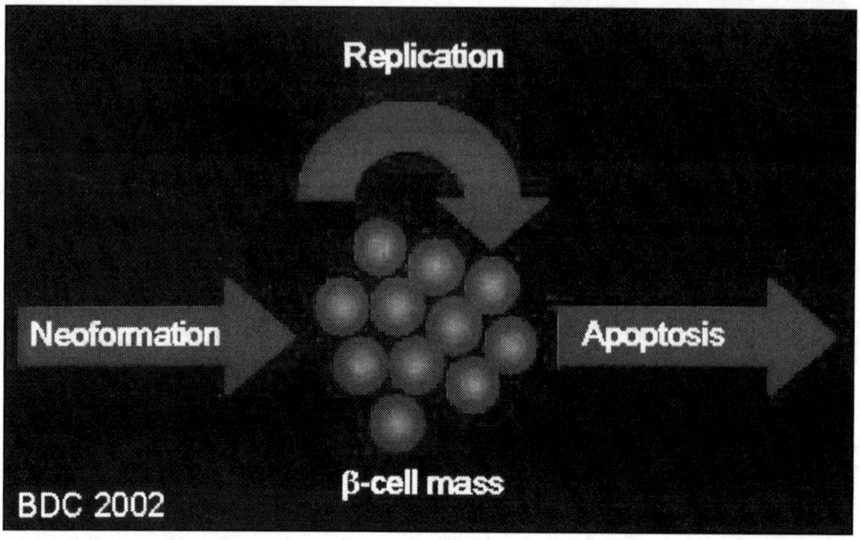

Figure 2. The β-cell mass is determined by the rate of newly differentiating cells (neoformation), the rate of cell division of the existing cells, minus the number of cells dying over time. During the initial embryological stages, neoformation account for most of the increase in β-cell mass. In the third trimester, high mitotic activity increases the β-cell mass (replication). In the early postnatal period, apoptosis reduces the amount of β-cells needed. Later demand for β-cell mass increase is generally observed to derive from replication, although it is not always easy to discriminate replication versus neoformation experimentally.

is a either a congenital disorder (diffuse form), or a loss-of-heterozygozity (LOH) (focal form) whereby the pancreatic β-cell mass is multiple fold upregulated. Mutations causing this disorder may fall in several genes, where the best known is the sulfonyl urea receptor-1 (SUR1).[12] This receptor associates to an ATP-sensitive K^+ inward rectifier channel (kir6.2), in which mutations associating to PHHI also have been found[13] that serves to normalize membrane polarity after stimulated insulin secretion. K_{ATP} channels thereby control the resting membrane potential of β-cells, and their loss results in a constitutive depolarization that allows voltage-gated Ca^{2+} channels to open spontaneously, increasing the cytosolic Ca^{2+} levels enough to trigger continuous release of insulin. If circulating glucose were sensed, and the β-cell mass would adjust accordingly, the hypoglycaemia of the PHHI child should cause an involution of the β-cell mass. However, exactly the opposite occurs.[11] Therefore, the sensing mechanism for a β-cell mass adjustment (via induction of either mitosis, or apoptosis) must rely on the individual cell's interpretation of integrated workload over time rather than being linked to the hyperglycaemic state. In the case of the PHHI β-cell, this is constantly asked for more insulin production and secretion – a "day-at-work" which may mimic the working characteristics of a β-cell in a progressive Type 2 diabetic patient. Supporting this view is the fact that during experimentally induced hyperglycaemia (48 h glucose infusion), which normally causes β-cell dysfunction manifested as impaired glucose-dependent insulin secretion, such deterioration can be prevented by infusion of the β-cell secretion inhibitor diazoxide.[14] For a review see Leahy et al.[15] Thus, reducing workload protects β-cell function.

Certain observations have led to a present view of the β-cell as a "fragile" cell, that may be more susceptible to damage than other cell types. For instance, the β-cells are more sensitive to cytokine-mediated destruction, notably IL-1b, than other islet cells.[16] Upon IL-1b exposure, β-cells express the inducible NO-synthetase (i-NOS),[17] and easily undergo apoptosis when the intracellular free radical NO is produced. Such increased sensitivity may help explain why a precipitating autoimmunity in Type 1 diabetes is specific for the β-cells in the islet, although

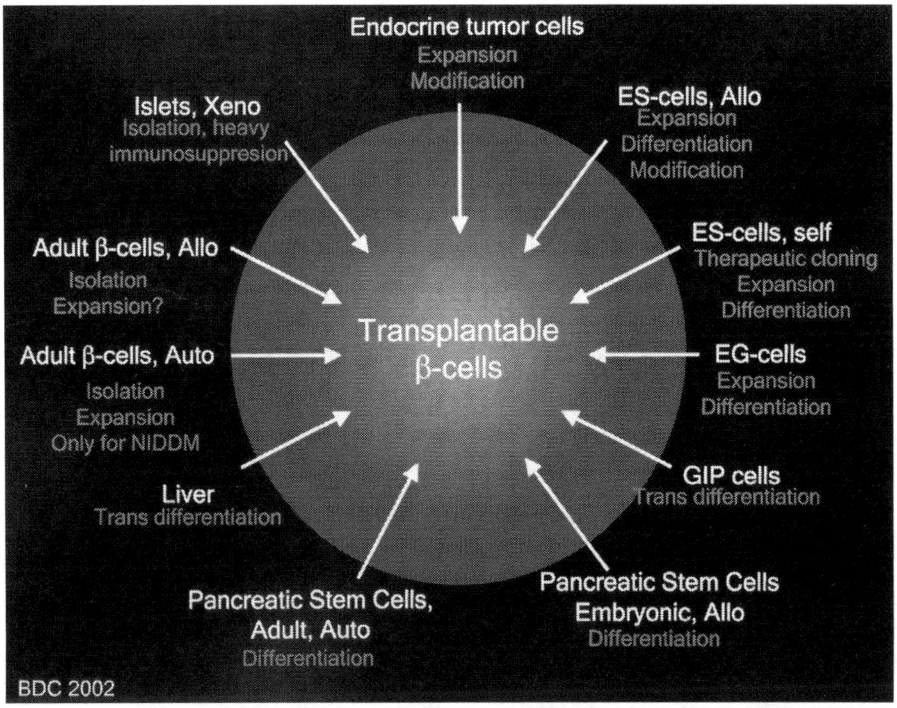

Figure 3. Various possibilities currently being investigated as sources of transplantable β-cells.

this has not been causally related at the moment. Also, β-cells are extremely sensitive to FAS-FAS-L interactions.[18] Normally, expression of FAS-L on immunoprivileged tissues renders cells resistant to FAS presented by activated T-cells, so that these are not eliminated. However, β-cells expressing FAS-L in transgenic NOD-mice, display increased sensitivity to the diabetogenic T-cells present in the NOD (non obese diabetic) mouse, and undergo programmed cell death.

Overview of Possible β-Cell Sources

Several possibilities exist for the generation of β-cells for transplantation (Fig. 3), and not all may be equally suited for clinical purposes due to either severe immunological compatibility issues, latent tumor-forming capability of the cells in question, a non, or aberrant glucose responsiveness, or simply due to an insufficient transplantable cell mass. When it comes to recurrent autoimmune destruction of a transplanted cell mass, this is difficult to avoid in the Type 1 diabetic patient, unless immune-suppression is maintained throughout life. Alternatively, cells transplanted will have to be modified in order to escape recognition by the memory T-cell population existing after Type 1 diagnosis. We can briefly divide up the possible reactions towards transplanted tissue based on known facts of immunity. First, transplantation of cells from a different species (Xenotransplantation) will lead to an acute response (see Chapter 6). This is a brisk immune destruction that is mediated both by the innate and adaptive immune system. For that reason, immuno-suppression is very difficult to invoke, in order to allow for host acceptance. Second, transplantation may be performed using cells from the same species, but different donor genotype (Allotransplantation), and this is the current applied transplantation method of the Edmonton protocol as the source of islets derive from human donors. Most approaches investigated today rely on cells being allo-source, including the generation of human β-cells from ES-cells (see later). We know that immunosuppression against allo-transplanted islets can be controlled, and the success rate of allo-transplatation is greatly increased if donor and host are matched HLA-haplotypes. Finally, we may view the possibility of auto-transplantation, where the derived islet cells initially originate from a certain cell type of the diabetic person, and will thus be of identical genotype. Several approaches consider the development of β-cells with a person-specific origin, and it is self-evident that such cells would be preferred over any of the other donor cells. These possibilities include therapeutic cloning of adult cells via oocyte nuclear transfer and transdifferentiation of other cell types e.g., pancreatic stem cells, liver cells, GIP-cells and hematopoeitic stem cells. However, these various approaches are also those where intricate knowledge of developmental processes and phenotype modulation are needed to be successful, as such cells are not initially of the β-cell type. In the following, we will briefly discuss these various approaches to generate β-cells.

β-Cells from β-Cells: Controlling Replication

The most readily available source of β-cells that may come to mind is the β-cells themselves. As shown above, replication of preexisting β-cells may account for most upward mass adjustments, so why not attempt to increase β-cell mass from a limited source via inducing a mitotic program? This is more easily suggested than done, primarily due to two problems. First, nontransformed β-cells are of "primary tissue" character, and thus susceptible to cell senescence like other primary cell cultures. Second, after decades of studies of β-cell mitogenic compounds, the conclusion is that it is very difficult to maintain primary islet cells in culture in a dividing state.[19] Thus, primary islet cell expansion has not been shown to be feasible. Some of the strongest mitogens for the β-cell are growth hormone and prolactin, which intracellularly signals through the JAK/STAT pathway.[20-22] IGFII is also a β-cell mitogen, which may be required for autocrine growth control during late embryonic and early postnatal stages, as this is expressed in the β-cells themselves.[23-25] Furthermore, upregulation of endogenous IGFII expression has been associated with insulinoma formation, and increased IGFII is also observed in the enlarged β-cell mass of PHHI patients.[26] A major downstream effector of the growth promoting effects of IGFII on islets is Akt/PKB, that is activated via the PI3 kinase pathway. Overexpression of Akt in the beta cell results in markedly increased β-cell mass.[27] Other β-cell mitogens include HGF,[28] GLP1[29,30] and GIP[31]. However, even with the identification of the above growth promoting agents, no treatment has yet proven capable of maintaining islet cells in a dividing state, allowing for significant expansion of the islet preparation. Finally, it has recently become clear that certain components of the cell cycle machinery has β-cell type specific functions – including the involvement of CDK4 and p16, where mutations in these genes specifically affect β-cell mass, leaving most other cell types in the organism fully functional.[32] These observations strongly emphasize that more work needs to be done in order to fully understand β-cell cycle control, and this area may well contain several unexpected findings relating to cell-type specific mitotic control in general.

Xenotransplantation

Xenotransplantation refers to cross-species transplantation. For the immunological aspects of xenotransplantation, please see Chapter 6, by M. Coulombe and R. Gill. Most efforts today evaluate the feasibility of pig islet cells, which are readily available. However, to use nonhuman islet cells for transplantation purposes has several pitfalls, although two different strategies may help towards this. First, if islet cell encapsulation can be performed, allowing sufficient oxygen and nutritional supply for continuous function and life for the pig islets, such islets represent an immediate source to be utilized. We still have to see much progress in the area of biomaterial encapsulation, before this is feasible for clinical use. Secondly, as transgenic techniques established in the mouse, are utilized to generate transgenic pigs, such technology may help for the generation of "humanized" islets. Even in this scenario, such islets will always be at minimum of the "allo"-type, and most likely still express certain pig-epitopes, that may trigger acute rejection. Therefore, therapy will require immunosuppression preceding transplantation. Notwithstanding, the use of pig islets, which have a very similar glucose responsive capability, as well as responsiveness to most types of human endocrine signaling, may appear to be our best bet for the time being in providing material for cell therapy of diabetics.

Embryonic Stem Cells – A Holy Grail or the Epitome of Man's Reduction of Life Through Science?

In the early 1980s the first permanent embryonic stem cell (ES) line from mice was established followed later in the 1990s by a pluripotent embryonic germ cell line (EG), reviewed in Wobus et al.[33] Now, primarily due to the ubiquitous use of transgenic animal generation relying on blastula injection of mouse ES-cells, rigid conditions for maintaining also human ES-cell lines has been achieved so that both human ES- and EG-cell lines are successfully maintained in culture over extended periods. Envisioning the use of ES-cells in any therapeutic strategy deriving β-cell from ES-cells, it is important to emphasize that most studies described so far are studies of animal stem cells, and we do not know if these results can be directly correlated to human stem cell research. Of most importance is the fact of long term ES-derived cell survival, as mice cells do not have the lifespan of human cells due to the earlier cell senescence, as the organism is shorter lived.

ES and EG cells have the capacity to self-renew and to differentiate into various somatic cell types (more than 200 different mature cell phenotypes). Studies in culture systems have shown that this process occurs in a presumably randomized manner, if the suppressive stimuli for differentiation, such as LIF, are removed from the culture. Because of the possibility for these cells to undergo differentiation into somatic cell types (Fig. 4), they are obvious targets for exploitation to derive either 'artificial' cells of interest or even tissue or organs of interest. Conversely, the talent and potential of stem cells, and future perspectives of how to obtain "personalized" human ES-cells also leads to major ethical discussions. Such cells are natural steps in the process of human cloning, and to allow such a procedure naturally causes concerns of where and how to set proper restrictions if this method is ever to be used.

Differentiation of Embryonic Stem Cells into Insulin Producing Cells

In contrast to mesodermally derived cardiomyocytes and ectodermally derived neural cells, no endoderm-derived cells (including the β-cell) differentiate spontaneously from in vitro culture of stem cells. Consequently, the ES cells require supplementary and generally unknown factors to be directed into a controlled pathway of endocrine development. Several approaches have been evaluated to establish endocrine lineages. One is to use selective markers together with a variety of added reagents. Alternatively, culture condition induced differentiation of ES-cells towards β-cells may be aided by following percentages of cells expressing markers that either label pancreatic precursor or endocrine derived stem cell lineages. An evaluation of both of these methods has been experimentally tested in mouse ES-lines, and these are described below.

The Active Selection Approach

Soria and coworkers used the advantage of transfection of a stem cell population with a resistance marker (Neo^R) construct controlled by the insulin gene promoter, which makes it possible to select only those cells that matures to activate this particular promoter (fate coupled transgene), and thereby allowing selection against unwanted cell lineages.[34,35] Taking this approach, Soria et al further designed the construct to coexpress a nontoxic, but easily detectable marker, green fluorescent protein (GFP). By so doing, the GFP marker should allow for selection of a phenotypic more pure cell population via cell sorting. This way, the cell line IB/3x-99 was derived directly from a differentiating mouse ES-cell culture and was found to have the capacity of forming cell aggregates that secreted insulin in a glucose dependent manner. To test for the functional capabilities of these cells in vivo, aggregates were implanted into the spleen of recipient syngeneic STZ diabetic mice. Encouragingly, most mice showed immediate improved

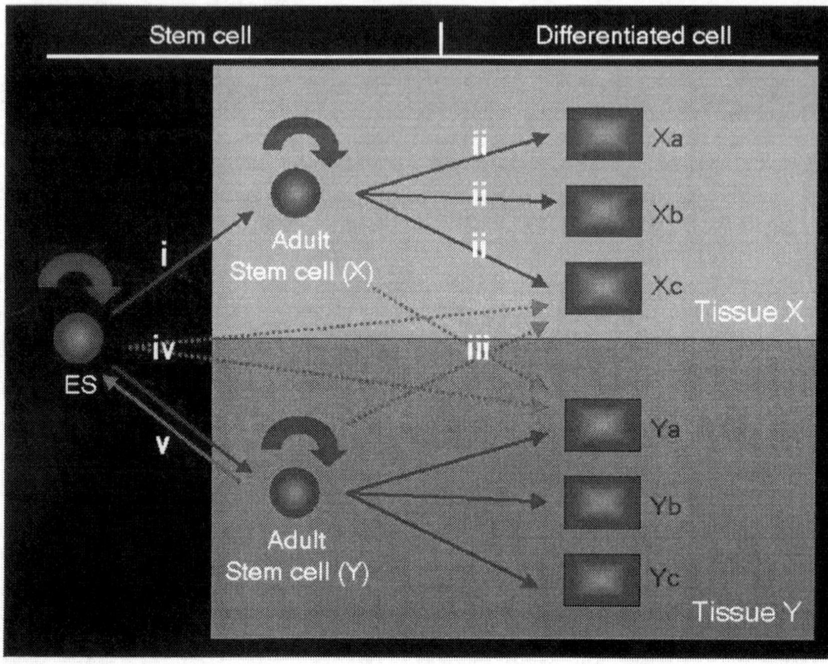

Figure 4. Differentation control of stem cells. Simplified version of cell differentiation, where two fully developed and independent tissues exist (tissue X and Y), each containing a pluripotent stem cell type (adult stem cells), such as e.g., the intestinal crypt cell, the basal cell type of the skin, or the hematopoeitic stem cell. The adult stem cell type is capable of self-renewal and can give rise to all more differentiated cells in the tissue (ii). At present, studies have shown that certain differentiation events may happen from such cells, where cell types normally not developing from the particular adult stem cell anyway will form (iii). However, we generally lack information of how such changes may be induced, if allowed at all, as certain transdifferentiation events simply can not happen due to loss of pluripotency from those lineages. Adult stem cells arise through gradual specification of previous precursor cells as embryogenesis proceeds (i). As ES cells represent early undifferentiated and omnipotent cells, these may differentiate into the various lineages of all adult tissues (iv) correctly in vivo, and aberrantly in vitro. However, such a differentiation event, especially if performed in vitro, will bypass most of the proper spatially and temporally defined signals that allow the normal adult stem cells to form (iv), and it is therefore questionable what exact phenotype may be obtained through such an aberrant differentiation procedure. An argument may be that only if we can recapitulate the signaling environment over time as it normally happens, we will be able to derive the correct phenotype. Countering this argument is the possibility that ES-cells may adapt functionally in vivo to the environment to which they have been introduced, based on the presence of local morphogenetic signals. Thus, it could be envisioned that when e.g., neuronal precursors are introduced into the adult brain, certain will respond by differentiation into becoming a fully mature cell type, and establish a new set of neuronal connections, according to its new phenotype – and thereby survive. Most likely, the performance and flexibility of ES-cells may lie within these views. At the moment, we need better understanding of how the plasticity of ES-cells is modulated by extra cellular factors to answer this question. Finally, we have the possibility of "reverse" the restricted potency of the adult stem cell, by performing nuclear transfer (v). Apparently, this procedure allows nuclei of adult stem cells to generate progeny that may contribute to all cell types in the organism, including the germ line. How the genomic restrictions within the adult stem cell is neutralized, and the potency thus gained is not well established. The fact that this is possible has shocked our understanding of irreversibility in gene transcription.

control of blood glucose. However, 40% of the transplanted mice developed hyperglycemia 12 weeks after transplantation, even though having regained normal blood glucose levels one week after surgery.[34] Thus, despite the positive finding that STZ mice can be treated with ES cell derived therapy, the study also pinpoints the difficulties of maintenance of cell clusters as physiological β-cells. Several reasons may account for the loss of the transplanted cells but were not investigated in detail. These include loss of differentiated characteristics in the absence of selective pressure, induced cell death perhaps due to the foreign environment, or targeted destruction by the host. Also, it remains a possibility that teratoma-like differentiation of certain cells in the aggregates would overgrow, and thereby adversely affect the function of adjacent cells continuously producing insulin.

The Culture Refinement Approach

Lumelsky et al tested the hypothesis that ES progeny cells expressing the gene nestin might undergo endocrine development.[36] Nestin is a filament protein that has been identified to be a marker for a multipotent stem cell in the central nervous system, and due the many similarities in development of the nervous system and the pancreas, nestin might be the marker for pancreatic endocrine precursor cells as well. At present, it is currently not known whether normal islet forms from nestin positive precursors. The protocol used by Lumelsky and coworkers to enhance nestin expressing cells within the ES cell-pool was based on selection pressure. Dividing up the culture phases of ES-cells into expansion and differentiation segments, where various components were present or absent (LIF deprivation, media supplements and bFGF), the investigators found that endocrine cell bodies were increased in their culture. In brief, the protocol enables ES cells to differentiate into cell aggregates that express not only insulin but also the endocrine markers glucacon, somatostatin and pancreatic polypeptide, which resample normal islet gene expression. No exocrine markers (amylase, carboxypeptidase A) were identified in the culture. Histologically, these cell clusters had significant common characteristics with normal islet structure, but only expressed insulin fifty times less than normal islet cells. Nevertheless, insulin secretion responded to physiological levels of glucose in the media and in a dose-dependent matter. Testing for in vivo functional capability, diabetic mice were implanted with these cultured cell aggregates and although the animals did show some improvement in general health, the surgery did not rescue the mice from the diabetic phenotype, most likely due to the low levels of secreted insulin (2% of normal). The primary note of caution in relation to the above study again goes towards the actual phenotype of the insulin-positive cells in the ES-culture. It is known that neuronal cells during embryogenesis are capable of expressing the insulin gene. As there is significant overlap in the critical transcription factor for neurogenesis and endocrinogenesis, it remains a possibility that the neuronal bHLH components such as neurogenin-1, 2, NeuroD, and NeuroD2 may help activate the insulin gene in a bona-fide neuronal cell, as these factors are either identical (NeuroD), or very similar (Ngn1 or 2) to those bHLH components necessary for pancreatic endocrine cell function and development (NeuroD and Ngn3). However, the endodermal-specific marker HNF3b was expressed in the derived ES-cells, and this is extremely encouraging. Further phenotypic characterization using more specific markers of the ES-derived cells should allow for a more comprehensible evaluation of how much these cells adhere to a β-cell phenotype.

Promises and Pitfalls on the Path to a Cure for Diabetes Using ES Cells

In addition to the known phenotypic plasticity, two of the more promising aspects of ES-cells is their capability for unlimited expansion in vitro, and their amenability for further modification prior to use. In fact, it can easily be envisioned that ES-cell clones can be established that carries immuno-modulatory genes, or can be modified so that they become "immuno-refractory", and will be more easily accepted in an allotransplantation procedure. On the other hand, if the

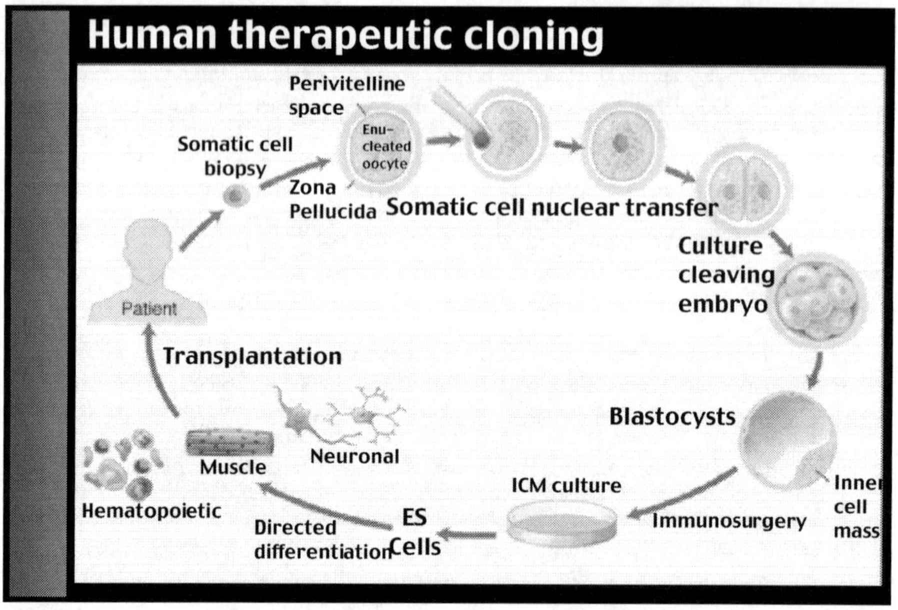

Figure 5. Overview of the therapeutic cloning procedure. In order to obtain self-derived stem cells, adult cell nuclei are transferred to human oocytes, obtained as excess material from in vitro fertilization. This procedure was used to create the sheep "Dolly", and is at present highly controversial, and ethically debated, as the resulting blastula from the procedure could be transferred to a human foster mother, and give rise to a complete human being. Figure reproduced by permission from Solter et al ScienceMag.

culture contains few undifferentiated cells, these may form teratocarcinomas, i.e., tumors, in the host.[37] For that reason, we need stable 'β-cell like' lineages, which are not tumorigenic, but still have the ability to proliferate at some level. For an ES-cell based therapy to be completely safe, it appears that we at minimum need to be able to control the biological equilibrium between proliferative capacity, cell death and latent teratogenesis post-transplantation in vivo, and also have a secure evaluation of the total lifespan of the implanted cells.

Therapeutic Cloning – Taking Advantage of Person-Specific Stem Cells

By the generation of the cloned sheep "Dolly" at the Roslin Institute (www.roslin.ac.uk), considerable interest has been placed in the possibility of making person-specific stem cells. Dolly was made through injection of an adult cell nucleus, into an enucleated, fertilized egg (Fig. 5). Any progeny from such transfer are genetically identical to the donor nuclei, and are thus exact clones of the adult sheep. The prevailing view in biology prior to this experiment was that it was not possible to do, as it was 'well known' that cells irreversibly loose immature characteristics, when maturing – recently Wakayami et al made 35 independent ES-cell lines from adult somatic nuclei, showing that such irreversibility does not really exist.[38] Apparently, the genetic program is apparently reset by this method – the possible mechanisms for this was currently reviewed by Surani.[39] Although human cloning is theoretically possible using this technique, to derive an identical human being from a living person is ethically controversial. Nevertheless, due to the possibility of taking advantage of the initial stages of the technique in order to make syngeneic stem cells, which will be readily accepted by the immune system (the host and donor are identical), this procedure is currently being intensely discussed. Artificial organ derivation, neural stem cell replenishment, or β-cell therapy, could all be helped through the use of this technique. If successful the most critical aspect is that of ethical considerations as the recipient fertilized eggs prior to enucleation (which have the capability of developing into a different human being), and the renucleated egg (which has the ability to develop into a clone of the donor patient) never will be allowed to do so. For a more elaborate discussion of this technique please see this review by Davor Solter in Science: www.sciencemag.org (1999; 283(5407):1468-70).

Adult Stem Cells a Resource for Tissue Homeostasis and Regeneration – And Perhaps Conversion

Stem cells are not only found during embryonic development, but also present in the adult organism, where these cells are as important for maintaining of the adult organ function and tissue maintenance, as embryonic stem cells are for the development. The best characterized adult stem cell is the hematopoietic stem cell, which give rise to a continuous generation of blood cells.[40,41] Adult stem cells have been verified in the liver, brain, skin, bone marrow, intestinal epithelium, myocardium and also suggested to exist in the pancreas (described later).[42-44] Although adult stem cells shares some of the same properties as embryonic stem cells, it has been generally observed that these have limited capabilities of differentiating into cells types other than that of the tissue they reside in. We can view these as undifferentiated cells that 'rest' in several if not all tissues, often characterized by a specific location in a proliferative unit, and sometimes these may be characterized by expression of certain markers. No common marker exist for adult stem cells, and in many cases, we even lack useful markers to identify cells in this state, depending on the tissue. The existence of adult stem cells is noteworthy in relation to cell-based therapies of disease. Due to their pluripotency, perhaps these may be converted into other cell types, such as the β-cells of the pancreas?

Plasticity of Adult Stem Cells

We do not know much of the mechanisms allowing for the plasticity of the adult stem cells to undergo transdifferentiation into different tissue than they originate from, or their ability to first undergo de-differentiation, and then further differentiate. The classical view has been that adult stem cells located in a specialized organ have certain limitations as to which cell fates they can give rise to. However, this interpretation is under heavy challenge these years, because it seems that even resting cells in a specialized organ are able to transdifferentiate into very different cell types, signifying that the adult stem cells are indeed multipotent. One example of this is bone marrow stem cells being capable of transdifferentiating into such diverse cells as brain-, liver- and muscle-cells.[41,43,45] Neural stem cells can undergo differentiation into muscle and hematopoeitic cell fates.[44-47] But more important for diabetes research is that hepatic stem cells which normally differentiate into morphologically and functionally mature hepatocytes, also have the capacity to transdifferentiated into pancreatic acinar-, ductal- and intestinal epithelial cells, if transplanted into pancreas or duodenal wall. Suzuki et al labeled hematopoeitic stem cells with enhanced-GFP and used these cells in transplantation studies.[48] Subsequent isolation from transplantations revealed expression of pancreatic specific markers, such as insulin, glucagon and somatostatin, detectable by RT-PCR when EGFP clones were cultured. This suggests that stem cells receive signals originating in the local environment, leading to a reprogramming of their cell fate choice. This opens an even broader perspective in the use of adult stem cells, because it seems that the preprogrammed fate can be reprogrammed depending upon the microenvironment to undergo an endodermal fate change.

Liver

Of the tissues outside the pancreas, the liver represents a very enticing source for generating β-cells. In a developmental perspective, the liver, ventral pancreas and gallbladder coordinately develop from the ventral side of the primitive gut tube. In fact, the ventral endoderm appears to be default programmed to develop as pancreas, which happens in absence of cues necessary to establish the liver fate.[49-51] These cues are primarily originating from the cardiac mesoderm, that helps pattern this region of the endoderm. This indicates that a common precursor state exist between hepatic and pancreatic development, which has to be rooted in sharing of genetic networks. Such sharing apparently continues even within the adult cells of the two organs. Rodents maintained on a copper-deficient diet display transdifferentiation of pancreas into liver[52-54] and embryonic rodent pancreas cultures undergo a similar conversion in high levels of glucocorticoid.[55] Liver-type carcinomas may also be found in the pancreas.[56] The genetic basis of the similarity between these organs is readily evident as several of the initially classified hepatocyte transcriptional regulators, such as HNF1, vHNF1, HNF3b, HNF3a, HNF4 and HNF6 are all expressed in the pancreas as well. Here, some of these have been linked to aspects of establishing the pancreatic fate,[57] as well as participating in the terminal gene activation of several genes expressed in the pancreas.[58-63] However, significant variance exists between these two tissues, which clearly display dramatic differences in organ formation. The pancreas coordinates growth as an interplay between a branching epithelial precursor sheet, supported by trophic signals from adjacent mesenchymal cells, and is capable of generating several independent cell types. In contrast, the liver primarily generates one cell type, where the major initial growth is conferred by immigration of hematopoeitic precursors, residing here through most of embryogenesis. However, the shared phenotypical basis between the pancreas and liver may be exactly what is needed for a pancreas-specific transcription factor to be activated, although transdifferentiation from liver to pancreas is seldom observed in adult pathological situations. Ferber et al hypothesized that the pancreas-specific, insulin transcription factor, PDX1 would be able to elicit a change in function of liver hepatocytes.[64] To test this, they injected PDX1-expressing adenovirus in the tail vein of mice, and observed the presence of adenovirus-encoded PDX1 mRNA primarily in the liver, but not pancreas. In all cases, mouse insulin II gene mRNA was detected in PDX1-expressing livers, which was confirmed by radioimmunoassay to be 20-fold elevated compared to controls. In addition, the presence of β-cell expressed prohormone convertase 1/3 was also noted, indicating that liver cells could be able to process insulin correctly. Importantly, mice rendered diabetic prior to administration of the virus with the β-cell toxin Streptozotocin (STZ), could be restored to normoglycaemia with the injection of the virus. Together, these are amazing findings. However, a certain amount of caution, and additional experimentation is needed to clarify the actual effect of the PDX1 ectopic expression. First, the authors have not shown what cells in liver in fact express the PDX1 gene, and what cell type that underwent a change in differentiation. It is argued that the cell type can not be the mature hepatocyte, as transduction of primary hepatocyte cultures failed to induce insulin. Second, we do not know how much of the exact identity and characteristics of the resulting transdifferentiated liver cell shares with normal β-cell function. Is it glucose-responsive? Is it capable of storing and releasing insulin at all? Does it respond to glucagon, as a native hepatocyte, by inducing gluconeogenesis and glycogenolysis? Along that line, how much of the original liver phenotype is retained? And when it comes to the delivery system is adenovirus a suitable transfer system that could be developed into a human therapy, or will manipulation of the tissue have to occur ex vivo?

Nevertheless, these are all issues that may be experimentally addressed and one of the more encouraging aspects of targeting liver tissue in a transdifferentation procedure ex vivo is that patient-derived liver biopsies may be readily available, due to the excellent regeneration potential of the liver.

Modified GIP Cells: Intestinal Rescue of a Pancreatic Problem

Several groups have successfully been able to ectopically express the insulin gene in various tissues as liver, muscle pituitary, hemapoietic stem cells, fibroblasts and the gastrointestinal tissue.[65-69] However, insulin was not secreted in a glucose-dependent manner. Looking for a possible β-cell replacement cell type, Cheung and coworkers rationalized that the intestinal glucose-dependent insulinotropic polypeptide (GIP)-cell could fulfill certain characteristics of the pancreatic β-cell, most importantly that of glucose-induced hormone secretion.[70] As intestinal GIP cells do not produce insulin, transgenic mice were produced by oocyte injection of a transgene incorporating the GIP-cell specific 5' regulatory region of the GIP gene coupled to the coding region of the human insulin gene. Also, such a strategy may prevent recurrent autoimmune destruction of the insulin-producing cells, as these cells are by all means besides the presence of insulin still of the GIP-type, and not a β-cell type. GIP cells are open-type cells located in the intestinal wall, capable of sensing increases in intestinal glucose after a meal. In the endocrine system, GIP is upstream of insulin, where GIP is one of the incretin hormones, acting on the β-cell directly. Thus, at least following a meal, the kinetics of GIP and insulin release has some resemblance. Importantly, it was found that ectopic insulin was expressed and fully processed in the K cells within the stomach and duodenum and its release was found to correlate with blood glucose concentrations. Furthermore, these mice were also protected against streptozotocin (STZ) induced diabetes, showing that the toxin does not affect the GIP-cell type. Encouraging, these animals displayed normal blood glucose tolerance, indicating that the mass of GIP cells is sufficient to clear a glucose bolus when compared to the normal β-cell mass. Finally, these cells are currently being tested whether they are resistant to autoimmune destruction by islet-reactive T-cells, (Dr. Tim Kieffer, personal communication).

If basing a β-cell replacement therapy in humans on this approach, some concerns arise. First, it is not possible to target human GIP cells in the same manner as when making a transgenic animal—instead, efficient gene delivery has to take place targeting the human GIP cells. This may not be possible at all—the GIP cells are a scarce population, and widely dispersed. Any targeting of this cell even with a viral delivery system will also target adjacent non GIP cells in the intestinal epithelium, which could have adverse affects due to the viral presence – although this effect will probably be of short duration, due to the rapid turnover of cells in the intestine. Rather, it is much more enticing to target the stem-cell population of the gut, which then would maintain the transgene, and GIP cells differentiating from this population will be able to express the transgene. In theory, this seems feasible, and this way also a continuous replacement of the modified GIP cells would be secured. However, this requires a viral delivery system that is optimized for use in the gut, and one that preferentially targets the crypt population. Weighing in the positive aspects of such a therapy including glucose-responsiveness, insulin processing capability, population renewal, and perhaps lack of recurrent autoimmune destruction, this is a highly promising approach. However, more studies are needed to evaluate the actual performance of the modified GIP cells in the mouse, as it is not known under which circumstances the GIP cells might not respond to the actual needs. As when elevation of blood glucose occurs by increased liver output, these will respond correctly. There may be metabolic or endocrine situations where these cells do not perform as a β-cell, and will therefore not be capable of maintaining normoglycaemia in all cases.

Endocrine Tumor Cells

Over the last decade significant research has been performed to evaluate the use of transformed β-cells for use in transplantation[71-73] and for a review see reference.[74] The reason for this is simple: such cells are already of the β-cell type, produce insulin, and has no problem of limited growth as these cells are transformed to immortality and easily expanded. However, some extremely annoying phenotype-linked problems have to be resolved if such cells are to be used for transplantation purposes. First, the benefit of unregulated growth is a major challenge. Only if this can be safely controlled, un-encapsulated tumor cells have a chance of being transferred to humans. Secondly, the transforming events leading to unlimited growth significantly destroys mature functions of the initial β-cell phenotype. Always, the glucose-response window is severely down-shifted, often due to expression of one or both of the low-Km hexokinase enzymes in the tumor cells. Presence of either hexokinase I or II leads to a blood-glucose independent glucose oxidation in the cell, with the effect of constitutive low-level insulin secretion. Storage of insulin does not occur, and phased insulin secretion is thereby impaired. Most likely, insulin secretory granule maturation is also affected. It is difficult to envision how to efficiently counter this effect. Selection for growth will always have the effect of negatively selecting for differentiated functions, such as insulin production and storage. These characteristics are energy-expensive, and not compatible with growth selection.

Due to these concerns, most research in this area presently focuses on applying gene regulatory systems that allows for a controlled growth phase entry/exit, hereby achieving a reversal to mature functions after the necessary expansion of the cells. At present, the tetracycline-based (Tet-OFF) gene regulatory technique is best characterized. Efrat and coworkers were first to establish conditionally transformed beta cells, naming these bTC-tet cells.[75,76] This β-cell line was established from tumors arising in double transgenic mice carrying both an insulin-promoter driven tet-repressor-VP16 fusion protein, and the tet-operon controlled SV40 Large T antigen. Thus, these cells were designed to be conditionally growing only in absence of tetracycline, or doxycycline (Tet/Dox), due to the combined presence of the chimeric activator which would drive expression of the tetracycline-controlled SV40 large T-antigen transforming oncogene. However, in presence of Tet, transcription of LargeT cease, as the TetR-VP16 fusion protein becomes impaired in DNA binding when tetracycline is bound to the factor. Thus, it was speculated that differentiated functions would reappear. Indeed, both in vitro, as well as in vivo, proliferation of the cells, or the resulting tumors, was inhibited by tetracycline. Accordingly, SV40 large T antigen was reduced in most cells, although a small number of cells displayed expression of LargeT even in presence of Tet. Importantly, when these cells were transplanted to STZ-rendered diabetic mice treated with Tet, these would alleviate diabetes, and not cause hypoglycaemia, as normal bTC cell lines do, due to their continuous growth. Even better, the cells were found to secrete insulin at a "normal" threshold above 2-3 mM glucose, even in the absence of Tet at least at lower passage numbers.

Despite the promising performance, these cells have a major drawback, signified by their latent transformed state. It is evident that heterogeneous LargeT antigen expression is already observed in the isolated culture, which is "bathed" in tetracycline. High-passage culturing of these cells seems to cause some of the same problems associated with regular bTC cells being a gradual loss of glucose responsiveness due to hexokinase expression.[77] This is most likely due to amplification of a culture heterogeneity, and is a phenomenon that is almost impossible to circumvent. Also, if clonal heterogeneity leads to loss of growth control in a few cells, this may have devastating effects over time, as this will cause insulinoma formation. For this reason, to apply such a genetic switch as the Tet-controlled LargeT system in cells for human transplantation may require the development of a fail-safe mechanism of inducible cellular elimination to provide for the safety needed.

Regeneration of β-Cells from Pancreatic Tissue

Normally, β-cells arise from pancreatic precursor cells during embryonic stages (Fig. 6). Thus, whatever specific changes in gene expression that has to happen during β-cell formation, these happen correctly when the starting point is the embryonic pancreas. We do not know whether this process can take place correctly from any other tissue besides pancreas, and we know even less of the required mechanisms to initiate such a process. Consequently, the pancreas represents the most promising starting material for the generation of fully functional β-cells. Unfortunately, the existence of precursor cells in the adult pancreas is debatable, as we lack usable markers describing the phenotype of such cells. The adult pancreas is densely packed with mature cells of various types, mostly pancreatic exocrine cells. In addition, the four islet cell types are organized into the endocrine cell mass, and although nestin-positive, hormone-negative, islet cells has been found, these have not been convincingly shown to be actual precursor cells of the islets.[78,79] In fact, to experimentally determine between β-cell neoformation versus replication is very difficult, and this concern has recently been discussed by Dr. Bonner Weir.[80] However, Krakowski and coworkers have found that over expression of

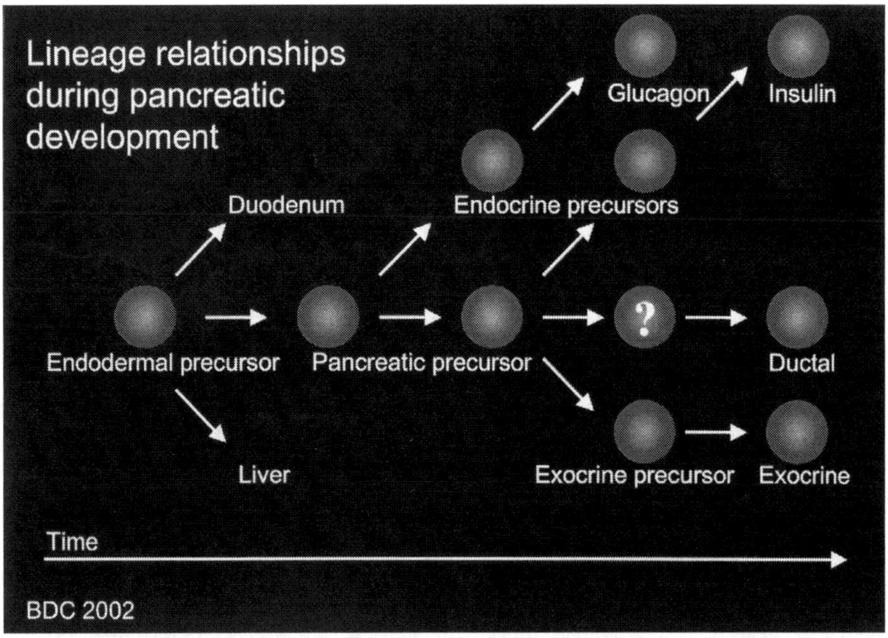

Figure 6. Lineage relationship diagram of the developing pancreas. Pancreatic precursor cells are established from endodermal precursors. These are initially specified at the early somite stages, at defined regions of the primitive gut. Differentiation of endocrine glucagon and insulin expressing cells is independently occurring from the precursor cells. These generally differentiate in the interior regions of the developing organ. More exterior regions are predisposed to develop as exocrine, which occurs concomitantly to the development of the mature β-cells at a timepoint known as the secondary transition. Ductal cells form most likely from remaining pancreatic precursors, although markers of their differentiation program is generally lacking. The cell type depicted with a "?" represent cells with a stem-cell like character in the adult organ. These have been suggested to be of the intercalated duct cell, or centroacinar cell type. However, markers characterizing this phenotype have not been found, and its existence is controversial. It is possible that adult exocrine cells may undergo a de-differentiation process towards this cell type during certain models of pancreatic regeneration.

FGF7 in β-cells lead to expansion of intra-islet ductal cells, which may suggest that an intra-islet precursor cell type may exist.[81] This suggests that stem cells in the pancreas may be of ductal-type morphology. Mature ductal cells generate the remaining endodermally derived components by forming the draining ductal tree for the exocrine acinar cells. None of these cell types resemble the pancreatic precursor cells from which the β-cells are generated through development, and it is therefore also a question of whether a β-cell differentiation process can occur from these.

However, under certain circumstances of adult pancreatic regeneration, de-novo differentiation of islet cells have been noted.[82-84] Thus, some pancreatic cells may under given conditions change phenotype into a possible reflection of the pancreatic precursor cell type. In most cases, such regeneration has been noted from pancreatic duct cells – although the issue of which subpopulation these may be, and where such ductal precursor cells may reside, is not clear. In the following, we will go through various models of pancreas regeneration and discuss finding of neodifferentiation of pancreatic cells. As these models differ significantly in the regenerative process generally in the aspect of (1) levels of inflammation, (2) presence or absence of systemic effects, such as hyperglycaemia, (3) exocrine/endocrine destruction, and (4) presence of local hypoxia and hemorrhage, we will briefly go through these individually.

Pancreatitis

Pancreatitis is an either acute or continuous inflammation of the organ. Most of the precipitating causes in humans are due to an obstruction of the pancreatic ductal system, leading to intra-organ retention of pancreatic juice, which secondarily causes autodigestion, and cell death. The main cause for pancreatitis is alcoholism. Several experimental animal models of pancreatic regeneration after pancreatitis fall into this category, including the duct-ligation model, the cellophane-wrap model, and supramaximal caerulein-treatment leading to exocrine hypersecretion. In early, or very severe cases as acute fulminant necrotizing pancreatitis, β-cell destruction occurs via inflammatory cytokines, leading to hyperglycaemia. However, the pancreas has an excellent recovery capability after pancreatitis, and a 100% recovery rate is observed in animals after a multiple dose caerulein injection. This regeneration is primarily due to massive proliferation of mature exocrine cells that survive the treatment. Normally, this treatment does not affect the pancreatic islets, unless the supramaximal dose is maintained until a necrotizing pancreatitis in induced. We have recently observed that only limited endocrine neoformation occurs during caerulein-induced regeneration. However, several markers of pancreatic precursors cells are reactivated within regenerating exocrine cells, including Pdx1 and components of the Notch-system (J.N. Jensen, unpublished). Such data holds promise for the possibility of converting such cells to a β-cell-type, which is presently being pursued in our laboratory.

Partial Pancreatectomy

Partial pancreatectomy (px) represents a highly used method of inducing pancreas regeneration in rodents. The method is performed by gently rubbing away pancreatic tissue, leaving the major vessels intact. The most common experimental use of px has been for studies of β-cell regeneration, where a 90% px procedure causes mild hyperglycaemia, and will thus trigger a β-cell response. Regeneration of islet cells from proliferating duct cells have also been noted.[84,85] The pancreatic regeneration occurs at the face of destruction, and this process is independent of the hyperglycaemia as normoglycaemic 50% px animals display a similar regenerative process. This process is different from that of pancreatitis as only little inflammation is observed. Some similarities to pancreatitis exist, as exocrine cells also proliferate strongly. However, it has been observed that these exocrine cells also may undergo a dedifferentiation process, and start to express the Pdx1 gene.[84] This could signify a reversal to the "protodifferentiated cell" phenotype as found in the pancreatic stem cells during embryogenesis.[86] Importantly, the regeneration is highly localized as no changes are observed in the remnant organ, and this is most likely caused by local hypoxia, blood-clotting from where wound-healing cytokines are produced, or both. Unfortunately, the molecular changes occurring over time during px have not been studied in detail, and we lack firm evidence that those exocrine cells found to de-differentiate has the ability like other embryonic stem cells to differentiate into all lineages of the pancreas.

Pancreas Toxins

Some pancreas toxins (ethionine, taurocholate) initiate a severe pancreatic destruction that may lead to a regenerative process where pancreas cells arise through differentiation. Ethionine (ethyl analog of the aminoacid methionine), given together with a choline-deficient diet cause both liver and pancreas destruction (diet-induced pancreatitis).[87-89] Ethionine inhibits intracellular S-adenosyl-methionine mediated methylation, and will eventually cause widespread liver and pancreatic necrosis. Very little information about the regenerative process following Ethionine treatment is available. However, although both caerulein and ethionine induce histological characterized pancreatitis, a major difference exists, as recovery is always observed after caerulein treatment, whereas ethionine triggers necrosis and death.[90] The reason for this difference is not clear, but may be rooted in irreversible changes to the exocrine cells in the ethionine procedure. Also, ethionine causes a block of exocrine secretion,[91] whereas caerulein has the opposite effect.

β-Cell Toxins

The β-cell toxins alloxan and Streptozotocin leads to destruction of the insulin producing cells, leaving the organ intact.[92] No regenerative process is initiated in the pancreas, except for induction of β-cell replication in surviving β-cells in submaximal doses of Streptozotocin. Furthermore, this is most likely an indirect result secondary to a hyperglycaemic state, that is initiated by a similar means as when the β-cell mass needs to expand during physiological processes such as pregnancy or body mass increases.

Reprogamming Cells – Where Do We Get a Grip?

Assuming that the adult pancreas does contain or can be made to contain cells of stem cell character that may differentiate into insulin producing cells. Or assuming that ES-cells do have the plasticity to convert into fully functional β-cells, then what control genes, or factors, do we need to express, or add, to elicit such a phenotype conversion? In this area, most can be learned from the developmental biology of the pancreas. Starting with the identification of the Pdx1 gene, which was the first gene shown to be required for pancreas development, a growing range of developmental regulatory genes have been identified that act at various levels in pancreatic and endocrine specification (Fig. 7). It falls outside the scope of this review to characterize these factors, suffice to say that unraveling the genetic network underlying pancreas development may give us excellent tools for converting cells towards a given lineage in the pancreas. In the case of the β-cell certain factors stand out. HNF6 and Ngn3 are both required for endocrine cell development – in fact, Ngn3 may by itself force an endocrine program in pancreatic precursor cells.[57,93,94] At present, the capacity of these factors in inducing an endocrine program in adult pancreatic cells have not been tested, although this appears to be likely at least under certain circumstances that would allow for a stem-cell-like reversal of more mature pancreatic cells. Pax4 is a gene that is required for proper β-cell specification.[95] In its absence, β-, and δ-cells do not form, although other endocrine cells do so. Therefore, Pax4 may be important in ES-cells conversion strategies, or adult stem cell conversion to make sure that β-cells, and not alpha-cells, are derived through the process. Other critical transcription factors for the β-cell are Pdx1, Nkx2.2 and Nkx6.1.[96-98] However, these genes are expressed both in the precursors of the β-cells, as well as in the mature cells themselves, and it seems more unlikely that these may trigger a differentiation process by themselves. Nonetheless, these are all required for proper establishment of β-cell phenotype, and are good markers for the fully mature β-cell. Pdx1 is a critical homeodomain factor that is involved in glucose-mediated insulin transcription, and most likely Nkx2.2 and Nkx6.1 play as critical roles in adult β-cell function as well, in addition to their role proper phenotype establishment.

In addition to the intracellular determination factors, we also need to understand how the immature pancreatic cells communicate in order to undergo normal development. Humans develop via the process of "regulative development" where cell fates are not predetermined, but given through each cell's continuous interpretation of its environment through embryogenesis. These signals originate from the cells themselves, and allow a constant sensing of neighbors, and organ size. Some of the signaling molecules that may relay such positional information include those of the BMP, hedgehog, FGF, Wnt and Notch genes. Several members of these various families are expressed in the developing pancreas, and play a role in organ development. For instance, FGF10 is required to allow for a proper expansion of pancreatic precursor cells[99] and Notch signaling is necessary for proper cell differentiation.[93,100] As these signaling systems act via the surface of the precursor cells, these are also accessible to the investigator. Therefore, we may in the future see these signaling networks exploited to govern fate allocation and tissue expansion in a controlled manner.

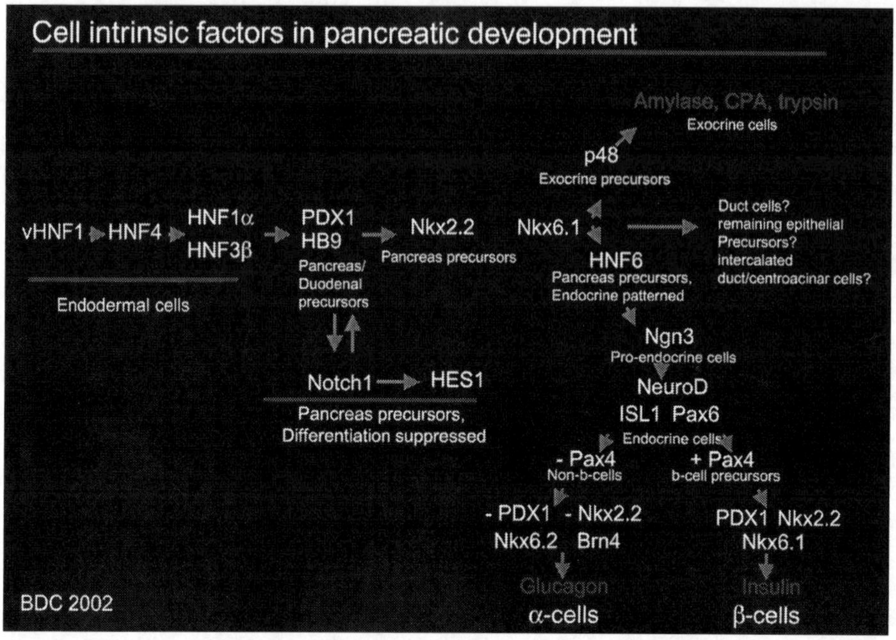

Figure 7. Diagram depicting the relationship between various transcription factors in the lineage tree of pancreatic development. A detailed description of these factors can be found in various reviews. The pancreatic state is signified by expression of a set of homeodomain factors (PDX1, HB9, Nkx2.2, Nkx6.1), which developed from endodermal precursors that expressed several of the hepatocyte nuclear factors (vHNF1, HNF4, HNF1a, HNF3b). Differentiation from the pancreatic precursors are controlled by bHLH factors, where p48 leads to exocrine development, and Ngn3 to endocrine development. The differentiation program is blocked by the Notch-system which is active in the pancreatic precursors prior to differentiation. HNF6 may play a role in the initation of the endocrine program by activating ngn3. Sub-specification of the endocrine cells is determined by several homeodomain factors, including PDX1, Nkx6.1, Nkx2.2, Pax4, Nkx6.2 and Brn4. The endocrine cell state is also signified by expression of the bHLH factor neuroD and the homeodomain factors Pax6 and Isl1.

What Technology Is Acceptable – And What Does the Patient with Diabetes Wish?

Here listing a plethora of pathways that may lead to solve the islet demand problem, maybe we should ask if all, or only some, may be acceptable for the diabetic person. It seems that in many cases the desire for the investigator to solve the β-cell problem does not take into account the problematic issues arising, if his/her method proves successful to create an "unlimited" source of glucose-responsive β-cells. For instance, the use of reengineered insulinoma cells may of little interest to the patient, simply due to the latent immortality of the cells and the loss of glucose-responsiveness that may occur post-transplantation. Such cells are always plastic, and very uncontrollable. Similarly, the tumor forming capability of the ES-lines, may hinder their use, simply due to fear by the person to receive such cells. Also, the actual cloning procedure to generate person-specific ES-cells may turn most people away from exploitation. This said, any technology harnessing new islet cells would also have to be of as little discomfort, and short hospitalization time for the patients as possible. The promises of an islet transplantation procedure in competition with present day insulin-therapy will have to include that of reliability, safety, ease, and the assurance that the worry of diabetic complications can be eliminated. This is a lot to ask for a therapy in its infancy, but it is certainly not an unrealistic demand to be met.

Perspectives

Banting and Best received in the shortest time in history the Nobel prize in 1923 for their discovery of insulin in 1921-22. Now, approaching a full century with the lifesaving insulin available for the world's diabetic population maybe the time for change is just around the corner. We have here compiled the expanding list of possible methods that is currently being investigated to generate the specific β-cell type. It is difficult to predict which, if any or even if more, of the present approaches will successfully provide islet material useful for transplantation of diabetic patients. However, the trend is strongly towards a scenario where insulin-therapy of Type 1 diabetic patients becomes obsolete. Perhaps both the basic science groundwork, the clinical proof-of-concept and the established treatment is already in place in the clinics by 2023? If so, we may possibly be able to marvel at the focused research effort that finally succeeded to provide a cure for diabetes, and simply in our daily lives view islet cell transplantation as common, and as safe as kidney transplantations are today.

Acknowledgements

We would like to acknowledge the support of the Childrens Diabetes Foundation, the University of Copenhagen, and the American Diabetes Association for helping establish the conditions for writing this chapter. We would also like to thank Dr. J. Hutton and Dr. Tobias Zahn at the Barbara Davis Center for Childhood Diabetes for helpful comments on the manuscript.

References

1. Vogel G. Can old cells learn new tricks? Science 2000; 287(5457):1418-1419.
2. Jones DB, Gill GV. Insulin dependent diabetes mellitus. In: Pickup J, Williams G, eds. Textbook of Diabetes. Blackwell Science, 1997.
3. Shapiro AM, Ryan EA, Lakey JR. Pancreatic islet transplantation in the treatment of diabetes mellitus. Best Pract Res Clin Endocrinol Metab 2001; 15(2):241-264.
4. Kruszynska YT. Normal metabolism: The physioogy of fuel homeostasis. In: Pickup J, Williams G, eds. Textbook of Diabetes. Blackwell Science, 1997.
5. Grupe A, Hultgren B, Ryan A et al. Transgenic knockouts reveal a critical requirement for pancreatic beta-cell glucokinase in maintaining glucose-homeostasis. Cell 1995; 83(1):-69.
6. Hanke J. Apoptosis and occurrence of Bcl-2, Bak, Bax, Fas and FasL in the developing and adult rat endocrine pancreas. Anat Embryol (Berl) 2000; 202(4):303-312.
7. Weir GC, Laybutt DR, Kaneto H et al. Beta-cell adaptation and decompensation during the progression of diabetes. Diabetes 2001; 50 Suppl 1:S154-S159.
8. Steil GM, Trivedi N, Jonas JC et al. Adaptation of beta-cell mass to substrate oversupply: Enhanced function with normal gene expression. Am J Physiol Endocrinol Metab 2001; 280(5):E788-E796.
9. Biarnes M, Montolio M, Nacher V et al. Beta-cell death and mass in syngeneically transplanted islets exposed to short- and long-term hyperglycemia. Diabetes 2002; 51(1):66-72.
10. Yki-Jarvinen H, Williams G. Insulin resistance in noninsulin dependent diabetes mellitus. In: Pickup J, Williams G, eds. Textbook of Diabetes. Blackwell Science, 1997.
11. Thomas PM, Cote GJ, Wohllk N et al. The molecular basis for familial persistent hyperinsulinemic hypoglycemia of infancy. Proc Assoc Am Physicians 1996; 108(1):14-19.
12. Thomas PM, Cote GJ, Wohllk N et al. Mutations in the sulfonylurea receptor gene in familial persistent hyperinsulinemic hypoglycemia of infancy. Science 1995; 268(5209):426-429.
13. Thomas P, Ye Y, Lightner E. Mutation of the pancreatic islet inward rectifier Kir6.2 also leads to familial persistent hyperinsulinemic hypoglycemia of infancy. Hum Mol Genet 1996; 5(11):1809-1812.
14. Sako Y, Grill VE. Coupling of beta-cell desensitization by hyperglycemia to excessive stimulation and circulating insulin in glucose-infused rats. Diabetes 1990; 39(12):1580-1583.
15. Leahy JL. b-cell dysfunction and chronic hyperglycaemia. In: Habener JF, Hussain MA, eds. Molecular basis of pancreas development and function. Kluwer Academic Press, 2001:35-46.
16. Johannesen J, Pociot F, Karlsen AE et al. Strain-dependent difference in inducible nitric oxide synthesis (iNOS) expression in rat pancreatic islets correlates with interferon regulating factor 1 (IRF-1) and heat shock protein 70 (HSP70) expression. Eur Cytokine Netw 2001; 12(3):501-509.
17. Nielsen K, Karlsen AE, Deckert M et al. Beta-cell maturation leads to in vitro sensitivity to cytotoxins. Diabetes 1999; 48(12):2324-2332.

18. Chervonsky AV, Wang Y, Wong FS et al. The role of Fas in autoimmune diabetes. Cell 1997; 89(1):17-24.
19. Nielsen JH, Svensson C, Galsgaard ED. Beta cell proliferation and growth factors. J Mol Med 1999; 77(1):62-66.
20. Stout LE, Svensson AM, Sorenson RL. Prolactin regulation of islet-derived INS-1 cells: Characteristics and immunocytochemical analysis of STAT5 translocation. Endocrinology 1997; 138(4):1592-1603.
21. Billestrup N, Nielsen JH. The stimulatory effect of growth hormone, prolactin, and placental lactogen on beta-cell proliferation is not mediated by insulin-like growth factor-I. Endocrinology 1991; 129(2):883-888.
22. Brelje TC, Sorenson RL. Role of prolactin versus growth hormone on islet B-cell proliferation in vitro: Implications for pregnancy. Endocrinology 1991; 128(1):45-57.
23. Hill DJ, Hogg J. Growth factor control of pancreatic B cell hyperplasia. Baillieres Clin Endocrinol Metab 1991; 5(4):689-698.
24. Miettinen PJ, Otonkoski T, Voutilainen R. Insulin-like growth factor-II and transforming growth factor-alpha in developing human fetal pancreatic islets. J Endocrinol 1993; 138(1):127-136.
25. Petrik J, Pell JM, Arany E et al. Overexpression of insulin-like growth factor-II in transgenic mice is associated with pancreatic islet cell hyperplasia. Endocrinology 1999; 140(5):2353-2363.
26. Kassem SA, Ariel I, Thornton PS et al. p57(KIP2) expression in normal islet cells and in hyperinsulinism of infancy. Diabetes 2001; 50(12):2763-2769.
27. Tuttle RL, Gill NS, Pugh W et al. Regulation of pancreatic beta-cell growth and survival by the serine/threonine protein kinase Akt1/PKBalpha. Nat Med 2001; 7(10):1133-1137.
28. Garcia-Ocana A, Takane KK, Syed MA. Hepatocyte growth factor overexpression in the islet of transgenic mice increases beta cell proliferation, enhances islet mass, and induces mild hypoglycemia. J Biol Chem 2000; 275(2):1226-1232.
29. Stoffers DA, Kieffer TJ, Hussain MA et al. Insulinotropic glucagon-like peptide 1 agonists stimulate expression of homeodomain protein IDX-1 and increase islet size in mouse pancreas. Diabetes 2000; 49(5):741-748.
30. Xu G, Stoffers DA, Habener JF et al. Exendin-4 stimulates both beta-cell replication and neogenesis, resulting in increased beta-cell mass and improved glucose tolerance in diabetic rats. Diabetes 1999; 48(12):2270-2276.
31. Trumper A, Trumper K, Trusheim H et al. Glucose-dependent insulinotropic polypeptide is a growth factor for beta (INS-1) cells by pleiotropic signaling. Mol Endocrinol 2001; 15(9):1559-1570.
32. Rane SG, Dubus P, Mettus RV et al. Loss of Cdk4 expression causes insulin-deficient diabetes and Cdk4 activation results in beta-islet cell hyperplasia. Nat Genet 1999; 22(1):44-52.
33. Wobus AM. Potential of embryonic stem cells. Mol Aspects Med 2001; 22(3):149-164.
34. Soria B, Roche E, Berna G et al. Martin F. Insulin-secreting cells derived from embryonic stem cells normalize glycemia in streptozotocin-induced diabetic mice. Diabetes 2000; 49(2):157-162.
35. Soria B, Skoudy A, Martin F. From stem cells to beta cells: New strategies in cell therapy of diabetes mellitus. Diabetologia 2001; 44(4):407-415.
36. Lumelsky N, Blondel O, Laeng P et al. Differentiation of embryonic stem cells to insulin-secreting structures similar to pancreatic islets. Science 2001; 292(5520):1389-1394.
37. Solter D, Gearhart J. Putting stem cells to work. Science 1999; 283(5407):1468-1470.
38. Wakayama T, Tabar V, Rodriguez I et al. Differentiation of embryonic stem cell lines generated from adult somatic cells by nuclear transfer. Science 2001; 292(5517):740-743.
39. Surani MA. Reprogramming of genome function through epigenetic inheritance. Nature 2001; 414(6859):122-128.
40. Kondo M, Weissman IL, Akashi K. Identification of clonogenic common lymphoid progenitors in mouse bone marrow. Cell 1997; 91(5):661-672.
41. Krause DS, Theise ND, Collector MI et al. Multi-organ, multi-lineage engraftment by a single bone marrow-derived stem cell. Cell 2001; 105(3):369-377.
42. Jones PH, Harper S, Watt FM. Stem cell patterning and fate in human epidermis. Cell 1995; 80(1):83-93.
43. Pittenger MF, Mackay AM, Beck SC et al. Multilineage potential of adult human mesenchymal stem cells. Science 1999; 284(5411):143-147.
44. Rietze RL, Valcanis H, Brooker GF et al. Purification of a pluripotent neural stem cell from the adult mouse brain. Nature 2001; 412(6848):736-739.
45. Mezey E, Chandross KJ, Harta G et al. Turning blood into brain: Cells bearing neuronal antigens generated in vivo from bone marrow. Science 2000; 290(5497):1779-1782.
46. Clarke DL, Johansson CB, Wilbertz J et al. Generalized potential of adult neural stem cells. Science 2000; 288(5471):1660-1663.

47. Mezey E, Chandross KJ. Bone marrow: A possible alternative source of cells in the adult nervous system. Eur J Pharmacol 2000; 405(1-3):297-302.
48. Suzuki A, Zheng Yw YW, Kaneko S et al. Clonal identification and characterization of self-renewing pluripotent stem cells in the developing liver. J Cell Biol 2002; 156(1):173-184.
49. Zaret KS. Liver specification and early morphogenesis. Mech Dev 2000; 92(1):83-88.
50. Zaret KS. Hepatocyte differentiation: From the endoderm and beyond. Curr Opin Genet Dev 2001; 11(5):568-574.
51. Deutsch G, Jung JN, Zheng MH et al. A bipotential precursor population for pancreas and liver within the embryonic endoderm. Development 2001; 128(6):871-881.
52. Rao MS, Subbarao V, Reddy JK. Induction of hepatocytes in the pancreas of copper-depleted rats following copper repletion. Cell Differ 1986; 18(2):109-117.
53. Rao MS, Dwivedi RS, Yeldandi AV et al. Role of periductal and ductular epithelial cells of the adult rat pancreas in pancreatic hepatocyte lineage. A change in the differentiation commitment. Am J Pathol 1989; 134(5):1069-1086.
54. Rao MS, Reddy JK. Hepatic transdifferentiation in the pancreas. Semin Cell Biol 1995; 6(3):151-156.
55. Shen CN, Slack JM, Tosh D. Molecular basis of transdifferentiation of pancreas to liver. Nat Cell Biol 2000; 2(12):879-887.
56. Paner GP, Thompson KS, Reyes CV. Hepatoid carcinoma of the pancreas. Cancer 2000; 88(7):1582-1589.
57. Jacquemin P, Durviaux SM, Jensen J et al. Transcription factor hepatocyte nuclear factor 6 regulates pancreatic endocrine cell differentiation and controls expression of the proendocrine gene ngn3. Mol Cell B 2000; 20(12):4445-4454.
58. Ben Shushan E, Marshak S, Shoshkes M et al. A pancreatic beta-cell-specific enhancer in the human PIX-1 gene is regulated by hepatocyte nuclear factor 3 beta (HNF-3 beta), HNF-1 alpha, and SPs transcription factors. J Biol Chem 2001; 276(20):17533-17540.
59. Jacquemin P, Durviaux SM, Jensen J et al. The transcription factor HNF-6 regulates pancreatic endocrine cell differentiation and controls the expression of the pro-endocrine gene *ngn3*. Molecular And Cellular Biology 2000; In press.
60. Gauthier BR, Schwitzgebel VM, Zaiko M et al. Hepatic Nuclear Factor-3 (HNF-3 or Foxa2) Regulates Glucagon Gene Transcription by Binding to the G1 and G2 Promoter Elements. Mol Endocrinol 2002; 16(1):170-183.
61. Pontoglio M, Sreenan S, Roe M et al. Defective insulin secretion in hepatocyte nuclear factor 1alpha- deficient mice. J Clin Invest 1998; 101(10):2215-2222.
62. Shen W, Scearce LM, Brestelli JE et al. Foxa3 (hepatocyte nuclear factor 3gamma) is required for the regulation of hepatic GLUT2 expression and the maintenance of glucose homeostasis during a prolonged fast. J Biol Chem 2001; 276(46):42812-42817.
63. Sladek FM, Dallas-Yang Q, Nepomuceno L. MODY1 mutation Q268X in hepatocyte nuclear factor 4alpha allows for dimerization in solution but causes abnormal subcellular localization. Diabetes 1998; 47(6):985-990.
64. Ferber S, Halkin A, Cohen H et al. Pancreatic and duodenal homeobox gene 1 induces expression of insulin genes in liver and ameliorates streptozotocin-induced hyperglycemia. Nat Med 2000; 6(5):568-572.
65. Bartlett RJ, Secore SL, Denis M et al. Toward the biologic release of human insulin from skeletal muscle. Transplant Proc 1997; 29(4):2199-2200.
66. Bochan MR, Sidner RA, Shah R et al. Stable transduction of human pancreatic adenocarcinoma cells, rat fibroblasts, and bone marrow-derived stem cells with recombinant adeno- associated virus containing the rat preproinsulin II gene. Transplant Proc 1998; 30(2):453-454.
67. Lipes MA, Davalli AM, Cooper EM. Genetic engineering of insulin expression in nonislet cells: Implications for beta-cell replacement therapy for insulin-dependent diabetes mellitus. Acta Diabetol 1997; 34(1):2-5.
68. Falqui L, Martinenghi S, Berra C et al. Human proinsulin production in primary rat hepatocytes after retroviral vector gene transfer. J Mol Med 1999; 77(1):250-253.
69. Goldfine ID, German MS, Tseng HC et al. The endocrine secretion of human insulin and growth hormone by exocrine glands of the gastrointestinal tract. Nat Biotechnol 1997; 15(13):1378-1382.
70. Cheung AT, Dayanandan B, Lewis JT et al. Glucose-dependent insulin release from genetically engineered K cells. Science 2000; 290(5498):1959-1962.
71. de la TD, Halvorsen T, Demeterco C et al. Beta-cell differentiation from a human pancreatic cell line in vitro and in vivo. Mol Endocrinol 2001; 15(3):476-483.
72. Demeterco C, Levine F. Gene therapy for diabetes. Front Biosci 2001; 6:D175-D191.

73. Itkin-Ansari P, Demeterco C, Bossie S et al. PDX-1 and cell-cell contact act in synergy to promote delta-cell development in a human pancreatic endocrine precursor cell line. Mol Endocrinol 2000; 14(6):814-822.
74. Levine F, Leibowitz G. Towards gene therapy of diabetes mellitus. Mol Med Today 1999; 5(4):165-171.
75. Efrat S, Linde S, Kofod H et al. Beta-cell lines derived from transgenic mice expressing a hybrid insulin gene-oncogene. Proc Natl Acad Sci USA 1988; 85(23):9037-9041.
76. Efrat S. Genetically engineered pancreatic beta-cell lines for cell therapy of diabetes. Ann NY Acad Sci 1999; 875:286-293.
77. Fleischer N, Chen C, Surana M et al. Functional analysis of a conditionally transformed pancreatic beta-cell line. Diabetes 1998; 47(9):1419-1425.
78. Hunziker E, Stein M. Nestin-expressing cells in the pancreatic islets of Langerhans. Biochem Biophys Res Commun 2000; 271(1):116-119.
79. Zulewski H, Abraham EJ, Gerlach MJ et al. Multipotential nestin-positive stem cells isolated from adult pancreatic islets differentiate ex vivo into pancreatic endocrine, exocrine, and hepatic phenotypes. Diabetes 2001; 50(3):521-533.
80. Bonner-Weir S. beta-cell turnover: Its assessment and implications. Diabetes 2001; 50 Suppl 1:S20-S24.
81. Krakowski ML, Kritzik MR, Jones EM et al. Transgenic expression of epidermal growth factor and keratinocyte growth factor in beta-cells results in substantial morphological changes. J Endocrinol 1999; 162(2):167-175.
82. Gu D, Lee MS, Krahl T et al. Transitional cells in the regenerating pancreas. Development 1994; 120(7):1873-1881.
83. Bonner-Weir S, Taneja M, Weir GC et al. In vitro cultivation of human islets from expanded ductal tissue. Proc Natl Acad Sci USA 2000; 97(14):7999-8004.
84. Sharma A, Zangen DH, Reitz P et al. The homeodomain protein IDX-1 increases after an early burst of proliferation during pancreatic regeneration. Diabetes 1999; 48(3):507-513.
85. Bonner-Weir S, Baxter LA, Schuppin GT et al. A second pathway for regeneration of adult exocrine and endocrine pancreas. A possible recapitulation of embryonic development. Diabetes 1993; 42(12):1715-1720.
86. Pictet R, Rutter WJ. Development of the embryonic endocrine pancreas. In: Steiner DF, Freinkel N, eds. Handbook of Physiology, section 7: Endocrinology, Vol. 1. Washington DC: American Physiological Society, 1972:25-66.
87. Niederau C, Niederau M, Luthen R et al. Pancreatic exocrine secretion in acute experimental pancreatitis. Gastroenterology 1990; 99(4):1120-1127.
88. Rodriguez AI, De DI, Garcia-Montero AC et al. Acute necrotizing pancreatitis induced by a CDE diet in the rat: A morphologic and functional study. J Lab Clin Med 1994; 123(5):693-700.
89. Walker NI, Winterford CM, Williamson RM et al. Ethionine-induced atrophy of rat pancreas involves apoptosis of acinar cells. Pancreas 1993; 8(4):443-449.
90. Kusske AM, Rongione AJ, Ashley SW et al. Interleukin-10 prevents death in lethal necrotizing pancreatitis in mice. Surgery 1996; 120(2):284-288.
91. Capdevila A, Decha-Umphai W, Song KH et al. Pancreatic exocrine secretion is blocked by inhibitors of methylation. Arch Biochem Biophys 1997; 345(1):47-55.
92. Portha B, Blondel O, Serradas P et al. The rat models of noninsulin dependent diabetes induced by neonatal streptozotocin. Diabete Metab 1989; 15(2):61-75.
93. Apelqvist A, Li H, Sommer L et al. Notch signalling controls pancreatic cell differentiation. Nature 1999; 400(6747):877-881.
94. Gradwohl G, Dierich A, LeMeur M et al. neurogenin3 is required for the development of the four endocrine cell lineages of the pancreas. Proc Natl Acad Sci USA 2000; 97(4):1607-1611.
95. Sosa-Pineda B, Chowdhury K, Torres M et al. The Pax4 gene is essential for differentiation of insulin- producing beta cells in the mammalian pancreas. Nature 1997; 386(6623):399-402.
96. Jonsson J, Carlsson L, Edlund T et al. Insulin-promoter-factor 1 is required for pancreas development in mice. Nature 1994; 371(6498):606-609.
97. Sander M, Sussel L, Conners J et al. Homeobox gene Nkx6.1 lies downstream of Nkx2.2 in the major pathway of beta-cell formation in the pancreas. Development 2000; 127(24):5533-5540.
98. Sussel L, Kalamaras J, Hartigan-O'Connor DJ et al. Mice lacking the homeodomain transcription factor Nkx2.2 have diabetes due to arrested differentiation of pancreatic beta cells. Development 1998; 125(12):2213-2221.
99. Bhushan A, Itoh N, Kato S et al. Fgf10 is essential for maintaining the proliferative capacity of epithelial progenitor cells during early pancreatic organogenesis. Development 2001; 128(24):5109-5117.
100. Jensen J, Pedersen EE, Galante P et al. Control of endodermal endocrine development by Hes-1. Nat Genet 2000; 24(1):36-44.

CHAPTER 2B

Proprotein Processing and Pancreatic Islet Function

John Hutton, Christina Wasmeier, Rodabe Amaria, Nicholas Bright and John Creemers

Overview and Introduction

In type 1A diabetes the insulin-secreting beta cells of the islets of Langerhans are selectively destroyed by an autoimmune process while other hormone secreting islet cells such as glucagon secreting A cell, somatostatin secreting D cell and pancreatic polypeptide secreting PP cells are relatively spared. Despite the specificity of beta cell destruction the targets of autoimmunity are not necessarily β-cell specific and are present as well in the non β-cells or more widely distributed in cells of neurondocrine origin. Such molecules include glutamic acid decarboxylase (GAD), carboxypeptidase E (CPE), ICA512 (IA-2) and phogrin (IA-2β) all of which are targets of humoral autoimmunity in man and experimental animals. A common feature of these molecules is their association with the regulated pathway of secretion and specifically the storage organelles of hormones and nonpeptide neurotransmitters. CPE, like insulin is a secreted lumenal protein, ICA512 and phogrin are transmembrane with the important B and T-cell epitopes localized on the cytoplasmic face of the organelle. GAD65 is a peripherally associated protein of the microvesicle surface; the related 67kD cytosolic form of GAD on the other hand is not a major autoantigen. There are marked differences between species in the expression of specific islet molecules and islet organization (e.g., almost no GAD in mouse islet cells, beta cell selective expression of GAD in rat islets, and GAD expression in all human islet cells) that may be important in diabetes pathogenesis. Likewise there is potential contribution to antigenicity of the post-translational modification of these molecules. Mature human insulin consists of individual A and B chains and is produced from a larger precursor proinsulin which incorporates the B and A-chains in a linear sequence separated by a connecting peptide (C-peptide). CPE, ICA512 and phogrin undergo glycoyslation and phosphorylation and GAD is modified by myristoylation. Thus it is timely to review aspects of islet biochemistry and physiology that relate to the processing and production of islet hormones and other potential islet autoantigens.

Islets, which are individual groups of 2000-5000 cells (Fig. 1), represent approximately 1-2% of pancreatic mass. They are readily isolated by intraductal injection of a protease mixture containing collagenase followed by mechanical agitation and density gradient centrifugation digestion. Suspensions of islets prepared in this manner can be introduced into the body by injection transcutaneously into the hepatic portal vein from where they disperse and lodge in capillary beds within the liver. A good islet preparation from a single donor is often sufficient to restore normoglycemia in a nonautoimmune diabetic person or experimental animal though preparations from multiple donors may be required to obtain true insulin independence and glucose tolerance.

Immunology of Type 1 Diabetes, Second Edition, edited by George S. Eisenbarth.

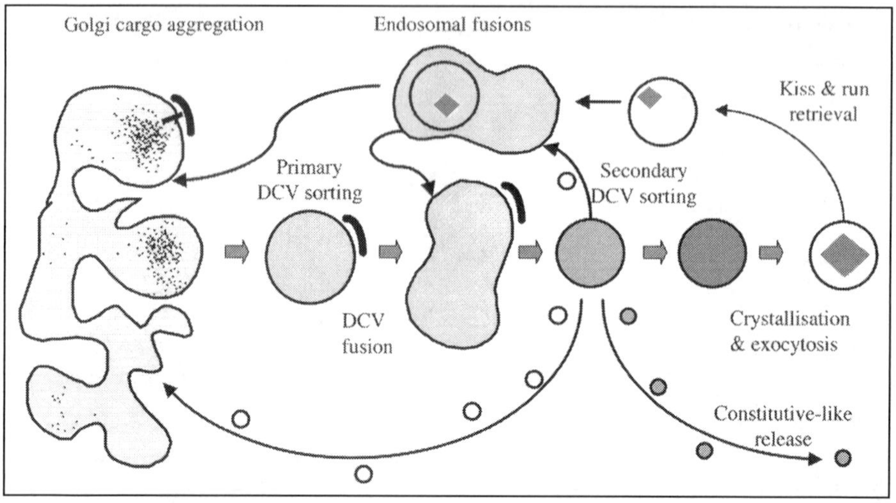

Figure 1. Post Golgi pathways of membrane traffic in the pancreatic β-cell. Granules appear to form from envelopment by lipid membrane of condensed material in the trans Golgi network. The spherical vesicles undergo a maturation process that includes proteolytic processing of the vesicle cargo, removal of missorted proteins by a clathrin-mediated mechanism and fusion of granules to one another. The proteins retrieved from the immature granules may be secreted (constitutive-like secretion). Following regulated secretion the granule membrane is retrieved possibly to an endosomal compartment from where it may be returned to the TGN or destroyed in lysosomes. DCV= dense core vesicle.

The combination of regulated exocytosis of insulin storage granules form the β-cells with the action of insulin on glucose transport into fat and muscle cells and insulin effects on the disposal and production of glucose in the liver provides the physiological basis of glucose homeostasis in the short term. Secretion can be sustained for hours with secretion rates of the order of 10-15% insulin content per hour. Glucose also has a profound effect on the rate of translation of preformed insulin mRNA such that at moderately elevated glucose concentrations (e.g., 8-11mM) there is no net loss of islet insulin content. Under more chronic hyperglycemic conditions transcriptional regulation of the insulin gene contributes to the secretory capacity and there is evidence in the long term that islet cell mass can respond to the hormonal or nutritional milieu. The mediators of the response varies in each case. Glucose provides a primary stimuli in the majority of monogastric mammals but there are responses to numerous physiologic and pharmacologic stimuli, including other islet hormones (e.g., glucagon stimulatory, somatostatin inhibitory). When glucose is administered orally (e.g., an oral glucose tolerance test) there is greater insulin secretion compared to an equivalent elevation of blood glucose achieved by administering intravenous glucose, a phenomenon that is attributable to the effects of glucagon-like peptide 1 and gastric insulinotropic peptide (GIP) both of which are released in a glucose-dependent manner from the gut. The secretory response to square wave pulse of glucose either in vivo or in vitro is biphasic with an initial first phase lasting approximately 5 minutes followed by a steadily increasing and sustained second phase. In vitro glucose stimulation produces a first phase and then second prolonged phase of insulin secretion. The first phase of insulin secretion is extremely sensitive to abnormalities of islet function and quantitatively sensitive to loss of islet beta cell mass. Thus with islet destruction, of essentially any etiology, there is lost first phase insulin secretion following glucose stimuli, while response to other stimuli such as glucagon or amino acids can be preserved or only modestly decreased. Loss of insulin secretion after intravenous glucose is used in clinical trials to stage the severity of islet dysfunction and approximate time to diabetes onset.

Islets secrete multiple soluble granule proteins and peptides in addition to insulin, including the C-peptide of the insulin molecule and another costored peptide amylin. C-peptide is produced on a 1:1 basis with insulin and serves as an important parameter for monitoring insulin secretion in individuals receiving insulin therapy. Therapy for type 1A diabetes utilizes insulin, not proinsulin, and thus any C-peptide in a diabetic individual is endogenously produced. The monitoring of C-peptide secretion is the usual outcome variable for trials for the prevention of further beta cell destruction in patients with new-onset type 1 diabetes.

Perspective

The processing of prohormones to bioactive peptides is highly conserved in evolutionary terms and appears to involve remarkably few gene products given the complexity and plasticity of the biological events that depend upon bioactive peptides. Seven enzymes of the eukaryote subtilase family account for the majority of endoproteolytic processing reaction and two carboxypeptidases for the removal of COOH terminal basic amino acids exposed by endopeptidase activity. Their substrate specificity is determined by simple primary sequence motifs; their biological specificity is gained through regulation of expression at the transcriptional and translational level, intracellular compartmentalization and intramolecular regulation. Some tissues appear to contain antiproteases which provide a further layer of regulation. Thus the combination of tissue-specific expression, intracellular trafficking, responsiveness to ionic conditions, and regulation of their activation state generates alternate pathways of proprotein processing. This allows cells to manufacture multiple possible bioactivities from the same precursor and at the same time permits the same enzymic machinery to act on hundreds of different substrates even within the same cell. Null mutations in PC1, PC2 and CPE have recently been described in rodents and man and the phenotypes broadly support the conclusions drawn from in vitro studies. They nevertheless indicate that further account needs to be given to redundancy in action between the candidate enzymes and the existence of further regulatory controls. They have provided further insight into how both the genetic and pathophysiological regulation of the proprotein processing pathway contributes to diseases such as diabetes and other endocrinopathies in man.

The Islet as a Neuroendocrine Organ

The pancreatic islet, alongside of the thyroid, adrenal medulla, adrenal cortex and diffuse neuroendocrine system are the major organs of metabolic regulation outside of the brain that determine the energetic balance between catabolism and anabolism. The islet can function autonomously of the brain as illustrated by the success of the transplantation of isolated islets in regulating blood glucose in an individual otherwise predisposed to diabetes. Nevertheless, its response to peptides more classically associated with the brain determines the functionality of the organ and the functional mass of β-cells in the body. The embryological derivation of the islet from foregut endoderm is in contrast to the nervous system's origin from neural ectoderm and yet islets expresses a large number of "nerve specific" proteins such as the granin family of secreted proteins,[1-3] neuropeptides such as NPY,[4,5] CCK,[6] POMC[7] and TRH,[8] glutamic acid decarboxylase, a critical biosynthetic enzyme in the production of the classical neurotransmitter gamma amino butyric acid.[9] The functional significance of most of these proteins in the islet remains obscure and their individual ablation is not often associated with an obvious endocrine pancreatic phenotype. Broader hints of a functional relationship is seen in the fact that the islet shares with the hypothalamus the ability to sense glucose concentrations through the possession of glucokinase, Glut-2 transporter[10] and translates these signals through regulated voltage gated K^+ channels[11,12] to induce electrical excitability and secretory responses (see Chapter 2C). Inactivation of the leptin signaling system produces pancreatic β-cell insufficiency possibly through the brain[13,14] as does pharmacological treatments such as gold thioglucose[15] which induces obesity via regions of the hypothalamus involved in appetite regulation. Local regulatory circuits with the same functional components conceivably provide part of the basis of parallel responses in islet and hypothalamus with respect to metabolites and effectors that cross the blood brain barrier. More direct communication is likely to be mediated by circulating polypeptides and most significantly those released by the pituitary. Islet β-cell mass is regulated, responds to prolactin and placental lactogen[16,17] and the secretion of insulin can be modulated directly by the pituitary peptide PACAP[18,19] and possibly other pituitary peptides (ACTH 22-39)[20] and downstream factors such as corticosteroids. The source of these peptides however is not necessarily the pituitary itself. There is increasing awareness that the brain and particularly cells of the hypothalamic/pituitary axis are sensitive to insulin[21] and that the islet itself receives information through the insulin signaling pathway that affects its growth and biosynthetic activity.[22] The major challenge is to put these studies into a physiological context; this however presumes knowledge of the regulatory components. The present chapter reviews what we have learnt in recent years about the production, storage and post-translational modification of polypeptide hormones and neurotransmitters that underlie metabolic regulation. The focus is principally the major secreted products of the beta, alpha, delta and PP cells namely the pancreatic hormones insulin, glucagon, somatostatin and pancreatic polypeptide, respectively.

The Regulated Pathway of Polypeptide Secretion

The storage of bioactive polypeptides in vesicles for rapid release by exocytosis is an evolutionary conserved mechanism present in most eukaryotes including single cell metazoans.[23] Peptides related to the major pancreatic endocrine polypeptides are found in primitive vertebrates in association with gastrointestinal tract without apparent relationship to digestive pancreatic organ as is evident in higher vertebrates.[24] Similar peptides are found in invertebrates, and evolutionary pressure, if anything has acted to decrease the diversity of peptides rather than increase it. Thus the nematode *C. elegans* has 37 insulin-like molecules with mainly neuronal localization[25] and a single homolog of the insulin receptor compared to the three major insulin related proteins in man (insulin, IGF I and IGF II). The function of insulin in nematodes is not clear beyond an apparent association of the insulin receptor molecule DAF-2 which plays a role in determination of life span and progression of aging.[26] In insects, insulin appears to play the role of a peptide neurotransmitter or neuromodulator[27] and already is associated with other proteins that characterize the insulin secretory granule in mammals, namely members of the IA-2 family of protein tyrosine phosphatases.[28,29]

The packaging of insulin into secretory granules follows a similar pathway of most regulated secreted proteins. It enters the secretory pathway at the level of the endoplasmic reticulum (ER) directed by the 16 aa hydrophobic leader (signal) sequence of the preproinsulin NH_2 terminus. The leader sequence is removed by signal peptidases to form proinsulin and the three disulphide bonds are formed assisted by the action of protein disulphide isomerase activity. As it progresses through the secretory pathway it becomes more resistant to reduction,[30] a phenomenon that may contribute to its ultimate condensation on the dense core of the insulin granule. It moves through the Golgi apparatus to the trans-most cisternae where it becomes concentrated as aggregates with the same approximate diameter as the mature secretory granule (Fig. 1) These are often juxtaposed to membrane bearing a partial clathrin coat and apparently give rise to immature secretory granule with an aggregate of similar morphology and a partial clalthrin coat.[31] Maturation is thought to involve the conversion of proinsulin to insulin and concomitant acidification of the granule interior and crystallization of the matrix and loss of the clathrin coat.[32] Recent tomographic reconstruction of the Golgi of rapidly frozen pancreatic β-cells[33] has called into question many aspects of this classical model that is based on ultrastructural analysis of chemically fixed islet tissue and the results of pulse-chase combined with autoradiography. The concept that the granule buds from elements of the TGN that bear a clathrin coat[34] has been set against a model which suggests that the granule can originate from more than one Golgi stack and by a process that uses a coat other than clathrin.[35] The question as to how cargo proteins are transported through the Golgi has also been subject to recent revision as has the concept that proteins undergoing regulated secretion are sorted away from other proteins that are constitutively released or are targeted to other subcellular compartments.[36,37]

It is clear that the trans-most cisternae of the Golgi gives rise to several distinct types of vesicles and that vesicle coat proteins play an important role in vesicle formation playing both a mechanical role and serving as a sorting mechanism.[32,38] The best documented case is the involvement of the heterotrimeric adaptor complex AP-1 binding to the cytoplasmic tail of the mannose-6-phosphate receptor which specifically recognizes lysosomal proteins via a mannose-6-phosphate modification received in the Golgi.[39] The types of Golgi-derived coat proteins[40] has expanded as has the number of members of the adaptor family which now includes 4 classical AP proteins and several related GGA proteins,[41] yet it is unclear which of these are important to regulated secretion. It should be borne in mind as well that major cargo such as the proinsulin molecule may rely on a different mechanism (e.g., aggregation)[42] to the minor granule constituents (e.g., association with lipids or other proteins) and that a third mechanism may be involved in the sorting of secretory granule membrane proteins (e.g., cytosolic factors).[43-45] The model whereby secreted proteins are sorted away from constitutively released proteins by a receptor-mediated mechanism has gained some support from the identification of putative sorting motifs and sortase molecules[46,47] and better definition of the mechanisms by which cargo aggregation or multimerization can occur.[31] On the other hand, the concept that constitutive secretion as the default pathway has been challenged by the finding that professional secretory cells as opposed to tumor cell lines show little constitutive release and can sort even cytosolic proteins into granules when they are fused to a signal sequence that directs them to the lumen of the endoplasmic reticulum.[37]

Our understanding of endocrine secretory granule biogenesis at the molecular level has borrowed extensively from genetic and biochemical studies of vacuolar sorting in yeast even though the yeast do not have the equivalent of dense core granules and regulated secretion. The application of this conceptual framework to in vitro studies of secretory granule biogenesis in neuroendocrine cell types, (PC12 phaeochromocytoma cell line[48] and stably transformed GH3 somatotrophs)[49] has led to the definition of roles for heterotrimeric G proteins,[50] tyrosine phosphorylation,[51] phospholipid transfer activity,[52] PI3 kinase, phospholipase D, syntaxin 6, Arf-1[53] and AP-1[54] as part of the generic membrane trafficking that supports the formation of the mature secretory granule. Rather than being a single sorting event, granule biogenesis can be seen as involving several stages of cargo and membrane remodeling (Fig. 1). This includes the removal of lysosomal enzymes by a mannose-6 phosphate mediated process,[55,56] the retrieval of Golgi resident proteins such as the endopeptidase furin,[48] and the subsequent homotypic fusion of immature granules and removal of excess membrane. The latter process is prominent in cells such as pituitary corticotrophs where the immature and mature granules vary markedly in size. In such models, the sorting of secretogranin II as a marker for the regulated pathway can be reconstituted in a vesicle population distinct from that containing constitutively secreted heparin sulphate proteoglycans, without significant colocalization observed in post-TGN compartments.[57]

There is evidence for the existence of similar pathways in pancreatic islet β-cells. Proinsulin is processed to mature insulin by removal of the connecting C-peptide subsequent to its arrival in the secretory granule. C-peptide is released in molar excess to insulin from unstimulated cells,[58] an observation which has led to a model of "sorting by retention". In this scheme, proteins not destined for the granule may initially be included in the immature granule but are removed by vesicular traffic during the maturation process, while regulated secretory proteins are retained due to their ability to condense into an insoluble aggregate,[59,60] a process that is enhanced by the conversion of proinsulin to insulin[61] and the uptake of Zn by the nascent granule.[62] Lysosomal hydrolase precursors are found in the immature but not in the mature granule, suggesting that they were selectively retrieved during maturation. Indeed, the inhibition of their glycosylation and hence their retrieval by the mannose 6 phosphate receptor leads to their retention and regulated secretion.[56] A vesicular trafficking pathway originating from immature granules could serve a number of roles, for instance an increase in the fidelity of

TGN-based sorting, or the recycling of membrane proteins with a role in granule budding back to the TGN for another round of vesicle formation.

Little information is available about integral membrane proteins specific to the secretory granule. Recently, though, two such molecules have been cloned from β-cells — the closely related protein tyrosine phosphatase homologues phogrin and IA-2.[44,63,64] IA-2 which reaches the cell surface following exocytosis has been shown to return to the Golgi and yet it appears that it may be cleaved in its cytosolic domain by μ calpain in the process.[65] The functional significance of this reaction is not clear; it would appear to eliminate the potential for the IA-2 molecule to interact with the cytoskeleton and thus may mark the granule as sensecent. Our own studies on the related phogrin molecule indicate that the membrane and possibly some cargo, is retrieved by a clathrin-mediated process post exocytosis and returns to an endosomal compartment (Bright, Wasmeier, Angelson et al, unpublished findings). Whether it is reincorporated into nascent granules or is destined lysosomal degradation is presently unclear. Their granule localization of these molecules has made them valuable markers for cell biological studies of secretory granule membrane traffic during granule biogenesis and secretion.[66] Both *Drosophila* and *C. elegans* possess homologues of the IA-2 family[28,29] which are localized in dense core vesicles (DCVs) of the nervous system which introduces the prospect of developing genetic analyses in multicellular organism similar to those in yeast.

The Enzymology of Pancreatic Hormone Proteolysis

Insulin, glucagon, somatostatin and pancreatic polypeptide, like the majority of polypeptide hormones and neuropeptides are produced from a larger, biologically-inactive precursors through intracellular post-translational proteolytic cleavage at sites marked by basic amino acid residues (see refs. 67-70 for review). The processing enzymes for the most part belong to the family of serine proteases which incorporate a catalytic domain with structural homology to bacterial subtilisins (Fig. 2). Seven mammalian gene products have so far been identified (furin, PACE4, PC1/3, PC2, PC4, PC5/6, PC7/8 (LPC)) and have a similar modular structure. They each possess signal sequence which precede a moderately conserved (30-50%) prodomain of 70-90 amino acids that terminates in a cluster of basic residues. This borders on the conserved (50-70%) catalytic domain of 320-340 amino acids which terminates with the topological equivalent of the COOH-terminus of subtilisin. All the eukaryote subtilisin-like convertases then incorporate a well conserved (35-50%) middle domain (also called P- or Homo B domain) of 110-130 amino-acids followed by the variable COOH terminus which ranges from 50 to several hundred amino acids and may incorporate other significant features that determine its cellular localization and trafficking in the cell.[71,72] CPE, although structurally unrelated to the subtilases also has a signal peptide, a short prodomain terminating in a cluster of basic amino acids, an extended catalytic domain with only weak homology to the digestive enzymes carboxypeptidases A and B and a short COOH terminal tail[73] which has some similarities in secondary though not primary structure to that of PC1 and PC2.[74]

Most members of the subtilase family are capable of cleaving generic substrates with a P1 Arg, a P2 Arg or Lys, a P4 Arg or Lys and a P6 Arg or Lys. A bulky hydrophobic residue is precluded in the P1' position whereas the P2' is usually hydrophobic in character. Where the individual members vary is in their capacity to accommodate nonbasic amino acids in the P4 and P6 position and their accommodation of Lys in the P1 position. Much has been learnt about their biological specificities from gene transfer experiments using vaccinia virus vectors though in some cases the results have been misleading with regard to in vivo function because of the level of overexpression achieved. Another major determinant of their biological function is their cellular localization and that of their substrates. For example, PC1 and PC2 which are involved in most prohormone cleavages need to be coordinately synthesized with their substrates and be copackaged into dense core secretory granules to exert their biological activity.[75] Furin, a Golgi-resident transmembrane protein[76-78] possibly only sees its substrates transiently as they pass through the TGN. The tissue distribution of the eukaryote subtilisins is variable; furin, PACE 4 and PC7/8 exhibit an almost ubiquitous localization, PC1 and PC2 are con-

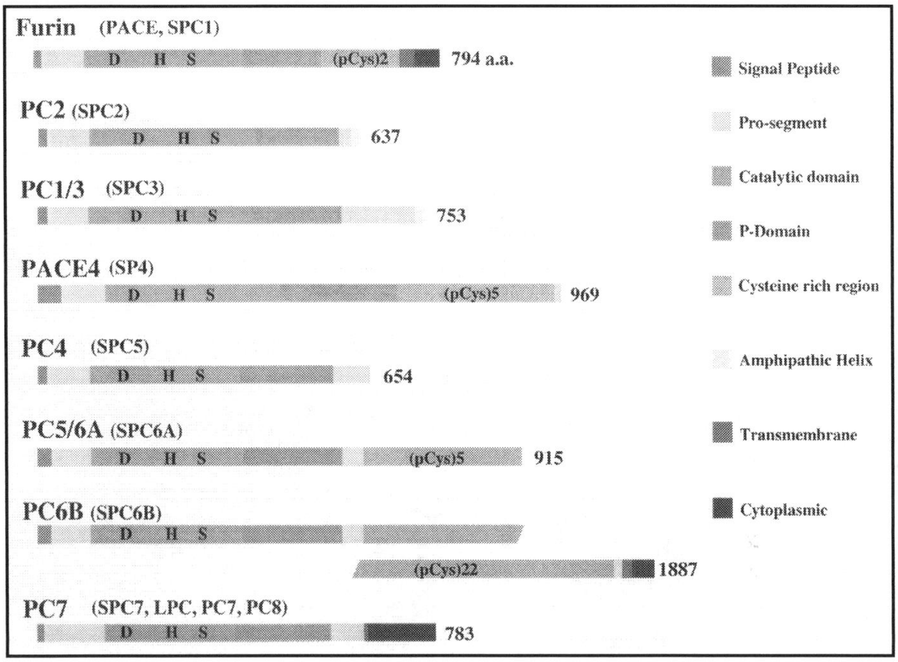

Figure 2. The Prohormone subtilase family members.

fined to tissues of neuroendocrine lineage as is PC5/6, PC4 is restricted to germ cells of the testis (see ref. 79). A further factor determining the activity of each member is the prevailing acidic pH and high Ca^{2+} concentration in the compartments through which they pass[80-86] and significant differences exist between family members in their requirements and sensitivity to ionic conditions.

Endoproteolytic cleavage by members of the mammalian subtilase family occurs COOH terminally to the ultimate basic amino acid in the cleavage motif and is followed by sequential removal of the Arg and Lys residues by carboxypeptidase E, a zinc metalloenzyme of the same class as exocrine pancreas carboxypeptidase A and B. CPE, like PC1 and PC2, is a soluble granule matrix protein with an acidic pH optimum suited to the intragranular environment.[87] Mice bearing a null mutation in the enzyme exhibit an obese and diabetic phenotype on some genetic backgrounds.[88] The fact that they are viable can be traced to the presence of another enzyme, carboxypeptidase D which can catalyze the same reaction though less efficiently by virtue of the fact that it is only transiently associated with the immature secretory granule.[89] Such redundancy however should remind us that the endoproteolytic reaction may not necessarily be the product of mammalian subtilases in all cases. Before the discovery of the subtilase family, many candidate prohormone convertase enzymes from all major classes were proposed. None of these were cloned at the time and few have stood the test of close scrutiny with regard to their physiological function. Remaining viable candidates are a thiol convertase,[90] an aspartyl enzyme homolog of yeast YAP.[91,92]

Their roles are still uncertain but it must be emphasized that while the subtilases may be sufficient to account for the majority of proprotein cleavage reactions that have been examined they have not always proven to be necessary. There are endoproteolytic cleavages of prohormones at single Arg residues and at sequences marked by Arg Lys and Lys Lys motifs that are not sites of efficient subtilase cleavage. There are also many instances where individual reactions within a pathway have been traced to the activities of a subtilase member but the entire sequence of

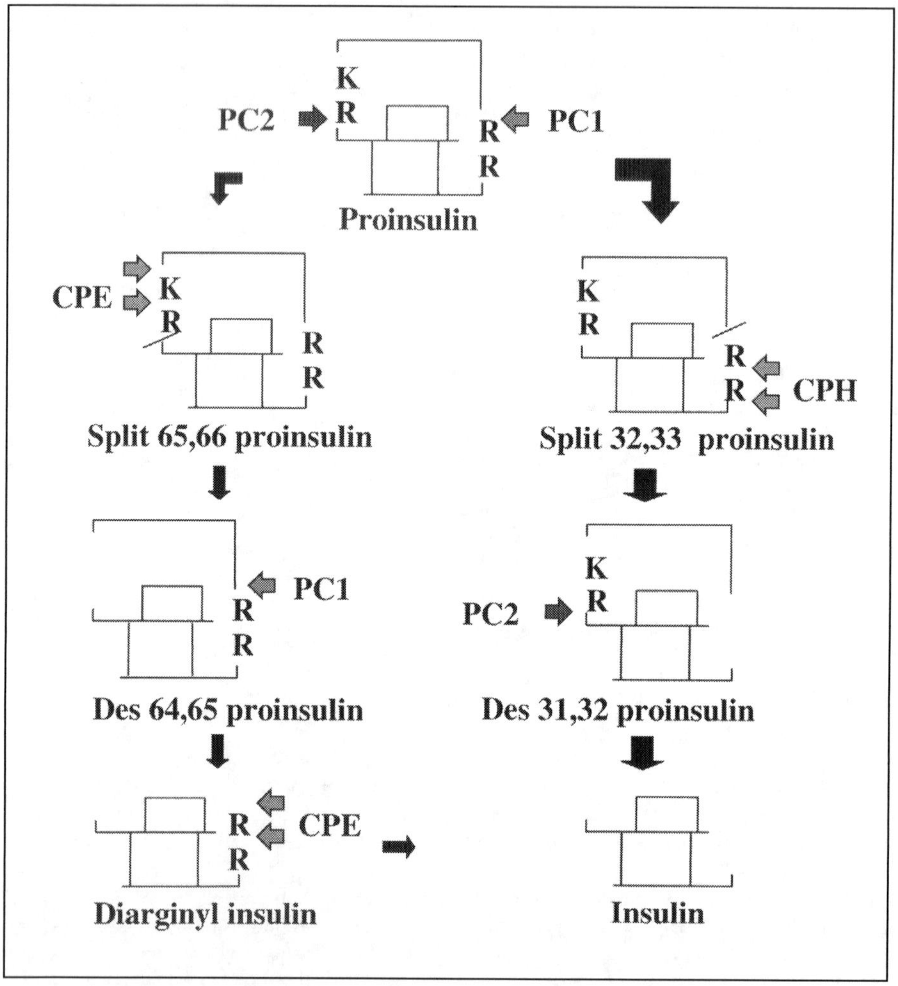

Figure 3. Pathway of processing of the human proinsulin molecule.

reactions leading from the proprotein precursor to the final biological product has not been fully elucidated. Novel endopeptidases could play a role here; on the other hand it may depend on a combination of known subtilases acting in concert or sequentially or the presence of a particular intraorganellar environment. The oft-adopted experimental approach of overexpressing the protease together with the substrate in a neuroendocrine cell line that normally possess neither component is obviously not the best approach to understanding the regulation and fine control of sequential reactions.[93]

Pathways of Prohormone Processing in the Endocrine Pancreas

Human proinsulin is cleaved at two sites marked by Arg Arg and Lys Arg in the molecule by PC1 and PC2 respectively (Figs. 3, 4). The enzymes appear to act sequentially as indicated by the finding that inactivation of PC1 in humans is manifest by the accumulation of high circulating concentrations of proinsulin only and not the des 64,65 proinsulin intermediate which might have been produced by PC2 action on proinsulin.[94] PC2 inactivation on the other hand leads to elevated circulating des 31,32 proinsulin.[95] The apparent preference of PC2 for des 31,32

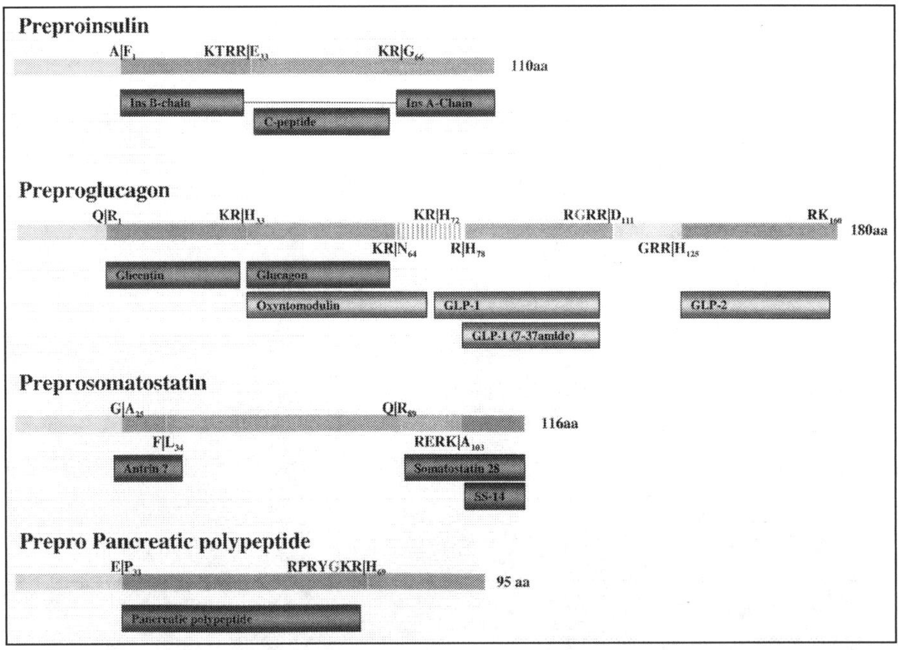

Figure 4. Sites of endoproteolytic processing of the major pancreatic hormones. For preproglucagon, the principal products found in the alpha cell and the intestinal L-cell are denoted. Glycine residues marked in red are substrates for peptide amidating mono-oxygenase activity and as amide donors to the carboxyl group of the preceding amino acid.

proinsulin as a substrate is manifest in cell free reactions and with the purified enzyme in vitro.[96] Immunohistochemical localization of the processing enzymes (Fig. 5) shows that both enzymes are present in the pancreatic β-cell.

Proglucagon has many more potential processing sites than proinsulin and is expressed in multiple neuroendocrine cell types, most prominently the alpha cell of the endocrine pancreas which produces only glucagon and the intestinal L-cell whose major products include glicentin, oxyntomodulin, GLP-1 (7-37 amide) and GLP-2[97] (Fig. 4) PC1 in transfection studies has been shown to generate GLP-1 whereas none of the subtilases either individually or in combination can generate glucagon from proglucagon.[97] Recombinant PC2 in vitro however can generate glucagon from oxytomodulin,[98] a partially processed form of the precursor that can be generated from proglucagon by PC1.[97] Animals with a PC2 null allele fail to produce glucagon[99] and a human with PC1 deficiency has normal circulating glucagon but low GLP-1.[94] The pancreatic alpha cell has high PC2 immunoreactivity and low PC1 (Fig. 5) whereas the converse is true for the intestinal L-cell.[100] These data fall short of defining the pathways of proglucagon conversion and one suspects as in the case of proinsulin processing that proteolytic events other than those catalyzed by PC1 and PC2 alone are involved or that the pathway is tightly regulated.[101] The favored substrate for PC2 may be the major glucagon related fragment produced by cleavage at residue 72 by another enzyme. Experiments with vaccinia viral transfection with a combination of PC2 and either furin, PACE4, PC5A or PC5B failed to define the pathway.[97] A satisfactory enzymological account of proglucagon processing however may be critical in the future for our understanding of the interaction between the various islets in adulthood and development. The observation that GLP-1 is not only a potent secretagogue but also involved in β-cell growth and differentiation hints at an important role for regulation of prohormone processing in regulation of β-cell mass. It is pro-

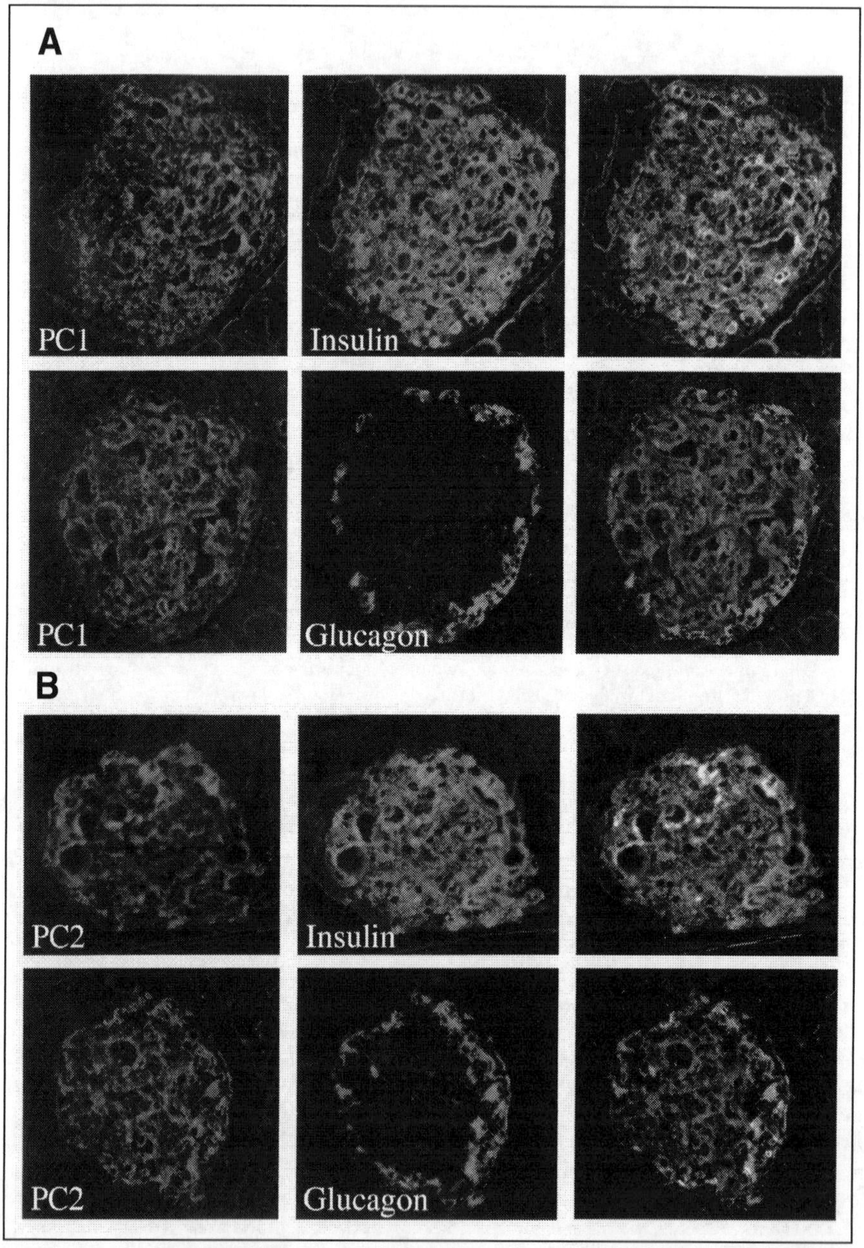

Figure 5A, B. Localization of prohormone converting enzymes in rat pancreatic islets. PC1 is found principally in the core of β-cells whereas PC2 is expressed more strongly in the peripheral cells alpha cells. Carboxypeptidase E while endocrine in localization is present in all islet cell types.

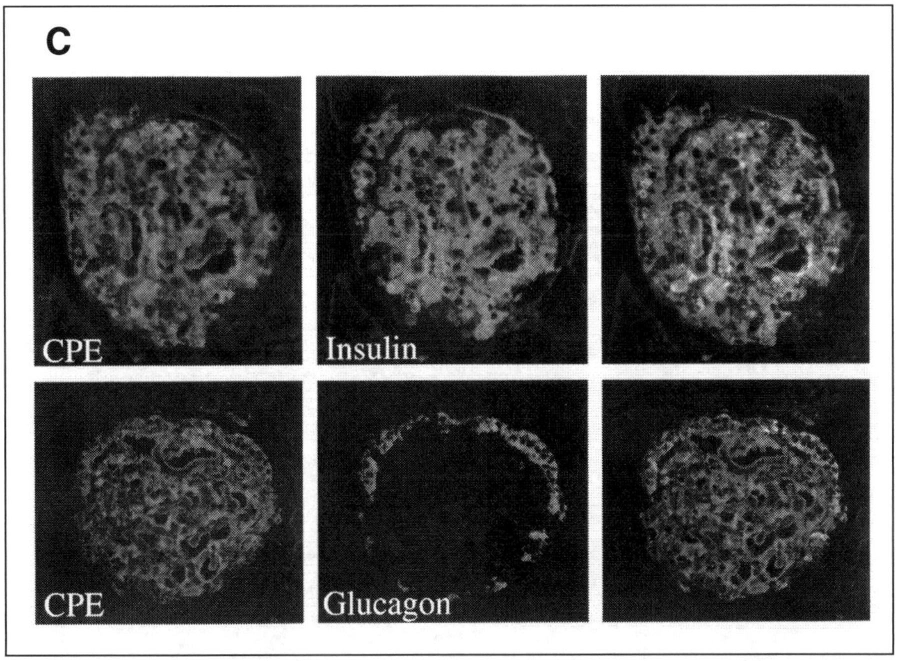

Figure 5C. Continued from previous page.

posed, for example, that the "alpha" cells that are the first differentiated endocrine cell type to appear during normal endocrine pancreatic development express PC1.[102,103] These cells are not mitotically active and thus probably are not the progenitors of mature alpha cells. The implication that they might produce GLP-1 because of their PC1 expression has important consequences to our understanding of how the β-cell precursor grow and differentiate. Any recapitulation of these developmental events in adulthood may be important in the response of β-cell mass to injury or physiological events such as pregnancy or obesity.

Somatostatin which is produced by the pancreatic D cell is the most broadly distributed of the endocrine hormones of the pancreas. Transfection studies in the constitutively secreting Lovo cell lines indicated that the somatostain SS-28 (single Arg site cleavage) could be produced by furin, SS-14 (Arg Lys site cleavage) with PC1 whereas PACE4 produced a mixture of peptides (Fig. 4). PC2 was inactive in these cells yet produced SS-14 in the AtT20 neuroendocrine cell line. PC5 was inactive in either cell line.[104,105] Immunohistochemical analysis shows that the pancreatic D-cell expresses PC2 but not PC1[106] indicating that SS-14 is possibly the physiological product of PC2. Processing of somatostatin is affected in PC2 null mice[95] in agreement with these findings.

Pancreatic polypeptide produced within the pancreatic PP cells has potential dibasic cleavage sites at Arg Arg (some species only) and Gly Lys Arg which produces the precursor of the mature amidated peptide[107] (Fig. 4). A single base cleavage results in further fragmentation of the COOH terminus to produce a 7aa fragment. No reports are available regarding the enzymes involved in these reactions but it is noted that PC2 and not PC1 appears to be localized in this cell type.[106,108]

The processing of the known major products of the pancreatic islets can thus broadly be accounted for by the activities of PC1 and/or PC2 within the secretory granule compartment with a less defined role for furin, PACE4 or PC7 at the level of the TGN. Islet cells express high levels of PC5 which can exist either in a TGN-localized membrane associated form or an

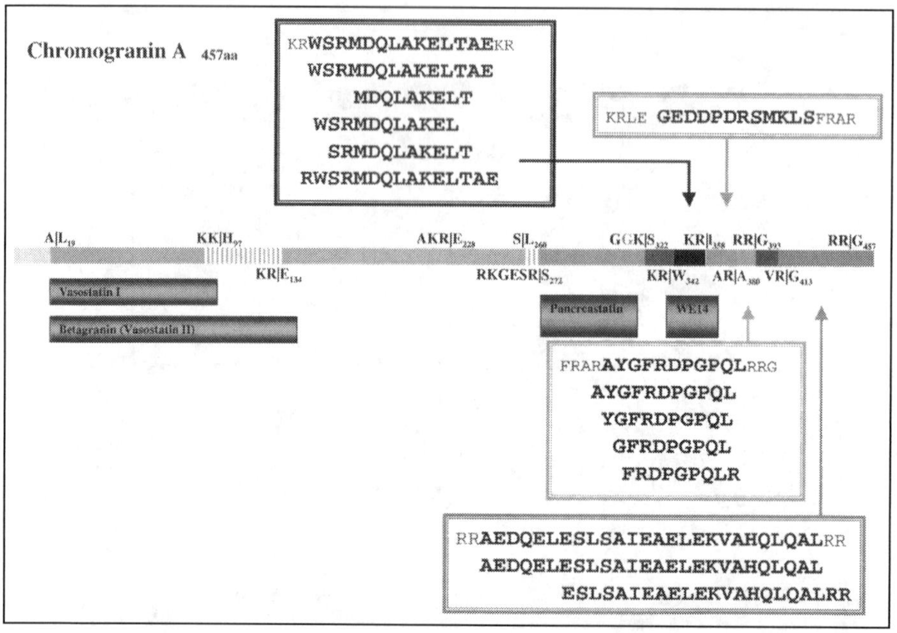

Figure 6. Pattern of processing of human chromogranin A. The marked cleavage sites have been deduced from the sequences of peptides found in a variety of neuroendocrine tissue. Peptides with documented bioactivity are identified. The boxed sequences are derived from HPLC MS MS proteomic analyses of rat insulinoma secretory granules and illustrate that besides the expected cleavages at pairs of basic amino acids, there appear to be a number of nonclassical endoproteolytic cleavages and/or aminopeptidase activity.

alternatively spliced variant that is localized to granules.[109] Given that the enzyme has a neuroendocrine distribution,[110] its presence in islets suggests a role though this has yet to be defined. It is reported to process procholeocystokinin to products resembling forms produced in the gastrointestinal tract[111] and it is noted that islet cell lines produce CCK and express the CCK receptor. Such findings highlight another feature of prohormone processing in the islet, namely that besides the major cargo there are a number of cosecreted proteins with the potential of producing bioactive peptides with hormonal, autocrine or paracrine activities. A particularly good example is the chromogranin A molecule that is widely distributed in neuroendocrine tissue but processed in a tissue specific manner (Fig. 6). Recent proteomic data indicates that not only is chromgranin A processed at the expected dibasic sites but the granule stores many variant forms of the peptides which suggest the activity of endoproteases acting at novel sequence motifs and the existence aminopeptidases in granules (Hutton and Duncan, unpublished findings).[112] Variant forms of the major secreted hormones have also been reported in a form of "mini glucagon" glucagon 9-29[113] which antagonizes the stimulatory effects of glucose, glucagon and GLP-1 on insulin secretion from the Min 6 cell line. The physiological function of most of these products remains to be defined but they should not be dismissed on the basis of being minor constituents.

Regulation of Prohormone Processing Activity within the Secretory Pathway

Given that prohormone processing usually proceeds over a time course of hours in vivo, it would appear that the level of endopeptidase activity in the regulated pathway of secretion is rate-limiting to prohormone conversion in the cell.[83] The control of gene expression at the

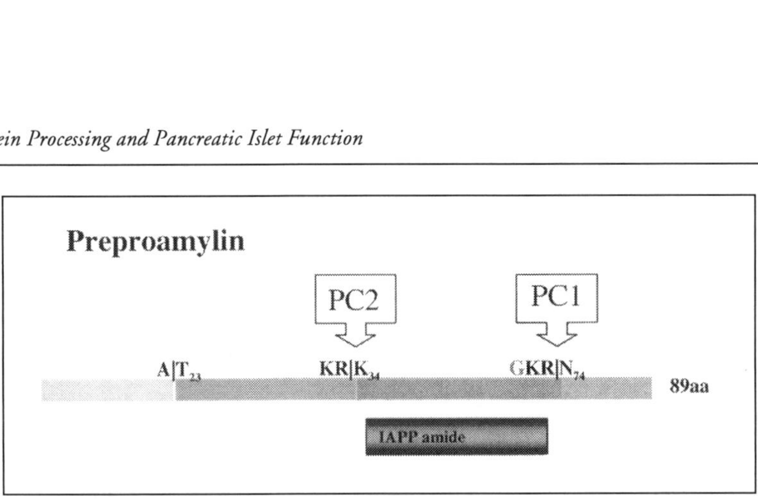

Figure 7. Pattern of processing of the amyloidogenic peptide, amylin (Islet amyloid polypeptide).

transcriptional and translational level is thus an important determinant of the rates of cleavage of proprotein substrates. The synthesis of PC1 and proinsulin in islets increases in response to secretory stimuli,[114] whereas PC2 and carboxypeptidase H appear to be synthesized constitutively.[115] Thus the relative composition of newly forming granules will likely change as a function of the stimulus received by the β-cell and its duration. A consequence is that the rate of conversion of proinsulin to insulin in newly formed granules will vary and thus more or less unprocessed proinsulin could be released depending on the average maturation state of the granule pool in the cell. There is evidence that the amyloidogenic peptide IAPP requires both PC1 for COOH terminal cleavage and PC2 cleavage at its NH2-terminal processing site and thus changes in the relative proportions of PC1 and PC2 could contribute to processing defects in this molecule[116-118] (Fig. 7). Changes in the relative proportion of proinsulin to CPH has a more hypothetical consequence arising from the suggestion that CPH acts as a sorting receptor.[47]

In vitro assays show that PC1 and PC2 are dependent of Ca^{2+} for activity and are active only in the acidic pH range (pH 4.5-6.5).[83] Both pH and Ca^{2+} are thought to play a role in restricting the action of these enzymes to the granule compartment and preventing their activity in earlier vesicular compartments of the secretory pathway. Cellular depletion of Ca^{2+} prevents proinsulin processing though it does so at several levels including the inhibition of the proteolytic activation of the enzymes themselves.[75] It remains an open question as to whether more subtle changes in intragranular Ca^{2+} concentration and pH which are thought to accompany stimulation of the β-cell would have any impact on the overall rates of conversion and maturation of enzymes and their catalytic activity.

The Structure of PC1, PC2 and CPE and Their Post-Translational Processing

The secretory granule proteins PC1, PC2 and CPE have many characteristic features in common with their substrates in that they are subject to many of the same post-translational modifications. They are sorted as efficiently into the regulated pathway as the major cargo[80,119] though not necessarily by the same mechanism that sorts proinsulin.

The prodomains of PC1, PC2 and CPE, as in other secretory proteins, presumably serve several functions: correct temporal and spatial activation of catalytic activity, promotion of folding and oligomerization, intracellular transport, stability and sorting. PC1 is cleaved rapidly via an intramolecular process[120] whereas pro-PC2 cleavage is slow and can occur in trans.[121,122] CPE prodomain cleavage is also rapid and presumably occurs in the ER, however the nature of the enzyme responsible remains obscure. In each case, removal of the prodomain affects the solubility of the proteins and their association with membranes. It has been postulated that this may be part of a sorting mechanism however it is not clear how and where this may operate in the cell.[122-124] The pro-forms of PC1 and PC2 do not appear to be catalytically active towards exogenous substrate, a feature that may be part of the mechanism by which their activities are confined to the dense core secretory granule. ProCPE in contrast is active, a feature which may be of consequence in the trimming of the COOH termini of proteins that are processed in the Golgi but not sorted to the regulated pathway.

Although they differ significantly in size and have little sequence homology, the COOH-terminal domains of PC1, PC2 and CPE terminate in putative amphipathic alpha-helices preceded by a stretch of charged amino acids.[125,126] The membrane association or aggregation of each of the proteins is promoted by calcium and acidic pH,[123,127] a phenomenon which is dependent upon the COOH terminal sequence being intact. The COOH-termini of PC1 (100aa), PC2 (< 20aa) and CPE (20aa) are truncated in a post-trans-Golgi compartment, usually with quite slow kinetics.[86,119,128] The truncated form of PC1 is reported to be more active but less stable and to exhibit changes in its pH and calcium requirements towards synthetic substrates.[129] Truncated CPE and PC2 on the other hand do not appear to be altered in their kinetic properties at least towards soluble substrates in dilute solutions. Deletion of the tails perturbs their proper sorting[71,130] and chimeric proteins, bearing the tails of CPE or PC2, can sort protein domains into the regulated pathway which would normally be committed to constitutive release. It has been postulated that CPE associates with glycosphingolipid-cholesterol rafts at the TGN to facilitate its sorting to the regulated secretory pathway.[131] Residues Arg 255 and Lys 260 within the core of CPE have been implicated as sites which interact with other proproteins (POMC, proenkephalin and proinsulin) to facilitate their sorting[132] through a piggyback mechanism. It is difficult to envisage how this might sort major cargo since the stoichiometry of CPE to cargo is far too low. CPE or the other processing enzymes could however act as "sortases" if they functioned as nucleation sites for aggregation of cargo or docking sites for preformed aggregates of cargo. Such a mechanism is attractive as it would guarantee that each nascent granule would contain a complement of processing enzymes as opposed to a model where the enzymes were sorted by a self-association mechanism similar to that postulated for the major cargo.[42]

Studies on the overexpression of the truncated form of PC1 and of the COOH terminal fragment alone in CHO cells transfected with prorenin or POMC[133] has pointed to an additional function of the carboxy terminus as an endogenous inhibitor of PC1. The evidence suggests that full length PC1 binds prorenin at its dibasic cleavage site but fails to cleave due to the inhibitory effect of the COOH-terminus of PC1. As a result, the PC1 / prorenin complex is targeted to the secretory granules, the COOH-terminal tail of PC1 is removed by proteolysis which in turn renders PC1 capable of processing prorenin. In many ways the model is similar to that formulated for CPE as a "sorting receptor" and encounters the same problems regarding stoichiometry.

Other models to account for the selective sorting of PC1, PC2 and CPE to dense core secretory granules focus on the pro-domain and cite the observation that the proproteins are invariably membrane associated in a pH and Ca^{2+} dependent fashion.[123] Cleavage of proPC1 and proCPE appear to occur at their point of exit from the endoplasmic reticulum and TGN sorting that depends upon the pro-domain would require that the mature enzyme and prodomain remain physically, if not covalently, coupled up until their delivery to the immature secretory granule. PC2 may not undergo early cleavage.[134] Furin, although requiring cleavage to exit the ER[135] remains associated with the prodomain until the prodomain undergoes later cleavage in the TGN at a second site.[136] Retention of binding of the cleaved furin propeptide to furin is postulated to provide a mechanism to suppress furin activity until it reaches the TGN. This may be a more general mechanism since the TGN-associated subtilase family members, furin, PC5 and PC7, are each inhibited by synthetic versions of their prodomain peptide in the nanomolar range. The COOH terminus of the prodomains is sufficient, though high affinity inhibition depends upon the NH2 terminus being present.[93] The prodomains of each protein however did not display selectivity for their cognate enzyme. It will be of interest to observe what happens with PC1 and PC2 in this respect since they too have prosequences containing two processing sites.

Antiproteases and Chaperones

Soon after the subtilase family was discovered, it was postulated that they may be associated with antiproteases if only on the basis of the existing paradigm for the mammalian serine protease family. The first convincing example came in the form of another secretory granule protein of neuroendocrine tissue 7B2.[137,138] 7B2 is evolutionarily conserved bearing a Pro-Pro-Asn-Pro-Cys-Pro motif and shares a weak homology with the chaperonin 60 class of molecules in its amino-terminus. It is postulated to either facilitate the folding of proPC2 to an active conformation or to stabilize it during its transport to the TGN.[139-141] The carboxyl-terminus of 7B2 in addition has a short segment of homology to the serpin class of serine proteinase inhibitors. Mutagenesis studies suggest that inhibitory activity results from the ability of the enzyme to recognize the dibasic motif in the sequence[169] VAKKSV but not to cleave it readily.[142] The 27kD proform of 7B2 is a substrate for PC1 and furin which cleave the sequence[150] RRKRR producing a NH2 terminal 21kD and COOH terminal 6kD fragment.[142] The cleaved molecule is inactive as a "chaperonin" and cleavage may serve to disengage 7B2 and proPC2 once they have reached the TGN or granule compartment. The resultant COOH terminal fragment is still inhibitory in the nM range but might be inactivated by cleavage of the internal $_{171}$KK site by PC2 and removal of the COOH terminal amino acids by CPE in the secretory granules.[143] Such a mechanism could act to delay or dampen PC2 activity however the need for such a mechanism is not obvious since it is generally held that the protein arrives in the granule as the inactive proform. That said, there is conflicting evidence that proPC2 is processed before or soon after leaving the ER[71,80] not least that it can cleave prosomatostatin in a pregranule compartment.[49] Knockout of 7B2 in mice results in the loss of PC2 activity but not PC2 gene expression, a finding that is consistent with its postulated role as a chaperonin. The 7B2 null phenotype however is more severe than the PC2 null and is characterized by Cushing's syndrome originating, not from the anterior pituitary but from the neurointermediate lobe thus indicating that the molecule may have functions unrelated to it interactions with PC2.[144]

PC1 also appears to be associated with a specific antiprotease, the 26 kDa granule protein proSAAS[145] previously identified as the secretory granule protein of neuroendocrine tissue, IA4.[146] proSAAS incorporates a proline-rich sequence within the first half of the molecule and a COOH-terminal 40aa inhibitory peptide separated by a furin consensus sequence. The COOH region sequence bears a hexapeptide (Leu-Leu- Arg-Val-Lys-Arg) previously identified by combinatorial peptide library screening as an inhibitor of PC1 in the nanomolar range.[147] Recombinant PC1 can cleave the inhibitory proSAAS CT peptide and presumably inactivate it. ProSAAS and 7B2 thus appear to be members of a functionally homologous family of convertase inhibitor proteins with specificity for PC1 and PC2 respectively. The sequence homology of the proteins is very weak and an essential chaperonin function for ProSAAS has not as yet been identified. 7B2 is localized relatively uniformly within the islet and thus is present in all cells that express its cognate inhibitory target, PC2 (Figs. 4, 8). ProSAAS on the other hand, is most highly expressed in the pancreatic alpha cell, a cell that has little if any PC1, its cognate target. Such localization suggests other possible functions for this inhibitor. Alternatively its presence in the alpha cell may be to ensure that any residual PC1 activity is truly inactivate.

Inherited Defects in Prohormone Processing in the Islet

Mutations of the proinsulin gene altering either one of the basic amino acids at a proteolytic cleavage site,[148,149] or more distally at the Zn^{2+}-binding His_{B10} residue (>Asp)[150] are the most common causes of frank proinsulinemia of genetic origin in man. Most of such cases are heterozygotes and the impact on glucose metabolism is relatively small as normal insulin is produced from the other allele. Homozygous null alleles have not been reported but would be expected to be neonatally lethal on the basis of experimental studies.[151] The $_{B10}$His>Asp mutation is less readily explained. Initial transfection experiments in AtT20 cells suggested that the

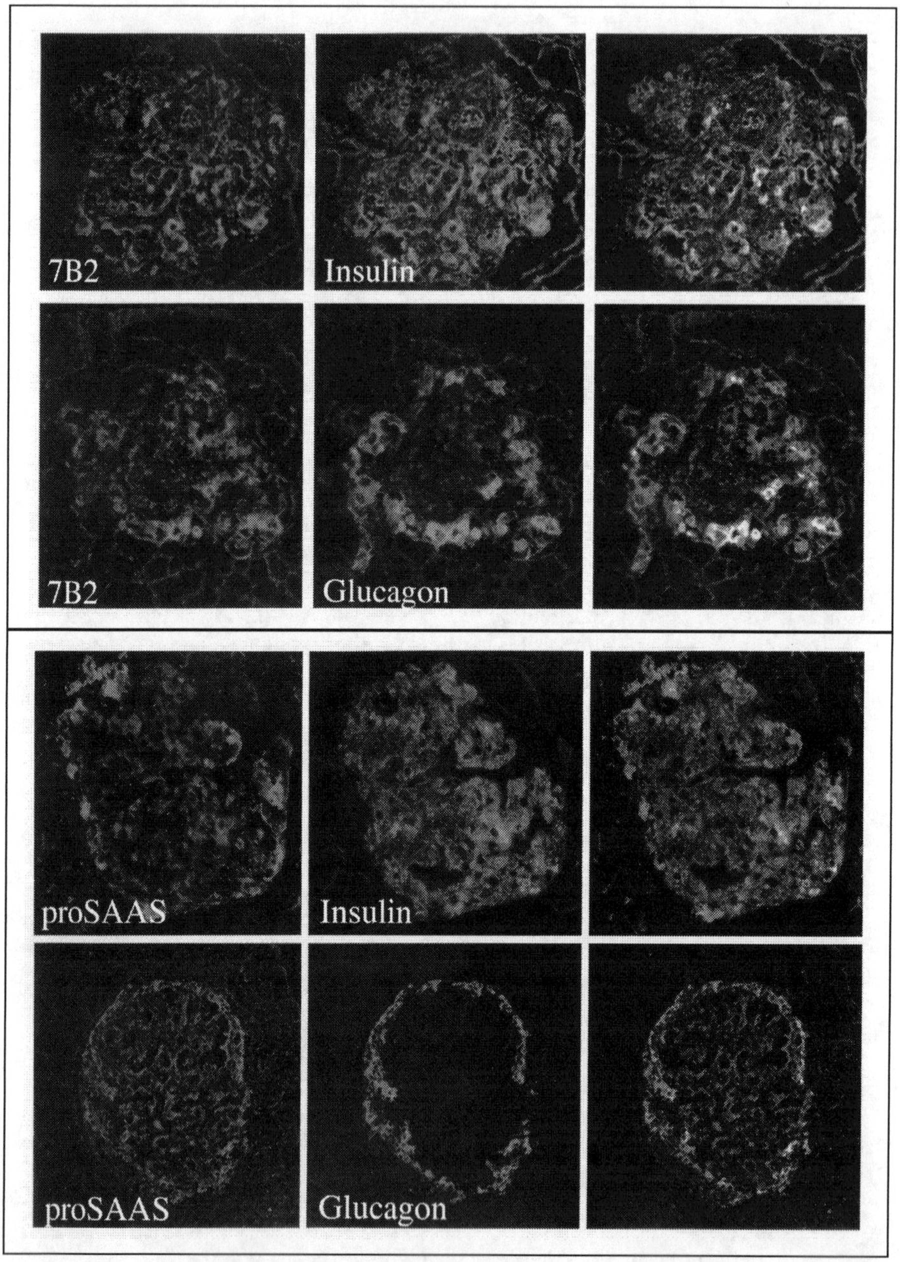

Figure 8. Immunohistochemical localization of the antiproteases 7B2 and ProSAAS in pancreatic islets. Double immunofluorescence labeling shows the presence of 7B2 in all major islet cell types Pro SAAS shows a similar localization but appears more concentrated in the alpha cell mantle surrounding the core of β-cells a region that shows only lowPC1 immunoreactivity (Fig. 4).

lack of a critical Zn^{2+} binding site impairs condensation of the prohormone in the TGN resulting in missorting and constitutive release. However, further studies implicate a kinetic defect in processing possibly brought about by alteration in secondary structural requirements for substrate recognition by the endopeptidases.[152] Another folding mutation has been identified in the mouse[153] in which a mutation affecting a Cys residue in one of the two mouse insulin genes causes the protein to aggregate in the endoplasmic reticulum which in turn disrupts or impedes the folding of the normal alleles resulting in a dominant mutation with insulinopenia.

Genetic deficiencies in the processing enzymes themselves have only recently been reported. The first of these, the fat/fat mouse[88] was traced to a point mutation (Ser_{202}>Pro) of the CPE gene. The enzyme is synthesized at a relatively normal rate but fails to fold and is degraded rapidly in the ER. In addition, it appears to be catalytically inactive. Homozygous animals show hyperproinsulinaemia, slowly developing obesity, infertility and glucose intolerance. This is associated with defects in proinsulin,[88] proenkephalin,[154] POMC[47] and CCK8[155] processing, most notably the accumulation of intermediate forms extended by COOH-terminal basic amino acids. A further feature of the fat/fat mutant is that the pro forms as well as the dibasic extended forms of hormones accumulate. This has been postulated as arising from the "sorting receptor" function[47] of CPE but could equally be accounted for by product inhibition or allosteric effects on the PC enzymes.[88] Secretion of proinsulin in the fat/fat animal is regulated by glucose a finding which suggests that sorting occurs normally. Moreover, diarginyl insulin which is secreted by these animals could only have been generated in a granule compartment thus arguing against a sorting defect.

A case of *PC1* inactivation in man was first identified[156] in an adult individual who presented with severe childhood obesity, developed primary amenorrhea secondary to idiopathic hypogonadotropic hypogonadism and exhibited a long-standing symptomatic reactive hypoglycaemia.[157] The defect resulted from a missense mutation in the middle domain ($_{483}$Gly>Arg) of PC1 combined with an exon 5 splice donor site mutation that results in truncation of the catalytic domain in the second allele (Table 1). Circulating insulin was undetectable in the proband but levels of intact proinsulin were elevated several hundred fold. POMC to ACTH ratios were also elevated however this was relatively mild suggesting that the mutant enzyme may still be active or perhaps compensated by another endopeptidase in this tissue.[90,91] The obesity might be attributable to deficient processing of hypothalamic satiety factors such as alpha MSH, GLP1 or CCK and a proLHRH processing defect might account for her reproductive problems though previous studies indicate that this propeptide is processed by PC2.[158] The site of the mutation is 2 amino acids from the end of the middle domain of PC1 and the affected Gly residue is conserved in all 7 members of the eukaryote subtilisin family. Recreation of the $PC1_{483}$Gly>Arg mutant in CHO cells shows that the protein is synthesized normally, translocated into the lumen of the endoplasmic reticulum but is retained there and not secreted.[156] It also fails to be processed and efficiently secreted in AtT20 cells and baculovirus expression systems (Creemers & Hutton, unpublished findings). There remains the possibility that the mutant in question retains some residual activity (<1% wild type) or that the other allele can be spliced to the wild type with low efficiency since a second reported *PC1* mutation affecting the catalytic domain was neonatally lethal (O'Rahilly, personal communication). A surviving mouse model with a *PC1* null allele whose phenotype resembles this second case will be informative in this regard (Steiner, personal communication).

PC2 mutations in man have not been reported, however a viable PC2 knockout in mice has been generated.[159] The mutant protein is synthesized in islets however it does not undergo post-translational processing and is not secreted. Homozygotes appear normal at birth, but grow slowly, exhibit chronic fasting hypoglycemia and a reduced rise in blood glucose levels following an intraperitoneal glucose tolerance test; features which are consistent with a deficiency of circulating glucagon. The processing of proglucagon, prosomatostatin, and proinsulin are severely impaired and the pancreatic islets show marked hyperplasia of alpha and delta cells and a reduction of beta cells. The phenotype is strikingly different from that of the fat/fat mouse and human PC1 mutant in not having diabetes or being associated with obesity. It can

Table 1. Known sites of mutation and polymorphisms in the prohormone convertase molecule

Site	Mutation		Reference
PRRS**R**RS	R>N 79	exon 2	Ohagi
Splice site donor AAC AA gtg**a**gt	a>c exon 5 skipped & stop	intron 5	Jackson (YC)
LILH**G**TSS	G>R 592	exon 13	Jackson (YC)
KKSP**S**AKLN	S>T 689	exon 14	Jackson (Ctrl)
GAPS**Q**AMLR	Q>E 664	exon 14	Ohagi

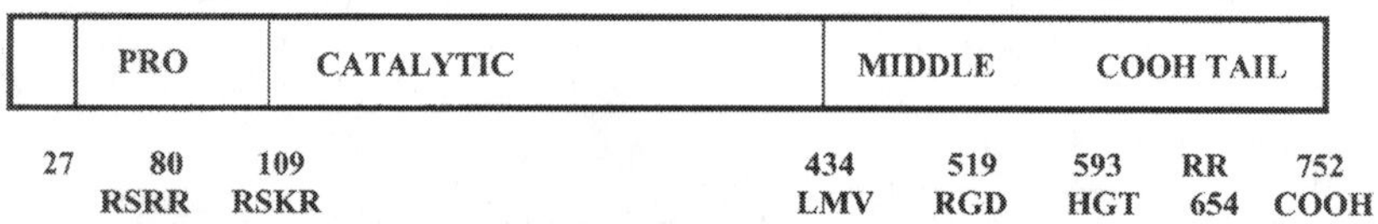

be explained to a large extent by the defect in glucagon production[99] and the phenotype is reversed by chronic glucagon infusion (D.F. Steiner, personal communication). The milder impairment in insulin production compared to the human case is accounted by the fact that rodent proinsulin conversion, unlike human can largely be achieved with PC1 alone.[160]

The consequences of the loss of PC1, PC2 or CPE function on the whole organism are less than might be predicted from in vitro studies using cultured cells and isolated enzymes.[161] This may be due to compensation by negative feedback regulation in the whole organism and redundancy of enzyme function, such that the loss of one convertase does not cause complete loss of function.[154,162] PC1, PC2 and CPE deficiency grossly elevate proinsulin concentration and PC1 defects also increase POMC levels. Elevation of the circulating proinsulin may in part compensate for the inactivation of the enzyme since the proinsulin has a few percent of the biological activity of insulin and is likely to control glycemia at the supraphysiological levels observed. The mouse knockout models will be essential to follow up the hypothesis that PC1 plays an important role in the production of neuropeptides in the hypothalamus and gut which mediate satiety since it is impossible at the moment to investigate the changes in hypothalamic neuropeptides noninvasively in humans.

Proprotein Processing Peturbations of Nongenetic Origin

Elevated circulating proinsulin (proinsulin and the intermediates des 31,32 proinsulin and des 64,65 proinsulin) occurs commonly in the prediabetic phase of type 1 diabetes, type 2 diabetes, impaired glucose tolerance, insulin resistance syndromes, hepatic cirrhosis, pancreatitis cardiovascular disorders and neoplasia.[79,163-168] Endocrine pancreatic tumors result in circulating levels one or more orders of magnitude greater than those found in normal individuals[169] which may be a reflection of the release of the prohormone by a constitutive mechanism or some form of processing defect.[170] In most cases, tumor resection is accompanied by a reduction of the level to the normal range and the follow-up measurement of proinsulin provides a useful prognostic index.

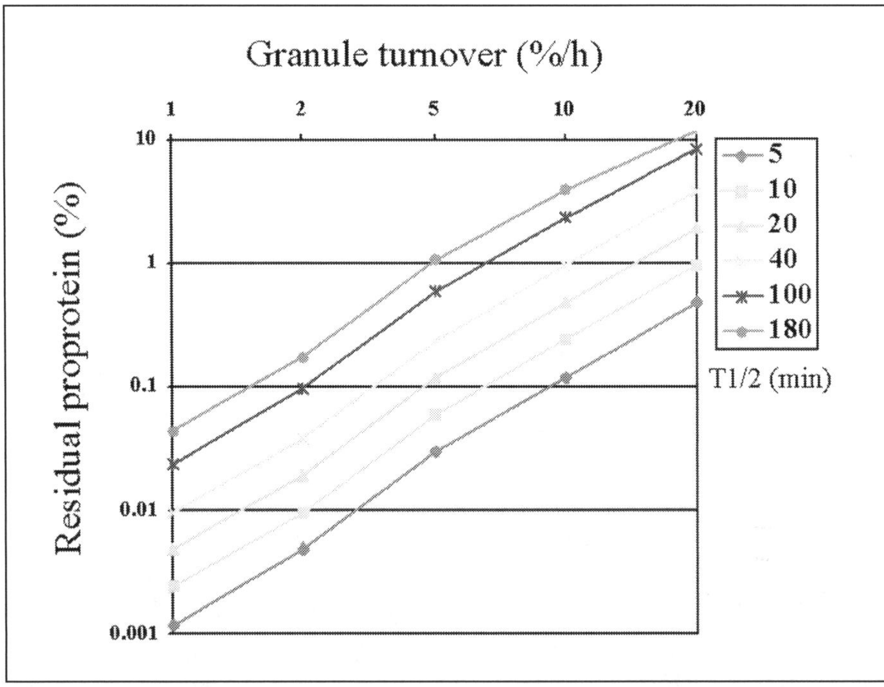

Figure 9. Impact of secretory rate of proprotein secretion. The mathematical model is based upon first order kinetics and steady state secretion balanced by de novo synthesis. It illustrates the impact of granule turnover on the release of unprocessed forms of proproteins and the potential effects of changes in endoprotease activity as reflected in the half lives of processing. Values of the latter range from the most rapid (PC1 t1/2 5min) to the slowest (PC2 $t_{1/2}$ 3h) and span a typical range for proinsulin (t1/2 40min (rat) –100min (human)).[174] The release of unprocessed proinsulin in the range of 0.5-1.5% characterizes the physiological situation whereas the 2-5% is the pathological range. Transition between this range occurs at rates of granule turnover of around 10%/h which approximates the maximal secretory rates of isolated islets. A two-fold reduction of the rate of processing would also result in proinsulinemia at high secretory rates.

In the case of diabetes or prediabetes, the changes are far less dramatic and for the most part appear to be attributable to the regulated release of proportionately more proinsulin and the intermediate des 31,32 proinsulin relative to insulin. The most probable origin of these changes is the decrease in functional β-cell mass and the proportionately greater release of immature secretory granules from the β-cell.[171,172] This hypothesis has gained some support from the finding that "resting" the pancreas in diabetic individuals by overnight somatostatin infusion restores subsequent proinsulin levels and the pulstile release of insulin.[173] A model predicting the effects of processing half-life and secretory granule turnover of processing (Fig. 9) suggests that at basal secretion rates (1-2%/h) all granule constituents would be fully processed (>99.9%). At an intermediate rate of secretion of 6% granule content per hour the amount of residual precursor would be around 1% for a protein like proinsulin with a half life of 100min. In the circulation this would translate into 10% of the circulating immunoreactive insulin because of the 10 fold difference in half lives of (pro)insulin and insulin in the circulation. If the granule turnover increased to 12% for example such as would occur if the storage pool were halved (degranulation of β-cell loss) or secretion rate doubled to match insulin resistance then the residual (pro)insulin would reach 5% in the granule and 50% in the circulation. It is exactly these sorts of quantitative changes that are seen in autoimmune prediabetes[163] and in glucose intolerant individuals.[165]

These observations do not discount the possibility of a more subtle defect exists in either prohormone conversion or in pathways of sorting proteins to the secretory granule. The GK rat, for example, which does not show marked changes in the circulating level of proinsulin (J.C. Hutton, C-G Ostenson, D. Gross, unpublished findings) nevertheless shows an impairment in chromogranin A, PC2 and CPH processing. An important difference between human and rat islets is the slower rate of processing ($t_{1/2}$=100 vs. 40min;[174] see Fig. 9) and the dependence of human proinsulin on PC2 because of differences in the sequence of the C-peptide/ A chain processing site. Likewise, the hyperlipidemia that often accompanies the onset of type 2 diabetes may itself impact negatively on proprotein processing.[175,176] Given that both PC1 and PC2 appear to be involved in the processing of proamylin in the pancreatic β-cell (Fig. 7), small changes in the processing and release of amyloidogenic peptides could have long term consequences for the loss of islet function and islet mass that accompanies type 2 diabetes.[118,177,178]

Processing of Islet Autoantigens and Type 1A Diabetes

The importance of proprotein processing in type 1A diabetes is relatively unexplored. There are suggestions that the initiating autoimmune reaction might be directed at the proinsulin molecule and that epitope spreading then extends this response to other regions, most notably the epitope represented by the B9-23 peptide. The proinsulin epitope shows significant homology to a region within another autoantigen GAD 65. The epitope would normally be destroyed in the post-translational processing reaction as it lies at the B-chain C-peptide boundary. A combination of alternative splicing and proteolysis with the secretory pathway might also contribute to the antigenicity of the autoantigen IA-2.[179,180] Normally the antigenic component of IA-2 resides on the intracytoplasmic face of the granule and would never gain access to the immune system components in the circulation short of cell lysis or phagocytosis. Nevertheless a common isoform of the molecule lacks the transmembrane portion of the molecule encoded within exon 13. In this circumstance the "cytoplasmic" portion of the molecule is translocated into the lumen of the endoplasmic reticulum and is subsequently processed to smaller secreted peptides. It is noteworthy that it is this part of the molecule that is immunogenic. Whether this is because the organism does not develop central tolerance to the piece of IA2 that is missing in the exon 13 variant or whether the exon 13 variant represents the only form of the antigen that circulating immune cells and antibodies "see" remains an intriguing question.[179,180]

Acknowledgements

The work from the author's laboratories was supported by the NIH, Juvenile Diabetes Foundation International, Wellcome Trust, and Fons voor Wetenschappelijk Onderzoek Vlaanderen. Carrie John is thanked for help in preparing the manuscript.

References

1. Hutton JC, Nielsen E, Kastern W. Molecular cloning of the chromogranin A-like precursor to β-granin and pancreastatin from the endocrine pancreas. FEBS Lett 1988; 236:269-274.
2. Hoflehner J, Eder U, Laslop A et al. Processing of secretogranin II by prohormone convertases: Importance of PC1 in generation of secretoneurin. FEBS Lett 1995; 360:294-298.
3. Laslop A, Weiss C, Savaria D et al. Proteolytic processing of chromogranin B and secretogranin II by prohormone convertases. J Neurochem 1998; 70:374-83.
4. Neophytou PI, Muir EM, Hutton JC. A subtractive cloning approach to the identification of mRNAs specifically expressed in pancreatic beta-cells. Diabetes 1996; 45:127-33.
5. Myrsen-Axcrona U, Ekblad E, Sundler F. Developmental expression of NPY, PYY and PP in the rat pancreas and their coexistence with islet hormones. Regul Pept 1997; 68:165-75.
6. Yoon J, Beinfeld MC. Prohormone convertase 1 is necessary for the formation of cholecystokinin 8 in Rin5F and STC-1 cells. J Biol Chem 1997; 272:9450-6.
7. Rhodes CJ, Thorne BA, Lincoln B et al. Processing of proopiomelanocortin by the insulin secretory granule proinsulin processing endopeptidases. J Biol Chem 1993; 268:4267-4275.

8. Wu P, Jackson IM. Identification, characterization and localization of thyrotropin- releasing hormone precursor peptides in perinatal rat pancreas. Regul Pept 1988; 22:347-60.
9. Christgau S, Aanstoot HJ, Schierbeck H et al. Membrane anchoring of the autoantigen GAD65 to microvesicles in pancreatic beta-cells by palmitoylation in the NH2-terminal domain. J Cell Biol 1992; 118:309-20.
10. Navarro M, Rodriquez de Fonseca F, Alvarez E et al. Colocalization of glucagon-like peptide-1 (GLP-1) receptors, glucose transporter GLUT-2, and glucokinase mRNAs in rat hypothalamic cells: Evidence for a role of GLP-1 receptor agonists as an inhibitory signal for food and water intake. J Neurochem 1996; 67:1982-91.
11. Ashford ML, Boden PR, Treherne JM. Glucose-induced excitation of hypothalamic neurones is mediated by ATP- sensitive K+ channels. Pflugers Arch 1990; 415:479-83.
12. Miki T, Liss B, Minami K et al. ATP-sensitive K+ channels in the hypothalamus are essential for the maintenance of glucose homeostasis. Nat Neurosci 2001; 4:507-12.
13. Spanswick D, Smith MA, Groppi VE et al. Leptin inhibits hypothalamic neurons by activation of ATP-sensitive potassium channels. Nature 1997; 390:521-5.
14. Konturek PC, Konturek SJ, Brzozowski T et al. Role of leptin in the stomach and the pancreas. J Physiol Paris 2001; 95:345-54.
15. Caterson ID, Taylor KW. Direct effects of gold thioglucose on insulin secretion by isolated rat islets of Langerhans. FEBS Lett 1979; 98:351-4.
16. Nielsen JH. Effects of growth hormone, prolactin, and placental lactogen on insulin content and release, and deoxyribonucleic acid synthesis in cultured pancreatic islets. Endocrinology 1982; 110:600-6.
17. Sorenson RL, Stout LE. Prolactin receptors and JAK2 in islets of Langerhans: An immunohistochemical analysis. Endocrinology 1995; 136:4092-8.
18. Arimura A, Shioda S. Pituitary adenylate cyclase activating polypeptide (PACAP) and its receptors: Neuroendocrine and endocrine interaction. Front Neuroendocrinol 1995; 16:53-88.
19. Filipsson K, Sundler F, Ahren B. PACAP is an islet neuropeptide which contributes to glucose-stimulated insulin secretion. Biochem Biophys Res Commun 1999; 256:664-7.
20. Humphries J, Nurse EF, Dunmore SJ et al. Beta-cell tropin: Synthesis and biological activity. Biochem Biophys Res Commun 1983; 114:763-6.
21. Spanswick D, Smith MA, Mirshamsi S et al. Insulin activates ATP-sensitive K+ channels in hypothalamic neurons of lean, but not obese rats. Nat Neurosci 2000; 3:757-8.
22. Burks DJ, White MF. IRS proteins and beta-cell function. Diabetes 2001; 50 Suppl 1:S140-5.
23. Turkewitz AP, Chilcoat ND, Haddad A et al. Regulated protein secretion in Tetrahymena thermophila. Methods Cell Biol 2000; 62:347-62.
24. Falkmer S, Dafgard E, el-Salhy M et al. Phylogenetical aspects on islet hormone families: A minireview with particular reference to insulin as a growth factor and to the phylogeny of PYY and NPY immunoreactive cells and nerves in the endocrine and exocrine pancreas. Peptides 1985; 6:315-20.
25. Pierce SB, Costa M, Wisotzkey R et al. Regulation of DAF-2 receptor signaling by human insulin and ins-1, a member of the unusually large and diverse C. elegans insulin gene family. Genes Dev 2001; 15:672-86.
26. Wolkow CA, Kimura KD, Lee MS et al. Regulation of C. elegans life-span by insulinlike signaling in the nervous system. Science 2000; 290:147-50.
27. Gorczyca M, Augart C, Budnik V. Insulin-like receptor and insulin-like peptide are localized at neuromuscular junctions in Drosophila. J Neurosci 1993; 13:3692-704.
28. Cai T, Krause MW, Odenwald WF et al. The IA-2 gene family: Homologs in Caenorhabditis elegans, Drosophila and zebrafish. Diabetologia 2001; 44:81-8.
29. Zahn TR, Macmorris MA, Dong W et al. IDA-1, a Caenorhabditis elegans homolog of the diabetic autoantigens IA- 2 and phogrin, is expressed in peptidergic neurons in the worm. J Comp Neurol 2001; 429:127-43.
30. Huang XF, Arvan P. Intracellular transport of proinsulin in pancreatic beta-cells. Structural maturation probed by disulfide accessibility. J Biol Chem 1995; 270:20417-23.
31. Tooze SA, Martens GJ, Huttner WB. Secretory granule biogenesis: Rafting to the SNARE. Trends Cell Biol 2001; 11:116-22.
32. Orci L, Halban P, Amherdt M et al. A clathrin-coated, Golgi-related compartment of the insulin secreting cell accumulates proinsulin in the presence of monensin. Cell 1984; 39:39-47.
33. Marsh BJ, Mastronarde DN, Buttle KF et al. Organellar relationships in the Golgi region of the pancreatic beta cell line, HIT-T15, visualized by high resolution electron tomography. Proc Natl Acad Sci USA 2001; 98:2399-406.

34. Molinete M, Dupuis S, Brodsky FM et al. Role of clathrin in the regulated secretory pathway of pancreatic beta- cells. J Cell Sci 2001; 114:3059-66.
35. Ladinsky MS, Mastronarde DN, McIntosh JR et al. Golgi structure in three dimensions: Functional insights from the normal rat kidney cell. J Cell Biol 1999; 144:1135-49.
36. Turner MD, Arvan P. Protein traffic from the secretory pathway to the endosomal system in pancreatic beta-cells. J Biol Chem 2000; 275:14025-30.
37. Molinete M, Lilla V, Jain R et al. Trafficking of nonregulated secretory proteins in insulin secreting (INS-1) cells. Diabetologia 2000; 43:1157-64.
38. Tooze J, Tooze SA. Clathrin-coated vesicular transport of secretory proteins during the formation of ACTH-containing secretory granules in AtT20 cells. J Cell Biol 1986; 103:839-50.
39. Pearse BM, Robinson MS. Clathrin, adaptors, and sorting. Annu Rev Cell Biol 1990; 6:151-71.
40. Ladinsky MS, Kremer JR, Furcinitti PS et al. HVEM tomography of the trans-Golgi network: Structural insights and identification of a lace-like vesicle coat. J Cell Biol 1994; 127:29-38.
41. Robinson MS, Bonifacino JS. Adaptor-related proteins. Curr Opin Cell Biol 2001; 13:444-53.
42. Chanat E, Huttner WB. Milieu-induced, selective aggregation of regulated secretory proteins in the trans-Golgi network. J Cell Biol 1991; 115:1505-1519.
43. Green SA, Setiadi H, McEver RP et al. The cytoplasmic domain of P-selectin contains a sorting determinant that mediates rapid degradation in lysosomes. J Cell Biol 1994; 124:435-48.
44. Wasmeier C, Hutton JC. Molecular cloning of phogrin, a protein tyrosine phosphatase homolgue localised to insulin secretory granule membranes. J Biol Chem 1996; In press.
45. Milgram SL, Mains RE, Eipper BA. Identification of routing determinants in the cytosolic domain of a secretory granule-associated integral membrane protein. J Biol Chem 1996; 271:17526-35.
46. Cool DR, Fenger M, Snell CR et al. Identification of the sorting signal motif within proopiomelanocortin for the regulated secretory pathway. J Biol Chem 1995; 270:8723-8729.
47. Cool DR, Normant E, Shen F et al. Carboxypeptidase E is a regulated secretory pathway sorting receptor: Genetic obliteration leads to endocrine disorders in Cpe(fat) mice. Cell 1997; 88:73-83.
48. Dittie AS, Klumperman J, Tooze SA. Differential distribution of mannose-6-phosphate receptors and furin in immature secretory granules. J Cell Sci 1999; 112:3955-66.
49. Stoller TJ, Shields D. Retrovirus-mediated expression of preprosomatostatin: Posttranslational processing, intracellular storage, and secretion in GH3 pituitary cells. J Cell Biol 1988; 107:2087-95.
50. Leyte A, Barr FA, Kehlenbach RH et al. Multiple trimeric G-proteins on the trans-Golgi network exert stimulatory and inhibitory effects on secretory vesicle formation. EMBO J 1992; 11:4795-804.
51. Austin CD, Shields D. Formation of nascent secretory vesicles from the trans-Golgi network of endocrine cells is inhibited by tyrosine kinase and phosphatase inhibitors. J Cell Biol 1996; 135:1471-83.
52. Ohashi M, Jan de Vries K, Frank R et al. A role for phosphatidylinositol transfer protein in secretory vesicle formation. Nature 1995; 377:544-7.
53. Donaldson JG, Klausner RD. ARF: A key regulatory switch in membrane traffic and organelle structure. Curr Opin Cell Biol 1994; 6:527-32.
54. Dittie AS, Hajibagheri N, Tooze SA. The AP-1 adaptor complex binds to immature secretory granules from PC12 cells, and is regulated by ADP-ribosylation factor. J Cell Biol 1996; 132:523-36.
55. Kornfeld S, Mellman I. The biogenesis of lysosomes. Annu Rev Cell Biol 1989; 5:483-525.
56. Kuliawat R, Klumperman J, Ludwig T et al. Differential sorting of lysosomal enzymes out of the regulated secretory pathway in pancreatic beta-cells. J Cell Biol 1997; 137:595-608.
57. Tooze SA, Huttner WB. Cell-free protein sorting to the regulated and constitutive secretory pathways. Cell 1990; 60:837-47.
58. Arvan P, Kuliawat R, Prabakaran D et al. Protein discharge from immature secretory granules displays both regulated and constitutive characteristics. J Biol Chem 1991; 266:14171-4.
59. Kuliawat R, Arvan P. Protein targeting via the "constitutive-like" secretory pathway in isolated pancreatic islets: Passive sorting in the immature granule compartment. J Cell Biol 1992; 118:521-9.
60. Kuliawat R, Arvan P. Distinct molecular mechanisms for protein sorting within immature secretory granules of pancreatic beta-cells. J Cell Biol 1994; 126:77-86.
61. Kuliawat R, Prabakaran D, Arvan P. Proinsulin endoproteolysis confers enhanced targeting of processed insulin to the regulated secretory pathway. Mol Biol Cell 2000; 11:1959-72.
62. Figlewicz DP, Formby B, Hodgson AT et al. Kinetics of 65zinc uptake and distribution in fractions from cultured rat islets of langerhans. Diabetes 1980; 29:767-73.
63. Lan MS, Lu J, Goto Y et al. Molecular cloning and identification of a receptor-type protein tyrosine phosphatase, IA-2, from human insulinoma. DNA Cell Biol 1994; 13:505-14.
64. Solimena M, Dirkx R Jr, Hermel JM et al. ICA 512, an autoantigen of type I diabetes, is an intrinsic membrane protein of neurosecretory granules. Embo J 1996; 15:2102-14.

65. Ort T, Voronov S, Guo J et al. Dephosphorylation of beta2-syntrophin and Ca2+/mu-calpain-mediated cleavage of ICA512 upon stimulation of insulin secretion. Embo J 2001; 20:4013-23.
66. Pouli AE, Emmanouilidou E, Zhao C et al. Secretory-granule dynamics visualized in vivo with a phogrin-green fluorescent protein chimaera. Biochem J 1998; 333:193-9.
67. Hutton JC. Subtilisin-like proteinases involved in the activation of proproteins of the eukaryotic secretory pathway. Curr Opin Cell Biol 1990; 2:1131-42.
68. Steiner DF, Smeekens SP, Ohagi S et al. The new enzymology of precursor processing endoproteases. J Biol Chem 1992; 276:23435-23438.
69. Halban PA, Irminger J-C. Sorting and processing of secretory proteins. Biochem J 1994; 299:1-18.
70. Seidah NG, Chrétien M, Day R. The family of subtilisin/kexin like pro-protein and pro-hormone convertases: Divergent or shared functions. Biochemie 1994; 76:197-209.
71. Creemers JWM, Usac EF, Bright NA et al. Identification of a transferrable sorting domain for the regulated pathway in the prohormone convertase PC2. (Submitted for publication) 1996; 271:25284-25291.
72. Zhou A, Martin S, Lipkind G et al. Regulatory roles of the P domain of the subtilisin-like prohormone convertases. J Biol Chem 1998; 273:11107-14.
73. Fricker LD, Adelman J, Douglas J et al. Isolation and sequence analysis of rat carboxypeptidase E cDNA. Mol Endocrinol 1989; 3:666-673.
74. Fricker LD, Das B, Hogue-Angeletti R. Idetification of the pH-dependent membrane anchor of carboxypeptidase E. J Biol Chem 1990; 265:2476-2482.
75. Guest PC, Bailyes EM, Hutton JC. Endoplasmic reticulum Ca2+ is important for the proteolytic processing and intracellular transport of proinsulin in the pancreatic beta-cell. Biochem J 1997; 323:445-50.
76. Bosshart H, Humphrey J, Deignan E et al. The cytoplasmic domain mediates localization of furin to the trans-Golgi Network en route to the endosomal/lysosomal system. J Cell Biol 1994; 126:1157-1172.
77. Molloy SS, Thomas L, Slyke JKV et al. Intracellular traffiking and activation of the furin proprotein convertase: Localisation to the TGN and recycling from the cell surface. EMBO J 1994; 13:18-33.
78. Schäfer W, Stroh A, Berghöfer S et al. Two independent targeting signals in the cytoplasmic domain determine trans-Golgi network localization and endosomal trafficking of the proprotein convertase furin. EMBO J 1995; 14:2424-2435.
79. Bennett D, C.Hutton J. Proinsulin-processing endopeptidases and diabetes. In: Draznin B, Leroith D, eds. Molecular Biology of Diabetes, Autoimmunity and Genetics Insulin Synthesis and Secretion. Vol 1. New York: Humana Press, 1994:213-248.
80. Guest PC, Arden SD, Bennett DL et al. The post-translational processing and intracellular sorting of PC2 in the islets of Langerhans. J Biol Chem 1992; 267:22401-22406.
81. Lindberg I, Ahn SC, Breslin MB. Cellular distributions of the prohormone processing enzymes PC1 and PC2. Mol Cell Neurosci 1994; 5:614-622.
82. Malide D, Seidah NG, Chrétien M et al. Electron microscopic immunocytochemical evidence for the involvement of the convertases PC1 and PC2 in the processing of proinsulin in pancreatic β-cells. Histochem Cytochem 1995; 43:11-19.
83. Davidson HW, Rhodes CJ, Hutton JC. Intraorganellar Ca and pH control proinsulin cleavage in the pancreatic β cell via two distint site-specific endopeptidases. Nature 1988; 333:93-96.
84. Bailyes EM, Hutton JC. Kinetic analysis of the type 1 proinsulin endopeptidase using a monoclonal antibody-based immunoabsorbent assay. Biochem J 1992; 286:223-229.
85. Bailyes EM, Shennan KIJ, Smeekens SP et al. A member of the eukaryotic subtilisin family (PC3) has the enzymic properties of the type 1 proinsulin-converting endopeptidase. Biochem J 1992; 285:391-394.
86. Bailyes EM, Shennan KIJ, Usac EF et al. Differences between the catalytic properties of recombinant human PC2 and endogenous rat PC2. Biochem J 1995; 309:587-594.
87. Docherty K, Hutton JC. Carboxypeptidase activity in the insulin secretory granule. FREBS Lett 1983; 137-141.
88. Naggert JK, Fricker LD, Varlamov O et al. Hyperproinsulinaemia in obese fat/fat mice associated with a carboxypeptidase E mutation which reduces enzyme activity. Nat Genet 1995; 10:135-42.
89. Dong W, Fricker LD, Day R. Carboxypeptidase D is a potential candidate to carry out redundant processing functions of carboxypeptidase E based on comparative distribution studies in the rat central nervous system. Neuroscience 1999; 89:1301-17.
90. Hook VY, Moran K, Kannan R et al. High-level expression of the prohormones proenkephalin, pro- neuropeptide Y, proopiomelanocortin, and beta-protachykinin for in vitro prohormone processing. Protein Expr Purif 1997; 10:80-8.

91. Azaryan AV, Wong M, Friedman TC et al. Purification and characterization of a paired basic residue-specific yeast aspartic protease encoded by the YAP3 gene. Similarity to the mammalian pro-opiomelanocortin-converting enzyme. J Biol Chem 1993; 268:11968-11975.
92. Cawley NX, Pu LP, Loh YP. Immunological identification and localization of yeast aspartic protease 3-like prohormone-processing enzymes in mammalian brain and pituitary. Endocrinology 1996; 137:5135-43.
93. Fugere M, Limperis PC, Beaulieu-Audy V et al. Inhibitory potency and specificity of subtilase-like pro-protein convertase (SPC) prodomains. J Biol Chem 2001; 26:26.
94. Jackson RS, Creemers JWM, Ohagi S et al. Obesity and impaired prohormone processing associated with mutations In the human prohormone convertase 1 gene. Nat Genet 1997; 16:303-306.
95. Furuta M, Yano H, Zhou A et al. Defective prohormone processing and altered pancreatic islet morphology in mice lacking active SPC2. Proc Natl Acad Sci USA 1997; 94:6646-51.
96. Rhodes CJ, Lincoln B, Shoelson SE. Preferential cleavage of des-31,32-proinsulin over intact proinsulin by the insulin secretory granule type II endopeptidase. Implication of afavored route for prohormone processing. J Biol Chem 1992; 267:22719-22727.
97. Dhanvantari S, Seidah NG, Brubaker PL. Role of prohormone convertases in the tissue-specific processing of proglucagon. Mol Endocrinol 1996; 10:342-55.
98. Rouille Y, Bianchi M, Irminger JC et al. Role of the prohormone convertase PC2 in the processing of proglucagon to glucagon. FEBS Lett 1997; 413:119-23.
99. Furuta M, Zhou A, Webb G et al. Severe defect in proglucagon processing in islet A-cells of prohormone convertase 2 null mice. J Biol Chem 2001; 276:27197-202.
100. Dhanvantari S, Izzo A, Jansen E et al. Coregulation of glucagon-like peptide-1 synthesis with proglucagon and prohormone convertase 1 gene expression in enteroendocrine GLUTag cells. Endocrinology 2001; 142:37-42.
101. Damholt AB, Buchan AM, Holst JJ et al. Proglucagon processing profile in canine L cells expressing endogenous prohormone convertase 1/3 and prohormone convertase 2. Endocrinology 1999; 140:4800-8.
102. Nie Y, Nakashima M, Brubaker PL et al. Regulation of pancreatic PC1 and PC2 associated with increased glucagon- like peptide 1 in diabetic rats. J Clin Invest 2000; 105:955-65.
103. Nielsen JH, Galsgaard ED, Moldrup A et al. Regulation of beta-cell mass by hormones and growth factors. Diabetes 2001; 50 Suppl 1:S25-9.
104. Brakch N, Galanopoulou AS, Patel YC et al. Comparative proteolytic processing of rat prosomatostatin by the convertases PC1, PC2, furin, PACE4 and PC5 in constitutive and regulated secretory pathways. FEBS Lett 1995; 362:143-6.
105. Galanopoulou AS, Seidah NG, Patel YC. Heterologous processing of rat prosomatostatin to somatostatin-14 by PC2: Requirement for secretory cell but not the secretion granule. Biochem J 1995; 311:111-8.
106. Tanaka S, Kurabuchi S, Mochida H et al. Immunocytochemical localization of prohormone convertases PC1/PC3 and PC2 in rat pancreatic islets. Arch Histol Cytol 1996; 59:261-71.
107. Boel E, Schwartz TW, Norris KE et al. A cDNA encoding a small common precursor for human pancreatic polypeptide and pancreatic icosapeptide. EMBO J 1984; 3:909-12.
108. Marcinkiewicz M, Ramla D, Seidah NG et al. Developmental expression of the prohormone convertases PC1 and PC2 in mouse pancreatic islets. Endocrinology 1994; 135:1651-60.
109. De Bie I, Marcinkiewicz M, Malide D et al. The isoforms of proprotein convertase PC5 are sorted to different subcellular compartments. J Cell Biol 1996; 135:1261-75.
110. Barbero P, Rovere C, De Bie I et al. PC5-A-mediated processing of pro-neurotensin in early compartments of the regulated secretory pathway of PC5-transfected PC12 cells. J Biol Chem 1998; 273:25339-46.
111. Cain BM, Vishnuvardhan D, Beinfeld MC. Neuronal cell lines expressing PC5, but not PC1 or PC2, process Pro-CCK into glycine-extended CCK 12 and 22. Peptides 2001; 22:1271-7.
112. Taylor CV, Taupenot L, Mahata SK et al. Formation of the catecholamine release-inhibitory peptide catestatin from chromogranin A. Determination of proteolytic cleavage sites in hormone storage granules. J Biol Chem 2000; 275:22905-15.
113. Dalle S, Blache P, Le-Nguyen D et al. Miniglucagon: A local regulator of islet physiology. Ann NY Acad Sci 1998; 865:132-40.
114. Alarcon C, Lincoln B, Rhodes CJ. The biosynthesis of the subtilisin-related proprotein convertase PC3, butno that of the PC2 convertase, is regulated by glucose in parallel to proinsulin biosynthesis in rat pancreatic islets. J Biol Chem 1993; 268:4276-4280.
115. Guest PC, Rhodes CJ, Hutton JC. Regulation of the biosynthesis of insulin-secretory-granule proteins: Coordinate translational control is exerted on some, but not all, granule matrix constituents. Biochem J 1989; 257:431-437.

116. Badman MK, Shennan KI, Jermany JL et al. Processing of pro-islet amyloid polypeptide (proIAPP) by the prohormone convertase PC2. FEBS Lett 1996; 378:227-31.
117. Higham CE, Hull RL, Lawrie L et al. Processing of synthetic pro-islet amyloid polypeptide (proIAPP) 'amylin' by recombinant prohormone convertase enzymes, PC2 and PC3, in vitro. Eur J Biochem 2000; 267:4998-5004.
118. Wang J, Xu J, Finnerty J et al. The prohormone convertase enzyme 2 (PC2) is essential for processing pro-islet amyloid polypeptide at the NH2-terminal cleavage site. Diabetes 2001; 50:534-9.
119. Guest PC, Pipeleers D, Rossier J et al. Cosecretion of carboxypeptidase H and insulin from isolated rat islets of Langerhans. Biochem J 1989; 264:503-508.
120. Goodman LJ, Gorman CM. Autoproteolytic activation of the mouse prohormone convertase mPC1. Biochem Biophys Res Commun 1994; 201:795-804.
121. Matthews G, Shennan KI, Seal AJ et al. Autocatalytic maturation of the prohormone convertase PC2. J Biol Chem 1994; 269:588-92.
122. Shennan KIJ, Taylor NA, Jermany JL et al. Differences in pH optima and calcium requirements for maturation of the prohormone convertases PC2 and PC3 indicates different intracellular locations for these events. J Biol Chem 1995; 270:1402-1497.
123. Shennan KI, Taylor NA, Docherty K. Calcium- and pH-dependent aggregation and membrane association of the precursor of the prohormone convertase PC2. J Biol Chem 1994; 269:18646-50.
124. Guest PC, Arden SD, Rutherford NG et al. The post-translational processing and intracellular sorting of carboxypeptidase H in the islets of Langerhans. Mol Cell Endocrinol 1995.
125. Smeekens SP, Avruch AS, LaMendola J et al. Identification of a cDNA encoding a second putative prohormone convertase related to PC2 in AtT-20 cells and islets of Langerhans. Proc Natl Acad Sci USA 1991; 88:340-344.
126. Mitra A, Song L, Fricker LD. The C-terminal region of carboxypeptidase E is involved in membrane binding and intracellular routing in AtT-20 cells. J Biol Chem 1994; 269:19876-19881.
127. Hill RM, Ledgerwood EC, Brennan SO et al. Comparison of the molecular forms of the Kex2/subtilisin-like serine proteases SPC2, SPC3, and furin in neuroendocrine secretory vesicles reveals differences in carboxyl-terminus truncation and membrane association. J Neurochem 1995; 65:2318-26.
128. Zhou A, Mains RE. Endoproteolytic processing of proopiomelanocortin and prohormone convertases 1 and 2 in neuroendocrine cells overexpressing prohormone convertases 1 or 2. J Biol Chem 1994; 269:17440-17447.
129. Zhou Y, Rovere C, Kitabgi P et al. Mutational analysis of PC1 (SPC3) in PC12 cells. 66-kDa PC1 is fully functional. J Biol Chem 1995; 270:24702-6.
130. Zhou A, Paquet L, Mains RE. Structural elements that direct specific processing of different mammalian subtilisin-like prohormone convertases. J Biol Chem 1995; 270:21509-16.
131. Dhanvantari S, Loh YP. Lipid raft association of carboxypeptidase E is necessary for its function as a regulated secretory pathway sorting receptor. J Biol Chem 2000; 275:29887-93.
132. Zhang CF, Snell CR, Loh YP. Identification of a novel prohormone sorting signal-binding site on carboxypeptidase E, a regulated secretory pathway-sorting receptor. Mol Endocrinol 1999; 13:527-36.
133. Jutras I, Seidah NG, Reudelhuber TL et al. Two activation states of the prohormone convertase PC1 in the secretory pathway. J Biol Chem 1997; 272:15184-8.
134. Muller L, Cameron A, Fortenberry Y et al. Processing and sorting of the prohormone convertase 2 propeptide. J Biol Chem 2000; 275:39213-22.
135. Creemers JWM, Vey M, Schäfer W et al. Endoproteolytic cleavage of its propeptide is a prerequisite for efficient transport of furin out of the endoplasmic reticulum. J Biol Chem 1995; 270:2695-2702.
136. Anderson ED, VanSlyke JK, Thulin CD et al. Activation of the furin endoprotease is a multiple-step process: Requirements for acidification and internal propeptide cleavage. EMBO J 1997; 16:1508-18.
137. Braks JA, Martens GJ. 7B2 is a neuroendocrine chaperone that transiently interacts with prohormone convertase PC2 in the secretory pathway. Cell 1994; 78:263-73.
138. Mbikay M, Seidah NG, Chretien M. Neuroendocrine secretory protein 7B2: Structure, expression and functions. Biochem J 2001; 357:329-42.
139. Martens GJ, Braks JA, Eib DW et al. The neuroendocrine polypeptide 7B2 is an endogenous inhibitor of prohormone convertase PC2. Proc Natl Acad Sci USA 1994; 91:5784-7.
140. Zhu X, Lamango NS, Lindberg I. Involvement of a polyproline helix-like structure in the interaction of 7B2 with prohormone convertase 2. J Biol Chem 1996; 271:23582-7.
141. Benjannet S, Mamarbachi AM, Hamelin J et al. Residues unique to the pro-hormone convertase PC2 modulate its autoactivation, binding to 7B2 and enzymatic activity. FEBS Lett 1998; 428:37-42.

142. Horssen AMV, Hurk WHVd, Bailyes EM et al. Identification of the region within the neuroendocrine polypeptide 7B2 responsible for the inhibition of prohormone convertase PC2. J Biol Chem 1995; 270:14292-14296.
143. Zhu X, Rouille Y, Lamango NS et al. Internal cleavage of the inhibitory 7B2 carboxyl-terminal peptide by PC2: A potential mechanism for its inactivation. Proc Natl Acad Sci USA 1996; 93:4919-24.
144. Westphal CH, Muller L, Zhou A et al. The neuroendocrine protein 7B2 is required for peptide hormone processing in vivo and provides a novel mechanism for pituitary Cushing's disease. Cell 1999; 96:689-700.
145. Fricker LD, McKinzie AA, Sun J et al. Identification and characterization of proSAAS, a granin-like neuroendocrine peptide precursor that inhibits prohormone processing. J Neurosci 2000; 20:639-48.
146. Donadel G, Marinos N, DeSilva MG et al. Molecular cloning and characterization of a highly basic protein, IA-4, expressed in pancreatic islets and brain. Neuroendocrinology 1998; 67:190-6.
147. Apletalina E, Appel J, Lamango NS et al. Identification of inhibitors of prohormone convertases 1 and 2 using a peptide combinatorial library. J Biol Chem 1998; 273:26589-95.
148. Shibasaki Y, Kawakami T, Kanazawa Y et al. Posttranslational cleavage of proinsulin is blocked by a point mutation in familial hyperproinsulinaemia. J Clin Invest 1985; 89:1902-1907.
149. Yano H, Kitano N, Morimoto M et al. A novel point mutation in the human insulin gene giving rise to hyperproinsulinaemia. Proc Natl Acad Sci USA 1987; 84:2194-2197.
150. Chan SJ, Seino S, Gruppuso PA et al. A mutation in the B-chain coding region is associated with impaired proinsulin conversion in a family with hyperproinsulinaemia. Proc Natl Acad Sci USA 1987; 84:2194-2197.
151. Duvillie B, Cordonnier N, Deltour L et al. Phenotypic alterations in insulin-deficient mutant mice. Proc Natl Acad Sci USA 1997; 94:5137-40.
152. Docherty KD, Rhodes CJ, Shennan KIJ et al. Proinsulin endopeptidase substrate specificities defined by site-directed mutagenesis of proinsulin. J Biol Chem 1989; 264:18335-18339.
153. Wang J, Takeuchi T, Tanaka S et al. A mutation in the insulin 2 gene induces diabetes with severe pancreatic beta-cell dysfunction in the Mody mouse. J Clin Invest 1999; 103:27-37.
154. Fricker LD, Berman YL, Leiter EH et al. Carboxypeptidase E activity is deficient in mice with the fat mutation. Effect on peptide processing. J Biol Chem 1996; 271:30619-24.
155. Cain BM, Wang W, Beinfeld MC. Cholecystokinin (CCK) levels are greatly reduced in the brains but not the duodenums of Cpe(fat)/Cpe(fat) mice: A regional difference in the involvement of carboxypeptidase E (Cpe) in pro-CCK processing. Endocrinology 1997; 138:4034-7.
156. Jackson RS, Creemers JW, Ohagi S et al. Obesity and impaired prohormone processing associated with mutations in the human prohormone convertase 1 gene. Nat Genet 1997; 16:303-6.
157. O'Rahilly S, Gray H, Humphreys PJ et al. Brief report: Impaired processing of prohormones associated with abnormalities of glucose homeostasis and adrenal function. N Engl J Med 1995; 333:1386-90.
158. Wetsel WC, Liposits Z, Seidah NG et al. Expression of candidate pro-GnRH processing enzymes in rat hypothalamus and an immortalized hypothalamic neuronal cell line. Neuroendocrinology 1995; 62:166-77.
159. Furuta M, Yano H, Zhou A et al. Defective Prohormone Processing and Altered Pancreatic Islet Morphology In Mice Lacking Active Spc2. Proc Natl Acad Sci USA 1997; 94:6646-6651.
160. Kaufmann JE, Irminger JC, Mungall J et al. Proinsulin conversion in GH3 cells after coexpression of human proinsulin with the endoproteases PC2 and/or PC3. Diabetes 1997; 46:978-82.
161. Rouille Y, Duguay SJ, Lund K et al. Proteolytic processing mechanisms in the biosynthesis of neuroendocrine peptides: The subtilisin-like proprotein convertases. Front Neuroendocrinol 1995; 16:322-61.
162. Rothenberg ME, Eilertson CD, Klein K et al. Evidence for redundancy in propeptide/prohormone convertase activities in processing proglucagon: An antisense study. Mol Endocrinol 1996; 10:331-41.
163. Heaton DA, Millward BA, Gray P et al. Evidence of β cell dysfunction which does not lead to diabetes: A study of identical twins of insulin dependent diabetes. Brit Med J 1987; 294:145-146.
164. Ward WK, LaCava EC, Paquette TL et al. Disproportionate elevation of immunoreactive proinsulin in type 2 (noninsulin dependent) diabetes mellitus and in experimental insulin resistance. Diabetologia 1987; 30:698-702.
165. Temple RC, Carrington CA, Luzio SD et al. Insulin deficiency in noninsulin dependent diabetes. Lancet 1989; 1:293-295.
166. Rhodes CJ, Alarcon C. What beta-cell defect could lead to hyperproinsulinemia in NIDDM? Some clues from recent advances made in understanding the proinsulin-processing. Diabetes 1994; 43:511-517.

167. Gadot M, Ariav Y, Cerasi E et al. Hyperproinsulinemia in the diabetic Psammomys obesus is a result of increased secretory demand on the beta-cell. Endocrinology 1995; 136:4218-4223.
168. Wareham NJ, Byrne CD, Williams R et al. Fasting proinsulin concentrations predict the development of type 2 diabetes. Diabetes Care 1999; 22:262-70.
169. Rubenstein AH, Block MB, Starr J et al. Proinsulin and C-peptide in blood. Diabetes 1972; 21:661-672.
170. Neerman-Arbez M, Sizonenko SV, Halban PA. Slow cleavage at the proinsulin B-chain/connecting peptide junction associated with low levels of endoprotease PC1/3 in transformed β cells. J Biol Chem 1993; 268:16098-16100.
171. Alarcon C, Leahy JL, Schuppin GT et al. Increased secretory demand rather than a defect in the proinsulin conversion mechanism causes hyperproinsulinemia in a glucose-infusion rat model of noninsulin-dependent diabetes mellitus. J Clin Invest 1995; 95:1032-9.
172. Roder ME, Porte Jr. D, Schwartz RS et al. Disproportionately elevated proinsulin levels reflect the degree of impaired B cell secretory capacity in patients with noninsulin- dependent diabetes mellitus. J Clin Endocrinol Metab 1998; 83:604-8.
173. Laedtke T, Kjems L, Porksen N et al. Overnight inhibition of insulin secretion restores pulsatility and proinsulin/insulin ratio in type 2 diabetes. Am J Physiol Endocrinol Metab 2000; 279:E520-8.
174. Sizonenko S, Irminger JC, Buhler L et al. Kinetics of proinsulin conversion in human islets. Diabetes 1993; 42:933-6.
175. Yudkin JS, Denver AE, Mohamed-Ali V et al. The relationship of concentrations of insulin and proinsulin-like molecules with coronary heart disease prevalence and incidence. A study of two ethnic groups. Diabetes Care 1997; 20:1093-100.
176. Furukawa H, Carroll RJ, Swift HH et al. Long-term elevation of free fatty acids leads to delayed processing of proinsulin and prohormone convertases 2 and 3 in the pancreatic beta- cell line MIN6. Diabetes 1999; 48:1395-401.
177. Westermark GT, Steiner DF, GebreMedhin S et al. Pro islet amyloid polypeptide (ProIAPP) immunoreactivity in the islets of Langerhans. Ups J Med Sci 2000; 105:97-106.
178. D'Alessio DA, Verchere CB, Kahn SE et al. Pancreatic expression and secretion of human islet amyloid polypeptide in a transgenic mouse. Diabetes 1994; 43:1457-61.
179. Park YS, Kawasaki E, Kelemen K et al. Humoral autoreactivity to an alternatively spliced variant of ICA512/IA- 2 in Type I diabetes. Diabetologia 2000; 43:1293-301.
180. Diez J, Park Y, Zeller M et al. Differential splicing of the IA-2 mRNA in pancreas and lymphoid organs as a permissive genetic mechanism for autoimmunity against the IA-2 type 1 diabetes autoantigen. Diabetes 2001; 50:895-900.

CHAPTER 2C

Stimulus-Secretion Coupling in the Pancreatic Beta-Cell

Kirstine Juhl and John Hutton

Introduction

The two most common forms of diabetes in man have very different etiologies and different clinical presentation. Nevertheless the underlying loss of β-cell function has similar consequences in terms of glycemic control and the emergence of devastating long-term complications. In type 1 autoimmune diabetes the loss of β-cells is usually absolute as reflected in the dependence on insulin therapy and the absence of circulating C-peptide. Its sudden onset belies the fact that the underlying loss of β-cell mass is the culmination of many years of gradual and progressive loss of β-cells in the face of autoimmune attack which is first evident with the appearance of autoantibodies to islet proteins in the preceding 1-10 years (see Chapter 11).[1,2] The infiltration of the islets with immune and inflammatory cells that initiates the disease first appears in the islet periphery, affects a subpopulation of islets and is possibly benign or at least kept in check by the presence of regulatory T cells.[3,4] The invasive insulitis seen in NOD mice closer to disease onset may reflect a change in the balance of destructive and protective responses in favor of the former. The histological changes in man are comparatively mild and may reflect the slower progression of the disease or possibly a different immune process. The islet tends to be viewed as the source of autoantigen that supports or initiates the immune attack and ultimately the victim of the crime. However, much more needs to be learned about the reaction of the islet to cytokine mediators of the immune response and about how the β-cell manages to survive so long or replenish its population from progenitor cells in the pancreas. Since the mechanism of autoimmune destruction by effector cells may be mediated by CD4+ cells, and thus indirect, there is also the question of whether the β-cell is uniquely susceptible to oxygen and nitrogen free radicals or cytokine mediators of cell death which may account for the fact that other islet cells exposed to same molecules survive while the β-cell dies. Although we have scant knowledge of β-cell biochemistry and physiology in this regard there is a wealth of information regarding the physiological and pathophysiological responses of the islet to nutrient secretagogues and pharmacological agents. The focus of the following review reflects this balance, however it also aims to emphasize how the β-cell differs from its neighbors and from other endocrine tissues and how it may participate in its own demise in type 1 diabetes. The review also illustrates the challenges faced by investigators wishing to genetically engineer non β cells for cellular therapy of type 1 diabetes or wanting to introduce specific genes into the β-cell population to afford it greater protection from autoimmune attack.

Physiology of the Islet of Langerhans

The endocrine pancreas is arranged in clusters of secretory cells the Islets of Langerhans scattered throughout the exocrine glandular tissue (Fig. 1). In man, the pancreas contains around one million islets that comprise 1-2% of the total mass of the gland. The islets are separated

Immunology of Type 1 Diabetes, Second Edition, edited by George S. Eisenbarth.

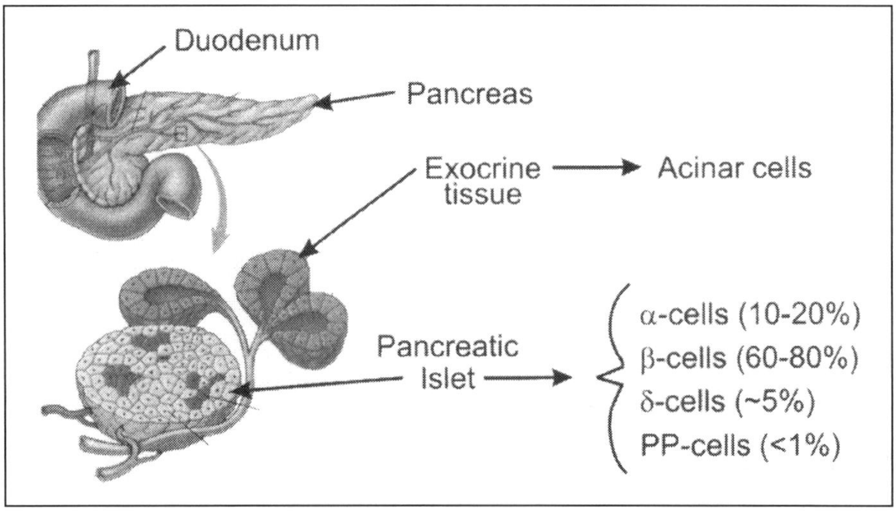

Figure 1. The pancreas is located close to the duodenum. The enlarged part of the pancreas shows the exocrine acinar cells next to the islet endocrine cells. The islets consist of four different cell types, which secrete different peptide hormones. The majority of the islets are β-cells that secrete insulin. The α-cells secrete glucagon, δ-cells secrete somatostatin and PP cells secrete pancreatic polypeptide.

from the exocrine tissue by a capsule made up of connective tissue fibers and human islets vary in size from less than 50 up to several thousand cells.[5] Four different endocrine cell types are contained in the islets; β-cells, which produce insulin and constitute 60-80% of the endocrine cell mass, glucagon secreting α-cells (10-20%), somatostatin producing δ-cells (~5%) and pancreatic polypeptide secreting PP-cells (<1%).[6]

Most mammalian islets have a β-cell rich core surrounded by a mantle of α-, δ- and PP-cells. Afferent arterioles enter the β-cell rich medulla where the arterioles split into a branching system of capillaries that traverse the β-cell mass before reaching the mantle of α- and δ-cells. Thus, the β-cells are the first endocrine islet cell type to sense blood-borne metabolic changes.[7] Insulin is known to inhibit glucagon secretion and glucagon to stimulate insulin secretion. Somatostatin will inhibit the secretion of both insulin and glucagon, hence the anatomical relationship of the islet cells is critical to the function of the organ. The islets are innervated by autonomic nerve fibers, containing both parasympathetic and sympathetic nerves that enter the islets and terminate in proximity to the endocrine cells.

The Biosynthesis of Insulin

The islet β-cell is the principal cell in the adult mammal able to transcribe the insulin gene. The insulin receptor by comparison is widely distributed even on cells that are not thought to be insulin responsive and in the case not even exposed to significant concentrations of the hormone. This may reflect the evolution of the molecule from primarily a neurotransmitter to an endocrine function.[8]

Insulin mRNA is translated to a prepro-insulin peptide, which is rapidly processed in the rough endoplasmic reticulum to pro-insulin by removal of the N-term signal peptide. Proinsulin contains the A and B chains of insulin linked by the C peptide, a structure that helps to align the disulfide bridges that are generated between the A and B chain. Proinsulin is transported through the Golgi apparatus and thereafter further processed in the maturing granule to insulin by excision of the C peptide by the endopeptidases, PC1 and PC2 (prohormone convertases), and carboxypeptidase H (for more details see Chapter 2B). The bioactive insulin molecule consists of an A and B chain (21 and 30 amino acids, respectively), linked intramolecularly by disulfide bridges. Most mammalian insulin bind Zn coordinately forming hexamers and can form large crystalline arrays which promote condensation and aggregation of the hormone in the acidic core of the secretory granule. Proinsulin although forming Zn hexamers is prevented from further aggregation by the highly charged C-peptide, a physical property that may be important in retaining its solubility in the proximal elements of the secretory pathway. In the mature secretory granule C peptide exists in equimolar amounts with insulin in the granule and is cosecreted thus making it a useful marker for β-cell function in diabetics receiving insulin therapy.

Transcription of the insulin gene is modulated by nutrient secretagogues, a variety of cytokines and by insulin itself. Insulin signal transduction enhances transcription of the insulin gene through phosphorylation of insulin receptor substrate (IRS)-1 and -2, and activation of phosphatidylinositol-3-kinase (PI_3K) and it has been postulated that insulin plays a feed forward role in insulin biosynthesis ensuring that the stores of insulin are always replenished.[9] Exogenous insulin is also shown to hyperpolarize β-cell plasma membrane in mice by insulin-induced activation of the K_{ATP} channels possible through the PI_3K thereby turning off the islet insulin secretion.[10] Glucose is also shown to stimulate the production of proinsulin through rapid activation of translation of a preformed pool of mRNA.[11]

The proinflammatory cytokines IL-1β, TNFα and INFγ modulate transcription of the insulin gene. Culturing human β-cells 48 or 72 h with single cytokines does not affect the cellular content in insulin or proinsulin whereas the combination of several cytokines (IL-1β + IFNγ) disproportionally elevates medium proinsulin levels.[12] This is caused by a conserved proinsulin synthesis and a slower hormone conversion possibly contributed to lower expression of PC1 and PC2 convertases. These cytokines mediates signal transduction involving binding to specific receptors, through MAPK and SAPK and mobilization of diverse transcription factors: NFκB, AP-1 and STAT-1 involved in β-cell apoptosis.[13]

Physiological Regulation of Islet Hormone Secretion

The blood glucose level is maintained within a narrow range around 5 to 7 mM in the fasting state mainly by the combined and reciprocal action of insulin and glucagons.[14] Essentially, when the blood glucose concentration is elevated after a meal, the β-cells are stimulated to secrete insulin. Conversely, when blood glucose levels are low the α-cells are stimulated to secrete glucagon, which leads to glycogenolysis and gluconeogenesis in the liver and therefore release of glucose to the blood. Between meals, when the plasma glucose decreases, the release of the neurotransmitter norepinephrine and the neuropeptide galanin from the sympathetic nerves will directly activate glucagon release and inhibit insulin release.[15] During a meal or the postprandial state, the parasympathetic nerves potentiate glucose-induced insulin release from islet β-cells. This is achieved by the release of the neurotransmitter acetylcholine and the neuropeptides pituitary adenylate cyclase activating polypeptide (PACAP), vasoactive intestinal polypeptide (VIP), GLP-1 and GIP.[15] The innervation of the islet and its microanatomy is probably important in the orchestration of these responses.

Somatostatin inhibits both insulin and glucagon secretion and potentially serves as a modulator of the islet hormone release. However, it is unclear whether it has a physiological role given the vascular anatomy of the islet in which the secreted products from the α- and δ-cell enter the blood circulation without reaching the β-cells. The physiological role of islet PP is unknown, but appears not to affect the secretion of the other islet hormones except a possible inhibition of islet insulin release.[16-18]

Much of the physiological regulation of insulin secretion can be reproduced in vitro and a wealth of information has been generated from the study of isolated islets, which appear to behave as autonomous organs. Such studies have led to the classification of insulin secretagogues into three groups, initiators, potentiators and inhibitors. Glucose is the most important initiator in monogastric animals, but other carbohydrates, amino acids and fatty acids alone can also initiate insulin secretion. Pharmacological agents like the clinically useful sulphonylureas

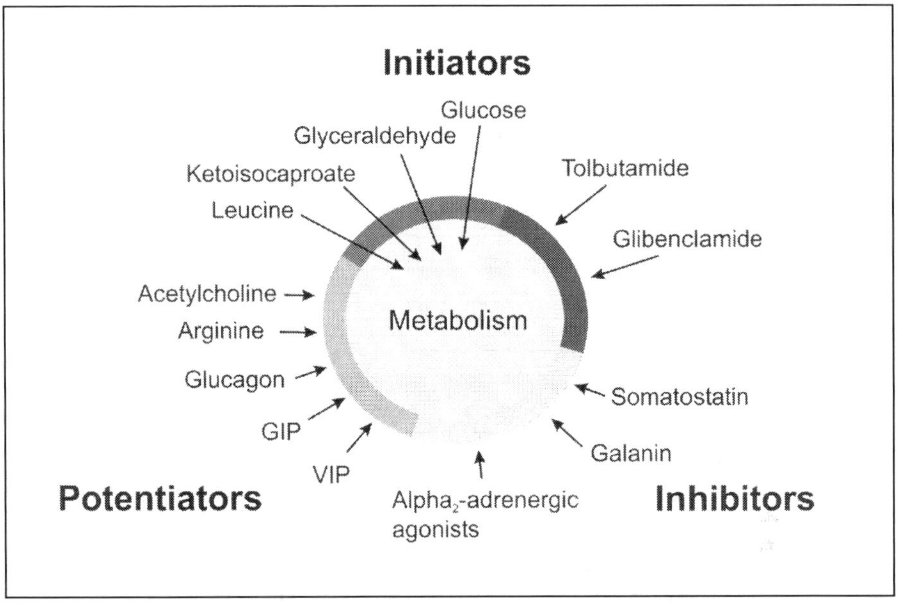

Figure 2. Examples of potentiators, initiators and inhibitors of insulin secretion from the pancreatic β-cell. Abbreviations used: GIP = gastric insulinotropic polypeptide; VIP = vasoactive intestinal polypeptide; GLP-1 = glucagon-like peptide 1.

tolbutamide and glibenclamide also initiate insulin secretion in a glucose-independent manner. Arginine initiates insulin secretion by depolarizing the plasma membrane because of its transport in a positive charge form. It triggers insulin release by increasing $[Ca^{2+}]_i$ through a K_{ATP} channel-independent pathway.[19] Potentiators of insulin secretion do not initiate insulin secretion by themselves but enhance stimulation of secretion in the presence of an initiator. Examples of potentiators are the intestinal glucoincretin hormones glucagon-like peptide 1 (GLP-1) and gastric insulinotropic polypeptide (GIP), which provide fine regulation of insulin secretion in the post prandial state along with neurotransmitters. Other neurotransmitters and hormones such as somatostatin, galanin and adrenalin reduce insulin secretion (Fig. 2).

Islet Pathophysiology in Type 1 Diabetes

Type 1 diabetes is characterized by the appearance of circulating antibodies targeted at β-cell proteins, which appear many years before the onset of clinical disease. Proteins such as insulin, GAD, IA2, phogrin, IGRP and CPH are specifically targeted but as yet there is little convincing evidence that any one of these molecules is primary or dominant in the autoimmune response. Insulin and IGRP are the only molecule in this group that is specific to the β-cell though the others have the common feature of being associated with regulated pathway of secretion. Insulin and CPH are secreted molecules and to some extent remain on the cell surface following exocytosis. By contrast the dominant epitopes seen by antibodies reactive to GAD, IA2 and phogrin are intracellular. It is presently unknown whether the β-cell is targeted directly or indirectly, whether the antigens are presented by the living cell or through death or antigen shedding and why the α-cell survives in spite of sharing many autoantigens with the β-cells and being bathed in the same milieu of soluble immune effector molecules. One hypothesis is that the β-cell participates in its own destruction by increased presentation of antigens in response to glucose, a situation that worsens as β-cell mass decreases and the remaining cells compensate by increased transcription and translation of secretory pathway proteins.[20] β and α-cells have a differential sensitivity to cytokines in that inhibition of glucose-stimulated insulin release by IL-1 is reversible, whereas the effect on glucose-modulated glucagon release is not.[21] The β-cells are more sensitive to cytokines than the other three endocrine cell types in the islets. Cytokine-induced free radicals in β-cells such as NO catalyzed by the inducible nitric oxide synthase (iNOS) that may be involved in β-cell-specific destruction in type 1 diabetes.[22] NO is an important mediator but not the sole mediator of cytokine-induced cytotoxicity in β-cells as shown by the fact that inhibition of NO production in human islets did not prevent cytokine-induced apoptosis, α-cells do not express NO in response to cytokines. β-cells but not α-cells induce the expression of heat shock protein 70 (hsp70), heme oxygenase and MnSOD (manganese superoxide dismutase) upon exposure to IL-1.[23] Native β-cells have been shown to possess low scavenging potential for oxygen-derived free radicals and overexpression in β-cells of scavenger or antioxidant enzymes such as MnSOD, catalase and glutathione peroxidase has been shown to increase their survival after exposure to NO and reactive oxygen species.[22] Production of oxygen free radicals mediated by macrophages can damage β-cells directly resulting in type 1 diabetes in NOD mice.[24] Superoxide dismutase and catalase protected isolated β-cells against alloxan-induced diabetes in vivo indicating a role for superoxide radicals and hydrogen peroxide in the toxicity of alloxan.[25]

Studies with isolated islets have provided models of how mediators of the immune response may damage the islet in the autoimmune response in type 1 diabetes. The macrophage cytokine IL-1β induces an initial phase of functional stimulation in rodent pancreatic islets, which is followed after 4-7 hours by a progressive inhibition of (glucose-induced) insulin release and eventually overt damage to the β-cell caused by production of NO.[26,27] The two cytokines, TNFα and INFγ potentiate the IL-1β-induced production of toxic NO and oxygen free radicals which inhibits insulin secretion.[28] This has led to the concept that IL-1β in combination with INFγ and TNFα plays an important role for β-cell dysfunction and death. The general process of cytokine-induced β-cell 'de-differentiation' with impairment of some of the most differentiated functions of β-cells, such as the preferential mitochondrial metabolism of glucose and the biosynthesis and release of insulin is paralleled by the activation of proteins related to cell survival, such as heat shock proteins and antioxidant enzymes.[29] There are important species differences in β-cell response to IL-1. IL-1 has stimulatory effects on human islets with no formation of NO compared to rodent islets, which produces NO when exposed to IL-1. Human islets cultured in the presence of IL-1β alone exhibit a more prolonged stimulatory phase, which may last up to 48 hours.[27] But again, with prolonged exposure of human islets to IL-1β + TNFα + INFγ an inhibition of human islet function is also observed. This difference between species is probably due to better capacity of human islets to scavenge oxygen free radicals and to the higher content of hsp70, catalase and superoxide dismutase in these cells.[13] Hsp70 has a direct anti-apoptotic role by inhibiting protein aggregation, decreasing formation of oxygen free radicals and blocking effector caspases and it could also decrease necrosis by preventing cellular ATP depletion.

The cytokine-induced β-cell de-differentiation is probably associated to cytokine-induced down-regulation of pancreas duodenum homeobox gene-1 (PDX-1) and Isl-1. PDX-1 is a crucial gene for β-cell development and for the maintenance of its differentiated phenotype (see Chapter 2A). The decreased expression of PDX-1 together with the inhibition of Isl-1 expression could contribute to the observed decreased expression of insulin and glucokinase mRNA which depend on PDX-1 for their regulated transcription. While IL-1β inhibits PDX-1 and Isl-1 expression and up-regulate *c-myc* it does not trigger β-cell apoptosis.[13]

Plasma type II secreted phospholipase A_2 (PLA_2) may be closely involved in the pathophysiology of acute pancreatitis. The serum levels of type II PLA_2 has been shown to correlate with the severity of the disease.[30] Also, an inhibitor of type II PLA_2 protects the pancreas against tissue damage when pancreatitis is induced in vitro.[31] Interestingly PLA_2 enzyme isoforms are also produced by inflammatory cells and by the islets themselves (see below). As in the case of the soluble immune effector molecules we have little information on the actual concentrations

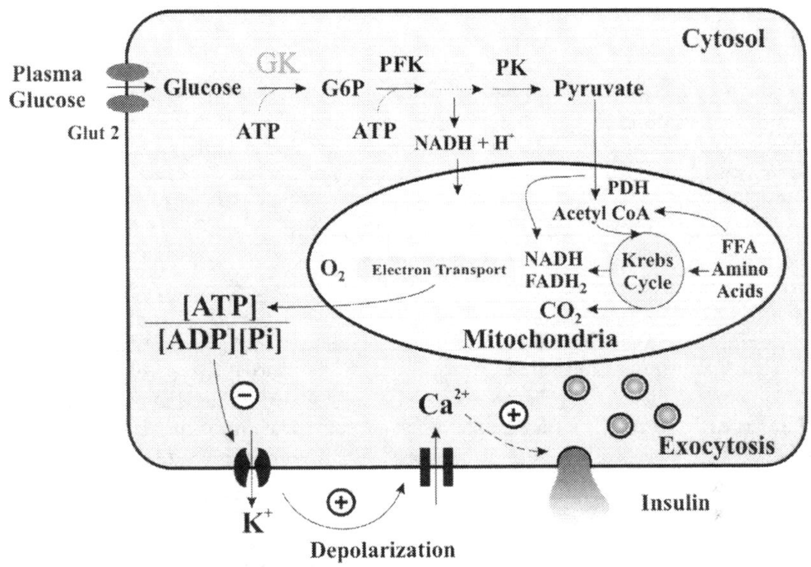

Figure 3. Stimulus-secretion coupling in the pancreatic β-cell. Glucose is transported across the plasma membrane through the GLUT2 transporter. Glucose is metabolized through the glycolysis and Krebs (TCA) cycle. The increased ATP/ADP ratio leads to closure of the ATP-sensitive K^+ channel in the plasma membrane, membrane depolarization and opening of the voltage-dependent Ca^{2+} channels. The resulting Ca^{2+} influx stimulates release of the insulin-containing granules by exocytosis.

to which the β-cells are exposed and the actual in vivo response in terms of regulation of receptors and signaling pathways.

The Stimulus-Secretion Coupling of the Pancreatic β-Cell

The pancreatic β-cell is unusual from the biochemical standpoint in that it has a high rate of glycolytic and oxidative metabolism and that its glucose metabolism is largely insulin-independent. Most-strikingly it responds to extracellular glucose concentrations by increased metabolic flux even at concentrations of the sugar as high as 25 mM. The latter provides a unique mechanism that couples the availability of metabolizable nutrients to the control of secretory, translational, and transcriptional processes without the mediation of ligand-specific cell surface receptors. It may also be the Achilles heel of the cell in that intracellular proteins are exposed to higher concentrations of glucose and reactive metabolites than would otherwise occur in a typical cell that restricts glucose entry and its metabolism to meet it energetic demands.

The β-Cell Glucose Sensor

The β-cell glucose sensor consists the combination of two molecules, which have a restricted tissue distribution, namely Glut2 and glucokinase (GK). Glucose enters the β-cell through the glucose transporter Glut2 (Km ~17 mM) and is quickly phosphorylated by GK to glucose 6 phosphate (G6P) (Fig. 3).[32] Glut2 expression changes in diabetes and hyperglycemic states, which seems to underline its importance in the response of the β-cell, however there is great species variation in the expression of this molecule and little evidence per se that glucose entry is rate limiting to glucose metabolism in the β-cell. GK on the other hand, although present in other tissues such as the liver and kidney appears to be driven by a pancreatic specific promoter.[33-35] Its high expression and accompanying downregulation of low Km forms of hexokinases[36-38] create a situation where kinetics of the enzyme dictates the kinetics of the generation of G6P in a cellular context. There has been considerable debate as to whether the β-cell has a specific glucose 6 phosphatase (G6Pase) and whether it participates with GK in a regulatory substrate cycling activity.[39-41] The β-cell expresses islet G6Pase related protein (IGRP) a β-cell specific homolog of the liver G6Pase catalytic subunit (55% identity) but it is not clear whether the molecule is catalytically active.[42] Other biochemical properties of the β-cell which have been implicated in its ability to sense metabolic fuels include the absence of lactate dehydrogenase and thus an inability to reoxidize cytoplasmically generated NADH[43] and the enhanced expression of the mitochondrial glycerophosphate shuttle[44,45] which may be involved in the transport of reducing equivalents in to and out of the mitochondria. Anaplerosis (the refilling of Krebs cycle intermediates) is shown to be implicated in the K_{ATP}-independent pathways of insulin secretion perhaps via malonyl-CoA formation and lipid esterification processes or through a pyruvate/malate shuttle generating NADPH.[46] Anaplerosis requires conversion of pyruvate into oxaloacetate by pyruvate carboxylase an enzyme that is abundant in islet tissue.[47] Consistent with the view that anaplerosis is an essential component of β-cell signaling is the fact that the dose dependence of insulin release highly correlates with the accumulation of citrate, malate, and citrate-derived malonyl-CoA in the cell.[46] The fact that the association of rise in citrate and α-ketoglutarate and initiation of insulin secretion does not require Ca^{2+} further indicates that anaplerosis is an early event of β-cell activation.[46] A net result of this biochemical machinery is that rates of glucose metabolism are strongly correlated to changes in the redox state of pyridine nucleotides (particularly NADPH)[48,49] and the ratio of ATP/ADP in the cell both which may act as intracellular signals for ionic and other events in the cell (Fig. 3).[50]

The β-cell is electrically excitable and when stimulated with glucose it shows oscillations in membrane potential characterized by bursts of 5-15 sec duration space by silent hyperpolarization, increasing glucose lengthen the burst and shortens the interval leading to continuous spiking over a base of $\Delta\psi$ <-40 mV.[51,52] The β-cell resting membrane potential of about -70 mV is mainly determined by the activity of ATP-sensitive K^+ channels (K_{ATP}).[53] When the ATP/ADP ratio increases as a result of glucose metabolism, the K_{ATP} channels in the plasma membrane close.[54-60] This causes the cell membrane to depolarize due to the presence of an inwardly directed cation current that is dominant in the absence of a reduced K_{ATP}-conductance. The nature of this current is currently unknown but may be a member of the trp family that carries Na^+ and Ca^{2+} ions.[61] The resulting membrane depolarization leads to opening of the voltage-dependent L-type Ca^{2+} channels in the plasma membrane. The resultant Ca^{2+} influx elevates the intracellular free Ca^{2+} concentration $[Ca^{2+}]_i$ and initiates exocytosis of insulin-containing secretory granules (Fig. 3).

Besides the effect of glucose on the K_{ATP} channels glucose is also shown to affect the stimulus-secretion coupling independently of the K_{ATP} channels thereby amplifying the pathway of glucose-induced insulin secretion.[62] This has been shown by the fact that glucose dose-dependently increases insulin secretion when the cell is depolarized and K_{ATP} channels cannot be closed (diazoxide plus high K^+).[63] Under conditions when the K_{ATP} channels are closed by sulfonylureas and therefore the β-cell plasma membrane is depolarized and insulin secretion is increased, glucose is shown to be able to further amplify insulin secretion dose-dependently. This K_{ATP} channel-independent (amplifying) effect of glucose on insulin secretion is Ca^{2+}-dependent but not mediated by any further rise in $[Ca^{2+}]_i$.[63] The amplifying pathway serves to augment the secretory response induced by the triggering signal and creates dose-dependent secretory response in the 5-25 mM concentration range.

Glucose may also regulate insulin secretion by a Ca^{2+}-independent mechanism.[64-66] Under stringent Ca^{2+}-deprived conditions when protein kinases A and C are activated simultaneously, glucose stimulates insulin release in the absence of an elevation of $[Ca^{2+}]_i$. This Ca^{2+}-independent amplification of glucose-induced insulin release may require GTP late in stimulus-secretion coupling.[67,68] This has been interpreted in terms of two different mechanism of exocytosis, one driven by Ca^{2+} and the other by GTP. Others argue that glucose-induced insulin release is mainly regulated by the two Ca^{2+}-dependent pathways (rise in β-cell $[Ca^{2+}]_i$ and increase in Ca^{2+} efficacy) without significant contribution of a Ca^{2+}-independent mechanism. In the following, some of the key molecular components in the β-cell stimulus-secretion coupling will be described.

The ATP-Sensitive K^+ Channel Complex

K_{ATP} channels are expressed in different cell types including islets, heart and ventromedial hypothalamus muscles (VMH) and serve to couple cell metabolism to membrane excitability. They are composed of a poreforming complex consisting of subunits, a specific K^+ channel (Kir6.2) surrounded by regulatory sulphonylurea (SUR) binding subunits. See recent reviews.[69,70]

SUR (Mw 140 kDa) is a member of the ABC-transporter superfamily and Kir6.2 (390 amino acids, 35 kDa) a member of the inward rectifier K^+ channel superfamily. Four subunits of Kir6.2 and SUR1 each constitute the functional channel complex in islets.[71,72] K_{ATP} channel activity is modulated by a range of compounds and intracellular metabolites. Channel activity is reduced by Mg-ATP and sulphonylureas, the most commonly used class of compounds in the treatment of type 2 diabetes. ADP and diazoxide conversely activate K_{ATP} channel activity. Changes in the β-cell ATP/ADP ratio are an important physiological regulator of channel activity with an increase in the blood glucose concentration leading to an enhanced ATP/ADP ratio and consequently channel closure. Recently, K_{ATP} channels have also been detected in α- and δ-cells.[73-76] Whereas the functional consequence of channel closure in the δ-cell is similar to that in the β-cell (increased hormone release), inhibition of channel activity in the α-cell leads to inhibition of glucagon release.[76]

The β-Cell Ca^{2+} Channels

Membrane depolarization, resulting from K_{ATP} channel closure leads to elevation of cytosolic Ca^{2+} via voltage-gated Ca^{2+} channels in the plasma membrane. The influx of Ca^{2+} in turn regulates several steps in exocytosis, such as the size of the readily releasable vesicle pool, the fusion event, and expansion of the fusion pore.[77] The Ca^{2+} concentration required for initiating insulin secretion in different experimental systems has been estimated in the range of 10-30 μM.[78,79] The resting level is in the submicromolar range.

Pancreatic β-cells are found to express N-, P/Q- and L-type Ca^{2+} channels, with the major contribution to Ca^{2+} influx mediated by the L-type channel (large and long-lasting).[53] The L-type channels are characterized by the requirement of a strong depolarization to activate. They inactivate slowly, which is seen in a patch-clamp experiment during a voltage-clamp depolarization to 0 mV where channel inactivation is maximal (Fig. 4).[80] The L-type channels are by themselves inactivated by Ca^{2+} ions at the inner side of the membrane, providing a negative feedback mechanism that limits Ca^{2+} entry into the cell.[81] There is also a voltage-dependent component in the inactivation.[82] The β-cell L-type Ca^{2+} channel activity is blocked by dihydropyridines e.g., nifedipine and stimulated by the cyclic AMP/protein kinase A pathway. Elevated cAMP levels lead to a reduced rate of Ca^{2+} channel inactivation whereas the peak current is only moderately increased.[83]

Molecular Motors and SNARES

When the Ca^{2+} channels in the plasma membrane open and Ca^{2+} flows in the cell, the onset of exocytosis follows less than 50 ms later. This latency is shorter than the time required for Ca^{2+} to equilibrate in the cytosol, which suggests that the granules are located in the vicinity of the Ca^{2+} channels and thus sensitive to local $[Ca^{2+}]_i$ changes. Indeed, areas with a high density of Ca^{2+} channels are found to colocalize with the granules and represent 'hot spots' of secretion.[78] A mouse β-cell contains about 13,000 secretory granules.[84] but only a small fraction

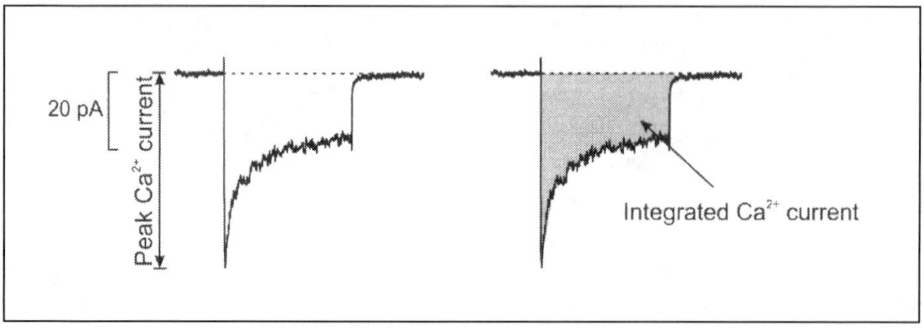

Figure 4. Ca^{2+} current in a mouse β-cell is mediated by L-type Ca^{2+} channels. The peak amplitude and the integrated Ca^{2+} current are depicted resulting from a membrane depolarization to 0 mV during a voltage-clamp experiment.

(50-75 granules[85]) in the readily releasable pool (RRP) are available for immediately release.[86,87] The remainder referred to as the reserve pool need to be mobilized into the RRP before they can undergo exocytosis. The number of granules released seems to be dependent on the phosphorylation state of key proteins. Activation of protein kinase A and C (PKA and PKC) or inhibition of protein phosphatases is required for a large exocytotic response.[88] The effects of these kinases are additive, which suggests that phosphorylation of different regulatory proteins are involved. Candidate proteins include the motor protein kinesin and the granule membrane protein phogrin.[89,90]

Glucose-stimulated insulin secretion in vivo is typically biphasic and characterized by a steep transient first phase that is followed by a gradually developing second phase (Fig. 5). In one model it is explained as the first phase represents release of granules in RRP, triggered by Ca^{2+} influx, and the second phase most likely reflects the ATP-dependent recruitment and release of granules from the reserve pool.[86] The electrophysiological response is also biphasic so this conclusion is open to debate.

Mobilization of granules may involve physical translocation within the β-cell and/or chemical modification[91] and docking of granules near the plasma membrane. Docked granules need to be primed in order to enter the RRP. The insulin granules are transported from the reserve pool to the plasma membrane initially along microtubules and then along the microfilament networks of the cytoskeleton associated with the plasma membrane.[92] An ATP-dependent motor protein such as kinesin is thought to provide the necessary motive force for transportation. However, the exact mechanism and its regulation involved in the transport of granules towards the plasma membrane remains largely unknown.[93]

The processes by which granules in close proximity to the plasma membrane bind and fuse to it can be divided into three phases docking, priming and fusion based on electrophysiological and secretion studies performed on various neurosecretory tissues and in combination with biochemical and molecular genetic analysis.

Docking describes the initial reversible interaction of granules with the plasma membrane and constitutes a pool of unprimed granules in close proximity to the exocytotic site. There is little morphological evidence of the existence of granule docking, but the presence of this pool in the β-cell is suggested from the speed by which RRP is replenished following addition of ATP in the cytosol. The next priming step involves ATP-hydrolysis mediated in part by the ATPase N-ethylmaleimide-sensitive factor (NSF), which is activated by the soluble NSF-attachment factor (α-SNAP).[94] NSF functions as a molecular chaperone to activate SNAp REceptor (SNARE) proteins destined for the fusion complex (Fig. 6).[93] Another possible ATP-dependent event in priming is the synthesis of phosphatidylinositol (4,5)-bisphosphate (PIP_2), which is thought to be important in the recruitment and modulation of proteins impli-

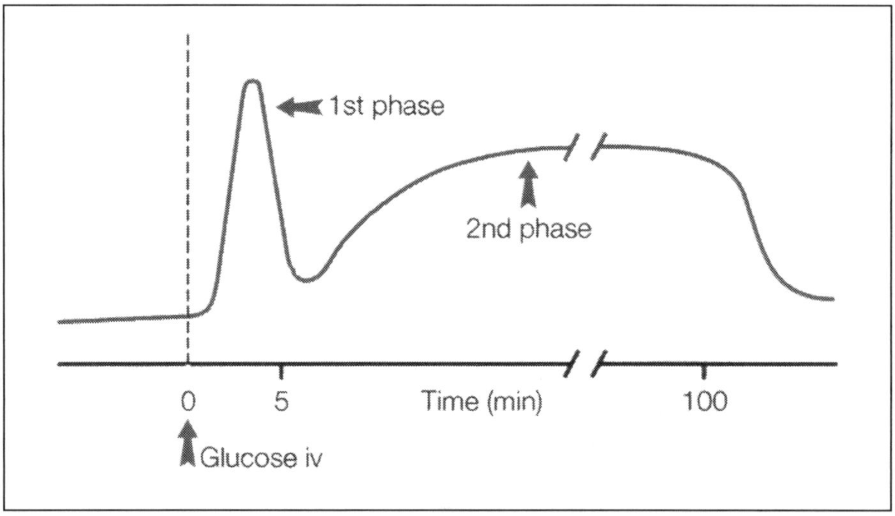

Figure 5. Glucose-induced insulin secretion is characterized by a rapid first phase and a slower second phase insulin released to the bloodstream. Abbreviation used: iv= intravenous.

cated in the fusion apparatus – namely the Ca^{2+} binding synaptotagmin and calcium-dependent activator protein for secretion (CAPS).[77,95]

Fusion of Granules

Conserved protein families involved in fusion include the SNAREs, the Rab GTPases, and the Secl/muncl8-related proteins (also referred to as SM proteins).[96] The SNARE proteins were originally identified as synaptic proteins, but are now generally accepted as universally involved in the core machinery in all intracellular fusion events. The primed granules are ready to fuse with the plasma membrane in an ATP-independent process when the intracellular Ca^{2+} concentration $[Ca^{2+}]_i$ is sufficiently elevated.

Synaptobrevin (VAMP-2, synaptic vesicle-associated membrane protein) is located on the β-cell granule as a *v*-SNARE, and syntaxin 1 and SNAP-25 (synaptosome-associated protein of 25 kDa) are located on the plasma membrane as *t*-SNARE (Fig. 6).[97,98] Synaptobrevin, SNAP-25 and syntaxin 1 assemble in a 1:1:1 stoichiometry to form a complex that drives membrane fusion. Helical portions in the SNARE complex bundle together (two helical domains from SNAP-25 and one from both syntaxin 1 and synaptobrevin) and form an extended coiled-coil rod-like structure, which brings the granule into close proximity with the plasma membrane, thereby overcoming the free energy barrier for fusion (Fig. 6).[99]

It has been postulated that muncl8 functions in a late stage in the fusion process in chromaffin cells, where its dissociation from syntaxin 1 determines the kinetics of postfusion events.[100] However, a muncl8 null mutation leads to a selective defect in the docking of LDCVs at the target membrane and overexpression of muncl8 leads to an increase in number of fusion-competent vesicles without affecting the kinetics of vesicle fusion. This suggests that muncl8 is also important in the docking step before trans-SNARE complexes have assembled.

The essential role of these proteins in neurosecretion is illustrated by the potent inhibition of insulin secretion by various botulinum and tetanus toxins that specifically cleave the different SNARE proteins. Studies with botulinum neurotoxin A which cleaves VAMP-2 and SNAP-25 demonstrate that VAMP-2 is absolutely necessary for insulin exocytosis whereas the action of SNAP-25 could be partially reversed by higher levels of Ca^{2+} or cAMP potentiation.[101] Botulinum neurotoxin C1 mainly cleaves syntaxin and reduces K^+-induced insulin

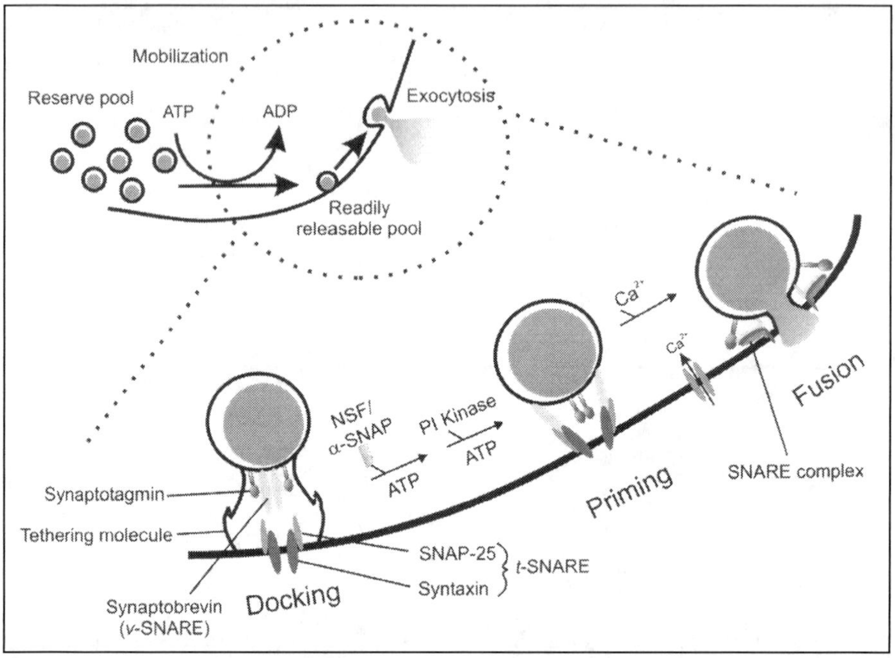

Figure 6. Docking, priming and fusion of insulin-containing granules. Granules from the reserve pool needs to be mobilized in order to enter RRP prior to release. The insert illustrates the three primary stages of insulin exocytosis at the plasma membrane, docking, priming and fusion. The granule is tethered at the plasma membrane during docking. Thereafter the a-SNAP activates NSF, which hydrolyses ATP thereby activating the SNARE proteins. A second ATP-dependent step in the priming process could be PI-4kinase phosphorylation in the generation of PIP_2 (after PI-5kinase phosphorylation). Synaptotagmin is proposed to act as a Ca^{2+} sensor and a fusion clamp preventing fusion until a Ca^{2+} signal arises. When the granule is primed, Ca^{2+} is the only requirement for granule fusion with the plasma membrane

release by 95% but only glucose stimulated insulin release by 25% pointing to further complexity in the exocytotic mechanism.[102]

The fusion event is highly Ca^{2+}-dependent (>10 μM), probably due to regulatory proteins that function as Ca^{2+} sensors. Synaptotagmin is the best characterized exocytotic Ca^{2+} sensor with two Ca^{2+}-binding regions, the C2 domains C2A and C2B, each sharing homology with the C2 regulatory domain of protein kinase C (PKC). The C2A domain is responsible for the Ca^{2+}-dependent insertion of synaptotagmin into the β-cell granule membrane at low micromolar (5 μM) free Ca^{2+} and the binding of synaptotagmin to the core complex protein syntaxin 1.[93,103] CAPS is another Ca^{2+}-binding protein thought to be important for Ca^{2+}-triggered fusion from neuroendocrine cells, including the β-cell.[93] It contains a pleckstrin homology domain that exhibits binding specificity towards PIP_2, which makes it a specific target for phospholipid mediators formed during priming. The lipid specificity of CAPS is switched at elevated Ca^{2+} concentrations and is thought to help disrupt the plasma membrane bilayer to permit fusion.[104] Synaptotagmin is also found to have an inhibitory function in exocytosis. It has been suggested to lock the fusion core complex by preventing α-SNAP from binding until a stimulatory Ca^{2+} signal arises.[98,105] Other components of the fusion process include syntaphilin that acts as a syntaxin clamp in regulating assembly of the SNARE complex during membrane fusion events[99] and NSF that may be necessary to dissociate SNARE complexes formed within a single membrane in favor of complexes between membranes (trans-SNARE complex).[106,107]

Rab GTPases represent a large family of homologous Ras-like GTP-binding proteins that direct the vectorial movement of secretory vesicles. Four isoforms of Rab3 (Rab3A, -B, -C, and -D) have been identified so far. Rab3A is associated with dense-core insulin-containing secretory granules and thought to play a negative role in the fusion process by preventing Ca^{2+}-dependent exocytosis.[108] Rab3A activation is shown to inhibit Ca^{2+}-evoked exocytosis in β-cells by possibly binding to calmodulin at low stimulatory $[Ca^{2+}]_i$.[109] Calmodulin is a key Ca^{2+} sensing protein and this may be a way for Ca^{2+} to act in concert with GTP in the insulin exocytotic mechanism (provides a downstream connection between an intracellular secondary messenger (Ca^{2+}) and the exocytotic machinery of the pancreatic β-cell). It is postulated[110] that it is a way to direct calmodulin to the cytoplasmic face of a β-granule membrane. When $[Ca^{2+}]_i$ increases, calmodulin binds CaMKII more readily than Rab3A and this activates CaMKII to phosphorylate effector exocytotic molecules and enhance either the transport of a β-granule to the plasma membrane and/or the mechanism of insulin exocytosis.[109]

Ca^{2+} plays another essential role in triggering endocytosis following secretion. This may be controlled by the Ca^{2+}/calmodulin-dependent serine/threonine protein phosphatase calcineurin, which is postulated to be a calcium sensor for endocytosis in synaptosomes. Calcineurin is activated by calcium and thereby leads to dephosphorylation of proteins involved in endocytosis like dynamin, amphiphysin 1, amphiphysin 2, and synaptojanin.[111,112]

Signaling Inside the β-Cell

Metabolizable secretagogues and extracellular molecules such as hormones, neurotransmitter that act by binding to receptors on the β-cell plasma membrane appear to mediate their effects through a common set of second messengers. These effects include (1) modulation of adenylate cyclase-cAMP system and activation of protein kinase A (PKA), (2) generation of inositol 1,4,5-triphosphate (IP_3) and *sn*-1,2-diacylglycerol (DAG) resulting in activation of the PKC, and finally (3) changes in intracellular free Ca^{2+} ions that function through Ca^{2+}-dependent regulatory proteins such as calmodulin and PKC.

Hormones binding to G-protein coupled receptors on the cell surface signal to a heterotrimeric seven transmembrane G-protein (inhibitory G-protein, G_i or stimulatory G-protein, G_s), which convey the signal to effector systems such as adenylate cyclase. When a stimulatory G-protein (G_s) is activated (e.g., glucagon stimulation of liver cells) the adenylate cyclase, which resides on the inner plasma membrane, converts ATP to the second messenger cyclic AMP (cAMP) that activates the cAMP-dependent serine/threonine kinase PKA. PKA phosphorylates several proteins at their serine or threonine residues causing a change in either enzyme activity or protein structure. An example is the activation of glycogen mobilization by activating glycogen phosphorylase.

cAMP has a short half-life in the cytoplasm and it is quickly hydrolyzed to AMP by the action of cAMP phosphodiesterases which is important for shutting down the stimulation by the second messenger. cAMP also potentiates Ca^{2+}-dependent insulin secretion.[83,113,114] Besides acting on the Ca^{2+} channels, the major action of cAMP on insulin release is by a direct effect on the secretory machinery itself.[83,115] This is thought to be a result of an increased Ca^{2+} sensitivity of exocytosis by cAMP via an extension of the distance over which the Ca^{2+} channels can recruit the secretory granules for release.[83] It has been suggested that cAMP acts by increasing the release probability of granules already in RRP (PKA-independent) and by accelerating granule mobilization thereby refilling the RRP (PKA-dependent).[113]

Other G-protein receptors (inhibitory and stimulatory G, G_o; stimulatory G, G_q) are coupled to phospholipase C (PLC). Acetylcholine stimulates exocytosis by activating the cholinergic muscarinic receptor that acts via a stimulating G-protein (G_q), which is bound to the Ca^{2+}-dependent membrane-bound PLC. PLC hydrolyses PIP_2 in the membrane to yield the two second messengers IP_3 and DAG. IP_3 diffuses to the endoplasmic reticulum (ER) where it binds to and opens a Ca^{2+} channel in the ER membrane thereby allowing Ca^{2+} to exit from the ER stores into the cytosol resulting in increased $[Ca^{2+}]_i$. The membrane bound DAG stimulates protein kinase C (PKC) by lowering its requirement for Ca^{2+}. DAG is also produced when

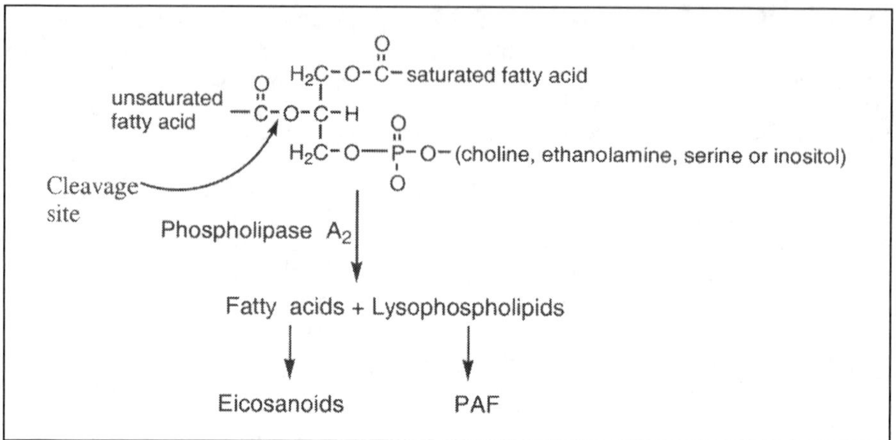

Figure 7. Illustration of cleavage sites for phospholipase A_2 in the glycerol backbone of a phospholipid. Phospholipase A_2 cuts the phospholipids in the sn-2 position resulting in formation of a fatty acid and a lysophospholipid, which can be further metabolized to eicosanoids and PAF, respectively. Abbreviations used: PAF= platelet-activating factor. (Modified from ref. 119).

PLC is activated following increases in $[Ca^{2+}]_i$, thus PKC will be activated under all conditions in which intracellular Ca^{2+} is elevated.[115] PKC is shown to modulate Ca^{2+}-dependent insulin secretion.[83,88] However, glucose-stimulated insulin secretion is unaffected in PKC-depleted islets, indicating that this enzyme is not essential for glucose stimulated insulin secretion. Rather, PKC appears to be involved in the stimulation of insulin secretion by potentiators of release e.g., acetylcholine and carbamylcholine by sensitizing the secretory machinery to Ca^{2+}.[88]

Glucose-induced insulin secretion is associated with inhibition of free fatty acid (FFA) oxidation, increased esterification and complex lipid formation by pancreatic β-cells. Increases in the mass of DAG, triglyceride and phosphatidic acid are observed. Exogenous FFA also potentiate glucose-stimulated secretion of insulin, possibly by providing additional acyl groups for long chain-CoA formation or complex lipid synthesis. FFA does not stimulate insulin secretion in the absence of glucose probably due to their rapid entry into the mitochondria when malonyl-CoA levels are low.[116] LC-CoA opens the ATP-sensitive K^+ channel whereas the corresponding FFA closes the channel. Long-chain fatty acids stimulate PKC, but the saturated ones are ineffective[117] or much less effective than the unsaturated ones.[118]

Phospholipase A_2 Superfamily and Pancreatic β-Cell Function

The phospholipase A_2 family plays an important role in the inflammatory response seen in IDDM by generating inflammatory molecules from lipids substrate. Phospholipase A_2 (PLA_2) constitutes a growing superfamily of enzymes that hydrolyze membrane phospholipids at the *sn*-2 position generating lipid second messengers as well as structural modulations of cellular membranes (Fig. 7). The fatty acid released from *sn*-2 is in most cases arachidonic acid.

Several different types of PLA_2 have been identified. They include: (1) cytosolic PLA_2 ($cPLA_2$), (2) secreted PLA_2 ($sPLA_2$), (3) Ca^{2+}-independent PLA_2 ($iPLA_2$) and (4) platelet-activating factor acetylhydrolase (PAF-AH). These groups are further subdivided according to structure, molecular weight, substrate specificity, cellular localization, and Ca^{2+} dependency of the enzyme.[120]

The cytosolic forms of PLA_2 (group IV PLA_2) are high molecular weight interfacial PLA_2 enzymes. They show a preference for phosphatidylcholine (PC) with arachidonic acid (AA) in the *sn*-2 position and are thought to play an important role in providing free AA for eicosaniod biosynthesis.[121-123] Type IV knock-out mice (IV $cPLA_2\alpha^{-/-}$ mice) show deficient inflammatory

Table 1. PLA_2 groups in the pancreatic islet

$sPLA_2$	$cPLA_2$	$iPLA_2$
Group IB	Group IVα	Group VIA
Group IIA	Group IVβ	
Group IIC [low]	Group IVγ [low]	

response[124] and peritoneal macrophages from IV $cPLA_2\alpha^{-/-}$ mice fail to produce prostaglandins and leukotrienes after stimulation.[125]

Group I, IIs (IIA, IIC, IID, IIE, IIF), III, V, IX, X and XI PLA_2 comprise a group of low molecular weight PLA_2 (13-18 kDa) interfacial enzymes which are either secreted or extracellular PLA_2 ($sPLA_2$). These enzymes have been implicated in various physiological and pathological functions, including lipid digestion, lipid mediator generation, cell proliferation, exocytosis, antibacterial defense, inflammatory diseases and cancer.[126]

The Ca^{2+}-independent PLA_2 ($iPLA_2$) is a group of cytosolic PLA_2 of 85-88 kDa whose major cellular function is the mediation of phospholipid remodelling and homeostasis.[127,128] This enzyme seems to regulate the incorporation of AA and other fatty acids into membrane phospholipids and as such may be a limiting factor for eicosanoid biosynthesis though it has been questioned in the case of the rat pancreatic islet group VIA enzyme.[129]

Platelet-activating factor acetylhydrolase (PAF-AH) is another group of Ca^{2+}-independent PLA_2s.[130] All of them have restricted substrate specificity and are selective for PAF and for phospholipids with a short (or intermediate) acyl chain at the *sn*-2 position that protects normal membrane lipids from hydrolysis by PAF-AH. A second group of substrates is oxidatively fragmented phospholipids. The main function of PAF-AH is to act as scavenger of bioactive phospholipids especially to regulate the level of PAF.[131]

Different Types of PLA_2 in Islets and Pancreatic β-Cells

Pancreatic islets contain at least five different types of PLA_2 (Table 1). Among the secreted form, both type IB and IIA are found with type IB more abundant. Type IB has been localized to the secretory granules of β-cells and stimulation with insulin secretagogues results in the cosecretion of insulin and the enzyme.[132] Type IB also appears to be regulated by fasting and ambient glucose.[132-134] Furthermore, mRNA for a $sPLA_2$ membrane receptor is found in rat pancreatic islets.

By use of different inhibitors of PLA_2, it has been demonstrated that Ca^{2+}-dependent PLA_2 may participate in acetylcholine-induced activation of the granule movement in pancreatic β-cells.[135] It is proposed that the type IV $cPLA_2$ couples the secretagogue induced increase in intracellular Ca^{2+} to the liberation of AA and the subsequent release of insulin.[136] Both exogenous $sPLA_2$ and products of its action have also been shown to suppress K_{ATP} channel activity in excised and cell-attached patches of plasma membranes from HIT insulinoma cells.[137]

In glucose-stimulated islets, hydrolysis of AA from membrane phospholipids appears in part to be mediated by a PLA_2 that is catalytically active in the absence of Ca^{2+} and inactivated by the $iPLA_2$ inhibitor HELSS (haloenollactone suicide substrate).[138] However, $iPLA_2$ is not required for AA incorporation or phospholipid remodelling in β-cells, suggesting that $iPLA_2$ plays another role in these cells.[129] $iPLA_2$ in rat and human β-cells is implicated as a putative glucose sensor, which can couple alterations in glycolytic metabolism to the generation of biologically active eicosanoids and thereby facilitate glucose-induced insulin secretion.[139] $iPLA_2$ is activated by ATP and other nucleotides and it has been postulated that it may respond to the increase in the ATP/ADP ratio after glucose stimulation, however this function has been questioned.

Besides playing an important role in the stimulus-secretion coupling in the β-cell different types of PLA_2 may be involved in inflammatory responses. Type II $sPLA_2$ and type IV $cPLA_2$ mRNA is upregulated in areas with histologic damage in pancreatitis suggesting that these isoforms might contribute to the morphologic changes that occur.[140] In addition, several proinflammatory cytokines such as IL-1, IL-6 and TNFα have been shown to induce transcription of the type IIA $sPLA_2$ gene in a variety of cells. TNF priming of human neutrophils (PMN) caused marked increase in fMLP stimulated AA release in parallel to enhanced activity and secretion of $sPLA_2$ and activity of $cPLA_2$.[141] Type I $sPLA_2$, on the contrary, is nontoxic to acinar cells in both severe and mild pancreatitis. This may indicate that type II but not type I $sPLA_2$ have pathophysiologic importance in severe acute pancreatitis associated with systemic inflammatory response syndrome.

Arachidonic Acid in Cell Signaling

Arachidonic acid is a major constituent of islet phospholipids comprising about 34-39% of the *sn*-2 fatty acyl mass in islet glycerophospholipids[139,142,143] compared to 13% in rat brain and heart and 1% in rat liver.[144] The intracellular level of free AA in islets is normally in the low micromolar range, but rises to a level of 50-200 μM during glucose stimulation.[145,146] Such changes could result by the action of PLA_2, but also by the action of diacylglycerol lipase and monoacylglycerol lipase on diacylglycerol (DAG), generated from the action of phospholipase C (Fig. 8).[147,148] Phosphatidic acid (PA), generated by phospholipase D hydrolysis of PL, is also a potential source of AA.[147]

Inhibition of AA release prevents insulin secretion.[149] Increases in the availability of AA (or its metabolites) by itself, however is probably insufficient to initiate insulin release, instead it seems that AA potentiates the action of other secretagogue.[150] In different cell types, AA has been associated with several specific cellular processes such as regulation of PKC and PLC and modulation of intracellular Ca^{2+} transients.[151-154] In β-cells, it has been suggested that both Ca^{2+}-dependent and Ca^{2+}-independent components are involved in PLA_2-induced insulin secretion. The Ca^{2+}-independent component could be a result of AA or a lipoxygenase product affecting late events in stimulus-secretion coupling.[155,156] The AA-induced Ca^{2+}-dependent event is biphasic with an early rise probably reflecting release of Ca^{2+} from intracellular stores and a sustained phase, which could reflect extracellular Ca^{2+} entry possible mediated by voltage-sensitive Ca^{2+} channels.[157] Using capacitance measurements it has been shown that recombinant type IV $cPLA_2$ stimulated the depolarization-evoked exocytosis without affecting the whole cell Ca^{2+} current or cytoplasmic Ca^{2+} level in isolated mouse pancreatic β-cells. This was mimicked by the combined action of arachidonic acid and lysophosphatidylcholine.[157a]

It is suggested that the ClC-3 Cl-channel is involved since antisense oligonucleotides against this channel and DIDS (a Cl-channel inhibitor) inhibited the stimulatory action of cPLA2. cPLA2 is proposed to play an important role in controlling the rate of exocytosis in β-cells by increasing the number of granules available for release.[157a]

Role of Arachidonic Acid Metabolites in Cell Signaling

When AA is generated, it can be reesterified into phospholipids by acyltransferases or metabolized by the cyclooxygenase (COX), lipoxygenase (LPX) or cytochrome P450 epoxygenase pathways generating prostaglandins and thromboxanes, leukotrienes and lipoxins, or epoxyeicosatrienoic acids.[158]

Prostaglandins made by the cyclooxygenase-pathway acting on AA have been demonstrated, in cell-attached patches on a β-cell line, to be involved in insulin secretion by increasing K_{ATP} channel activity.[137] Such action in vivo would result in inhibition of insulin secretion and may explain the observation that the COX inhibition potentiates glucose-induced insulin release in some circumstances.

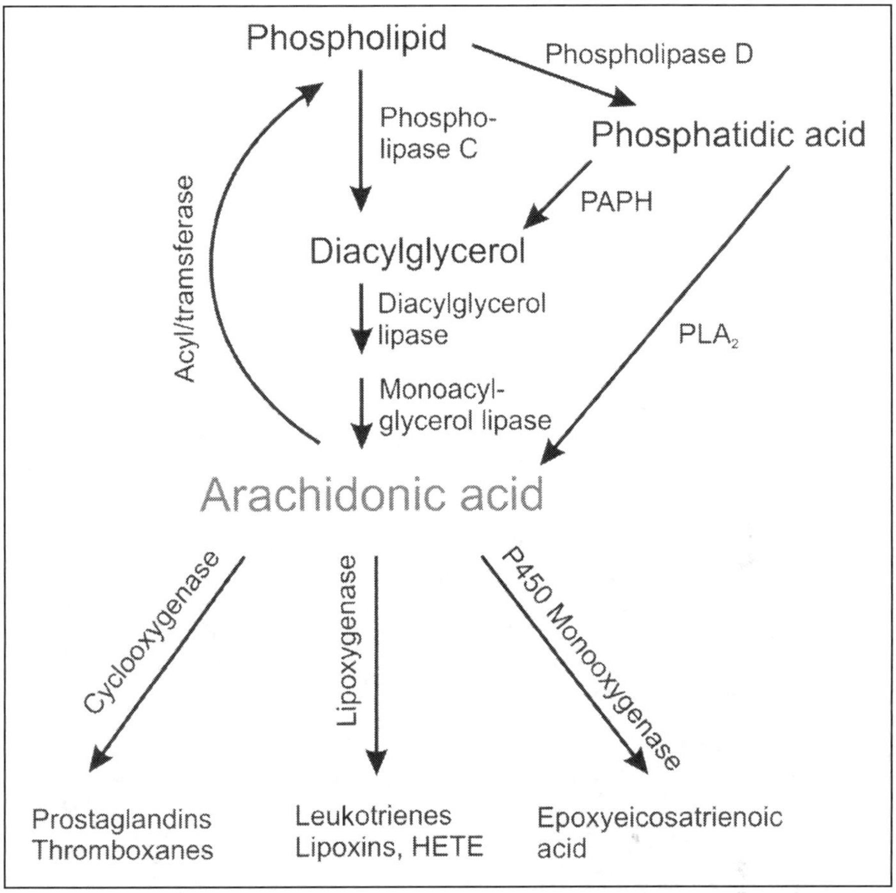

Figure 8. Arachidonic acid can be made in different ways, but also degraded in several ways. AA can be reesterified into membranes or metabolized by the lipoxygenase pathway to leukotrienes, by the cyclooxygenase pathway to prostaglandins and thromboxanes or by the cytochrome P450 monooxygenase enzyme to epoxyeicosatrienoic acid. Abbreviations used: PAPH= phosphatidic acid phosphohydrolase.

The lipoxygenase metabolites of AA are also potential mediators of insulin secretion. During nutrient-induced insulin secretion 12-hydroxyeicosatetraenoic acid (12-HETE; a product of 12-lipoxygenase metabolized AA) is augmented[150] and a lipoxygenase inhibitor (nordihydroguaiaretic acid, NDGA) inhibits glucose-induced insulin secretion.[159] Products of the lipoxygenase pathway have also been demonstrated to play a role in apoptosis induced by Ca^{2+} store depletion in β-cells by using the inhibitor NDGA.[160]

The cytokine IL-1 also augments islet production of 12-HETE and PGE_2.[161] This is not caused by an increased production of AA but by suppression of reesterification of AA into islet phospholipids. This might occur as a result of the IL-1-induced production of NO and inhibition of ATP-dependent conversion of the fatty acid to a coenzyme A thioester intermediate.[161] 12-HPETE, a precursor of 12-HETE, has been shown to induce apoptosis of cells with reduced levels of glutathione peroxidase. Thus the effects of IL-1 to increase substrate flux through the lipoxygenase pathway and to augment NO production may interact cooperatively to inflict injury on β-cells, and at the same time stimulate insulin release in the short term.

Possible Effects of Lysophospholipids on β-Cell Signaling

PLA_2 hydrolysis of membrane phospholipids also generates a lysophospholipid (lysoPL). Several types of lysoPLs exist in the islet membrane: lysophosphatidylcholine (lysoPC), -ethanolamine (lysoPE), -serine (lysoPS) and -inositol (lysoPI). LysoPLs and phosphatidic acid (PA) are shown to promote insulin release when the cellular content of either lipid increases.[156] LysoPLs affect stimulus-secretion coupling at a number of points.[162,163] They have been observed to inhibit directly the K_{ATP} channel activity in the pancreas.[164] LysoPC and lysophosphatidylglycerol can also effectively mobilize Ca^{2+} from intracellular stores.[163] LysoPAF has been demonstrated to circumvent the inhibition of glucose-induced insulin release caused by phospholipase inhibitors.[165]

Pathophysiogical Role of Lipid Metabolites

Inflammatory response in the pancreas not only plays an important role in the destruction of islet tissue in autoimmune diabetes (see Chapter 5), but also in the ability of the pancreatic β-cell to recover from such insult.

Pancreatitis induced by cellophane wrapping (CW), surgery, coxsackievirus B4, cerulein treatment or with other chemicals[166-168] for example can lead to islet regeneration. Locally produced INFγ causes lymphocyte infiltration and islet cell destruction and while IFN-γ gene linked to an insulin promoter produces diabetes new cells are formed from duct epithelial cells and they differentiate into endocrine cells.[169] The proinflammatory cytokines, IL-1, TNF and perhaps IL-6 is shown to induce PLA_2 expression and release.[170] The proinflammatory action of TNF depends in part on activation of PLA_2.[171] In humans with acute pancreatitis, there is a correlation between elevated serum PLA_2 levels and inflammation and lipid metabolites are also potential mediators in these events. In this regard it will be of interest to see whether selective inhibition of PLA_2 can be used as a therapeutic approach against type 1 diabetes.

Concluding Remarks

While it is not year clear as exactly how the generation of lipid metabolites is essential to or contributes to the physiological response of the β-cell, one can envisages many ways in which they interact in a pathophysiological setting. The fact that the pancreatic β-cell is an excitable cell that is dependent on the influx of Ca^{2+} through gated channels and its subsequent removal from the cell by intracellular sequestration or plasma membrane pumping leaves it in a precarious position when faced with agents that disrupt Ca^{2+} homeostasis in the cell. In a similar way, its susceptibility to drugs such as the DNA alkylating agent streptozotocin or the free radical generator alloxan may be the consequence of the mechanism of sugar transport, which can carry these agents into the cell. Such phenomenon may explain in part why so many of the agents that one associates with pancreatic β-cell damage and death turn out to be specific secretagogues at least in the short-term. The challenge is to discover which of these interactions are part of the normal responsiveness of the cell to metabolizable substrates, peptides and neurotransmitters and which are truly deleterious and thus a legitimate target for pharmacological intervention. It is likely that the concepts that there are "good" or a "bad" cytokines in the context of autoimmune diabetes is no more valid than that of glucose being "good' or "bad" in the context of glucotoxicity and type 2 diabetes. The short-term effects of secretagogues on islet function have been extensively documented but there is also the need to consider the broader perspective of how the cell responds in the longer term to inflammatory stimuli or physiological changes in energy homeostasis such as accompany pregnancy and starvation (Fig. 9).

Acknowledgements

We thank Jesper Gromada for valuable comments and suggestions to the manuscript.

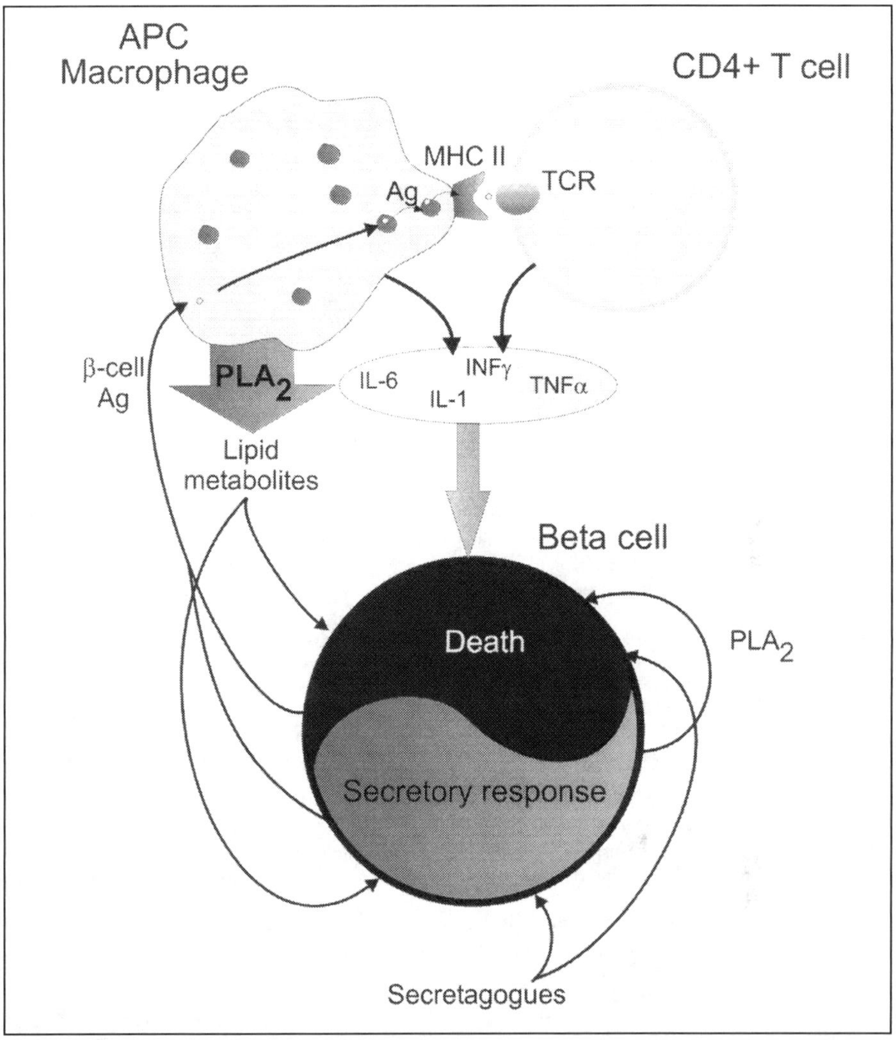

Figure 9. Yin and Yan representation of the pancreatic β-cell in terms of induction of either cell death or secretory response to secretagogues such as inflammatory cytokines and PLA_2.

Reference

1. Yu L, Rewers M, Gianani R et al. Antiislet autoantibodies usually develop sequentially rather than simultaneously. J Clin Endocrinol Metab 1996; 81(12):4264-4267.
2. Rewers M, Norris JM, Eisenbarth GS et al. Beta-cell autoantibodies in infants and toddlers without IDDM relatives: Diabetes autoimmunity study in the young (DAISY). J Autoimmun 1996; 9(3):405-410.
3. von Herrath MG, Oldstone M, Homann D et al. Is activation of autoreactive lymphocytes always detrimental? Viral infections and regulatory circuits in autoimmunity. Curr Dir Autoimmun 2001; 4:91-122.
4. von Herrath MG, Holz A, Homann D et al. Role of viruses in type I diabetes. Semin Immunol 1998; 10(1):87-100.

5. Bishop, Polak. The Anatomy, Organization and Ultrastructure of the Islets of Langerhans. Textbook of Diabetes. San Diego: Blackwell Science, 1997.
6. Ashcroft FM, Rorsman P. Electrophysiology of pancreatic islet cells. CRC Press Inc., 1995:207-242.
7. Burkitt HG, Young B, Heath JW. Wheater's Functional Histology. 3rd ed. Edinburgh: Churchill Livingstone, 1993.
8. Pierce SB, Costa M, Wisotzkey R et al. Regulation of DAF-2 receptor signaling by human insulin and ins-1, a member of the unusually large and diverse C. elegans insulin gene family. Genes Dev 2001; 15(6):672-686.
9. Leibiger IB, Leibiger B, Moede T et al. Exocytosis of insulin promotes insulin gene transcription via the insulin receptor/PI-3 kinase/p70 s6 kinase and CaM kinase pathways. Mol Cell 1998; 1(6):933-938.
10. Khan FA, Goforth PB, Zhang M et al. Insulin activates ATP-sensitive K(+) channels in pancreatic beta-cells through a phosphatidylinositol 3-kinase-dependent pathway. Diabetes 2001; 50(10):2192-2198.
11. Schuit FC, In't Veld PA, Pipeleers DG. Glucose stimulates proinsulin biosynthesis by a dose-dependent recruitment of pancreatic beta cells. Proc Natl Acad Sci USA 1988; 85(11):3865-3869.
12. Hostens K, Pavlovic D, Zambre Y et al. Exposure of human islets to cytokines can result in disproportionately elevated proinsulin release. J Clin Invest 1999; 104(1):67-72.
13. Eizirik DL, Mandrup-Poulsen T. A choice of death - the signal-transduction of immune-mediated beta- cell apoptosis. Diabetolog 2001; 44(12):2115-2133.
14. Matschinsky FM, Glaser B, Magnuson MA. Pancreatic beta-cell glucokinase: Closing the gap between theoretical concepts and experimental realities. Diabetes 1998; 47(3):307-315.
15. Schuit FC, Huypens P, Heimberg H et al. Glucose sensing in pancreatic beta-cells: A model for the study of otherglucose-regulated cells in gut, pancreas and hypothalamus. Diabetes 2001; 50(1):1-11.
16. Ahren B, Alumets J, Hakanson R et al. Somatostatin, pancreatic polypeptide, substance P and neurotensin. Cellular distribution and effects on stimulated insulin secretion in the mouse. Endocrinology 1979; 104(3):832-838.
17. Szecowka J, Tatemoto K, Rajamaki G et al. Effects of PYY and PP on endocrine pancreas. Acta Physiol Scand 1983; 119(2):123-126.
18. Degano P, Peiro E, Miralles P et al. Effects of rat pancreatic polypeptide on islet-cell secretion in the perfused rat pancreas. Metabolism 1992; 41(3):306-309.
19. Henquin JC. Triggering and amplifying pathways of regulation of insulin secretion by glucose. Diabetes 2000; 49(11):1751-1760.
20. Aguilar-Diosdado M, Parkinson D, Corbett JA et al. Potential autoantigens in IDDM. Expression of carboxypeptidase-H and insulin but not glutamate decarboxylase on the beta-cell surface. Diabetes 1994; 43(3):418-425.
21. Rabinovitch A, Pukel C, Baquerizo H. Interleukin-1 inhibits glucose-modulated insulin and glucagon secretion in rat islet monolayer cultures. Endocrinology 1988; 122(6):2393-2398.
22. Nielsen K, Karlsen AE, Deckert M et al. Beta-cell maturation leads to in vitro sensitivity to cytotoxins. Diabetes 1999; 48(12):2324-2332.
23. Strandell E, Buschard K, Saldeen J et al. Interleukin-1 beta induces the expression of hsp70, heme oxygenase and Mn-SOD in FACS-purified rat islet beta-cells, but not in alpha-cells. Immunol Lett 1995; 48(2):145-148.
24. Horio F, Fukuda M, Katoh H et al. Reactive oxygen intermediates in autoimmune islet cell destruction of the NOD mouse induced by peritoneal exudate cells (rich in macrophages) but not T cells. Diabetolog 1994; 37(1):22-31.
25. Jorns A, Tiedge M, Lenzen S et al. Effect of superoxide dismutase, catalase, chelating agents and free radical scavengers on the toxicity of alloxan to isolated pancreatic islets in vitro. Free Radic Biol Med 1999; 26(9-10):1300-1304.
26. Eizirik DL, Sandler S, Welsh N et al. Interleukin-1 beta-induced stimulation of insulin release in mouse pancreatic islets is related to diacylglycerol production and protein kinase C activation. Mol Cell Endocrinol 1995; 111(2):159-165.
27. Chen MC, Paez-Espinosa V, Welsh et al. Interleukin-1beta regulates phospholipase D-1 expression in rat pancreatic beta-cells. Endocrinology 2000; 141(8):2822-2828.
28. Karlsen AE, Sparre T, Nielsen K et al. Proteome analysis—A novel approach to understand the pathogenesis of Type 1 diabetes mellitus. Dis Markers 2001; 17(4):205-216.
29. Eizirik DL. Interleukin-1 induced impairment in pancreatic islet oxidative metabolism of glucose is potentiated by tumor necrosis factor. Acta Endocrinol (Copenh) 1988; 119(3):321-325.

30. Miura M, Endo S, Kaku LL et al. Plasma type II phospholipase A2 levels in patients with acute pancreatitis. Res Commun Mol Pathol Pharmacol 2001; 109(3-4):159-164.
31. Uhl W, Schrag HJ, Schmitter N et al. Experimental study of a novel phospholipase A2 inhibitor in acute pancreatitis. Br J Surg 1998; 85(5):618-623.
32. Matschinsky FM. Banting Lecture 1995. A lesson in metabolic regulation inspired by the glucokinase glucose sensor paradigm. Diabetes 1996; 45(2):223-241.
33. Watada H, Kajimoto Y, Miyagawa J et al. PDX-1 induces insulin and glucokinase gene expressions in alphaTC1 clone 6 cells in the presence of betacellulin. Diabetes 1996; 45(12):1826-1831.
34. Leibiger B, Leibiger IB, Moede T et al. Selective insulin signaling through A and B insulin receptors regulates transcription of insulin and glucokinase genes in pancreatic beta cells. Mol Cell 2001; 7(3):559-570.
35. Magnuson MA. Tissue-specific regulation of glucokinase gene expression. J Cell Biochem 1992; 48(2):115-121.
36. German MS. Glucose sensing in pancreatic islet beta cells: The key role of glucokinase and the glycolytic intermediates. Proc Natl Acad Sci USA 1993; 90(5):1781-1785.
37. Halban PA, Praz GA, Wollheim CB. Abnormal glucose metabolism accompanies failure of glucose to stimulate insulin release from a rat pancreatic cell line (RINm5F). Biochem J 1983; 212(2):439-443.
38 Ferber S, BeltrandelRio H, Johnson JH et al. GLUT-2 gene transfer into insulinoma cells confers both low and high affinity glucose-stimulated insulin release. Relationship to glucokinase activity. J Biol Chem 1994; 269(15):11523-11529.
39. Ling ZC, Khan A, Delauny F et al. Increased glucocorticoid sensitivity in islet beta-cells: Effects on glucose 6-phosphatase, glucose cycling and insulin release. Diabetolog 1998; 41(6):634-639.
40. Khan A, Chandramouli V, Ostenson CG et al. Glucose cycling in islets from healthy and diabetic rats. Diabetes 1990; 39(4):456-459.
41. Giroix MH, Sener A, Malaisse WJ. Hexose metabolism in pancreatic islets. Absence of glucose-6-phosphatase in rat islet cells. Mol Cell Endocrinol 1987; 49(2-3):219-225.
42. Ebert DH, Bischof LJ, Streeper RS et al. Structure and promoter activity of an islet-specific glucose-6- phosphatase catalytic subunit-related gene. Diabetes 1999; 48(3):543-551.
43. Sekine N, Cirulli V, Regazzi R et al. Low lactate dehydrogenase and high mitochondrial glycerol phosphate dehydrogenase in pancreatic beta-cells. Potential role in nutrient sensing. J Biol Chem 1994; 269(7):4895-4902.
44. Giroix MH, Rasschaert J, Sener A et al. Study of hexose transport, glycerol phosphate shuttle and Krebs cycle in islets of adult rats injected with streptozotocin during the neonatal period. Mol Cell Endocrinol 1992; 83(2-3):95-104.
45. Sener A, Malaisse WJ. Hexose metabolism in pancreatic islets. Ca(2+)-dependent activation of the glycerol phosphate shuttle by nutrient secretagogues. J Biol Chem 1992; 267(19):13251-13256.
46. Farfari S, Schulz V, Corkey B et al. Glucose-regulated anaplerosis and cataplerosis in pancreatic beta- cells: Possible implication of a pyruvate/citrate shuttle in insulin secretion. Diabetes 2000; 49(5):718-726.
47. Schuit F, De Vos A, Farfari S et al. Metabolic fate of glucose in purified islet cells. Glucose-regulated anaplerosis in beta cells. J Biol Chem 1997; 272(30):18572-18579.
48. Malaisse WJ, Hutton JC, Kawazu S et al. The stimulus-secretion coupling of glucose-induced insulin release. XXXV. The links between metabolic and cationic events. Diabetolog 1979; 16(5):331-341.
49. Newgard CB, McGarry JD. Metabolic coupling factors in pancreatic beta-cell signal transduction. Annu Rev Biochem 1995; 64:689-719.
50. Defronzo RA. Pathogenesis of type-2 diabetes - metabolic and molecular implications for identifying diabetes genes. Diabet Rev 1997; 5(3):177-269.
51. Williams JA. Electrical correlates of secretion in endocrine and exocrine cells. Fed Proc 1981; 40(2):128-134.
52. Rorsman P, Bokvist K, Ammala C et al. Ion channels, electrical activity and insulin secretion. Diabete Metab 1994; 20(2):138-145.
53. Ashcroft FM, Rorsman P. Electrophysiology of the pancreatic beta-cell. Prog Biophys Mol Biol 1989; 54(2).
54. Cook DL, Satin LS, Ashford ML et al. ATP-sensitive K+ channels in pancreatic beta-cells. Spare-channel hypothesis. Diabetes 1988; 37(5):495-498.
55. Loussouarn G, Pike LJ, Ashcroft FM et al. Dynamic sensitivity of ATP-sensitive K(+) channels to ATP. J Biol Chem 2001; 276(31):29098-29103.
56. Song DK, Ashcroft FM. ATP modulation of ATP-sensitive potassium channel ATP sensitivity varies with the type of SUR subunit. J Biol Chem 2001; 276(10):7143-7149.

57. Ashcroft FM, Gribble FM. ATP-sensitive K+ channels and insulin secretion: Their role in health and disease. Diabetolog 1999; 42(8):903-919.
58. Ashcroft FM, Rorsman P. ATP-sensitive K+ channels: A link between B-cell metabolism and insulin secretion. Biochem Soc Trans 1990; 18(1):109-111.
59. Ashcroft SJ, Ashcroft FM. Properties and functions of ATP-sensitive K-channels. Cell Signal 1990; 2(3):197-214.
60. Bryan J, Aguilar-Bryan L. The ABCs of ATP-sensitive potassium channels: More pieces of the puzzle. Curr Opin Cell Biol 1997; 9(4):553-559.
61. Sakura H, Ashcroft FM. Identification of four trp1 gene variants murine pancreatic beta-cells. Diabetolog 1997; 40(5):528-532.
62. Gembal M, Detimary P, Gilon P et al. Mechanisms by which glucose can control insulin release independently from its action on adenosine triphosphate-sensitive K+ channels in mouse B cells. J Clin Invest 1993; 91(3):871-880.
63. Gembal M, Gilon P, Henquin JC. Evidence that glucose can control insulin release independently from its action on ATP-sensitive K+ channels in mouse B cells. J Clin Invest 1992; 89(4):1288-1295.
64. Komatsu M, Noda M, Sharp GW. Nutrient augmentation of Ca2+-dependent and Ca2+-independent pathways in stimulus-coupling to insulin secretion can be distinguished by their guanosine triphosphate requirements: Studies on rat pancreatic islets. Endocrinology 1998; 139(3):1172-1183.
65. Komatsu M, Schermerhorn T, Noda M et al. Augmentation of insulin release by glucose in the absence of extracellular Ca2+: New insights into stimulus-secretion coupling. Diabetes 1997; 46(12):1928-1938.
66. Komatsu M, Sharp GW, Aizawa T et al. Glucose stimulation of insulin release without an increase in cytosolic free Ca2+ concentration: A possible involvement of GTP. Jpn J Physiol 1997; 47 Suppl 1:S22-S24.
67. Regazzi R, Li G, Ullrich S et al. Different requirements for protein kinase C activation and Ca2+-independent insulin secretion in response to guanine nucleotides. Endogenously generated diacylglycerol requires elevated Ca2+ for kinase C insertion into membranes. J Biol Chem 1989; 264(17):9939-9944.
68. Li GD, Luo RH, Metz SA. Effects of inhibitors of guanine nucleotide synthesis on membrane potential and cytosolic free Ca2+ levels in insulin-secreting cells. Biochem Pharmacol 2000; 59(5):545-556.
69. Babenko AP, Aguilar-Bryan L, Bryan J. A view of sur/KIR6.X, KATP channels. Annu Rev Physiol 1998; 60:667-687.
70. Gribble FM, Reimann F, Ashfield R et al. Regulation of KATP channels by nucleotides and drugs. Meeting of the physiological society, Cambridge, England, UK, July 21-22, 1989. J Physiol (London) 2000; 525P:27S-28S.
71. Suzuki M, Fujikura K, Inagaki N et al. Localization of the ATP-sensitive K+ channel subunit Kir6.2 in mouse pancreas. Diabetes 1997; 46(9):1440-1444.
72. Miki T, Nagashima K, Seino S. The structure and function of the ATP-sensitive K+ channel in insulin-secreting pancreatic beta-cells. J Mol Endocrinol 1999; 22(2):113-123.
73. Bokvist K, Olsen HL, Hoy M et al. Characterisation of sulphonylurea and ATP-regulated K+ channels in rat pancreatic A-cells. Pflug Arch Eur J Physiol 1999; 438(4):428-436.
74. Suzuki M, Fujikura K, Kotake K et al. Immuno-localization of sulphonylurea receptor 1 in rat pancreas. Diabetolog 1999; 42(10):1204-1211.
75. Gopel SO, Kanno T, Barg S et al. Patch-clamp characterisation of somatostatin-secreting delta-cells in intact mouse pancreatic islets. J Physiol (London) 2000; 528(3):497-507.
76. Gopel SO, Kanno T, Barg S et al. Regulation of glucagon release in mouse alpha-cells by K-ATP channels and inactivation of TTX-sensitive Na+ channels. J Physiol -London 2000; 528(3):509-520.
77. Ann K, Kowalchyk JA, Loyet KM et al. Novel Ca2+-binding protein (CAPS) related to UNC-31 required for Ca2+-activated exocytosis. J Biol Chem 1997; 272(32):19637-19640.
78. Bokvist K, Eliasson L, Ammala C et al. Colocalization of L-type Ca2+ channels and insulin-containing secretory granules and its significance for the initiation of exocytosis in mouse pancreatic B-cells. EMBO J 1995; 14(1):50-57.
79. Wiser O, Trus M, Hernandez A et al. The voltage sensitive Lc-type Ca2+ channel is functionally coupled to the exocytotic machinery. Proc Natl Acad Sci USA 1999; 96(1):248-253.
80. Nilius B, Hess P, Lansman JB et al. A novel type of cardiac calcium channel in ventricular cells. Nature 1985; 316(6027):443-446.
81. Ashcroft FM. Voltage-gated Ca^{2+} Channels. Ion Channels and Disease. San Diego: Academic Press, 2000:161-183.
82. Gutnick MJ, Lux HD, Swandulla D et al. Voltage-dependent and calcium-dependent inactivation of calcium channel current in identified snail neurones. J Physiol (Lond) 1989; 412:197-220.

83. Ammala C, Ashcroft FM, Rorsman P. Calcium-independent potentiation of insulin release by cyclic AMP in single beta-cells. Nature 1993; 363(6427):356-358.
84. Dean PM. Ultrastructural morphometry of the pancreatic -cell. Diabetolog 1973; 9(2):115-119.
85. Eliasson L, Renstrom E, Ding WG et al. Rapid ATP-dependent priming of secretory granules precedes Ca(2+)- induced exocytosis in mouse pancreatic B-cells. J Physiol (Lond) 1997; 503(Pt2):399-412.
86. Rorsman P, Eliasson L, Renstrom E et al. The cell physiology of biphasic insulin secretion. News Physiol Sci 2000; 15:72-77.
87. Gromada J, Hoy M, Renstrom E et al. CaM kinase II-dependent mobilization of secretory granules underlies acetylcholine-induced stimulation of exocytosis in mouse pancreatic B-cells. J Physiol (Lond) 1999; 518(Pt 3):745-759.
88. Ammala C, Eliasson L, Bokvist K et al. Activation of protein kinases and inhibition of protein phosphatases play a central role in the regulation of exocytosis in mouse pancreatic beta cells. Proc Natl Acad Sci USA 1994; 91(10):4343-4347.
89. Wasmeier C, Hutton JC. Secretagogue-dependent phosphorylation of the insulin granule membrane protein phogrin is mediated by cAMP-dependent protein kinase. J Biol Chem 2001; 276(34):31919-31928.
90. Wasmeier C, Hutton JC. Secretagogue-dependent phosphorylation of phogrin, an insulin granule membrane protein tyrosine phosphatase homologue. Biochem J 1999; 341(Pt3):563-569.
91. Rorsman P. The pancreatic beta-cell as a fuel sensor: An electrophysiologist's viewpoint. Diabetolog 1997; 40(5):487-495.
92. Huang JD, Brady ST, Richards BW et al. Direct interaction of microtubule- and actin-based transport motors. Nature 1999; 397(6716):267-270.
93. Easom RA. beta-granule transport and exocytosis. Semin Cell Dev Biol 2000; 11(4):253-266.
94. Sollner T, Bennett MK, Whiteheart SW et al. A protein assembly-disassembly pathway in vitro that may correspond to sequential steps of synaptic vesicle docking, activation, and fusion. Cell 1993; 75(3):409-418.
95. Rupnik M, Kreft M, Sikdar SK et al. Rapid regulated dense-core vesicle exocytosis requires the CAPS protein. Proc Natl Acad Sci USA 2000; 97(10):5627-5632.
96. Jahn R, Sudhof TC. Membrane fusion and exocytosis. Annu Rev Biochem 1999; 68:863-911.
97. Jacobsson G, Bean AJ, Scheller RH et al. Identification of synaptic proteins and their isoform mRNAs in compartments of pancreatic endocrine cells. Proc Natl Acad Sci USA 1994; 91(26):12487-12491.
98. Lang J. Molecular mechanisms and regulation of insulin exocytosis as a paradigm of endocrine secretion. Eur J Biochem 1999; 259(1-2):3-17.
99. Lao GF, Scheuss V, Gerwin CM et al. Syntaphilin: A syntaxin-1 clamp that controls SNARE assembly. Neuron 2000; 25(1):191-201.
100. Fisher RJ, Pevsner J, Burgoyne RD. Control of fusion pore dynamics during exocytosis by Munc18. Science 2001; 291(5505):875-878.
101. Huang X, Kang YH, Pasyk EA et al. Ca(2+) influx and cAMP elevation overcame botulinum toxin A but not tetanus toxin inhibition of insulin exocytosis. Am J Physiol Cell Physiol 2001; 281(3):C740-C750.
102. Land J, Zhang H, Vaidyanathan VV et al. Transient expression of botulinum neurotoxin C1 light chain differentially inhibits calcium and glucose induced insulin secretion in clonal beta-cells. Febs Lett 1997; 419(1):13-17.
103. Li C, Ullrich B, Zhang JZ et al. Ca(2+)-dependent and -independent activities of neural and nonneural synaptotagmins. Nature 1995; 375(6532):594-599.
104. Loyet KM, Kowalchyk JA, Chaudhary A et al. Specific binding of phosphatidylinositol 4,5-bisphosphate to calcium-dependent activator protein for secretion (CAPS), a potential phosphoinositide effector protein for regulated exocytosis. J Biol Chem 1998; 273(14):8337-8343.
105. Sollner T, Bennett MK, Whiteheart SW et al. A protein assembly-disassembly pathway in vitro that may correspond to sequential steps of synaptic vesicle docking, activation and fusion. Cell 1993; 75(3):409-418.
106. Xu T, Rammner B, Margittai M et al. Inhibition of SNARE complex assembly differentially affects kinetic components of exocytosis. Cell 1999; 99(7):713-722.
107. Xu T, Ashery U, Burgoyne RD et al. Early requirement for alpha-SNAP and NSF in the secretory cascade in chromaffin cells. EMBO J 1999; 18(12):3293-3304.
108. Iezzi M, Escher G, Meda P et al. Subcellular distribution and function of Rab3A, B, C, and D isoforms in insulin-secreting cells. Mol Endocrinol 1999; 13(2):202-212.

109. Kajio H, Olszewski S, Rosner PJ et al. A low-affinity Ca2+-dependent association of calmodulin with the Rab3A effector domain inversely correlates with insulin exocytosis. Diabetes 2001; 50(9):2029-2039.
110. Balsinde J, Balboa MA, Insel PA et al. Regulation and inhibition of phospholipase A2. Annu Rev Pharmacol Toxicol 1999; 39:175-189.
111. Lai MM, Hong JJ, Ruggiero AM et al. The calcineurin-dynamin 1 complex as a calcium sensor for synaptic vesicle endocytosis. J Biol Chem 1999; 274(37):25963-25966.
112. Bauerfeind R, Takei K, De Camilli P. Amphiphysin I is associated with coated endocytic intermediates and undergoes stimulation-dependent dephosphorylation in nerve terminals. J Biol Chem 1997; 272(49):30984-30992.
113. Renstrom E, Eliasson L, Rorsman P. Protein kinase A-dependent and -independent stimulation of exocytosis by cAMP in mouse pancreatic B-cells. J Physiol (Lond) 1997; 502(Pt 1):105-118.
114. Holz GG, Leech CA, Scott-Heller R et al. cAMP-dependent mobilization of intracellular Ca2+ stores by activation of ryanodine receptors in pancreatic beta-cells. J Biol Chem 1999; 274(20):14147-14156.
115. Ashcroft FM, Proks P, Smith PA et al. Stimulus-secretion coupling in pancreatic beta cells. J Cell Biochem 1994; 55(Suppl):54-65.
116. Corkey BE, Deeney JT, Yaney GC et al. The role of long-chain fatty acyl-CoA esters in beta-cell signal transduction. J Nutr 2000; 130(2S Suppl):299S-304S.
117. McPhail LC, Clayton CC, Snyderman R. A potential second messenger role for unsaturated fatty acids: Activation of Ca2+-dependent protein kinase. Science 1984; 224(4649):622-625.
118. Murakami K, Chan SY, Routtenberg A. Protein kinase C activation by cis-fatty acid in the absence of Ca2+ and phospholipids. J Biol Chem 1986; 261(33):15424-15429.
119. Prescott SM. A thematic series on phospholipases. J Biol Chem 1997; 272(24):15043.
120. Tischfield JA. A reassessment of the low molecular weight phospholipase A2 gene family in mammals. J Biol Chem 1997; 272(28):17247-17250.
121. Clark JD, Lin LL, Kriz RW et al. A novel arachidonic acid-selective cytosolic PLA2 contains a Ca(2+)-dependent translocation domain with homology to PKC and GAP. Cell 1991; 65(6):1043-1051.
122. Sharp JD, White DL, Chiou XG et al. Molecular cloning and expression of human Ca(2+)-sensitive cytosolic phospholipase A2. J Biol Chem 1991; 266(23):14850-14853.
123. Leslie CC. Properties and regulation of cytosolic phospholipase A2. J Biol Chem 1997; 272(27):16709-16712.
124. Sapirstein A, Bonventre JV. Specific physiological roles of cytosolic phospholipase A2 as defined by gene knockouts. Biochim Biophys Acta 2000; 1488(1-2):139-148.
125. Bonventre JV, Huang Z, Taheri MR et al. Reduced fertility and postischaemic brain injury in mice deficient in cytosolic phospholipase A2. Nature 1997; 390(6660):622-625.
126. Murakami M, Nakatani Y, Kuwata H et al. Cellular components that functionally interact with signaling phospholipase A2s. Biochim Biophys Acta 2000; 1488(1-2):159-166.
127. Balsinde J, Dennis EA. Function and inhibition of intracellular calcium-independent phospholipase A2. J Biol Chem 1997; 272(26):16069-16072.
128. Winstead MV, Balsinde J, Dennis EA. Calcium-independent phospholipase A(2): Structure and function. BBA-Mol Cell Biol Lipids 2000; 1488(1-2):28-39.
129. Ramanadham S, Hsu FF, Bohrer A et al. Studies of the role of group VI phospholipase A2 in fatty acid incorporation, phospholipid remodeling, lysophosphatidylcholine generation, and secretagogue-induced arachidonic acid release in pancreatic islets and insulinoma cells. J Biol Chem 1999; 274(20):13915-13927.
130. Dennis EA. The growing phospholipase A2 superfamily of signal transduction enzymes. Trends Biochem Sci 1997; 22(1):1-2.
131. Stafforini DM, McIntyre TM, Zimmerman GA et al. Platelet-activating factor acetylhydrolases. J Biol Chem 1997; 272(29):17895-17898.
132. Ramanadham S, Ma Z, Arita H et al. Type IB secretory phospholipase A2 is contained in insulin secretory granules of pancreatic islet beta-cells and is co secreted with insulin from glucose-stimulated islets. Biochim Biophys Acta 1998; 1390(3):301-312.
133. Metz S, Holmes D, Robertson RP et al. Gene expression of type I phospholipase A2 in pancreatic beta cells. Regulation of mRNA levels by starvation or glucose excess. Febs Lett 1991; 295(1-3):110-112.
134. Loweth AC, Scarpello JH, Morgan NG. Phospholipase A2 expression in human and rodent insulin-secreting cells published erratum appears in Mol Cell Endocrinol 1995 Oct 30; 114(1-2):227-8. Mol Cell Endocrinol 1995; 112(2):177-183.

135. Niwa T, Matsukawa Y, Senda T et al. Acetylcholine activates intracellular movement of insulin granules in pancreatic beta-cells via inositol trisphosphate-dependent correction of triphosphate-dependent mobilization of intracellular Ca2+ [published erratum appears in Diabetes 1999; 48(1):233]. Diabetes 1998; 47(11):1699-1706.
136. Parker KJ, Jones PM, Hunton CH et al. Identification and localisation of a type IV cytosolic phospholipase A2 in rat pancreatic beta-cells. J Mol Endocrinol 1996; 17(1):31-43.
137. Eddlestone GT. ATP-sensitive K channel modulation by products of PLA2 action in the insulin-secreting HIT cell line. Am J Physiol Cell Physiol 1995; 268(1 37-1):C181-C190.
138. Ramanadham S, Gross RW, Han X et al. Inhibition of arachidonate release by secretagogue-stimulated pancreatic islets suppresses both insulin secretion and the rise in beta-cell cytosolic calcium ion concentration. Biochemistry (Mosc) 1993; 32(1):337-346.
139. Gross RW, Ramanadham S, Kruszka KK et al. Rat and human pancreatic islet cells contain a calcium ion independent phospholipase A2 activity selective for hydrolysis of arachidonate which is stimulated by adenosine triphosphate and is specifically localized to islet beta- cells. Biochemistry (Mosc) 1993; 32(1):327-336.
140. Kashiwagi M, Friess H, Uhl W et al. Phospholipase A2 isoforms are altered in chronic pancreatitis. Ann Surg 1998; 227(2):220-228.
141. Seeds MC, Jones DF, Chilton FH et al. Secretory and cytosolic phospholipases A2 are activated during TNF priming of human neutrophils. Biochim Biophys Acta 1998; 1389(3):273-284.
142. Turk J, Wolf BA, Lefkowith JB et al. Glucose-induced phospholipid hydrolysis in isolated pancreatic islets: Quantitative effects on the phospholipid content of arachidonate and other fatty acids. Biochim Biophys Acta 1986; 879(3):399-409.
143. Ma Z, Ramanadham S, Kempe K et al. Pancreatic islets express a Ca2+-independent phospholipase A2 enzyme that contains a repeated structural motif homologous to the integral membrane protein binding domain of ankyrin. J Biol Chem 1997; 272(17):11118-11127.
144. Ramanadham S, Hsu FF, Bohrer A et al. Electrospray ionization mass spectrometric analyses of phospholipids from rat and human pancreatic islets and subcellular membranes: Comparison to other tissues and implications for membrane fusion in insulin exocytosis. Biochemistry (Mosc) 1998; 37(13):4553-4567.
145. Wolf BA, Turk J, Sherman WR et al. Intracellular Ca2+ mobilization by arachidonic acid. Comparison with myo- inositol 1,4,5-trisphosphate in isolated pancreatic islets. J Biol Chem 1986; 261(8):3501-3511.
146. Wolf BA, Pasquale SM, Turk J. Free fatty acid accumulation in secretagogue-stimulated pancreatic islets and effects of arachidonate on depolarization-induced insulin secretion. Biochemistry (Mosc) 1991; 30(26):6372-6379.
147. Bell RL, Kennerly DA, Stanford N et al. Diglyceride lipase: A pathway for arachidonate release from human platelets. Proc Natl Acad Sci USA 1979; 76(7):3238-3241.
148. Konrad RJ, Major CD, Wolf BA. Diacylglycerol hydrolysis to arachidonic acid is necessary for insulin secretion from isolated pancreatic islets: Sequential actions of diacylglycerol and monoacylglycerol lipases. Biochemistry (Mosc) 1994; 33(45):13284-13294.
149. Konrad RJ, Jolly YC, Major C et al. Inhibition of phospholipase A2 and insulin secretion in pancreatic islets. Biochim Biophys Acta 1992; 1135(2):215-220.
150. Metz SA. Pancreatic islet phospholipase A2: Differential, Ca2+-dependent effects of lysophospholipids, arachidonic acid, and its lipoxygenase-derived metabolites on insulin release. Adv Prostaglandin Thromboxane Leukot Res 1987; 17B:668-676.
151. Shuttleworth TJ. Arachidonic acid activates the noncapacitative entry of Ca2+ during Ca2+i oscillations. J Biol Chem 1996; 271(36):21720-21725.
152. Damron DS, Van-Wagoner DR, Moravec CS et al. Arachidonic acid and endothelin potentiate Ca2+ transients in rat cardiac myocytes via inhibition of distinct K+ channels. J Biol Chem 1993; 268(36):27335-27344.
153. Graber MN, Alfonso A, Gill DL. Ca2+ pools and cell growth: Arachidonic acid induces recovery of cells growth-arrested by Ca2+ pool depletion. J Biol Chem 1996; 271(2):883-888.
154. McPhail LC, Clayton CC, Snyderman R. A potential second messenger role for unsaturated fatty acids: Activation of Ca2+-dependent protein kinase. Science 1984; 224(4649):622-625.
155. Metz SA. Exogenous arachidonic acid promotes insulin release from intact or permeabilized rat islets by dual mechanisms. Putative activation of Ca2+ mobilization and protein kinase C. Diabetes 1988; 37(11):1453-1469.
156. Metz SA. The pancreatic islet as Rubik's Cube. Is phospholipid hydrolysis a piece of the puzzle? Diabetes 1991; 40(12):1565-1573.

157. Ramanadham S, Gross R, Turk J. Arachidonic acid induces an increase in the cytosolic calcium concentration in single pancreatic islet beta cells. Biochem Biophys Res Commun 1992; 184(2):647-653.

157a. Juhl K, Hoy M, Olsen HL et al. cPLA2alpha-evoked formation of arachidonic acid and lysophospholipids is required for exocytosis in mouse pancreatic beta-cells. Am J Physiol Endocrinol Metab 2003; 285(1):E73-81.

158. Hirabayashi T, Shimizu T. Localization and regulation of cytosolic phospholipase A2. Biochim Biophys Acta 2000; 1488(1-2):124-138.

159. Kato R, Yamamoto S, Nakadate T et al. Possible involvement of phospholipase A2 activation and lipoxygenase product(s) in the mechanism of insulin secretion. Adv Prostaglandin Thromboxane Leukot Res 1983; 12:265-270.

160. Zhou YP, Teng D, Dralyuk F et al. Apoptosis in insulin-secreting cells. Evidence for the role of intracellular Ca2+ stores and arachidonic acid metabolism. J Clin Invest 1998; 101(8):1623-1632.

161. Ma Z, Ramanadham S, Corbett JA et al. Interleukin-1 enhances pancreatic islet arachidonic acid 12- lipoxygenase product generation by increasing substrate availability through a nitric oxide-dependent mechanism. J Biol Chem 1996; 271(2):1029-1042.

162. Metz SA. Mobilization of cellular Ca2+ by lysophospholipids in rat islets of Langerhans. Biochim Biophys Acta 1988; 968(2):239-252.

163. Rustenbeck I, Lenzen S. Effect of lysophospholipids, arachidonic acid and other fatty acids on regulation of Ca2+ transport in permeabilized pancreatic islets. Cell Calcium 1992; 13(4):193-202.

164. Gasser KW, Holda JR. ATP-sensitive potassium transport by pancreatic secretory granule membrane. Am J Physiol 1993; 264(1 Pt 1):G137-G142.

165. Metz SA. Ether-linked lysophospholipids initiate insulin secretion. Lysophospholipids may mediate effects of phospholipase A2 activation on hormone release. Diabetes 1986; 35(7):808-817.

166. Frossard JL, Bhagat L, Lee HS et al. Both thermal and nonthermal stress protect against caerulein induced pancreatitis and prevent trypsinogen activation in the pancreas. Gut 2002; 50(1):78-83.

167. Ramsingh AI, Chapman N, Tracy S. Coxsackieviruses and diabetes. Bioessays 1997; 19(9):793-800.

168. Sevilla N, Homann D, von Herrath M et al. Virus-induced diabetes in a transgenic model: Role of cross-reacting viruses and quantitation of effector T cells needed to cause disease. J Virol 2000; 74(7):3284-3292.

169. Gu D, Sarvetnick N. Epithelial cell proliferation and islet neogenesis in IFN-g transgenic mice. Development 1993; 118(1):33-46.

170. Pruzanski W, Vadas P. Phospholipase A2—a mediator between proximal and distal effectors of inflammation. Immunol Today 1991; 12(5):143-146.

171. van Puijenbroek AA, Wissink S, van der Saag PT et al. Phospholipase A2 inhibitors and leukotriene synthesis inhibitors block TNF-induced NF-kappaB activation. Cytokine 1999; 11(2):104-110.

172. Roe MW, Worley JF, Qian F et al. Characterization of a Ca2+ release-activated nonselective cation current regulating membrane potential and [Ca2+]i oscillations in transgenically derived beta-cells. J Biol Chem 1998; 273(17):10402-10410.

173. Nalefski EA, Falke JJ. The C2 domain calcium-binding motif: Structural and functional diversity. Protein Sci 1996; 5(12):2375-2390.

174. Bottomley MJ, Salim K, Panayotou G. Phospholipid-binding protein domains. Biochim Biophys Acta 1998; 1436(1-2):165-183.

175. Six DA, Dennis EA. The expanding superfamily of phospholipase A2 enzymes: Classification and characterization. Biochim Biophys Acta 2000; 1488(1-2):1-19.

CHAPTER 3

Animal Models of Type 1 Diabetes:
Genetics and Immunological Function

Julie Lang and Donald Bellgrau

The ancient Egyptians described diabetes several millennia ago, but the first major break through in the understanding of the physiopathology of the disease came in 1921 with the discovery of insulin by Banting, Macleod, and Best. Because of the limitations of performing pathological studies in the human pancreas, the field has turned to multiple animal models of type 1 diabetes. Spontaneous mutations in several inbred rodent lines led to sublines which succumb to type 1 diabetes with varying incidences. These type 1 diabetes-susceptible models include the nonobese diabetic (NOD) mouse, the BioBreeding Diabetes-Prone (BB-DP) rat and the Komeda Diabetes-Prone (KDP) subline of the Long-Evans Tokushima Lean rat. In addition, other experimentally-induced models of type 1 diabetes were established as T cell receptor (TCR) transgenic (Tg) mice generated from diabetogenic T cell clones as well as neo-antigen (Ag) expression under the control of the rat insulin promoter (RIP) to establish neo-self antigen pancreatic expression. With these various animal models numerous studies have addressed the genetic susceptibility and immunology of the autoimmune attack on the pancreatic beta cells. A major conclusion from these models, as well as studies of human type 1 diabetics, is that type 1 diabetes is not the result of a well-defined pathway but can be the result of numerous distinct mechanisms; i.e., all type 1 diabetics do not get the disease in the same way. Nonetheless, some generalities do exist including the polygenic nature of the disease, the involvement of T cells in the destruction of pancreatic beta cells, and the incomplete penetrance of the disease implying that environmental factors influence disease susceptibility. These models provide useful tools for studying the autoimmune process as well as establishing and testing numerous provocative treatments for the prevention and possible cure of type 1 diabetes.

The Nonobese Diabetic Mouse

We will begin our discussion with the NOD mouse – the most-characterized animal model of type 1 diabetes.[1,2] The NOD inbred mouse line was established at the Shionogi Research Laboratories in Japan through sib breedings of a hyperglycemic female mouse from the CTS subline.[3] NOD mice show islet infiltration around 6-7 weeks of age followed by the spontaneous development of overt diabetes in approximately 70% of the females and 40% of the males by 30 weeks of age.[4,5] Similar to humans, NOD mice transiently express anti-insulin autoantibodies in their serum prior to hyperglycemia.[6-8] Although all NOD mice develop insulitis, this is not always followed by diabetes. Diabetes incidence varies among established NOD colonies although in all cases, unlike humans, gender influences disease incidence with higher disease frequencies in female mice than males.[5] Type 1 diabetes incidence is also affected by environmental conditions with higher disease frequencies in sterile conditions[9] and lower disease penetrance with infectious agents.[10,11] The disease can be prevented in numerous ways (>134) including immunological or genetic manipulations of NOD mice.[12] This model system enables testing of experimental preventative approaches before starting clinical trials in humans.

Immunology of Type 1 Diabetes, Second Edition, edited by George S. Eisenbarth.

At least four studies currently in clinical trials were initiated in the NOD mouse – oral nicotinamide, oral insulin (DPT-1 Trial), anti-CD3 (hOKT3gamma1 Ala-ala) antibody (Ab) treatment and ingestion of IFN-alpha (http://clinicaltrials.gov, see Chapter 12). In this chapter we will summarize the current available knowledge on the NOD mouse, concentrating on its immunological and genetic features.

Genetics of Diabetogenesis in the NOD Mouse

Like the human disease, diabetes susceptibility in the NOD is inherited through multiple genes, with a strong predisposing effect deriving from genes within the major histocompatibility complex (MHC).[13-15] A great deal of effort in the last decade has been involved in mapping the diabetes-associated alleles in both NOD mice and humans. In both cases, numerous loci have been identified (>20) although very few genes have been found to be clearly associated with the disease.[15,16] Genetic analysis has involved microsatellite mapping of the mouse genome with NOD backcrosses to nonautoimmune-prone strains. Loci (identified as Iddn in the mouse and IDDMn in the human) with significant LOD scores for association have been analyzed for candidate genes. Multiple congenic strains containing Idd locus(i) from disease-resistant strains have been generated and analyzed for disease incidence, insulitis, and more recently immunological phenotypes associated with diabetes in the NOD.[16-18] This approach has not only confirmed association with disease but also aided in further mapping of the locus. A general finding is that several identified regions actually contain multiple-disease associated loci which alone rarely confer protection but in multitudes of two or more can nearly completely block disease.[16] Positional cloning of numerous Idd loci is underway with few clear results to date. Newer technologies including Tg mice with human YACs, Chromosomal Substitution Strains (CSS) of mice, SNP markers every 10 kb for positional cloning and knowledge of human and mouse genomic sequences should enhance the speed at which disease-associated genes are discovered.[17] Nonetheless the process still necessitates verification of the locus's effect on disease incidence, a process requiring considerable effort.

A list of the identified Idd loci, their positions on human and mouse chromosomes, identified phenotypes and candidate genes are presented in Table 1. The discussion below will highlight current knowledge of the numerous Idd loci. These loci can either confer susceptibility or resistance to disease development. Idd1 has been mapped to the MHC locus on mouse chromosome 17.[5,13] Although the class II allele within the MHC does not completely explain the association of disease to this locus (i.e., other nonMHC genes within this locus likely contribute to disease susceptibility), the MHC class II locus is the most studied gene within this locus and will therefore be the focus of our discussion.[16] The I-A^{g7} molecule behaves as a recessive allele, normally required in the homozygous state for diabetes development in the NOD model.[13] Of note, both human DQ and mouse IA diabetes-associated sequences (DQB1*0302 and IA^{g7}) carry a serine at position 57 of their β-chain instead of the aspartic acid residue conserved on other mouse strains and on other DQB1 alleles conferring low diabetes risk in humans.[14,19] These findings have suggested that this particular residue may influence diabetes susceptibility, but it is likely that other amino acids within DQ or IA as well as other MHC genes may be important both in humans and mice. If fact, population studies,[20] and experiments in transgenic mice,[21,22] have failed to show position 57 alone as a susceptibility factor and rather support a significant role for alleles at the DRB1 locus in humans and for the corresponding IE molecule in the mice. The NOD mouse lacks surface expression of the IE molecule and its expression as a transgene prevents diabetes.[21-23] These data must be interpreted with caution since control experiments with transgenic expression of the diabetogenic IA^{g7} molecule also protect NOD mice from diabetes.[24] The Idd1 locus may act as a gene complex with at least two susceptibility loci, IA and IE.[22] It must be noted that, as with any one identified Idd locus, the combination of the IA and IE NOD alleles is not sufficient for diabetes development as CTS mice share the NOD class II alleles but are disease-resistant.[25,26] In addition, the requirement for IA^{g7} homozygosity is not absolute as models of mice heterozygous for IA^{g7} have been found to be susceptible to type 1 diabetes.[27]

Table 1. Identified Idd loci in NOD mouse

Locus	Chromosome	NOD Allele	Analysis	Phenotype of NOD Congenics	Disease Protection of Congenic Strains	Candidate Genes	References
Idd1	17	Susceptible	Backcross/ congenic	Disease protection however some signs of inflammation	100%	MHC I and II (IAg7); HSP70, Bat5, TNFα TNFβ, Tap1, Tap2	LS Wicker (1995)
Idd2	9	Susceptible	Backcross and F_2			Thy1	LS Wicker (1995) and MA McAleer (1995)
Idd3*	3	Susceptible	Backcross/ congenic	Delayed onset and protection; insulitis; Auto Abs; protects from EAE	69%	IL2; Fgf2; IL21	PL Podolin (1997), NJ Hill (2000) and PA Lyons (2000 Genome Res)
Idd4	11	Susceptible	Backcross				LS Wicker (1995)
Idd5*	1	Susceptible	Backcross/ congenic	Delayed onset; insulitis; Auto Abs; apoptosis resistance	45%		F Colucci (1997) and NJ Hill (2000)
Idd5.1	1	Susceptible	Congenic	Delayed onset; insulitis	26%	Casp8; FLIP, CD28, CD152, ICOS	NJ Hill (2000)
Idd5.2	1	Susceptible	Congenic	No protection	0%	Nramp1; CXCR2	NJ Hill (2000)
Idd6	6	Susceptible xB10; resistant xNON	Backcross and F_2				LS Wicker (1995) and MA McAleer (1995)
Idd7	7	Resistance	Backcross	Involved in NOD BDC2.5 TCR Tg disease resistance			A Gonzalez (1997) and MA McAleer (1995)
Idd8	14	Resistance	Backcross				LS Wicker (1995)
Idd9	4	Susceptible	Backcross, F_2 and congenic	Insulitis; High AutoAb titer; high CD30 expression and IL4 secretion; heterozygotes intermediate protection	90%		PA Lyons (2000-Immunity)
Idd9.1	4	Susceptible	Congenic			Jak1, Lck	PA Lyons (2000-Immunity)
Idd9.2	4	Susceptible	Congenic			CD30 - strain variations but same expression; Tnfr2	PA Lyons (2000-Immunity) and T Siegmund (2000)

Locus	Chromosome	NOD Allele	Analysis	Phenotype of NOD Congenics	Disease Protection of Congenic Strains	Candidate Genes	References
Idd9.3	4	Susceptible	Congenic	Delayed onset; insulitis	20%	CD137	PA Lyons (2000-Immunity)
Idd9.2/9.3	4	Susceptible	Congenic	Delayed onset; insulitis	62%		PA Lyons (2000-Immunity)
Idd10*	3	Susceptible	Backcross/ congenic	Small protection from disease	36%	Nras ruled out	PL Podolin (1998)
Idd11	4	Susceptible	Backcross	Protection; heterozygotes intermediate protection	62%	May be same as Idd9.1; not slc9a1	TC Brodnicki (2000)
Idd12	14	Susceptible	Backcross				LS Wicker (1995)
Idd13	2	Susceptible	F_2 and congenic	Insulitis; delayed onset	74%		D Serreze (1998)
Idd13a	2	Susceptible	Congenic	Insulitis; delayed onset	38%	beta 2-microglobulin	D Serreze (1998)
Idd13b	2	Susceptible	Congenic	Insulitis; delayed onset	25%		D Serreze (1998)
Idd14	13	Susceptible	F_2				M McAleer (1995)
Idd15	5	Susceptible	F_2			Xmv65	M McAleer (1995)
Idd16	17	Susceptible	Congenic	Delayed onset	52%	Hsp70; TNFα (identical gene sequence); class I	H Ikegami (1995)
Idd17	3	Susceptible	Congenic				PL Podolin (1997)
Idd18	3	Susceptible	Congenic		9%	Csf1 ruled out	PL Podolin (1997) and PL Podolin (1998)
Combined congenic strains (reviewed in JA Todd and LS Wicker - 2001)							
Idd10/18	3	Susceptible	Congenic		38%		PL Podolin (1998)
Idd3/10/18	3	Susceptible	Congenic	No Insulitis	92%		PA Lyons (2000-Immunity)
Idd3/10	3	Susceptible	Congenic		93%		PL Podolin (1997)
Idd3/5.1/5.2	1/3	Susceptible	Congenic	No insulitis; no Auto Abs	97%		NJ Hill (2000)
Idd3/10/18/9	3/4	Susceptible	Congenic	Complete protection; small number develop insulitis (10%)	100%		PA Lyons (2000-Immunity)
Idd10/17	3	Susceptible	Congenic		58%		PL Podolin (1997)
Idd3/10/17	3	Susceptible	Congenic		98%		PL Podolin (1997)

*Loci involved in insulitis formation (defective tolerance)

Recent crystallography studies of the IA^{g7} molecule showed this MHC protein is structurally stable. In comparison to other class II molecules, this diabetes-associated allele contains an altered peptide binding groove which allows more promiscuous binding of numerous peptides,[28,29] confirming earlier studies of weak peptide binding.[30] Furthermore, diabetes-associated DQ alleles in humans were also found to have similarly altered peptide-binding grooves in crystallography studies.[31] One possible role of weak presentation of self-peptides by MHC in autoimmunity could be inefficient thymic deletion. Support for this concept comes from limiting dilution studies which showed a high frequency of autoreactive T cells in IA^{g7} mice.[32]

Investigations into the role of MHC on thymic tolerance processes provide evidence for both a lack of thymic deletion in IA^{g7} homozygous mice and positive selection of regulatory T cells by protective MHC alleles. The CD4+ anti-islet 4.1 TCR Tg is positively selected by IA^{g7} but negatively selected with coexpression of other MHC class II alleles.[33] In addition, Ridgway and coworkers showed that dosing the IA^{g7} allele correlates with the degree of insulitis and autoreactive repertoire.[34] On the other hand, BDC2.5 TCR CD4+ anti-islet Tg mice show selection of regulatory subsets by protective alleles.[35] Thus, the current data suggest altered MHC alleles result in inefficient thymic deletion and altered positive selection in the NOD mouse. However, there is also evidence for a role of the IA^{g7} molecule in the peripheral activation of autoreactive T cells.[36] A mutated form of the IA^{g7} molecule replacing the His and Ser at positions 56 and 57 with the more common Pro and Asp residues generates a Tg MHC molecule known as BALBg7PD. This MHC Tg mouse mediates positive selection of BDC2.5 TCR Tg cells however these mature T cells are unable to mount an anti-islet response with BALBg7PD antigen-presenting cells (APCs) while responses with NOD WT APCs are adequate.[37] In conclusion, the Idd1 MHC locus may contribute to type 1 diabetes both in altering thymic selection processes and facilitating activation of autoreactive T cells in the periphery.

Idd2 has been assigned to mouse chromosome 9 and linked to the T lymphocyte marker thy1, although a significant association with diabetes has not been found in other studies.[38,39] Besides Idd1, most progress in gene identification and contribution to disease has been made with the Idd3 locus. NOD mice congenic for the Idd3 region of chromosome 3 from B6 mice show a low incidence of diabetes (25% compared to 80%).[40] The Idd3 locus contains the IL2 gene which has a unique glycosylation form in the NOD;[41] however, the protein appears to have normal function. In a recent oral presentation, Wicker reported that the IL2 gene is indeed not the Idd3 locus (Wicker, personal communication). Other identified candidate genes include Fgf2 and IL21.[16] Idd4 maps on chromosome 11 and may influence the frequency and severity of insulitis and progression to diabetes.[5,9] In particular, Idd4 homozygosity determines the age of onset of diabetes and Idd2, Idd3, and Idd4 together may accelerate progression to overt diabetes.[42]

The Idd5 locus is located on chromosome 1 and alone gives 50% protection from diabetes and in combination with Idd3 provides nearly complete protection from infiltration of the pancreas, thyroid and salivary glands.[43] Therefore, these genes are likely to influence tolerance processes in the animal.[44] The synergistic effect follows a model of additivity rather than multiple epistasis.[45] Two loci have been identified within the Idd5 region which result in delayed onset of disease. Idd5.1 includes casp8, flip, CD28 and CD152 as candidate genes and overlaps with the IDDM12 locus in humans (candidate gene is CD152 or $CTLA_4$).[43] Idd5 F_2 mice show a resistance to gamma-induced apoptosis in the NOD and NOR strains while T cells from C57Bl/6 and DBA/2 strains show a "high-apoptosis" phenotype.[46] The $CTLA_4$ candidate gene is intriguing because $CTLA_4$-/- mice are also resistant to apoptosis.[47] Candidate genes for the Idd5.2 locus include cxcr2, IL8R and Nramp1 which contains polymorphisms between NOD and B10.[43]

The Idd6 locus contains the NK cell cluster and NOD.Idd6 (containing NK1.1) congenic mice have been shown to have reduced disease incidence.[48] The importance of NK cell function in reducing disease in NOD mice has been a story of great interest this past year[47a,48-52] although the complete localization to Idd6 is still unknown. F_2 intercrossing between NOD and C57Bl/6 mice showed an association between the Idd6 locus on chromosome 6 and decreased proliferation of immature thymocytes in NOD mouse[53] although the contribution of this phenotype to disease is still unknown. Little information is available on Idd7 and Idd8 except that these loci appear to be protective in the homozygous state.[42]

The Idd9 locus on chromosome 4 is now known to contain 3 distinct loci. NOD.B10 Idd9.1/9.2/9.3 triple congenic mice are almost completely protected from diabetes yet still show insulitis, suggesting these loci are involved in regulating autoreactive T cells.[40] A change in cytokine production from IFNγ– to IL4 of infiltrating cells in triple congenic mice compared to wild-type NOD supports this interpretation. The presence of insulitis and salivary gland infiltrates in the Idd9 congenic mice suggest that tolerance defects still exist. Genes of the TNFR superfamily with polymorphisms between NOD and B10 mice include candidate genes CD30 and TNFR2 for Idd9.2[54] and CD137 for Idd9.3.[40]

Idd10 and Idd18 are closely linked on mouse chromosome 3. NOD.B610/18 congenic mice show reduced incidence of diabetes (50% vs 80%) and include candidate genes Csfm, CD53, KCNA3, Nras and Rap1a.[55] Adding another locus from chromosome 3 to create NOD.B6 3/10/18 congenic mice results in almost complete loss of diabetes as well as insulitis.[56] The few animals that do develop diabetes have a delayed-onset. Idd11 has been localized to one of the three Idd9 loci. Idd13, like Idd5, is involved in the regulation of T cell progression from benign to destructive insulitis. It contains at least two identified loci, Idd13a and Idd13b which contains the candidate gene β2-microglobulin.[18] The role of the majority of Idd loci on diabetes development is still largely unknown, with the generation of NOD congenic strains quite time-consuming. Immunological phenotypes associated with the NOD mouse have been described in the last decade (see section below). The role or relevance of these phenotypes in disease development is still largely unknown although investigators are beginning to study the association of various Idd loci with phenotypes to determine a correlation.

One concept emerging from genetic studies of numerous animal models of autoimmunity is the existence of autoimmune-prone genotypes, i.e., certain individuals inherit genes which predispose them to autoimmunity. Numerous autoimmune-associated loci in mouse, human and rat studies of type 1 diabetes, multiple sclerosis, collagen-induced arthritis and systemic lupus erythematosus (SLE) have been mapped to the same chromosomal sites, suggesting a common genetic basis.[17,57,58] In addition, families, and even individuals, predisposed to autoimmunity can have multiple immune-mediated diseases.[59-62] Indeed the NOD mouse itself develops multiple autoimmunities in addition to type 1 diabetes including thyroiditis, lymphocyte infiltration of the salivary and lacrimal glands,[15] and has been shown to be susceptible to both experimental autoimmune encephalitis (EAE — an animal model for multiple sclerosis)[63] and experimental autoimmune prostatis.[64] Therefore certain Idd genes or combinations thereof may alter the immune system such that tolerance processes are defective. The form of autoimmunity may then depend on other genetic (e.g., MHC and antigens) or environmental factors. One recent study shows that autoimmunity in the NOD switches from pancreas-specific to destruction of the peripheral nervous system by altering the costimulatory molecules present on dendritic cells.[65] This disease induction is not possible in mouse strains lacking "autoimmunity" genes.

Immunology of the NOD Mouse

Along with work in other animal models of type 1 diabetes, studies in the NOD mouse clearly showed the beta cell destruction was lymphocyte dependent, requiring CD4 T cells, CD8 T cells[66-68] and B cells for its spontaneous occurrence.[69] This model allowed studies of the progression of the disease including the timing of insulitis and the nature of insulitis i.e., cytokine production during the course of the disease. Prior to beta cell damage a relatively long period of "benign" insulitis is observed followed by a destructive form in which Th1 type cytokines (IFNγ– and TNF-α) dominate. There remains a debate among investigators whether the beta-cell loss is a gradual or acute process, occurring suddenly once regulation of autoreactive T cells wanes.

In NOD mice, transient expression of insulin autoantibodies (IAA) is noted prior to onset of hyperglycemia.[70] However, it was recently reported that NOR mice, a strain closely related to the NOD with the IA^{g7} MHC haplotype, also develop transient IAA antibodies but not type 1 diabetes, illustrating the incompleteness of this assay for disease prediction.[71] Genetic studies using congenic NOD strains show the expression of IAA Abs is controlled by multiple loci.[71a] In addition the amount of IAA was found to associate with degrees of insulitis more than disease incidence.[71a]

The role of various cell populations in the beta cell damage have been addressed using knockout animals, T cell clones and adoptive transfer models of the NOD mouse and are discussed in the following chapter. To summarize numerous findings, spontaneous disease requires CD4 and CD8 T cells as well as B cells, which are thought to be involved in Ag presentation at onset of insulitis. However some CD4 clones and even some CD8 clones, and now TCR Tg cells, have been isolated which do not require other cell types for disease transfer. Whether a CD8 clone requires CD4 "help" appears to be associated with the TCR's affinity for autoAg.

The confirmation that lymphocytes are required for the beta cell destruction in type 1 diabetes led to numerous studies on the immune system in NOD mice compared to non autoimmune-prone strains. A deficiency in the ability of NOD APCs to mount equivalent T cell responses has been reported.[72,73] These NOD APCs have been found to have low CD86 expression[74] and recent studies have reported phenotypic and functional defects in bone-marrow-derived dendritic cells (DCs) in response to GM-CSF.[75] Wong et al have found a deficiency in priming of NOD T cell responses to both endogenous and exogenous antigens.[76] Similarly, defects in TCR-mediated signaling resulting in inferior NOD T cell responses have been linked to alterations in $p21^{ras}$ signaling.[77] Neonatal CD28 costimulation has been recently found to restore this signaling as well as protect NOD mice from diabetes, suggesting a relationship between this T cell hyporesponsiveness and disease onset.[78] A corresponding defect in T cell responses to anti-TCR or ConA were found in a French group of type 1 diabetics. In contrast to differences in the $p21^{ras}$ pathway determined in NOD mice, the hyporesponsive T cell phenotye in humans was associated with reduced $p56^{lck}$ expression.[79] We have recently found a similar T cell hyporesponsiveness among numerous strains of autoimmune-prone mice susceptible to EAE, SLE, rheumatoid arthritis and autoimmune hemolytic anemia. The inferior T cell response in animals susceptible to autoimmunity, which is not commonly seen in normal strains of mice, suggests this phenotype may be useful in predicting individuals prone to autoimmune diseases. The prevention of diabetes resulting from administration of immune adjuvants[80,81] or DNA vaccinations to self-Ags[82] may be related to reversing this hyporesponsive immunity observed in the NOD. Further work is required to clarify the mechanism of protection and possible role of a hyporeactive immune response in autoimmunity.

One possible mechanism by which a hyporeactive T cell response may contribute to autoimmunity is through defective negative selection in the thymus. Similar to peripheral T cell activation, negative selection in the thymus requires signaling from the APC presenting self-Ag to the TCR. The combination of a poor-Ag presenting MHC molecule (IA^{g7}), poor Ag presentation by the APC itself (low costimulatory molecules) and inferior responsiveness of the T cell itself would be an overall weakened response to Ag (self or foreign). Thus, negative selection in the thymus would occur in the NOD to only higher affinity Ags compared to negative selection in a normal strain. Evidence for poor central tolerance in the NOD includes direct demonstration of inferior thymic deletion in NOD compared to Balb/c mice upon administration of anti-TCR Abs, increased frequency of autoreactive T cell responses to pancreatic antigens,[32,83-86] and a recent demonstration that epitope spreading in NOD follows a hierarchy from highest affinity TCR:self-Ag interactions to lower affinities suggesting the spreading hierarchy is determined by the extent of negative selection.[87] The finding that a subset of thymic cells present peripheral Ags,[88] and that a mutation in this process may be responsible for a monogenic syndrome of numerous autoimmunities,[89,90] supports a central role for thymic tolerance in the prevention of autoimmunity. The contribution of defective central tolerance in type 1 diabetes induction in NOD mice is supported by the suppression of diabetes with transgenic thymic expression of proinsulin,[91] and the lower expression of pancreatic proteins GAD67 and prosomatostatin in NOD compared to C57Bl/6 thymi as determined by real-time RT-PCR analyses.[92] Other thymic defects in the NOD include abnormal corticomedullary environments,[93] a deficiency in the number of TCR^+ CD4-, CD8- population in NOD mice, and low proliferation of immature thymocytes.[53] Whether these phenotypes have relevance to autoimmunity remains to be determined although the deficiency in TCR^+ DN thymocytes does not appear to be related as it is seen in a number of congenic Idd mouse strains which are not diabetic.[94]

Inefficient thymic tolerance leads to a peripheral T cell repertoire of higher affinity autoreactive cells. It is well established that autoreactive T cells can be controlled by regulatory cells in the periphery. Numerous studies have identified a defect in peripheral tolerance induction in the NOD[2] including a resistance to apoptosis induction in NOD peripheral T cells.[95] Examples for the role of regulatory T cells in delaying or preventing overt diabetes are numerous, including prevention of adoptive transfer of diabetes by mixing splenocytes from prediabetic NOD mice with the disease-penetrating splenocytes from diabetic donors as an example.[96] More than a decade of work attempting to define these regulatory T cells has provided evidence that the suppressor cells are heterogeneous in nature. A few distinct populations are beginning to be elucidated.

A critical role of NK T cells in protection from diabetes has recently been demonstrated in NOD mice. A deficiency in numbers and function (IL4 secretion) of CD1-restricted NK T cells has been observed for many years;[97,98] however, the contribution of this phenotype to diabetes development was unclear. Several recent studies demonstrate the protection offered by increased numbers or function of CD1-restricted NK T cells.[48-52] NK T cells show a restricted TCR repertoire with a predominance of Vα14/Jα281 chains. These cells are specific for lipid Ags presented in the CD1 MHC molecule. Upon activation, which can be achieved with alpha-galactosylceramide (αGalCer), NK T cells produce large amounts of cytokines, most notably IL-4 and IL-10. The Idd6 locus contains the NKR-P1 gene cluster which includes the NK1.1 gene. The NK1.1 gene is absent in NOD mice, making NK T cells more difficult to analyze. Carnaud and coworkers demonstrated a contribution of the Idd6 locus to disease incidence, as NOD.NK1.1 congenic mice have reduced diabetes incidence which correlated with improved NK cell function.[48] The contribution of NK T cell defects to diabetes incidence was also demonstrated in NOD CD1 KO mice which showed an exacerbated disease course.[47a,50] In addition, increasing the numbers of NK T cells in the NOD by transgenic expression of the Vα14Jα281 receptor resulted in reduced disease.[49] The protection appears to be related to increased IL4 production observed in the pancreas as IL12 or anti-IL4 Abs were shown to abolish this protection.[49] It appears that increasing numbers of NK T cells is not necessary for protection as activation of the existing NK T cells with αGalCer alone protects diabetes in CD1-sufficient NOD[50-52] but not CD1-deficient NOD mice.[52] This CD1-restricted NK activating reagent also enhances the survival of transplanted islets.[51] The ability to protect NOD mice from disease by this simple protocol is intriguing. Future studies in other animal models and humans should establish whether this NK cell defect observed in the NOD is a general feature of type 1 diabetes.

Another population of regulatory T cells has been shown to be important in type 1 diabetes induction in an accelerated model of NOD diabetes. NOD mice genetically-deficient for CD28 expression succumb to diabetes with faster kinetics than wildtype NOD mice.[99] This accelerated disease incidence was found to be associated with a loss of CD25+CD4+ regulatory T cells.[100] The role of the regulatory cells appears to keep the autoreactive cells in check and is responsible for the long period of insulitis in this model, and most likely the lack of disease in some females and the majority of males. In conclusion, alterations in immune responses in NOD mice can lead to loss of thymic and peripheral T cell tolerance allowing autoimmune disease development. The loss of tolerance is evident in the transient production of autoAbs, most notably anti-insulin (IAA) Abs, in NOD mice just prior to disease onset. Although IAA

Abs can be induced in other strains of mice (Balb/c) by immunization, the amount of IAA Abs in the non autoimmune-prone strain appears lower than in the NOD mouse, which produces substantial amounts of IAA Abs upon immunization.[101] These data suggest tolerance processes in the Balb/c mice are more potent than in the NOD. However, it is noteworthy that the NOD defects are not absolute losses of tolerance but rather reduced tolerance induction occurring at altered TCR affinities.

TCR Tg Mouse Models of Type 1 Diabetes

A number of T cell clones have been isolated from the spleens of diabetic NOD (BDC2.5, BDC6.9),[102] and pancreas of prediabetic[103-106] NOD or RIP-B7.1 NOD Tg[107] mice and islet-transplanted diabetic[108] NOD mice. Both CD4 and CD8 clones have been generated which have varying abilities to induce or suppress insulitis and diabetes.[109] A few of these clones have been constructed into TCR Tg mice on various backgrounds. A summary of these TCR Tg diabetic mice is listed in Table 2. Early studies breeding nonpancreatic specific TCR Tgs, either class-II restricted (D011.10 anti-OVA) or class I-restricted (2C anti-L^d) onto the NOD background resulted in no change in disease incidence.[110] Although the transgenes showed good allelic exclusion at the beta locus, the alpha locus was more leaky allowing endogenous receptor expression. This finding supported a selection for pancreatic-specific T cells in a system skewed with innocuous receptors.

The BDC2.5 TCR Tg mouse was generated from the BDC2.5 CD4 T cell clone specific for pancreatic islet Ag.[111] The Tg mouse gets a uniform insulitis at 3 weeks of age followed by a low incidence of overt diabetes. Surprisingly, BDC2.5 Tg mice on the C57Bl/6.IA^{g7} background showed a higher incidence of diabetes than on the NOD background. The difference in disease incidence was mapped to the Idd7 disease-resistance interval in the NOD.[15] The BDC2.5 TCR transgene bred onto the NOD.scid background resulted in a very accelerated onset of type 1 diabetes, suggesting regulatory cells prevent diabetes induction in the normal NOD setting.[111a] The immunology of this mouse is described in depth in Chapter 4 of this book. Recently another CD4+ clone, BDC6.9, isolated in Kathryn Haskin's laboratory has been made into a TCR Tg mouse. Unlike the BDC2.5 Tg mice, BDC6.9 Tg mice develop early diabetes by the 3rd and 4th generation backcross to the NOD background.[70] The advantage to this clone is the ability to manipulate the presence and absence of self-Ag, as the stimulatory Ag is found in pancreas from NOD and SWR mice, but not other common strains of mice.

Both Class II-restricted CD4+ and class-I restricted CD8+ T cell clones isolated in the laboratory of P. Santamaria have been developed into TCR Tg mice with either beta chains alone or in combination with alpha chains. A majority of CD8+ T cell clones isolated from the pancreas of prediabetic NOD mice show the same TCR Vα17/Jα42 with N region sequence M-R-D/E, which has been used to generate 8.3 TCRαβ Tg mice. The Vβ8.1 CD8+ TCR Tg mice, transgenic for the TCR beta chain only, and 8.3 CD8+ Tg mice show an accelerated incidence of disease with a reduced period of benign insulitis.[103] In the 8.1 beta-chain only Tg mice, the islet-reactive T cells were found to have the same alpha chain as the original clone, suggesting a selection of this clone in diabetogenesis.[103] The 4.1 Tg mouse is from a CD4 clone. This TCR Tg mouse has the same disease incidence on RAG2 KO or normal NOD background, confirming the ability of some CD4+ cells to induce diabetes in the absence of CD8+ T cells.[103] However, breeding the Vβ8.3 CD8+ TCRαβ Tg mice onto the RAG2 KO background results in greatly reduced disease onset and demonstrates a role for CD4 cells in the efficient recruitment of CD8+ T cells to islets.[104] The Ag for this clone is IGRP and a mimotope which stimulates the clone has been found by screening combinatorial peptide libraries.[112,112a] The transition from benign insulitis to overt diabetes appears to be a loss in the ability to regulate autoreactive cells. Epitope spreading and affinity maturation have been shown to be important in this process. Islet-reactive CD8+ T cells increase their avidity binding to the 8.3-reactive peptide presented in K^d as shown through tetramer analysis.[113]

Table 2. TCR Tg diabetic mice

T Cell Receptor Transgenic	Vbeta	Valpha	CD4/ CD8	Antigen	Disease Phenotype	References
D011.10	8.2	Endogeneous	CD4	OVA/IA^d	No change in diabetes frequency	Lipes MA (1993)
2C	8.2	3.1	CD8	L^d	No change in diabetes frequency	Lipes MA (1993)
BDC2.5	4		CD4	Unknown AutoAg in pancreas	Rapid, uniform insulitis at 3 weeks of age; reduced diabetes incidence; higher incidence of diabetes in BDC2.5 C57Bl.IA^{g7} congenic mice which maps to Idd7 resistance in NOD mice; accelerated diabetes on NOD SCID background	Katz JD (1993) Suri A (1999)
BDC6.9			CD4	Unknown AutoAg on chr 6 in NOD and SWR pancreas	Accelerated diabetes on NOD background positively selected on IA^{g7}	Haskins K (personal communication)
4.1-NOD			CD4	Unknown AutoAg in pancreas	Positively selected with IA^{g7}; negatively selected with other class II molecules Increased diabetes incidence; same frequencyRag-/- background -CD8 independent	Verdaguer J (1997)
8.3-TCRβ	8.1	Endogeneous	CD8		Accelerated diabetes with reduced period of benign insulitis	Verdaguer J (1996)
8.3-NOD	8.1	Vα17/Jα42	CD8	Unknown AutoAg in pancreas	Accelerated diabetes with reduced period of benign insulitis; On rag-/- background get decreased disease incidence - requires CD4 help	Verdaguer J (1997)
9.33	6	19	CD8	Unknown AutoAg in pancreas	Accelerated diabetes and increased incidence; requires CD4 help; No diabetes on scid background - not due to lack of CD4 but rather lack ofescape from negative selection with only one TCR	Kanagawa O (2000)
NOD.AI4	2	8	CD8	Unknown AutoAg in pancreas	Accelerated diabetes in absence of CD4 help or B7.1 expression	Graser RT (2000)

Previous reports demonstrated that a CD8+ T cell clone (G9C8) isolated from the pancreas of RIP-B7.1 Tg mice did not require CD4 cells for disease transfer.[107] In a more recent study a TCR Tg mouse (NOD.AI4) was generated from a CD8 clone which was isolated from an unmanipulated NOD pancreas. The transgenic T cells were capable of inducing diabetes free of CD4 help or transgenic costimulation, supporting the idea that high-affinity CD8 T cells are independently capable of beta cell destruction.[105] However, another CD8+ islet-reactive TCR Tg, the 9.3 Tg, showed accelerated diabetes compared to normal NOD mice but no diabetes on the NOD.scid background.[114] The unusual finding in this transgenic system was the necessity for two dissimilar receptors on the cell surface for disease incidence as identified by costaining with antibodies specific for the transgene and a mix of other Vbeta-specific Abs. The authors conclude that the lowered expression of the pancreatic-specific receptor due to coexpression of other TCRs allowed escape of the Tg receptors from negative selection. Adoptive transfer studies showed these cells did require CD4 help for diabetes induction.[114]

Transgenic Expression of NeoSelf-Antigens in Pancreas and Diabetes Incidence

A limitation for many of the studies using TCRs from pancreatic-reactive clones is the lack of knowledge of the Ag. Therefore, several groups have generated Tg models which express neo self-Ags in the pancreas through use of the rat insulin promoter (RIP). TCR Tg mice specific for these well-defined Ags facilitate studies of tolerance and disease in these models. A list of models is present in Table 3.

Transgenic expression of viral LCMV under the RIP promoter induces ignorance of the T cells to the LCMV antigen unless prompted by LCMV systemic infection which leads to diabetes.[115,116] The diabetes in this model always requires CD8+ T cells whereas CD4+ T cells are only required in some lines.[117-119] When the RIP-LCMV is crossed with a TCR Tg mouse specific for LCMV, the animal still requires systemic LCMV infection for diabetes induction, although in this model the disease onset is accelerated compared to the non TCR Tg model.[119a] In the LCMV Tg model, the CD8+ T cells do not require CD4+ help as suggested by disease incidence with LCMV infection in class II-deficient[118] and CD40L-deficient animals.[120] As with other systems, crossing the RIP-LCMV with RIP-B7.1 Tg mice, which provides costimulation directly on islet cells, results in spontaneous diabetes.[121,122] Recent work with a defined LCMV protein (NP) crossed with a RIP-IL4 Tg mouse showed protection from diabetes following LCMV-infection.[123] The IL4 was shown to inhibit the acquisition of cytolytic activity yet promote expansion and survival of CD8+ T cells through an increased expression of B7.2 on DCs.[123]

Another neo self-Ag model targets expression of flu hemagglutinin (HA) Ag to pancreatic beta cells using the RIP promoter. Unlike the LCMV protein, the Ins-HA mice appear tolerant to the HA Ag following flu immunization if administered in adulthood.[121,124] A CD8+ TCR Tg specific for HA (CL4) can adoptively transfer diabetes in Ins-HA recipients if the Tg T cells are preactivated in vitro.[125] In addition, CL4xIns-HA double Tg mice are not tolerant and get spontaneous diabetes by 8 days of age with few T cells in the islet infiltrate.[124] Further work on the CL4 TCR showed this TCR possessed an unusually high affinity to HA/K^d. This high affinity is evident in the activation of the TCR in the absence of costimulation or with soluble monomeric complexes.[125] Like this CD8+ TCR Tg mouse, tolerance was also not established to the HA Ag in mice doubly transgenic for the Ins-HA Tg and a CD4+ anti-HA TCR Tg.[126] Other HA-specific CD4+ TCR Tg mice crossed with Ins-HA developed varying degrees of diabetes depending on the background strain.[126,127] Recently, the Ins-HA model has been used in studies of peripheral tolerance of CD8+ T cells in various mouse studies. Defects in peripheral tolerance in NOD mice compared to Balb/c and B10.D2 were confirmed by showing higher avidity binding of NOD T cells to K^d-HA tetramers.[86]

Similar to the Ins-HA model, RIP-OVA Tg mice can become diabetic when transferred with preactivated anti-OVA CD8+ T cells.[128] In addition, early diabetes is observed in OT-1

Table 3. TCR Tg mice specific for well-defined Ags

Pancreatic Specific Ag	TCR Tg	Tg Phenotypes	References
LCMV	None	Ignorance; diabetes upon viral infection	Oldstone M (1991) Ohashi P (1991)
	CD8	Accelerated disease with infection when crossed with TCR Tg CD8 T cells required but not CD4;spontaneous disease when crossed with RIP-B7.1 Tg; crossed with RIP-IL4 get protection disease following infection	
Hemagglutinin (HA)	None	Tolerant to HA; no diabetes following infection	Lo D (1992)
	CD8 (CL4)	Not tolerant (get diabetes) when crossed with CL4 TCR Tg Can transfer diabetes with preactivated TCR Tg cells	Morgan DJ (1996)
	CD4	Not tolerant (get diabetes) when crossed with CD4 TCR Tg although amount of diabetes varies with strain and receptor	Degerman S (1994) Scott B (1994)
OVA	None	Tolerant	Kurts C (1998)
	CD8(OT1)	Can transfer diabetes with preactivated TCR Tg cells Early diabetes when crossed with OT1 TCR Tg	Blanas E (1996)
MHC class I K^b	CD8(Des)	Tolerance (thymic deletion and anergy) to RIP-K^b; Cross with RIP-IL2 mice results in loss of tolerance (diabetes)	Miller JF (1991) Heath WR (1992)
Hen Egg Lysozyme (HEL)	CD4	Periinsulitis on C57Bl/6 background; Diabetes on NOD.IA^k background	Lesage S (2001)

(TCR Tg anti-OVA)xRIP-OVA double Tg mice, suggesting a lack of tolerance and ignorance in this model.[129] Another model places the class I K^b molecule under control of the RIP-promoter.[130] Crossing the RIP-K^b mice with a CD8+ TCR anti-K^b Tg mouse shows tolerance to this pancreatic islet which is associated with both thymic deletion of high-affinity clones by residual thymic expression and functional unresponsiveness of remaining clones. This tolerance can be broken, resulting in diabetes induction, with RIP expression of IL2, a cytokine which has been shown to reverse T cell anergy.[131]

Transgenic expression of the exogeneous hen egg lysoszyme (HEL) protein by the RIP promoter, together with an anti-HEL TCR Tg, shows thymic deletion and peripheral anergy of Tg cells on the C57Bl/6 background. However when bred onto the NOD background, maintaining the necessary H-2^k MHC, the HEL model showed defects in both thymic deletion and T cell anergy induction leading to overt diabetes in the NOD RIP-HEL/TCR double-Tg mice. This model clearly demonstrates the influence of NOD nonMHC genes on tolerance processes.[132]

As mentioned above, the Tg mouse expressing the costimulatory molecule B7.1 on pancreatic islet cells (RIP-B7.1) has been useful in studying "spontaneous" diabetes by crossing this mouse with models which otherwise do not show disease incidence. This mouse on the C57Bl/

6 background does not get diabetes but can become diabetic when crossed with RIP-LCMV, RIP-TNFα, or RIP-IE or with mice expressing human HLA molecules associated with type 1 diabetes. The usefulness of the RIP-B7.1 model is to address the sensitivity of disease induction allowing the systems to be stressed to determine if diabetes can occur. RIP-B7.1 Tg expression on the NOD background results in diabetes incidence as early as the first backcross although IA^{g7} homozygosity is required.[125] The largest effect of B7.1 expression is on the CD8 cells, thereby negating the necessity of CD4+ help although the presence of CD4+ cells accelerates the disease progression by aiding in migration of cells to the pancreas. Unlike spontaneous NOD diabetes, B cells are not required for disease in the RIP-B7.1 model.[133]

Human Autoimmunity Genes Expressed as Transgenes in Mice

Another approach undertaken in the past decade in the study of type 1 diabetes is the transgenic expression in mice of human genes associated with autoimmunity. The class II HLA molecule DQB1 chain with serine, alanine or valine at P57 of the β chain, similar to the NOD MHC class II haplotype, confers the greatest risk for type 1 diabetes in humans. The DQ8 (DQA*0301 plus DQB1*0302 (alanine at P57)) haplotype has been expressed in transgenic mice.[134] The DQ8 Tg mice do not develop diabetes even when backcrossed to the NOD background. However, when these mice are crossed with the RIP-B7.1 Tg mice, 80% of the mice develop insulitis and type 1 diabetes whereas control human DQ6 (diabetes-resistant haplotype)xRIP-B7.1 double-Tg mice do not develop insulitis or diabetes. The DQ8xRIP-B7.1 Tg mice have been used to map approximately 10 different T cell epitopes in human GAD65, one of which is also a dominant T cell epitope in NOD mice restricted to IA^{g7}.[135,136] Transgenic expression of the human disease-associated DR4 allele in mice showed surprisingly that this allele, in combination with the highly susceptible DQ8 allele, actually protected mice from disease. Whereas 25% of DR4/RIP-B7 mice developed diabetes compared to >80% of DQ8/RIP-B7 mice, the triple transgenic DR4/DQ8/RIP-B7 had only a 23% disease incidence, showing a dominant protective effect for the DR4 allele.[137] Thus, the diabetes incidence afforded by the immunostimulatory RIP-B7.1 transgene enables studies of human autoimmunity genes in mice.

BioBreeding Diabetes-Prone (BB-DP) Rat Model of Spontaneous Type 1 Diabetes

The first inbred animal model for spontaneous, type 1 diabetes evolved from the discovery of diabetic animals in an outbred colony of Wistar rats maintained at the Bio Breeding Research Laboratories in Ottawa, Canada. Selective breeding led to the first paper on the Bio Breeding (BB) rat in 1975 wherein the clinical description of insulin-dependent diabetes was documented.[138] Prevention of diabetes with antiserum to rat lymphocytes suggested an autoimmune pathogenesis for disease.[139] Further support for this hypothesis was provided by studies describing the linkage of the disease with the MHC.[140] The disease is now known to have a polygenic basis for susceptibility[141] and abnormalities in T cell function have been described.[142-145] In comparison to the NOD animal model for type 1 diabetes, the BB model has certain advantages including higher and more uniform disease penetrance and no gender bias as diabetes is observed in >90% of both male and female BB-DP rats when maintained in pathogen-free conditions.[146,147] The time between insulitis and overt diabetes in the BB-DP rat is much shorter than in the NOD mouse, as diabetes develops between 2-4 weeks after insulitis in the rat model. However, the BB-DP model requires a unique lymphopenia (lyp) locus for spontaneous disease which is not present in the NOD mouse or the vast majority of humans with type 1 diabetes.[142] This lyp locus is required in the homozygous state for disease and results in a profound loss of peripheral T cells including a >85% reduction in CD8+ T cells and a 50-80% loss of CD4+ T cells. In this section we will focus on the genetics and immune status of the BB rat, with a special emphasis on the relationship between T cell lymphopenia, T cell function and disease susceptibility.

Derivation of BB Rats

All BB rat colonies worldwide were derived from a progenitor stock in Ottawa. As there was considerable genetic heterogeneity in the original Ottawa stock, inbreeding at other locations has predictably led to genetically distinct but related, inbred BB rat lines. In a study of 11 different BB rat colonies describing over 20 different BB sublines, inbred lines could be classified as belonging to one of four groups, distinguished by eight protein markers from at least three linkage groups.[148] Selected for sensitivity or resistance to diabetes, BB rats can be further subdivided into diabetes-prone (BB-DP) or diabetes-resistant (BB-DR).[149] As described later the BB-DR subline progeny are designated by their ability to develop diabetes following experimental manipulations that do not induce diabetes in other animals.

Genetics of BB Rats

As in the NOD and humans, genetic susceptibility for development of type 1 diabetes in the BB-DP rat is polygenic. At present, five disease-associated loci (iddmn) have been reported.[150,151] The remarkably lower number of loci in the rat model compared to the NOD may reflect the lack of investigations on genetics in this model or reduced disease complexity due to the large influence of the lymphopenia locus. The lyp locus has been mapped to the Ian gene of rat chromosome 4,[151a] is designated iddm1, and is responsible for the severe T cell loss when present in the homozygous state.[142] Similar to NOD and human type 1 diabetics, as well as the majority of other autoimmune-induced diseases, linkage to the MHC constitutes yet another disease-associated loci-iddm2. Linkage to the MHC was first defined by Colle and coworkers in an analysis of diabetes incidence among F_2 animals from a cross of BB-DP with Lewis strain rats.[140] Lewis rats are not prone to diabetes and are incompatible with the BB-DP at the MHC. All diabetic F_2 animals were shown to be homozygous for the BB-DP MHC. In the rat model the MHC haplotype susceptible to disease is $RT1^u$.

Although the contributions of the iddm1 (lyp) and idddm2 (MHC) loci to disease are quantitatively large, they are not, in and of themselves, sufficient for disease. Fixing the $RT1^u$ and lyp alleles when breeding the BB-DP rat to diabetes-resistant ACI[152] and PVG[153] strains did not result in diabetes. Therefore, other loci are clearly involved in rendering the animal susceptible to autoimmunity. Recent reports link iddm4 with diabetes susceptibility in crosses between disease-resistant Wistar-Furth (WF) and BB-DR animals, which do not normally develop spontaneous disease.[150] The BB-DR animals are genetically similar to the BB-DP rat yet are resistant to spontaneous type 1 diabetes due to the lack of the iddm1 locus. Thus the BB-DR animals have normal peripheral T cell numbers. However, autoimmune-susceptibility genes are clearly present in the BB-DR as type 1 diabetes can be induced in BB-DR animals with immune manipulations if administered in a short time window of 25-35 days of age. These manipulations include viral infections,[154] poly IC injections[155] and depletion of RT6+ T cells along with an environmental trigger.[156] Backcrosses of $(WFxDR\text{-}BB)F_1$ to WF rats resulted in approximately one-half of the pups susceptible to type 1 diabetes induction. Furthermore, this susceptibility mapped strongly to the iddm4 locus on chromosome 4.[150] This locus contains several major autoimmunity loci including aia2, aia3, and cia3. These data, as well as studies in the BB/OK strain,[151,157] crosses between BB-DP and nonautoimmune-prone rats,[158] and $RT1^u$ congenic strains[159] support the contribution of nonMHC loci in general susceptibility to autoimmunity. Two other loci have been found to be associated with type 1 diabetes in the BB-DP rat, including the diabetes-susceptible iddm3 and diabetes-resistant iddm5 loci.[151,157]

Immune Dysfunction in the BB-DP Rat and Its Relationship to Lymphopenia

Lymphopenia was first defined in 1981 by Jackson and colleagues.[160] An early important characteristic distinguishing lymphopenic from nonlymphopenic animals was the absence of the RT6+ subset of peripheral T cells.[161,162] The under-representation of the CD45R+ T cell subset isoform was also documented.[163] The absence of RT6+CD45R+ T cells correlated well with the overall reduction in the numbers of peripheral T cells found in the BB-DP. Since thymocytes are both RT6- [161] and 98% CD45R-,[164] a plausible role of the lymphopenia gene is to retard T cell maturation to a stage prior to the expression of these peripheral T cell antigens. The RT6+ population is known to contain regulatory function since (1) depletion of the RT6+ population along with poly I:C injection induces type 1 diabetes in BB-DR rats,[165] (2) diabetes can be induced by adoptive transfer of BB-DR lymphocytes into athymic nude WAG rats only if the BB-DR lymphocytes are pretreated with anti-RT6 antibodies while cotransfer of RT6+ T cells prevents diabetes,[166] and (3) injection of RT6+ cells from BB-DR rats into BB-DP lymphopenic rats prevents their spontaneous development of type 1 diabetes.[167,168] However, it seems unlikely that the association of RT6 with lymphopenia indicates that a defect in RT6 gene transcription is the cause of the lymphopenia, as the BB-DP RT6 allele is expressed in F_1s between BB-DP and other diabetes-resistant strains expressing the other RT6 allele.[169] Further, there is significant RT6 expression on intraepithelial lymphocytes[170] as well as low-level expression on T cells in BB-DP rats,[171] confirming a functional RT6 gene. Taken together these results indicate that the RT6 phenotype leading to a paucity in RT6+ T cells is due to the impact of the lymphopenia on the capacity of BB-DP T cells to mature to the point where RT6 would normally be significantly expressed.

Studies into the immune function of lyp T cells from BB-DP rats report contradictory results of poor in vitro responses and hyper-activated phenotypes. Early studies report weak proliferative[172] and cytotoxic responses[144] to alloantigen as well as altered CD4-mediated signaling.[173] Given the reduced numbers of T cells in the lyp animals, especially in the CD8 compartment, the differences in responses in these bulk read-out in vitro assays may be due to differences in functional T cells present. It is now known that many of the peripheral T cells in the BB-DP are undergoing apoptosis and therefore may not be functional, although present, in these assays. More recent studies suggest that lyp T cells show increased signs of activation ex vivo, including increased expression of CD25[174] and OX40[175] activation markers, increased INFγ- production[176] and increased mRNA of T cell signaling adapter protein vav.[177] The lyp cells also show increased signs of spontaneous proliferation ex vivo with (1) 2X number of cycling cells as determined by increased DNA content,[178] (2) a 90% reduction in levels of cell-cycle inhibitor $p27^{kip}$, increased levels of PCNA and phosphorylated Rb,[179] and (3) incorporation of BrdU into >90% of lyp T cells compared to ~30% of normal T cells in a 13-day period.[180] Collectively these data suggest that lyp T cells go through DNA synthesis prior to their imminent demise although no studies have investigated progression through mitosis or cell division. The resultant loss of T cells, as opposed to an accumulation, suggests this proliferative state is nonproductive.

Recent data from our laboratory shows a hyperresponsive phenotype of lyp T cells activated in vitro with anti-TCR Ab stimulation. The percentage of cells upregulating OX40 activation marker and proliferating as determined by loss of CFSE is greater at lower stimulating doses in lyp T cells compared to normal T cells. This phenotype was dependent only on the lyp locus and not on other background genes as Fisher lyp/lyp T cells show an identical phenotype to BB-DP lyp/lyp T cells. Furthermore, these assays measure activation on a per cell basis as opposed to responses of bulk cultures in which T cell purities are more critical. A hyperresponsive T cell phenotype could be due to (1) previous activation of the T cell, (2) a lowered activation threshold due to the lyp gene or (3) both. The high proliferative rate, increased percentages of activated T cells, and hyperresponsive phenotype cumulatively suggest an inherent global activation of lyp T cells as opposed to weakened immune responses.

In addition to T cell abnormalities associated with the lyp gene, studies of nonT cell subsets of the BB-DP immune system suggest thymic and APC defects similar to the NOD mouse model. Recent studies report a lower number of splenic DCs with decreased expression of MHC class II and costimulatory ligand CD80[181] and decreased clustering which improves stimulating capacity of DCs.[182] In addition defects in the NK cell population[183] as well as thymic B cells[184] have been observed. The data support an overwhelming defect in peripheral

T cell regulation as a mechanism for disease and reports of enhanced proliferative responses to superAgs following in vivo administration[145] support the conclusion of defects in peripheral tolerance in lyp rats. Few studies on thymic tolerance have been reported in the BB-DP model although alterations in thymic architecture, which have been associated with thymic tolerance defects, both in the medulla[142] and cortex[185] were noted.

Activation and Death of T Cells Homozygous for lyp Mutation

A remarkable phenotype afforded by the lyp mutation is the severely shortened lifespan of the lyp T cells. Although thymic development appears normal in lyp animals, with an almost normal distribution of double-positive and CD4+ and CD8+ single-positive subsets which vary depending on background strain,[153] peripheral T cell numbers are drastically reduced. The CD8+ T cell subset is nearly absent in spleen and lymph nodes of lyp animals, with a few cells expressing lower levels of CD8,[186] suggesting a faster death following thymic selection than CD4+ T cells.[187,188] Following the export of T cells from the thymus with fluorescent-labeling, Zadeh and coworkers measured the lifespan of CD4+ lyp T cells to be less than one week.[178] In addition, only few labeled thymocytes expressed RT6 which is normally upregulated 1-2 weeks following thymic export. The majority of the labeled thymocytes expressed the Thy1 Ag which is characteristic of recent thymic emigrants (RTEs) in the rat.[178] The thymus of the lyp rat also showed reduced thymic export of RTEs, data which is consistent with increased death of thymocytes in adult thymic organ cultures (ATOCs) from lyp rats.[189] Further data supporting an overwhelmingly shortened lifespan of lyp T cells is the rapid loss of T cells from lymph node and spleen following thymectomy in lyp rats compared to a fairly stable T cell pool in thymectomized normal rats.[190,191]

The efficient clearance of apoptotic cells by phagocytosis has made detection of apoptosis difficult in vivo. In vitro, death by apoptosis in lyp T cells is rapid[153,180] with a majority (~80%) of the cells dying in overnight cultures compared to negligible death (~20%) observed in cultured T cells from nonlymphopenic rats. Advances in detection of apoptotic cells allowed identification of these dying cells in vivo through TUNEL and FITC-Annexin V staining assays.[192] Further characterization identified the clearance of the apoptotic T cells in lyp rats in the liver.[183] In mice, dying apoptotic CD8+ T cells are cleared in the liver whereas CD4+ T cells remain in the LN even when apoptotic.[193] The dying cells were difficult to define as T cells due to their detectable but low expression of TCR, CD8, CD4, B220 and HSA proteins on their surface.[183]

The collective data point to a rigorous life for the lyp T cell. Those T cells which survive thymic selection become spontaneously activated following their export from the thymus. The extent of activation and proliferation is not well-characterized nor is the degree of functionality (i.e., cytotoxicity and cytokine secretion) well understood. The majority of the lyp T cells commit suicide for some unknown reason a few days later. There is evidence that increased activation may rescue these cells from death, as cells expressing CD25 and OX40 appear to have a survival advantage in that they accumulate over time[180] (our unpublished observations). In addition, a higher proportion of CD25+ cells is seen in the TUNEL-negative (live) population in vivo.[183] The absence of RT6+ and CD8+ cells is not complete suggesting a minority of cells are able to proceed through the "developmental block".[171] The lyp T cells do show normal responses (upregulation of CD25 and OX40, cell growth and proliferation) to the T cell mitogen, ConA, as well as to TCR and CD28 Ab-mediated signaling in vitro. In addition, in vitro activation reduces the number of lyp T cells undergoing cell death[153] (our unpublished observations). This is in contrast to increased apoptosis in culture of normal T cells with exogenous stimulation (our unpublished observations). Furthermore, activation by alloantigen appears to rescue alloreactive lyp T cells in vivo such that allografts can be destroyed.[180] These data, and the destruction of islet beta cells by T cells, illustrate that the functional capacity of lyp T cells is adequate to elicit immune responses.

The vast reduction in RT6+ cells along with a preponderance of Thy1+ cells suggest lyp impacts T cell development at the RTE stage. Our studies on the spontaneous entry into cell cycle of lyp T cells show the RTE subset has low levels of cell cycle inhibitor p27^{kip}, whereas the medullary thymocytes show a normal expression of this protein.[179] The data suggest lyp T cells proceed through normal thymic development followed by activation and death at the RTE stage. The effects of the lyp mutation appear to be cell autonomous and restricted to bone marrow cells as demonstrated through lyp → normal bone marrow chimeras.[194] In addition, transplants of normal thymus into lyp recipients do not reverse the lymphopenia, pointing to the developing T cells themselves harboring the genetic disturbance.[195] It is noteworthy that the T cell loss, activation and death phenotype observed in the BB-DP rat is also observed in lyp congenic rats from diabetes-resistant strains (Fisher-lyp and PVG-lyp) which show T cell lymphopenia but not diabetes.[153,179]

The lyp Paradox

The reduced number and early death of the T cells and their necessary role in type 1 diabetes induction present a paradox in the BB-DP model. Although a lack of regulation in the BB-DP rat is well-documented, less is known about the properties of the effector T cells responsible for the beta cell destruction. There is disagreement whether the short-lived RTEs are responsible for disease, as suggested by disease prevention with adult thymectomy.[191] However, the thymectomy is required during a distinct time period for protection – when performed at 8 weeks of age animals still develop diabetes even though the vast majority of RTEs still undergo rapid death. Combined with newer studies showing Ag activation can rescue lyp T cells from death both in vivo[180] and in vitro,[153] another model suggests that higher affinity autoreactive T cells are activated by autoantigen, survive and serve the effector functions necessary for immune clearance. Isolation and characterization of the effector T cells responsible for beta cell destruction are still necessary to differentiate among these models.

Disease Prevention in the BB-DP Rat

As in the NOD mouse model, numerous treatments have been shown to prevent diabetes in the BB-DP rat model. Early immunosuppression to prevent effector T cell function, including T cell depletion with Abs,[196-198] thymectomy[199,199a,191] or FK506,[200,201] prevents diabetes in the BB-DP rat. Similar to the effects in the NOD model, complete Freund's adjuvant[202] and viral infections[203,204] prevent disease in the BB-DP model as well. Reversing the defect in the regulatory population also prevents diabetes. This reversal has been shown to be successful in disease prevention with the addition of RT6+ cells[168,205-207] or with human TNFα, a cytokine which influences regulatory properties of T cells.[208] Although alterations in peripheral tolerance are well-documented in the BB-DP, little is known about alterations in thymic tolerance in this model. However a report that in ATOCs of BB-DP thymocytes, coculture with islets, but not thyrocytes, prevents diabetes and reduces insulitis upon adoptive transfer.[209] These data suggest diabetes can be reversed with increased thymic tolerance. The nonMHC and nonlyp loci may indeed influence the repertoire of autoreactive T cells and distinguish lyp animals susceptible or resistant to autoimmunity.

Concluding Remarks

Although the BB-DP rat model differs from type 1 diabetes in humans in its necessity for the lyp locus, important information regarding mechanisms of autoimmunity can be found in this model. For one, as in humans, there is no gender bias in disease incidence. In addition, this model genetically separates regulatory and autoreactive cells with the lyp mutation, affording easier studies on the role of these distinct populations. Finally, given the observation that type 1 diabetes can occur through multiple mechanisms in various animal models as well as humans, it is important to test preventative treatments in multiple animal models to assess its possible utility in humans. For example, the finding that CFA prevents diabetes in both the NOD and BB-DP[202] models suggests it may work in altering the autoimmune potential of the animal. Studies into the mechanism of treatment in both models should provide important information as to possible benefits for humans.

Long Evans Tokushima Lean (LETL) Rat Model

An additional rat model of spontaneous development of type 1 diabetes was described in 1991.[210] Inbreeding of a rat showing signs of diabetes in 1983 resulted in the LETL strain maintained at the Tokushima Research Institute. These rats have the advantage of spontaneous disease incidence without a gender bias or a requirement for lymphopenia. Like the NOD mouse, infiltration of the pancreas, salivary and lacrimal glands is observed.[210] The disease shows a polygenic mode of inheritance including a large contribution of the MHC RT-1^u haplotype shared with the BB-DP rat.[210] Selection of the Komeda Diabetes-Prone rat substrain improved disease penetrance, with 100% insulitis and 70% diabetes by 120 days of age.[211] A genome-wide scan of this strain identified a novel nonMHC associated disease gene (Cblb) on rat chromosome 11 which was shown to be essential for development of insulitis.[211,211a] An additional Komeda NonDiabetic (KND) substrain was established which, although genetically very similar to the KDP strain, does not develop type 1 diabetes.[212] This strain serves as an important control in both genetic and immunological studies of the KDP rat. This rat model thus most closely resembles human type 1 diabetes in its spontaneous onset, lack of gender bias, lack of lymphopenia and association with MHC class II.

Through studies of diabetes development in the three animal models in which diabetes occurs spontaneously (NOD mice, BB-DP and LETL rats) as well as the numerous models in which diabetes is experimentally-induced through transgenic expression of either islet cell neoantigens or diabetogenic T cell receptors, we can begin to understand mechanisms of disease induction. Similarities among the multiple models should be emphasized as potential targets for treatments. The differences in disease mechanism in these three spontaneous models alone illustrate the complexity of disease. In addition, the variation in disease penetrance among and within colonies add to this complexity. Each of these genetically-inbred models can be viewed as representing an individual's disease such that the numerous cases of type 1 diabetes among humans may represent as many distinct mechanisms for disease. Therefore one must use caution as well as multiple models in developing potential therapies useful for numerous patients. The use of multiple animal models can only enhance our understanding of this complex disease and aid in developing therapies useful for the treatment and cure of type 1 diabetes in humans.

References

1. Tisch R, McDevitt H. Insulin-dependent diabetes mellitus. Cell 1996; 85:291-297.
2. Delovitch TL, Singh B. The nonobese diabetic mouse as a model of autoimmune diabetes: Immune dysregulation gets the NOD. Immunity 1997; 7:727-738.
3. Makino S, Kunimoto K, Muraoka Y et al. Breeding of a nonobese, diabetic strain of mice. Exp Anim 1980; 29:1-13.
4. Kanazawa Y, Komeda K, Sato S et al. Nonobese-diabetic mice: Immune mechanisms of pancreatic B-cell destruction. Diabetologia 1984; 27:113-115.
5. Wicker LS, Miller BJ, Coker LZ et al. Genetic control of diabetes and insulitis in the nonobese diabetic (NOD) mouse. J Exp Med 1987; 165:1639-1654.
6. Maruyama T, Takei I, Yanagawa T et al. Insulin autoantibodies in nonobese diabetic (NOD) mice and streptozotocin induced diabetic mice. Diabetes Res 1988; 7:93-96.
7. Ziegler AG, Vardi P, Ricker AT et al. Radioassay determination of insulin autoantibodies in NOD mice. Correlation with increased risk of progression to overt diabetes. Diabetes 1989; 38:358-363.
8. Gottlieb PA, Eisenbarth GS. Mouse and man: Multiple genes and multiple autoantigens in the aetiology of type I DM and related autoimmune disorders. J Autoimmun 1996; 9:277-281.
9. Todd JA, Aitman TJ, Cornall RJ et al. Genetic analysis of autoimmune type I diabetes mellitus in mice. Nature 1991; 351:542-547.
10. Oldstone MBA. Prevention of type 1 diabetes in nonobese diabetic mice by virus infection. Science 1988; 239:500-502.
11. Wilberz S, Partke HJ, Dagnaes-Hansen F et al. Persistent MHV (mouse hepatitis virus) infection reduces the incidence of diabetes mellitus in nonobese diabetic mice. Diabetologia 1991; 34:2-5.
12. Atkinson MA, Leiter EH. The NOD mouse model of type 1 diabetes: As good as it gets? Nat Med 1999; 5:601-604.

13. Hattori M, Buse JB, Jackson RA et al. The NOD mouse: Recessive diabetogenic gene within the major histocompatibility complex. Science 1986; 231:733-735.
14. Todd JA, Bell JI, McDevitt HO. HLA-DQB gene contributes to susceptibility and resistance to insulin-dependent diabetes mellitus. Nature 1987; 329:599-604.
15. Wicker LS, Todd JA, Peterson LB. Genetic control of autoimmune diabetes in the NOD mouse. Ann Rev Immunol 1995; 13:179-200.
16. Todd JA, Wicker LS. Genetic protection from the inflammatory disease type 1 diabetes in humans and animal models. Immunity 2001; 15:387-395.
17. Encinas JA, Kuchroo VK. Mapping and identification of autoimmunity genes. Curr Opin Immunol 2000; 12:691-697.
18. McDuffie M. Genetics of autoimmune diabetes in animal models. Curr Opin Immunol 1998; 10:704-709.
19. Acha-Orbea H, McDevitt HO. The first external domain of the nonobese diabetic mouse class II I-A chain is unique. Proc Natl Acad Sci USA 1987; 84:2435-2439.
20. Nepom GT, Erlich H. MHC class-II molecules and autoimmunity. Ann Rev Immunol 1991; 9:493-525.
21. Lund T, O'Reilly L, Hutchings P et al. Prevention of insulin-dependent diabetes mellitus in nonobese diabetic mice by transgenes encoding modified I-A beta-chain or normal I-E alpha-chain. Nature 1990; 345:727-729.
22. Miyazaki T, Uno M, Uehira M et al. Direct evidence for the contribution of the unique I-ANOD to the development of insulitis in nonobese diabetic mice. Nature 1990; 345:722-724.
23. Nishimoto H, Kikutani H, Yamamura K et al. Prevention of autoimmune insulitis by expression of I-E molecules in NOD mice. Nature 1987; 328:432-434.
24. Wherrett DK, Singer SM, McDevitt HO. Reduction in diabetes incidence in an I-A^{g7} transgenic nonobese diabetic mouse line transgenic non. Diabetes 1997; 46:1970-1974.
25. Ikegami H, Makino S, Harada M et al. The cataract Shionogi mouse, a sister nonobese diabetic mouse similar class II but different class I gene products. Diabetologia 1988; 31:254-258.
26. Ikegami H, Eisenbarth GS, Hattori M. Major histocompatibility complex-linked diabetogenic gene of the nonobese diabetic mouse. Analysis of genomic DNA amplified by polymerase chain reaction. J Clin Invest 1990; 85:18-24.
27. Wicker LS, Miller BJ, Fischer PA et al. Genetic control of diabetes and insulitis in the nonobese diabetic mouse. Pedigree analysis of a diabetic H-2nod/b heterozygote. J Immunol 1989; 142:781-784.
28. Corper AL, Stratmann T, Apostolopoulos V et al. A structural framework for deciphering the link between I-Ag7 and autoimmune diabetes. Science 2000; 288:505-511.
29. Latek RR, Suri A, Petzold SJ et al. Structural basis of peptide binding and presentation by the type I diabetes-associated MHC class II molecule of NOD mice. Immunity 2000; 12:699-710.
30. CarrascoMarin E, Kanagawa O, Unanue ER. Insights into the chemistry and biology of the I-A^{g7} class II molecule. Res Immunol 1997; 148:291-301.
31. Lee KH, Wucherpfennig KW, Wiley DC. Structure of a human insulin peptide/HLA-DQ8 complex and susceptibility to type 1 diabetes. Nat Immunol 2001; 2:501-507.
32. Kanagawa O, Martin SM, Vaupel BA et al. Autoreactivity of T cells from nonobese diabetic mice: An I-Ag7-dependent reaction. Proc Natl Acad Sci USA 1998; 95:1721-1724.
33. Schmidt D, Verdaguer J, Averill N et al. A mechanism for the major histocompatibility complex-linked resistance to autoimmunity. J Exp Med 1997; 186:1059-1075.
34. Ridgway WM, Ito H, Fasso M et al. Analysis of the role of variation of major histocompatibility complex class II expression on nonobese diabetic (NOD) peripheral T cell response. J Exp Med 1998; 188:2267-2275.
35. Luhder F, Hoglund P, Allison JP et al. Cytotoxic T lymphocyte-associated antigen 4 (CTLA-4) regulates the unfolding of autoimmune diabetes. J Exp Med 1998; 187:427-432.
36. Kanagawa O, Vaupel BA, Xu G et al. Thymic positive selection and peripheral activation of islet antigen-specific T cells: Separation of two diabetogenic steps by an I-A(g7) class II MHC beta-chain mutant. J Immunol 1998; 161:4489-4492.
37. Suri A, Katz JD. Dissecting the role of CD4+ T cells in autoimmune diabetes through the use of TCR transgenic mice. Immunol Rev 1999; 169:55-65.
38. Ikegami H, Jackson RA, Makino S et al. Homozygosity for two genes (H-2: Chromosome 17 and Thy-1: Chromosome 9) linked to development of type I diabetes of the NOD mouse. Clin Res 1986; 34:683A.
39. Prochazka M, Leiter EH, Serreze DV et al. Three recessive loci required for insulin-dependent diabetes in nonobese diabetic mice. Science 1987; 237:286-289.

40. Lyons PA, Hancock WW, Denny P et al. The NOD Idd9 genetic interval influences the pathogenicity of insulitis and contains molecular variants of Cd30, Tnfr2, and Cd137. Immunity 2000; 13:107-115.
41. Podolin PL, Wilusz MB, Cubbon RM et al. Differential glycosylation of interleukin 2, the molecular basis for the NOD Idd3 type 1 diabetes gene? Cytokine 2000; 12:477-482.
42. Ghosh S, Palmer SM, Rodrigues NR et al. Polygenic control of autoimmune diabetes in nonobese diabetic mice. Nat Genet 1993; 4:404-409.
43. Hill NJ, Lyons PA, Armitage N et al. NOD Idd5 locus controls insulitis and diabetes and overlaps the orthologous CTLA4/IDDM12 and NRAMP1 loci in humans [In Process Citation]. Diabetes 2000; 49:1744-1747.
44. Fox CJ, Paterson AD, Mortin-Toth SM et al. Two genetic loci regulate T cell-dependent islet inflammation and drive autoimmune diabetes pathogenesis. Am J Hum Genet 2000; 67:67-81.
45. Cordell HJ, Todd JA, Hill NJ et al. Statistical modeling of interlocus interactions in a complex disease: Rejection of the multiplicative model of epistasis in type 1 diabetes. Genetics 2001; 158:357-367.
46. Colucci F, Bergman ML, Penha-Goncalves C et al. Apoptosis resistance of nonobese diabetic peripheral lymphocytes linked to the Idd5 diabetes susceptibility region. Proc Natl Acad Sci USA 1997; 94:8670-8674.
47. Bergman ML, Cilio CM, Penha-Goncalves C et al. CTLA-4-/- mice display T cell-apoptosis resistance resembling that ascribed to autoimmune-prone nonobese diabetic (NOD) mice. J Autoimmun 2001; 16:105-113.
47a. Shi FD, Flodstrom M, Balasa B et al. Germ Line deletion of the CD1 locus exacerbates diabetes in the NOD mouse. Proc Natl Acad Sci USA 2001; 98:6777-6782.
48. Carnaud C, Gombert J, Donnars O et al. Protection against diabetes and improved NK/NKT cell performance in NOD.NK1.1 mice congenic at the NK complex. J Immunol 2001; 166:2404-2411.
49. Laloux V, Beaudoin L, Jeske D et al. NK T cell-induced protection against diabetes in V alpha 14-J alpha 281 transgenic nonobese diabetic mice is associated with a Th2 shift circumscribed regionally to the islets and functionally to islet autoantigen. J Immunol 2001; 166:3749-3756.
50. Wang B, Geng YB, Wang CR. CD1-restricted NK T cells protect nonobese diabetic mice from developing diabetes. J Exp Med 2001; 194:313-320.
51. Sharif S, Arreaza GA, Zucker P et al. Activation of natural killer T cells by alpha-galactosylceramide treatment prevents the onset and recurrence of autoimmune Type 1 diabetes. Nat Med 2001; 7:1057-1062.
52. Hong S, Wilson MT, Serizawa I et al. The natural killer T-cell ligand alpha-galactosylceramide prevents autoimmune diabetes in nonobese diabetic mice. Nat Med 2001; 7:1052-1056.
53. Bergman ML, Penha-Goncalves C, Lejon K et al. Low rate of proliferation in immature thymocytes of the nonobese diabetic mouse maps to the Idd6 diabetes susceptibility region. Diabetologia 2001; 44:1054-1061.
54. Siegmund T, Armitage N, Wicker LS et al. Analysis of the mouse CD30 gene: A candidate for the NOD mouse type 1 diabetes locus Idd9.2. Diabetes 2000; 49:1612-1616.
55. Heath VL, Moore NC, Parnell SM et al. Intrathymic expression of genes involved in organ specific autoimmune disease. J Autoimmun 1998; 11:309-318.
56. Podolin PL, Denny P, Lord CJ et al. Congenic mapping of the insulin-dependent diabetes (Idd) gene, Idd10, localizes two genes mediating the Idd10 effect and eliminates the candidate Fcgr1. J Immunol 1997; 159:1835-1843.
57. Griffiths MM, Encinas JA, Remmers EF et al. Mapping autoimmunity genes. Curr Opin Immunol 1999; 11:689-700.
58. Becker KG. Comparative genetics of type 1 diabetes and autoimmune disease: Common loci, common pathways?. [Review]. Diabetes 1999; 48:1353-1358.
59. Ginn LR, Lin JP, Plotz PH et al. Familial autoimmunity in pedigrees of idiopathic inflammatory myopathy patients suggests common genetic risk factors for many autoimmune diseases. Arthritis Rheum 1998; 41:400-405.
60. Sels F, Westhovens R, Emonds MP et al. HLA typing in a large family with multiple cases of different autoimmune diseases. J Rheumatol 1997; 24:856-859.
61. Todd JA. From genome to aetiology in a multifactorial disease, type 1 diabetes. Bioessays 1999; 21:164-174.
62. Lorber M, Gershwin ME, Shoenfeld Y. The coexistence of systemic lupus erythematosus with other autoimmune diseases: The kaleidoscope of autoimmunity. Semin Arthritis Rheum 1994; 24:105-113.
63. Winer S, Astsaturov I, Cheung R et al. Type I diabetes and multiple sclerosis patients target islet plus central nervous system autoantigens; nonimmunized nonobese diabetic mice can develop autoimmune encephalitis. J Immunol 2001; 166:2831-2841.

64. Rivero VE, Cailleau C, Depiante-Depaoli M et al. Nonobese diabetic (NOD) mice are genetically susceptible to experimental autoimmune prostatitis (EAP). J Autoimmun 1998; 11:603-610.
65. Salomon B, Rhee L, Bour-Jordan H et al. Development of spontaneous autoimmune peripheral polyneuropathy in b7-2-deficient nod mice. J Exp Med 2001; 194:677-684.
66. Bendelac A, Carnaud C, Boitard C et al. Syngeneic transfer of autoimmune diabetes from diabetic NOD mice to healthy neonates. Requirement for both L3T4+ and Lyt-2+ T cells. J Exp Med 1987; 166:823-32.
67. Wicker LS, Miller BJ, Mullen Y. Transfer of autoimmune diabetes mellitus with splenocytes from nonobese diabetic (NOD) mice. Diabetes 1986; 35(8):855-60.
68. Charlton B, Mandel TE. Progression from insulitis to Beta cell destruction in NOD mouse requires L3T4+T-lymphocytes. Diabetes 1988; 37:1108-12.
69. Serreze DV, Chapman HD, Varnum DS et al. B lymphocytes are essential for the initiation of T cell-mediated autoimmune diabetes: Analysis of a new "speed-congenic" stock of NOD.Igu^{null} mice. J Exp Med 1996; 184:2049-2053.
70. Haskins K, Atkinson M. Prediction of Diabetes and Response to Therapy in Individual Animals: Are there Lessons for Man? Report on the JDF/Snow Mountain Meeting, 5-7 October 2000 Winter Park CO. USA. J Autoimmun 2001; 16:15-19.
71. Abiru N, Yu L, Miao D et al. Transient Insulin Autoantibody Expression Independent of Development of Diabetes: Comparison of NOD and NOR Strains. J Autoimmun 2001.
71a. Robles DT, Eisenbarth GS, Dailey NJM et al. Insulin autoantibodies are associated with islet inflammation but not always related to diabetes progression in NOD congenic mice. Diab 2002; 52:882-886.
72. Serreze DV, Gaskins HR, Leiter EH. Defects in the differentiation and function of antigen presenting cells in NOD/Lt mice. J Immunol 1993; 150:2534-2543.
73. Piganelli JD, Martin T, Haskins K. Splenic macrophages from the NOD mouse are defective in the ability to present antigen. Diabetes 1998; 47:1212-1218.
74. Dahlen E, Hedlund G, Dawe K. Low CD86 expression in the nonobese diabetic mouse results in the impairment of both T cell activation and CTLA-4 up-regulation. J Immunol 2000; 164:2444-2456.
75. Strid J, Lopes L, Marcinkiewicz J et al. A defect in bone marrow derived dendritic cell maturation in the nonobesediabetic mouse. Clin Exp Immunol 2001; 123:375-381.
76. Wong FS, Karttunen J, Dumont C et al. Identification of an MHC class I-restricted autoantigen in type 1 diabetes by screening an organ-specific cDNA library. Nat Med 1999; 5:1026-1031.
77. Buchs AE, Rapoport MJ. T cell signaling and autoimmune diabetes. J Pediatr Endocrinol Metab 2000; 13:1549-1554.
78. Zhang J, Salojin KV, Delovitch TL. CD28 costimulation restores T cell responsiveness in NOD mice by overcoming deficiencies in Rac-1/p38 mitogen-activated protein kinase signaling and IL-2 and IL-4 gene transcription. Int Immunol 2001; 13:377-384.
79. Nervi S, Atlan-Gepner C, Kahn-Perles B et al. Specific deficiency of p56lck expression in T lymphocytes from type 1 diabetic patients. J Immunol 2000; 165:5874-5883.
80. McInerney MF, Pek SB, Thomas DW. Prevention of insulitis and diabetes onset by treatment with complete Freund's adjuvant in NOD mice. Diabetes 1991; 40:715-725.
81. Guytingco R, Quddus J, Richardson B et al. Single administration of bacillus Calmette-Guerin (BCG) vaccine in nonobese diabetic (NOD) mice results in suppression of in vitro T-lymphocyte proliferation and regression of insulitis. Diabetes 1991; 40:152A.
82. Singh B. Stimulation of the developing immune system can prevent autoimmunity. J Autoimmun 2000; 14:15-22.
83. Ridgway WM, Fassò M, Lanctot A et al. Breaking self-tolerance in nonobese diabetic mice. J Exp Med 1996; 183:1657-1662.
84. Boyton RJ, Lohmann T, Londei M et al. Glutamic acid decarboxylase T lymphocyte responses associated with susceptibility or resistance to type I diabetes: Analysis in disease discordant human twins, nonobese diabetic mice and HLA-DQ transgenic mice. Int Immunol 1998; 10:1765-1776.
85. Dosch HM Becker DJ. Measurement of T-cell autoreactivity in autoimmune diabetes [letter]. Diabetologia 2000; 43:386-387.
86. Kreuwel HT, Biggs JA, Pilip IM et al. Defective CD8+ T cell peripheral tolerance in nonobese diabetic mice. J Immunol 2001; 167:1112-1117.
87. Tian J, Gregori S, Adorini L et al. The frequency of high avidity T cells determines the hierarchy of determinant spreading. J Immunol 2001; 166:7144-7150.
88. Hanahan D. Peripheral-antigen-expressing cells in thymic medulla: Factors in self- tolerance and autoimmunity. Curr Opin Immunol 1998; 10:656-662.

89. Aaltonen J, Björses P, Sandkuijl L et al. An autosomal locus causing autoimmune disease: Autoimmune polyglandular disease type I assigned to chromosome 21. Nat Genet 1994; 8:83-87.
90. Heino M, Peterson P, Kudoh J et al. Autoimmune regulator is expressed in the cells regulating immune tolerance in thymus medulla. Biochem Biophys Res Commun 1999; 257:821-825.
91. French MB, Allison J, Cram DS et al. Transgenic expression of mouse proinsulin II prevents diabetes in nonobese diabetic mice. Diabetes 1997; 46:34-39.
92. Pleau JM, Esling A, Geutkens S et al. Pancreatic hormone and glutamic acid decarboxylase expression in the mouse thymus: A real-time PCR study. Biochem Biophys Res Commun 2001; 283:843-848.
93. Naquet P, Naspetti M, Boyd R. Development, organization and function of the thymic medulla in normal, immunodeficient or autoimmune mice. Semin Immunol 1999; 11:47-55.
94. Lord CJ, Howlett S, Lyons PA et al. The murine type 1 diabetes loci, Idd1, Idd3, Idd5, Idd9, and Idd17/10/18, do not control thymic CD4-CD8-/TCRalphabeta+ deficiency in the nonobese diabetic mouse. Mamm Genome 2001; 12:175-176.
95. Lamhamedi-Cherradi SE, Luan JJ, Eloy L et al. Resistance of T-cells to apoptosis in autoimmune diabetic (NOD) mice is increased early in life and is associated with dysregulated expression of Bcl-x. Diabetologia 1998; 41:178-184.
96. Boitard C, Yasunami R, Dardenne M et al. T cell-mediated inhibition of the transfer of autoimmune diabetes in NOD mice. J Exp Med 1989; 169:1669-1680.
97. Gombert JM, Herbelin A, Tancrede-Bohin E et al. Early quantitative and functional deficiency of NK1+-like thymocytes in the NOD mouse. Eur J Immunol 1996; 26:2989-2998.
98. Falcone M, Yeung B, Tucker L et al. A defect in interleukin 12-induced activation and interferon gamma secretion of peripheral natural killer T cells in nonobese diabetic mice suggests new pathogenic mechanisms for insulin-dependent diabetes mellitus. J Exp Med 1999; 190:963-972.
99. Lenschow DJ, Herold KC, Rhee L et al. CD28/B7 regulation of Th1 and Th2 subsets in the development of autoimmune diabetes [published erratum appears in Immunity 1997 Feb;6(2):following 215]. Immunity 1996; 5:285-293.
100. Salomon B, Lenschow DJ, Rhee L et al. B7/CD28 costimulation is essential for the homeostasis of the CD4+CD25+ immunoregulatory T cells that control autoimmune diabetes. Immunity 2000; 12:431-440.
101. Abiru N, Maniatis AK, Yu L et al. Peptide and MHC specific breaking of humoral tolerance to native insulin with the B:9-23 peptide in diabetes prone and normal mice. Diabetes 2001; 50:1274-1281.
102. Haskins K, Portas M, Bergman B et al. Pancreatic islet-specific T cell clones from nonobese diabetic mice. Proc Natl Acad Sci USA 1989; 86:8000-8004.
103. Verdaguer J, Yoon J-W, Anderson B et al. Acceleration of spontaneous diabetes in TCR-β-transgenic nonobese diabetic mice by β-cell cytotoxic $CD8^+$ T cells expressing identical endognous TCR-α chains. J Immunol 1996; 157:4726-4735.
104. Verdaguer J, Schmidt D, Amrani A et al. Spontaneous autoimmune diabetes in monoclonal T cell nonobese diabetic mice. J Exp Med 1997; 186:1663-1676.
105. Graser RT, DiLorenzo TP, Wang F et al. Identification of a CD8 T cell that can independently mediate autoimmune diabetes development in the complete absence of CD4 T cell helper functions. J Immunol 2000; 164:3913-3918.
106. Wegmann DR, Norbury-Glaser M, Daniel D. Insulin-specific T cells are a predominant component of islet infiltrates in prediabetic NOD mice. Eur J Immunol 1994; 24:1853-1857.
107. Wong FS, Visintin I, Wen L et al. CD8 T cell clones from young nonobese diabetic (NOD) islets can transfer rapid onset of diabetes in NOD mice in the absence of CD4 cells. J Exp Med 1996; 183:67-76.
108. Wegmann DR, Shehadeh N, Lafferty KJ et al. Establishment of islet-specific T cell lines and clones from islet isografts placed in spontaneously diabetic NOD mice. J Autoimmun 1993; 6:517-527.
109. Haskins K, Wegmann D. Diabetogenic T-cell clones. Diabetes 1996; 45:1299-1305.
110. Lipes MA, Rosenzweig A, Tan KN et al. Progression to diabetes in nonobese diabetic (NOD) mice with transgenic T cell receptors. Science 1993; 259:1165-1169.
111. Katz JD, Wang B, Haskins K et al. Following a diabetogenic T cell from genesis through pathogenesis. Cell 1993; 74:1089-1100.
111a. Suri A, Katz JD. Dissecting the role of CD4+ T cells in autoimmune diabetes through the use of TCR transgenic mice. Immunol Rev 1999; 169:55-65.
112. Anderson B, Park BJ, Verdaguer J et al. Prevalent $CD8^+$ T cell response against one peptide/MHC complex in autoimmune diabetes. Proc Natl Acad Sci USA 1999; 96:9311-9316.

112a. Lieberman SM, Evans AM, Han B et al. Identification of the {beta} cell antigen targeted by a prevalent population of pathogenic CD8+ T cells in autoimmune diabetes. Proc Natl Acad Sci USA 2003; 100(14)8384-8388.
113. Amrani A, Verdaguer J, Serra P et al. Progression of autoimmune diabetes driven by avidity maturation of a T-cell population. Nature 2000; 406:739-742.
114. Kanagawa O, Shimizu J, Vaupel BA. Thymic and postthymic regulation of diabetogenic CD8 T cell development in TCR transgenic nonobese diabetic (NOD) mice. J Immunol 2000; 164:5466-5473.
115. Oldstone MBA, Nerenberg M, Southern P et al. Virus infection triggers insulin-dependent diabetes mellitus in a transgenic model: Role of anti-self (virus) immune response. Cell 1991; 65:319-331.
116. Ohashi PS, Oehen S, Buerki K et al. Ablation of "tolerance" and induction of diabetes by virus infection in viral antigen transgenic mice. Cell 1991; 65:305-317.
117. von Herrath MG, Coon B, Lewicki H et al. In vivo treatment with a MHC class I-restricted blocking peptide can prevent virus-induced autoimmune diabetes. J Immunol 1998; 161:5087-5096.
118. Laufer TM, von Herrath MG, Grusby MJ et al. Autoimmune diabetes can be induced in transgenic major histocompatibility complex class II-deficient mice. J Exp Med 1993; 178:589-596.
119. von Herrath MG, Dockter J, Oldstone MBA. How virus induces a rapid or slow onset insulin-dependent diabetes mellitus in a transgenic model. Immunity 1994; 1:231-242.
119a. Ohashi PS, Ochen S, Buerki K et al. Ablation of "tolerance" and induction of diabetes by virus infection in viral antigen transgenic mice. Cell 1991; 65:305-317.
120. Garza KM, Chan SM, Suri R et al. Role of antigen-presenting cells in mediating tolerance and autoimmunity. J Exp Med 2000; 191:2021-2027.
121. Lo D, Freedman J, Hesse S et al. Peripheral tolerance to an islet cell-specific hemagglutinin transgene affects both CD4+ and CD8+ T cells. Eur J Immunol 1992; 22:1013-1022.
122. Harlan DM, Hengartner H, Huang ML et al. Mice expressing both B7-1 and viral glycoprotein on pancreatic beta cells along with glycoprotein-specific transgenic T cells develop diabetes due to a breakdown of T-lymphocyte unresponsiveness. Proc Natl Acad Sci USA 1994; 91:3137-3141.
123. King C, Mueller HR, Malo CM et al. Interleukin-4 acts at the locus of the antigen-presenting dendritic cell to counter-regulate cytotoxic CD8+ T-cell responses. Nat Med 2001; 7:206-214.
124. Morgan DJ, Liblau R, Scott B et al. CD8(+) T cell-mediated spontaneous diabetes in neonatal mice. J Immunol 1996; 157:978-983.
125. Vizler C, Bercovici N, Cornet A et al. Role of autoreactive CD8+ T cells in organ-specific autoimmune diseases: Insight from transgenic mouse models. Immunol Rev 1999; 169:81-92.
126. Degermann S, Reilly C, Scott B et al. On the various manifestations of spontaneous autoimmune diabetes in rodent models. Eur J Immunol 1994; 24:3155-3160.
127. Scott B, Liblau R, Degermann S et al. A role for nonMHC genetic polymorphism in susceptibility to spontaneous autoimmunity. Immunity 1994; 1:73-83.
128. Kurts C, Heath WR, Kosaka H et al. The peripheral deletion of autoreactive CD8+ T cells induced by cross-presentation of self-antigens involves signaling through CD95 (Fas, Apo-1). J Exp Med 1998; 188:415-420.
129. Blanas E, Carbone FR, Allison J et al. Induction of autoimmune diabetes by oral administration of autoantigen. Science 1996; 274:1707-1709.
130. Miller JF, Morahan G, Allison J et al. A transgenic approach to the study of peripheral T-cell tolerance. Immunol Rev 1991; 122:103-116.
131. Heath WR, Allison J, Hoffmann MW et al. Autoimmune diabetes as a consequence of locally produced interleukin-2. Nature 1992; 359:547-549.
132. Lesage S, Hartley S, Wilson JD et al. T cell tolerance in a transgenic mouse model for autoimmune diabetes. Keystone Symposia Regulation of Immunity and Autoimmunity 2001; 115-
133. Wong FS, Dittel BN, Janeway CAJ. Transgenes and knockout mutations in animal models of type 1 diabetes and multiple sclerosis. Immunol Rev 1999; 169:93-104.
134. Fugger L. Human autoimmunity genes in mice. Curr Opin Immunol 2000; 12:698-703.
135. Herman AE, Tisch RM, Patel SD et al. Determination of glutamic acid decarboxylase 65 peptides presented by the type I diabetes-associated HLA-DQ8 class II molecule identifies an immunogenic peptide motif. J Immunol 1999; 163:6275-6282.
136. Liu J, Purdy LE, Rabinovitch S et al. Major DQ8-restricted T-cell epitopes for human GAD65 mapped using human CD4, DQA1*0301, DQB1*0302 transgenic IA(null) NOD mice. Diabetes 1999; 48:469-477.
137. Wen L, Chen NY, Tang J et al. The regulatory role of DR4 in a spontaneous diabetes DQ8 transgenic model. J Clin Invest 2001; 107:871-880.
138. Nakhooda AF, Like AA, Chappel CI et al. The spontaneously diabetic Wistar rat: Metabolic and morphologic studies. Diabetes 2001; 26:100-112.

139. Like AA, Rossini AA. Spontaneous autoimmune diabetes mellitus in the BioBreeding/Worcester rat. Surv Synth Path Res 1984; 3:131-138.
140. Colle E, Guttmann RD, Seemayer T. Spontaneous diabetes mellitus syndrome in the rat. I. Association with the major histocompatibility complex. J Exp Med 1981; 154:1237-1242.
141. Guttmann RD, Colle E, Michel F et al. Spontaneous diabetes mellitus syndrome in the rat. II. T lymphopenia and its association with clinical disease and pancreatic lymphocytic infiltration. J Immunol 1983; 130:1732-1735.
142. Elder ME, Maclaren NK. Identification of profound peripheral T lymphocyte immunodeficiencies in the spontaneously diabetic BB rat. J Immunol 1983; 130:1723-1731.
143. Bellgrau D, Naji A, Silvers WK et al. Spontaneous diabetes in BB rats: Evidence for T cell dependent immune response defect. Diabetologia 1982; 23:359-364.
144. Woda BA, Like AA, Padden C et al. Deficiency of phenotypic cytotoxic-suppressor T lymphocytes in the BB/W rat. J Immunol 1986; 136:856-859.
145. Sellins KS, Gold DP, Bellgrau D. Resistance to tolerance induction in the diabetes-prone biobreeding rat as one manifestation of abnormal responses to superantigens. Diabetologia 1996; 39:28-36.
146. Crisa L, Mordes JP, Rossini AA. Autoimmune diabetes mellitus in the BB rat. Diabetes Metab Rev 1992; 8:4-37.
147. Mordes JP, Bortell R, Doukas J et al. The BB/Wor rat and the balance hypothesis of autoimmunity. Diabetes Metab Rev 1996; 12:103-109.
148. Prins JB, Herberg L, den Bieman M et al. Genetic variation within and between lines of diabetes-prone and nondiabetes-prone BB rats: Allele distribution of 8 protein markers. Lab Anim 2001; 25:207-211.
149. Butler L, Guberski DL, Like A. Changes in Penetrance and Onset of Sontaneous Diabetes in the BB/Wor Rat 1990:50-53.
150. Martin AM, Maxson MN, Leif J et al. Diabetes-prone and diabetes-resistant BB rats share a common major diabetes susceptibility locus, iddm4: Additional evidence for a "universal autoimmunity locus" on rat chromosome 4. Diabetes 1999; 48:2138-2144.
151. Kloting II, van den BJ, Kovacs P. Quantitative trait loci for blood glucose confirm diabetes predisposing and protective genes, Iddm4 and Iddm5r, in the spontaneously diabetic BB/OK rat. Int J Mol Med 1998; 2:597-601.
151a.Hornum L, Romer J, Markholst H. The diabetes-prone BB rat carries a frameshift mutation in Ian4, a positional candidate of Iddm1. diab 2002; 51(6):1972-1979.
152. Colle E, Fuks A, Poussier P et al. Polygenic nature of spontaneous diabetes in the rat: Permisive MHC haplotype and presence of the lymphopenic trait of the BB rat are not sufficient to produce susceptibility. Diabetes 1992; 41:1617-1623.
153. Hernandez-Hoyos G, Joseph S, Miller NG et al. The lymphopenia mutation of the BB rat causes inappropriate apoptosis of mature thymocytes. Eur J Immunol 1999; 29:1832-1841.
154. Guberski DL, Thomas VA, Shek WR et al. Induction of Type 1 diabetes by Kilham's rat virus in diabetes-resistant BB/Wor rats. Science 1991; 254:1010-1013.
155. Sobel DO, Newsome J, Ewel CH et al. Poly I:C induces development of diabetes mellitus in BB rat. Diabetes 1992; 41:515-520.
156. Like AA. Depletion of RT6.1+ lymphocytes alone is insufficient to induce diabetes in diabetes-resistant BB/WOR rats. Am J Pathol 1990; 136:565-574.
157. Kloting I, Schmidt S, Kovacs P. Mapping of novel genes predisposing or protecting diabetes development in the BB/OK rat. Biochem Biophys Res Commun 1998; 245:483-486.
158. Jackerott M, Hornum L, Andreasen BE et al. Segregation of autoimmune type 1 diabetes in a cross between diabetic BB and brown Norway rats. J Autoimmun 1997; 10:35-41.
159. Ellerman KE, Like AA. Susceptibility to diabetes is widely distributed in normal class IIu haplotype rats. Diabetologia 2000; 43:890-898.
160. Jackson R, Rassi N, Crump A et al. The BB diabetic rat. Profound T-cell lymphocytopenia. Diabetes 1981; 30:887-889.
161. Ely JM, Greiner DL, Lubaroff DM et al. Characterization of monoclonal antibodies that define rat T cell alloantigens. J Immunol 1983; 130:2798-2803.
162. Greiner DL, Handler ES, Nakano K et al. Absence of the RT-6 T cell subset in diabetes-prone BB/W rats. J Immunol 1986; 136:148-151.
163. Groen H, Van Der Berk J, Nieuwenhuis P et al. Peripheral T cells in diabetes prone (DP) BB rats are CD45R-negative. Thymus 1989; 14:145-150.
164. Spickett GP, Brandon MR, Mason DW et al. MRC OX-22, a monoclonal antibody that labels a new subset of T lymphocytes and reacts with the high molecular weight form of the leukocyte-common antigen. J Exp Med 1983; 158:795-810.

165. Thomas VA, Woda BA, Handler ES et al. Altered expression of diabetes in BB/Wor rats by exposure to viral pathogens. Diabetes 1991; 40:255-258.
166. Whalen BJ, Greiner DL, Mordes JP et al. Adoptive transfer of autoimmune diabetes mellitus to athymic rats: Synergy of CD4+ and CD8+ T cells and prevention by RT6+ T cells. J Autoimmun 1994; 7:819-831.
167. Rossini AA, Mordes JP, Greiner DL et al. Spleen cell transfusion in the bio-breeding/worcester rat. J Clin Invest 1986; 77:1399-1401.
168. Ramanathan S, Poussier P. T cell reconstitution of BB/W rats after the initiation of insulitis precipitates the onset of diabetes. J Immunol 1999; 162:5134-5142.
169. Angelillo M, Greiner DL, Mordes JP et al. Absence of RT6 T cells in diabetes prone biobreeding/Worcester rats is due to genetic and cell developmental defects. J Immunol 1988; 141:4146-4151.
170. Fangmann J, Schwinzer R, Hedrich H-J et al. Diabetes-prone BB rats express the RT6 alloantigen on intestinal intraepithelial lymphocytes. Eur J Immunol 1991; 21:2011-2015.
171. Crisa L, Sarkar P, Waite DJ et al. An RT6 gene is transcribed and translated in lymphopenic diabetes prone BB rats. Diabetes 1993; 42:688-695.
172. Georgiou HM, Lagarde AC, Bellgrau D. T cell dysfunction in the diabetes-prone BB rat: A role for thymic migrants that are not T cell precursors. J Exp Med 1988; 167:132-148.
173. Bellgrau D, Redd JM, Sellins KS. Peculiar T-cell signaling does not preclude positive selection in the diabetes-prone BB rat. Diabetes 1994; 43:47-52.
174. Hosszufalusi N, Chan E, Granger G et al. Quantitative analyses comparing all major spleen cell phenotypes in BB and normal rats: Autoimmune imbalance and double negative T cells associated with resistant, prone and diabetic animals. J Autoimmun 1992; 5:305-318.
175. Moore JK, Bellgrau D. Promiscuous activation and cell cycle entry in T cells from autoimmune animals. Transplant Proc 1999; 31:1606-1610.
176. Bieg S, Möller C, Olsson T et al. The lymphopenia (*lyp*) gene controls the intrathymic cytokine ratio in congenic BioBreeding rats. Diabetologia 1997; 40:786-792.
177. Bieg S. Differential expression of p95vav in primary lymphoid tissue of BB rats congenic for the lymphopenia gene. Autoimmunity 1999; 30:37-42.
178. Zadeh HH, Greiner DL, Wu DY et al. Abnormalities in the export and fate of recent thymic emigrants in diabetes-prone BB/W rats. Autoimmunity 1996; 24:35-46.
179. Moore JK, Scheinman RI, Bellgrau D. The identification of a novel T cell activation state controlled by a diabetogenic gene. J Immunol 2001; 166:241-248.
180. Ramanathan S, Norwich K, Poussier P. Antigen activation rescues recent thymic emigrants from programmed cell death in the BB rat. J Immunol 1998; 160:5757-5764.
181. Delemarre FG, Simons PJ, de Heer HJ et al. Signs of immaturity of splenic dendritic cells from the autoimmune prone biobreeding rat: Consequences for the in vitro expansion of regulator and effector T cells. J Immunol 1999; 162:1795-1801.
182. Delemarre FG, Hoogeveen PG, Haan-Meulman M et al. Homotypic cluster formation of dendritic cells, a close correlate of their state of maturation. Defects in the biobreeding diabetes-prone rat. J Leukoc Biol 2001; 69:373-380.
183. Iwakoshi NN, Greiner DL, Rossini AA et al. Diabetes prone BB rats are severely deficient in natural killer T cells. Autoimmunity 1999; 31:1-14.
184. Tullin S, Farris P, Petersen JS et al. A pronounced thymic B cell deficiency in the spontaneously diabetic BB rat. J Immunol 1997; 158:5554-5559.
185. Rozing J, Coolen C, Tielen FJ et al. Defects in the thymic epithelial stroma of diabetes prone BB rats. Thymus 1989; 14:125-135.
186. Bellgrau D, Lagarde A-C. Cytotoxic T-cell precursors with low-level CD8 in the diabetes-prone biobreeding rat: Implications for the generation of an autoimmune T-cell repertoire. Proc Natl Acad Sci USA 1990; 87:313-317.
187. Groen H, Klatter FA, Brons NH et al. Abnormal thymocyte subset distribution and differential reduction of CD4+ and CD8+ T cell subsets during peripheral maturation in diabetes-prone BioBreeding rats. J Immunol 1996; 156:1269-1275.
188. Plamondon C, Kottis V, Brideau C et al. Abnormal thymocyte maturation in spontaneously diabetic BB rats involves the deletion of CD4-8+ cells. J Immunol 1990; 144:923-928.
189. Whalen BJ, Weiser P, Marounek J et al. Recapitulation of normal and abnormal BioBreeding rat T cell development in adult thymus organ culture. J Immunol 1999; 162:4003-4012.
190. McKeever U, Mordes JP, Greiner DL et al. Adoptive transfer of autoimmune diabetes and thyroiditis to athymic rats. Proc Natl Acad Sci USA 1990; 87:7618-7622.
191. Sarkar P, Crisa L, McKeever U et al. Loss of RT6 message and most circulating T cells after thymectomy of diabetes prone BB rats. Autoimmunity 1994; 18:15-22.

192. Jung CG, Kamiyama T, Agui T. Elevated apoptosis of peripheral T lymphocytes in diabetic BB rats. Immunology 1999; 98:590-594.
193. Crispe IN, Dao T, Klugewitz K et al. The liver as a site of T-cell apoptosis: Graveyard, or killing field? Immunol Rev 2000; 174:47-62.
194. Scott J. The spontaneously diabetic BB rat: Sites of the defects leading to autoimmunity and diabetes mellitus. A review. Curr Top Microbiol Immunol 1990; 156:1-14.
195. Francfort JW, Naji A, Silvers WK et al. The influence of T-lymphocyte precursor cells and thymus grafts on the cellular immunodeficiencies of the BB rat. Diabetes 1985; 34:1134-1138.
196. Like A, Rossini AA, Guberski DL et al. Spontaneous diabetes mellitus: Reversal and prevention in the BB/W rat with antiserum to rat lymphocytes. Science 1979; 206:1421-1423.
197. Like AA, Biron CA, Weringer EJ et al. Prevention of diabetes in BioBreeding/Worcester rats with monoclonal antibodies that recognize T lymphocytes or natural killer cells. J Exp Med 1986; 164:1145-1159.
198. Barlow AK, Like AA. Anti-CD2 monoclonal antibodies prevent spontaneous and adoptive transfer of diabetes in the BB/Wor rat. Am J Pathol 1992; 141:1043-1051.
199. Like AA, Kislauskis E, Williams RR et al. Neonatal thymectomy prevents spontaneous diabetes mellitus in the BB/W rat. Science 1982; 216:644-646.
199a. McKeever U, Mordes JP, Greiner DL et al. Adoptive transfer of autoimmune diabetes and thyroiditis to athymic rats. Proc Natl Acad Sci USA 1990; 87:7618-7622.
200. Murase N, Lieberman I, Nalesnik M et al. Prevention of spontaneous diabetes in BB rats with FK 506. Lancet 1990; 336:373-374.
201. Nicoletti F, Meroni PL, Barcellini W et al. FK-506 prevents diabetes in diabetes-prone BB/Wor rats. Int J Immunopharmacol 1991; 13:1027-1030.
202. Sadelain MW, Qin HY, Sumoski W et al. Prevention of diabetes in the BB rat by early immunotherapy using Freund's adjuvant. J Autoimmun 1990; 3:671-680.
203. Like AA, Guberski DL, Butler L. Influence of environmental viral agents on frequency and tempo of diabetes mellitus in BB/Wor rats. Diabetes 1991; 40:259-262.
204. Dryberg T, Schwimmbeck PL, Oldstone MBA. Inhibition of diabetes in BB rats by virus infection. J Clin Invest 1988; 81:928-931.
205. Greiner DL, Malkani S, Kanaitsuka T et al. The T cell marker RT6 in a rat model of autoimmune diabetes. Adv Exp Med Biol 1997; 419:209-216.
206. Rossini AA, Faustman D, Woda BA et al. Lymphocyte transfusions prevent diabetes in the Bio-Breeding/ Worcester rat. J Clin Invest 1984; 74:39-46.
207. Burstein D, Mordes JP, Greiner DL et al. Prevention of diabetes in BB/Wor rat by single transfusion of spleen cells. Parameters that affect degree of protection. Diabetes 1989; 38:24-30.
208. Satoh J, Seino H, Shintani S et al. Inhibition of type 1 diabetes in BB rats with recombinant human tumor necrosis factor-alpha. J Immunol 1990; 145:1395-1399.
209. Whalen BJ, Marounek J, Weiser P et al. BB rat thymocytes cultured in the presence of islets lose their ability to transfer autoimmune diabetes. Diabetes 2001; 50:972-979.
210. Kawano K, Hirashima T, Moris S et al. New inbred strain of Long-Evans Tokushima lean rats with IDDM without lymphopenia. Diabetes 1991; 40:1375-1381.
211. Yokoi N, Kanazawa M, Kitada K et al. A nonMHC locus essential for autoimmune type I diabetes in the Komeda diabetes-prone rat. J Clin Invest 1997; 100:2015-2021.
211a. Yokoi N, Komeda K, Wang HY et al. Cblb is a major susceptibility gene for rat type 1 diabetes mellitus. Nat Genet 2002; 31(4):391-394.
212. Komeda K, Noda M, Terao K et al. Establishment of two substrains, diabetes-prone and nondiabetic, from Long-Evans Tokushima Lean (LETL) rats. Endocr J 1998; 45:737-744.

Chapter 4

The Role of T Cells in Beta Cell Damage in NOD Mice and Humans

Katalin Kelemen

Type 1A diabetes is a chronic disorder that results from destruction of the insulin-producing beta cells of the pancreatic islets. In its initial phase, which is clinically silent, T lymphocytes and other inflammatory cells invade the islets and eventually destroy them. The disease becomes clinically evident with the pathological consequences (hyperglycemia, ketosis and long-term complications) of the inability to maintain glucose homeostasis.

The NOD Mouse

The study of Type 1A diabetes has been greatly facilitated through the use of the Biobreeding rat and the nonobese diabetic (NOD) mouse.[1] The NOD mouse in particular has been the subject of extensive studies over a number of years. The NOD mice spontaneously develop diabetes as a result of destruction of pancreatic beta cells, associated with heavy cellular infiltrates surrounding the individual islets beginning at least 7 weeks prior the onset of overt diabetes.[2,3] Through the use of transgenic technology, several new mouse models of human Type 1A diabetes has been generated and are now used extensively to understand the complex nature of Type 1A diabetes.

Adoptive Transfer of Diabetes by T Cells

There is extensive evidence supporting a crucial role for T cells and MHC-restricted self antigen recognition for diabetes pathogenesis, as disease can be transferred by T cell populations or T cell clones.[4-6] Haskins and colleagues isolated clones of CD4+ T cells capable of conferring disease. These clones recognized pancreas-specific antigens presented by the NOD MHC class II molecule, I-A^{g7}.[7,8] Recently, a line of transgenic mice expressing the T Cell Receptor (TCR) of one such clone, BDC2.5 was generated to facilitate the analysis of the selection and the tolerance induction forces operating on such cells.[9] Evaluation of the lymphoid compartment showed an absence of tolerance induction; there was no sign of negative selection in the thymus and the T cell reactivity was normal in the periphery. However, autoimmune pathogenesis followed a particular course. There was no manifestation in the first 3 weeks of life with pancreatic islets completely free of infiltration. Insulitis sets in abruptly at 3 weeks, quickly becoming massive and involving essentially all islets. Yet, in spite of this overwhelming insulitis, diabetes only appeared much later and onset was widely spread over 10-25 weeks within a cohort of transgenic animals in a manner not fundamentally different from that of disease in nontransgenic NOD mice. In the NOD mouse, long-term insulitis does not always progress to disease. For example, in male mice, insulitis is as prevalent as in females, but males less often develop hyperglycemia.[1]

Immunology of Type 1 Diabetes, Second Edition, edited by George S. Eisenbarth.

Checkpoints in the Development of Autoimmune Diabetes

The BDC2.5 TCR transgenic mouse allowed investigators to define two major checkpoints in the pathogenesis of Type 1A diabetes. Checkpoint 1 controls the onset of insulitis, checkpoint 2 controls the switch to overt diabetes and in spite of extensive and active insulitis, intact beta cells persist for long periods of time and no diabetes occurs. Eventually, this balance is lost and the insulitis becomes terminally aggressive. Both checkpoints also exist in regular NOD mice but they are more clearly delineated in the transgenic mouse. The molecular and cellular mechanisms that underlie these checkpoints are not just of academic interest since therapeutic interventions against Type 1A diabetes may stem from influencing such control mechanisms. The onset of insulitis coincides with weaning and the large immunological changes that take place at that time in which major shifts in food intake and in intestinal flora result in a novel array of antigens. The broad T cell stimulation that takes place at weaning modifies the homing potential of T cells and endows them with the ability to migrate into pancreatic connective spaces. There are documented changes in the expression of the addressins (MadCAM and PNAd) on the pancreatic blood vessel endothelium around the onset of insulitis.[10-12] Adhesion molecules, in particular alpha4 integrin has been shown to be important in the insulitic process.[13-15] It is not completely understood whether these changes in the expression of the adhesion molecules by the endothelium play an integral role in the process or they are merely secondary to the local inflammatory process. Checkpoint 1 would correspond to increased homing potential of T cells coupled with increased attractiveness of the endothelium. Transfer experiments indicate that the transition between tolerated insulitis and full blown disease involves a switch in the pathogenic potential of the lymphoid populations. Splenocytes from BDC2.5 mice that have insulitis but not diabetes given to neonatal recipients do not transfer disease, whereas those from overtly diabetic mice do. In early studies on the genetics of Type 1A diabetes in NOD mice, Todd and colleagues showed[16] that some idd loci affect only the progression to diabetes and not the incidence of insulitis. The genetic background seems to have a strong influence on the outcome -mild and tolerated insulitis or aggressive disease.[17] Possible events that may underlie the transition through checkpoint 2 include cell-intrinsic (originating in the pathogenic T cell itself) or cell-extrinsic mechanisms. Potential cell-intrinsic changes may occur through acquisition of new effector capabilities (Th1/Th2 balance, cytokines, expression of FasLigand) or through loss of sensitivity to negative signaling (CTLA4, IL-2R, NK-like receptors). Cell-extrinsic changes may manifest through recruitment of new antigen specificities involving epitope spreading or recruitment of CD8+ cells. Abrogation of negative control, such as inhibitory cytokines or loss of regulatory cell populations may contribute to regulative switch, together with the recruitment of accessory cell-types into the lesion. In conclusion, the autoreactive TCR transgenics simplify to a degree the analysis of the autoimmune process but many questions need to be answered before we reach a good understanding of what these checkpoint represent.

T Cell Subsets in Type 1A Diabetes

In both NOD mice and humans, the infiltrating cells circumscribing the islets are composed of CD4+ and CD8+ lymphocytes, B lymphocytes, macrophages, dendritic cells and other bone-marrow-derived cells.[18,19] T lymphocytes play an obligatory, if not primary role in the development of Type 1A diabetes and both CD4+ and CD8+ T cells are needed to generate destruction of islet beta cells in most models.[20-22] However, both CD4+ and CD8+ T cell clones specific to beta cell antigens were shown to be able to transfer diabetes without the presence of the other T cell subset[20,23] The CD4+ T cell subset can be broadly classified into two major subgroups on the basis of cytokine secretion profile and effector function: T helper1 (Th1) cells, which secrete IFNγ, TNFα and IL-2, are primarily associated with cellular immunity. Th2 cells, which secrete IL-4, IL-5, IL-6, IL-10 and IL-13 are mainly involved in humoral immune responses and allergy.[24] A number of studies have implicated Th1 cells in the development of diabetes[25-27] and some of the signature cytokines such as IFNγ and TNFα have been associated with promoting the development of disease. In addition, the systemic treatment of

NOD mice with IL-12, a potent inducer of Th1 cells, led to accelerated diabetes.[28,29] On the other hand, Th2 cells seem either relatively innocuous[25] or seem to offer protection from diabetes, in part by their distinct cytokine profile.[30-37] However, this putative protective role for Th2 cells must be tempered by the observation that NOD mice deficient in IL-4, which is the primary polarizing cytokine for Th2 development, exhibit similar disease incidence as regular NOD mice.[38] Although NOD mice that express IL-4 under the rat insulin promoter in the pancreas are protected from diabetes,[30] NOD mice with ectopic expression of IL-10, an immunosuppressive Th2 cytokine in islets demonstrate an accelerated onset of Type 1A diabetes.[39] Transfer experiments with islet-reactive Th2 cells generated from BDC2.5 transgenic mice, provided contradictory results. When the T cells were cultured in presence of IL-4 and IFNγ monoclonal antibody for a short time (4-6 days) and then transferred into neonatal NOD recipients, Th2 cells invaded the islets but did not provoke disease, neither did they provide protection.[25] When T cells from the BDC2.5 transgenic mouse were cultured long term with IL-4 and IFNγ monoclonal antibody, they were capable of accelerating disease in neonatal NOD recipients and also of transferring disease to NODscid recipients.[40] Furthermore, Th2 cells have been shown to cause diabetes in immunocompromised NODscid mice although the kinetics of the disease and the inflammatory lesion were markedly different from those induced by Th1 cells.[41] These results suggest that Th2 cells do indeed possess the potential to mediate Type 1A diabetes. The development of diabetes in the NOD mouse may be modulated by yet another thymus-derived T cell population that is often referred to as the regulatory T cell subset. The phenotypic nature of these cells remain elusive. However, CD4+ regulatory T cells, referred to by some as Th3 cells, which secrete TGFβ and IL-10 have been recently described and hypothesized to play a regulatory role in many autoimmune disorders.[42] Studies involving adoptive transfer of splenocytes from diabetic NOD mice into irradiated NOD recipients demonstrated that the kinetics of diabetes was both delayed and reduced when lymphoid cells from prediabetic NOD mice were cotransferred with the cells from diabetic NOD donors.[43] Additional reports have since confirmed the role of CD4+ regulatory T cells in both mouse and rat models of Type 1A diabetes, although there is still no evidence for the primary mechanism of regulation.[44-48] Bluestone and coworkers have provided evidence that CD4+ CD25+ T regulatory cells require costimulatory molecules and in their absence diabetes is greatly accelerated.[49]

Mechanism and Mediators of Beta Cell Death

The hallmark of Type 1A diabetes is the selective destruction of insulin-producing beta cells in the pancreatic islets. There is substantial amount of evidence that the primary mechanism of beta cell death is apoptosis.[50-53] Beta cell apoptosis is difficult to detect in vivo in NOD mice due to the slow and complicated kinetics of disease process, the asynchronous loss of islet beta cells and the rapid clearance of cell debris by host phagocytes. What are the factors responsible for mediating beta cell apoptosis? Utilizing diverse transgenic models, recent reports have suggested the role for CD8+ or CD4+ T cells in β cell apoptosis via the involvement of distinct death effector molecules. NOD lpr/lpr mice (deficient in Fas expression) are protected from spontaneous diabetes. The engagement of Fas on β cells by Fas ligand (FasL) on CD8+ T cells was suggested.[54,55] It is not yet clear if the lack of disease in the NOD lpr/lpr mice is primarily the result of a block in the end-stage killing of beta cells or the mutation's abnormal effect on peripheral T cells. Perforin-dependent cytotoxicity was also shown to be a crucial effector mechanism for beta cell elimination by cytotoxic T cells in autoimmune diabetes. In perforin-deficient NOD mice, the incidence of diabetes was reduced and the onset of the disease was significantly delayed.[56] However, perforin-deficient 8.3-TCR transgenic NOD mice (expressing TCR of a diabetogenic CD8+ T cell clone) developed diabetes even more frequently than their perforin-competent littermates.[57] Other studies implicate CD4+ T cells in promoting beta cell death by another distinct effector mechanism, via the involvement of the TNFα/TNF Receptor molecules. In this model system, when islets deficient in TNFα Receptor type 1 were transplanted into NODscid recipients, the graft was protected from destruction by BDC2.5 T cell.[58] An important issue arising from the model system described above pertains to whether the destruction of transplanted islets is dependent upon the direct action of BDC2.5 T cells or are they in fact killed by the secondary action of BDC 2.5 cells on other cell types, such as Natural Killer (NK) cells. The NODscid recipients are devoid of endogenous T and B lymphocytes but do possess NK cells. It has been reported that NOD NK cells exhibit poor effector responses,[59] but in this model it remains a formal possibility that NK cells could lyse beta cells upon local activation by infiltrating BDC2.5 T cells. However, antibody-mediated depletion of NK cells had no impact on the onset of diabetes in the NODscid recipients, thereby establishing that beta cell destruction depends on the presence of BDC2.5 T cells but not on NK cells.[60]

The Autoantigens

As T cells are thought to play a major role in the pathogenesis of diabetes, many studies have been performed to try to identify the islet-specific autoantigens to which T cells respond and the antigens identified in the human antibody studies have been investigated. Most work to date has focused on insulin and its precursors as well as on the enzyme GAD (glutamic acid decarboxylase) and more recently the protein tyrosine-like proteins IA-2 (ICA512) and phogrin (IA-2b) and IGRP.[60a] To examine the role of these antigens, a number of approaches have been used which investigate the spontaneous responses to the antigens, the ability to obtain immune responses and the use of these various antigens in tolerization protocols to prevent disease. In addition, in the NOD mouse model, it has been possible to use transgenic technology to try to examine the role of important putative autoantigens to the development of diabetes.

Insulin

Insulin or its precursors are obvious target autoantigens. Insulin is abundantly expressed in the pancreatic islets. Its expression was thought to be islet-specific but recently it has been shown that there are cells within the thymus that express peripheral antigens, such as insulin and GAD in humans[61-63] and in rats and mice.[64-66] Within the islets of NOD mice there are clearly documented CD4+ T cells which recognize insulin[67] specifically as a peptide consisting of amino acids 9-23 of the B chain and when cloned these cells are capable of causing disease on adoptive transfer.[68] Most recently it was shown that a highly diabetogenic CD8+ T cell clone recognizes a peptide consisting of amino acids 15-23 of the B chain of insulin.[69] Thus, the insulin B chain is clearly a target for both CD4+ and CD8+ T cells in the NOD mouse. This finding has important implications for immunoprevention. Previous studies using the NOD mouse have shown that insulin or insulin B chain peptide 9-23 given orally[68,70] subcutaneously[71] or intranasally[72] protect from diabetes. Insulin that has been metabolically inactivated changing one amino acid in the B chain is also effective at diabetes prevention.[73] Certainly CD4+ T cells that have a protective function and recognize insulin as their target antigen have also been isolated.[74] In addition, it has been shown that regulatory γδ CD8+ cells are induced after nasal administration of insulin.[72] It remains to be seen whether success in modulating diabetes in the NOD mouse using various modes of delivering of insulin/proinsulin or nonmetabolically active insulin can be translated similarly to humans.[72a]

Glutamic Acid Decarboxylase (GAD)

The other major putative autoantigen in humans, GAD, exists in two independently encoded forms, GAD-65 and GAD-67. The two forms differ mainly in the amino acid sequence at the amino-terminal end. GAD is of major importance in the central nervous system where it catalyses the formation of the inhibitory neurotransmitter, g-amino-butyric acid (GABA). GAD is mainly found in neuroendocrine tissues and the distribution of the isoforms of GAD differs between man and the mouse models of diabetes. GAD-65 is more predominant in the pancreas in humans whereas GAD-67 is expressed in the islets of mice, however the level of expression is very low.[75,76] The physiological role of GAD and GABA within the pancreas is still uncertain.

The earliest responses to GAD were found in young NOD mice at a time when reactivity to other antigens was not detectable.[77,78] A number of studies have shown in the NOD mouse that it is possible to induce tolerance to GAD and prevent diabetes by administration of GAD in a number of ways.[35,78-80] Although there is a report of CD4+ T cells that are responsive to GAD and are diabetogenic in NOD mice,[81] to date no CD8+ T cells have been isolated that respond to GAD in mice although they have been found in humans.[82] Several transgenic models have now been developed in NOD mice in order to attempt to further clarify the role GAD in diabetes. Under normal circumstances in the NOD mouse GAD-65 is expressed at very low levels in the pancreas. One of the engineered models had a mutated GAD expressed on a MHC class I promoter such that GAD-65 was expressed highly in all tissues.[83] It was found that the expression of the mutated GAD in this manner in one out of three transgenic lines resulted in a greater degree of insulitis in the males and a greater final incidence of diabetes. Of note, diabetes was not prevented while Harrison and coworkers, producing a similar transgenic with wide expression of proinsulin, prevented both NOD diabetes and insulitis.[84]

Another transgenic model expressed GAD-65 on the Rip(rat insulin promoter)-7 promoter, such that the GAD-65 was expressed at high levels in the pancreatic islets.[85] Two lines were generated, one of which expressing higher levels of GAD was protected from diabetes. It was suggested that this was due to an increase in the production of IL-10. The third and most recently reported transgenic model investigating GAD as an autoantigen used an antisense construct to prevent the expression of both GAD-65 and GAD-67.[86] In one of three transgenic lines, with apparently the greatest reduction in GAD expression, not only were diabetes and insulitis abolished but also the cells developing in these transgenic mice were unable to transfer diabetes. In addition, the islets were protected from adoptive transfer of diabetic spleen cells. These data are interesting and perhaps might have relevance for use in engineered islets in transplants. However, it raises a number of questions as to exactly how the antisense construct is working because the results presented make it unlikely that protection is solely due to the immunological effects of suppression of GAD and with so little GAD in normal mouse islets and the lack of "true" congenic strains the report has generated considerable controversy.[87,88]

Protein Tyrosine Phosphatases

The role of the PTP members IA-2 and phogrin as T cell autoantigens has not been extensively explored, yet these molecules are expressed at much higher levels than GAD-65 in islets and like so many of the candidate autoantigens are found exclusively in dense core secretory granules.[89,90] Both proteins are type 1 transmembrane receptor-like glycoproteins of similar size, domain structure and topology and show 80% and 26% identity, respectively, in their cytosolic and lumenal domains. IA-2 and phogrin were initially described as autoantibody-reactive 40 and 37 kDa tryptic fragments derived from a nonGAD 64 kDa precursor immunoprecipitated from radiolabeled β cells.[91,92] They were subsequently cloned from various tissue sources by subtractive and expression screening strategies.[90,93,94] Approximately 60-70% of sera of new-onset Type 1A diabetes patients[92,95,96] react to both molecules, as do 40-50% of sera from first-degree relatives starting from an early stage in the disease process. The role of IA-2 and phogrin as autoantigenic targets in the NOD mouse was discounted over a period of time. Later, autoantibodies have been detected in sera of NOD mice and they appeared close to the onset of hypergycemia and were usually transient.[97] Recently, we generated a series of phogrin-specific T cell clones from prediabetic NOD female mice immunized with recombinant phogrin in CFA. All of these T cell clones are of CD4+/Th1 phenotype. Some of these clones were able to destroy islet transplants and cause diabetes in vivo in a disease recurrence model.[98] Epitopes of these destructive T cell clones have been identified as two distinct peptides on the C terminus represented by amino acids (aa) 640-659 and amino acids 755-777. Spontaneous T cell proliferative responses have been detected to the epitope peptide amino acid 755-777 in spleens of prediabetic NOD mice, but not in those of Balb/c mice of the same age. When NOD mice were primed with the epitope peptides in CFA (complete Freund's adjuvant), an intramolecular epitope spreading appeared from the epitope aa 640-659 to the peptide aa 755-777, but never to the other direction. Determinant spreading is supposed to play an important role in the amplification of the T cell mediated inflammatory process and might be an important factor in the transition of benign insulitis into destructive process and clinical diabetes. The mechanism of a potential intermolecular epitope spreading between phogrin and IA-2 and the question whether this mechanism can be effectively targeted in prevention strategies, needs to be further analyzed.

Results from Investigations of Human Subjects

Although there is a considerable amount of direct evidence that T cells mediate beta cell destruction in the NOD mouse, there is much less information available with regard to human subjects. The major obstacles to obtaining islet-specific T cells from humans are due to practical and technical problems. The only practical source of starting material from human subjects is peripheral blood. It is likely that the frequency of islet-specific T cells in the peripheral blood of prediabetic or newly diabetic patients is very low and therefore the likelihood of isolating these cells is also low. Although, as will be reviewed below, a number of investigators have obtained evidence that newly diabetic subjects have T cell reactivity to beta cell associated antigens, evidence that such cells play a role in beta cell destruction is lacking.

Histological Investigations

The first indication that Type 1A diabetes was an immune-mediated phenomenon was provided by a comprehensive histological examination of pancreata of recently diagnosed diabetics.[99] These investigators observed that most of these subjects had significant infiltration of the islets and loss of beta cell mass. At the time of this report (1965), cellular immunology as we know it today was in its infancy and the observation of these infiltrates provided little insight into the mechanism of beta cell destruction. In a subsequent report published in 1981, it was suggested that these lymphocytic infiltrates accumulated as a result of attraction by antigens present in the beta cells of the islets.[100] The first indication that T cells were involved in beta cell destruction was provided by a detailed immunohistochemical examination of the pancreas from a newly diabetic subject in which it was observed that T cells, particularly CD8+ T cells were a predominant component of the islet infiltrates present.[101]

Demonstration of in Vitro T Cell Reactivity to Islet Cell Antigens

The mere presence of T cells in infiltrates is only suggestive of a direct role in development of diabetes, but these histological findings were soon followed by reports of T cell reactivity to beta cell products. The first evidence that T cells reactive to beta cell products were present in the peripheral blood of newly diabetic subjects was provided by Van Vliet and coworkers in 1989.[102] These investigators established T cell lines and clones from the peripheral blood of new-onset diabetics by stimulation with rat insulinoma cells. They found that these cells were CD4+ and restricted to HLA-DR. Of the eight clones examined, five appeared to be reactive with the membrane fraction of insulinoma cells. Despite the considerable difficulties in detecting islet cell specific autoreactive T cells in peripheral blood, several studies have reported T cell responses to more or less defined islet antigens.[103-108] The antigens to which such T cell responses were measured include insulin secretory granules, RIN (rat insulinoma) cells, insulinoma membranes, unidentified determinants of fetal pig proislets, human adult islets and the enzyme GAD. The identification of ICA512 (IA-2) as a humoral autoantigen was followed by reports of T cell proliferative responses to this molecule in the peripheral blood of new onset Type 1A diabetes patients.[109,110] It has been suggested that pancreatic islet autoimmunity leading to Type 1A diabetes could be triggered by viruses in genetically susceptible individuals. The possible mechanism could be the peripheral activation of T cells by homologous peptide sequences shared between enteroviruses and certain beta cell antigens ("molecular mimicry"). The relation between T cell responses to GAD and coxsackie virus B4,[111] GAD and rubella virus[112] IA-2 and rotavirus[113] has been investigated. However, the direct involvement of enteroviral infections in the pathogenesis of Type 1A diabetes still remains to be proven.

Problems Associated with the Detection of T Cell Reactivity in Diabetes – Conclusions from the First International T Cell Workshop

Evaluation of T cell autoreactivity in Type 1A diabetes is hampered by several features. First, it is apparent that the precursor frequency of autoreactive T cells in the periphery is low.[114] In addition, it is conceivable that regulatory T cells may modulate the activity of such T cells.[115] Newly diagnosed Type 1A diabetes patients are hyperglycemic and often lymphopenic.[116] There is relatively high percentage of monocytes, which in turn may affect the in vitro proliferation assays.[117] It has been recently demonstrated that the autoimmune T cell responses were primarily measured in the subset of activated T cells (CD45RA+RO+) in newly diagnosed Type 1A diabetes patients, whereas those with longer disease duration reacted to autoantigens with memory cells (CD45RO+).[118] The separation of the different T cell subsets might help to detect differences between T cell responses of different clinical groups of patients. The first international workshop for the standardization of the T cell assays organized by the Immunology of Diabetes Society in 1999 aimed to identify the problems associated with the autoreactive T cell assays in Type 1A diabetes.[119] The project included the analysis of a series of candidate autoantigens for quality by reference laboratories. Subsequently, T cell proliferative responses of diabetic and nondiabetic individuals to these antigen preparations were analyzed worldwide in 26 laboratories. The results can be summarized as followed: (1) The quality of the recombinant antigens required improvement; (2) All laboratories were able to detect T cell responses to tetanus toxoid although with significant inter-laboratory variation in sensitivity and in the extent of response measured; and (3) Differences in T cell proliferative responses could not be detected between diabetics and nondiabetic individuals. The T cell workshop of T cell autoreactivity was helpful at interpreting previously reported conflicting results and provided future directions for the use of more suitable antigens and standardized assays.

T Cell Assay Development

The difficulties with current T cell assays have stimulated the development of alternative procedures with considerable effort in developing ELISPOT and tetramer based assays. Aliva and coworkers have recently reported that insulin B:9-23 reactive T cells were present in new onset type 1 and prediabetic patients. Nepom and coworkers isolated a naturally processed GAD peptide[120] and have developed a DRB1*0401 tetramer using this peptide with encouraging preliminary studies in patients. Susan Wong and coworkers and Santamaria and coworkers have developed tetramers reacting with autoreactive T cells of the NOD mouse. For the Wong tetramer, the target peptide is insulin B:15-23. For Santamaria, the natural ligand is IGRP (islet glucose related phosphatase).[121,60a] Using a mimotope peptide Santamaria has provided intriguing data that progression to diabetes of the NOD mouse may be dependent upon T cell receptor "affinity maturation."[121a]

Summary

The NOD mouse has been available to research community for approximately 15 years and in that time tremendous advances have been made in understanding the role of the immune system in development of Type 1A diabetes. It is clear that beta cell destruction is mediated primarily by T cells with both CD4+ and CD8+ T cells playing an important role. It should be kept in mind that development of diabetes in the NOD mouse differs from that in humans in several respects, including a marked sex bias in females and very high concordance (80-90% in NOD female mice) compared with the 30-50% observed in human identical twins. Nevertheless, the NOD mouse provides a very important model for understanding diabetes pathogenesis in humans.

The data obtained thus far from investigation of human subjects are much less compelling with regard to a direct role for T cells in beta cell destruction than data from the NOD mouse. Nevertheless, it appears that T cells are involved and it is probably only a matter of time before advance in techniques will allow a definitive assessment of the participation of T cells in human type I diabetes.

References

1. Makino S. Genetic analysis of IDDM in NOD mice. Exp Anim 1998; 47:suppl-9.
2. Miyazaki A, Hanafusa T, Yamada K et al. Predominance of T lymphocytes in pancreatic islets and spleen of prediabetic nonobese diabetic (NOD) mice: A longitudinal study. Clin Exp Immunol 1985; 60:622-630.
3. Kanazawa Y, Komeda K, Sato S et al. Nonobese-diabetic mice: Immune mechanisms of pancreatic B-cell destruction. Diabetologia 1984; 27:113-115.
4. Wicker LS, Miller BJ, Mullen Y. Transfer of autoimmune diabetes mellitus with splenocytes from nonobese diabetic (NOD) mice. Diabetes 1986; 35(8):855-60.
5. Bendelac A, Boitard C, Bedossa P et al. Adoptive T cell transfer does not require recruitment of host B lymphocytes. J Immunol 1988; 141:2625-2628.
6. Bendelac A, Carnaud C, Boitard C et al. Syngeneic transfer of autoimmune diabetes from diabetic NOD mice to healthy neonates. Requirement for both L3T4+ and Lyt-2+ T cells. J Exp Med 1987; 166:823-32.
7. Haskins K, Portas M, Bergman B et al. Pancreatic islet-specific T cell clones from nonobese diabetic mice. Proc Natl Acad Sci USA 1989; 86:8000-8004.
8. Bergman B, Haskins K. Islet-specific T-cell clones of the NOD mouse respond to B-granule antigen. Diabetes 1994; 43:197-203.
9. Katz JD, Wang B, Haskins K et al. Following a diabetogenic T cell from genesis through pathogenesis. Cell 1993; 74:1089-1100.
10. Hanninen A, Taylor C, Streeter PR et al. Vascular addressins are induced on islet vessels during insulitis in nonobese diabetic mice and are involved in lymphoid cell binding to islet endothelium. J Clin Invest 1993; 92:2509-2515.
11. Linn T, Strate C, Federlin K et al. Intercellular adhesion molecule-1 (ICAM-1) expression in the islets of the nonobese diabetic and low-dose streptozocin-treated mouse. Histochemistry 1994; 102:317-321.
12. Faveeuw C, Gagnerault MC, Lepault F. Expression of homing and adhesion molecules in infiltrated islets of Langerhans and salivary glands of nonobese diabetic mice. J Immunol 1994; 152:5969-5978.
13. Hasegawa Y, Yokono K, Taki T et al. Prevention of autoimmune insulin-dependent diabetes in nonobese diabetic mice by anti-LFA-1 and anti-ICAM-1 mAb. Int Immunol 1994; 6:831-838.
14. Yang XD, Karin N, Tisch R et al. Inhibition of insulitis and prevention of diabetes in nonobese diabetic mice by blocking L-selectin and very late antigen 4 adhesion receptors. Proc Natl Acad Sci ΥΣΑ 1993; 90:10494-10498.
15. Baron JL, Reich EP, Visintin I et al. The pathogenesis of adoptive murine autoimmune diabetes requires an interaction between alpha 4-integrins and vascular cell adhesion molecule-1. J Clin Invest 1994; 93:1700-1708.
16. Wicker LS, Todd JA, Peterson LB. Genetic control of autoimmune diabetes in the NOD mouse. Ann Rev Immunol 1995; 13:179-200.
17. Scott B, Liblau R, Degermann S et al. A role for nonMHC genetic polymorphism in susceptibility to spontaneous autoimmunity. Immunity 1994; 1:73-83.
18. Miller BJ, Appel MC, O'Neil JJ et al. Both the Lyt-2+ and L3T4+ T cell subsets are required for the transfer of diabetes in nonobese diabetic mice. J Immunol 1988; 140:52-58.
19. O'Reilly LA, Hutchings PR, Crocker PR et al. Characterization of pancreatic islet cell infiltrates in NOD mice: Effect of cell transfer and transgene expression. Eur J Immunol 1991; 21:1171-1180.
20. Haskins K, McDuffie M. Acceleration of diabetes in young NOD mice with a CD4+ islet- specific T cell clone. Science 1990; 249:1433-1436.
21. Palmer JP, Lernmark Å. Diabetes Mellitus 1990; 4:414-435.
22. Wang B, Gonzales A, Benoist C et al. The role of CD8+ T cells in the initiation of insulin-dependent diabetes mellitus. Eur J Immunol 1996; 26:1762-1769.
23. Wong FS, Visintin I, Wen L et al. CD8 T cell clones from young nonobese diabetic (NOD) islets can transfer rapid onset of diabetes in NOD mice in the absence of CD4 cells. J Exp Med 1996; 183:67-76.
24. Abbas AK, Murphy KM, Sher A. Functional diversity of helper T lymphocytes. Nature 1996; 383:787-793.
25. Katz JD, Benoist C, Mathis D. T helper cell subsets in insulin-dependent diabetes. Science 1995; 268:1185-1188.

26. Healey D, Ozegbe P, Arden S et al. In vivo activity and in vitro specificity of CD4+ Th1 and Th2 cells derived from the spleens of diabetic NOD mice. J Clin Invest 1995; 95:2979-2985.
27. Liblau RS, Singer SM, McDevitt HO. Th1 and Th2 CD4+ T cells in the pathogenesis of organ-specific autoimmune diseases [see comments]. Immunol Today 1995; 16:34-38.
28. Trembleau S, Germann T, Gately MK et al. The role of IL-12 in the induction of organ-specific autoimmune diseases. Immunol Today 1995; 16:383-386.
29. Trembleau S, Penna G, Bosi E et al. Interleukin 12 administration induces T helper type 1 cells and accelerates autoimmune diabetes in NOD mice. J Exp Med 1995; 181:817-821.
30. Mueller R, Krahl T, Sarvetnick N. Pancreatic expression of interleukin-4 abrogates insulitis and autoimmune diabetes in nonobese diabetic (NOD) mice. J Exp Med 1996; 184:1093-1099.
31. Sarvetnick N, Shizuru J, Liggitt D et al. Loss of pancreatic islet tolerance induced by β-cell expression of interferon-y. Nature 1990; 346:844-847.
32. Campbell IL, Kay TWH, Oxbrow L et al. Essential role for interferon-γ and interleukin-6 in autoimmune insulin-dependent diabetes in NOD/Wehi mice. J Clin Invest 1991; 87:739-742.
33. Rapoport MJ, Jaramillo A, Zipris D et al. Interleukin 4 reverses T cell proliferative unresponsiveness and prevents the onset of diabetes in nonobese diabetic mice. J Exp Med 1993; 178:87-99.
34. Shimada A, Charlton B, Rohane P et al. Immune regulation in type 1 diabetes. J Autoimmun 1996; 9:263-269.
35. Tian J, ClareSalzler M, Herschenfeld A et al. Modulating autoimmune responses to GAD inhibits disease progression and prolongs islet graft survival in diabetes-prone mice. Nat Med 1996; 2:1348-1353.
36. Rabinovitch A. Immunoregulatory and cytokine imbalances in the pathogenesis of IDDM. Therapeutic intervention by immunostimulation? Diabetes 1994; 43:613-621.
37. Cameron MJ, Arreaza GA, Zucker P et al. IL-4 prevents insulitis and insulin-dependent diabetes mellitus in nonobese diabetic mice by potentiation of regulatory T helper-2 cell function. J Immunol 1997; 159:4686-4692.
38. Wang B, Gonzalez A, Höglund P et al. Interleukin-4 deficiency does not exacerbate disease in NOD mice. Diabetes 1998; 47:1207-1211.
39. Wogensen L, Lee M-S, Sarvetnick N. Production of interleukin 10 by islet cells accelerates immune-mediated destruction of β cells in nonobese diabetic mice. J Exp Med 1994; 179:1379-1384.
40. Poulin M, Haskins K. Induction of diabetes in nonobese diabetic mice by Th2 T cell clones from a TCR transgenic mouse. J Immunol 2000; 164:3072-3078.
41. Pakala SV, Kurrer MO, Katz JD. T helper 2 (Th2) T cells induce acute pancreatitis and diabetes in immune-compromised nonobese diabetic (NOD) mice. J Exp Med 1997; 186:299-306.
42. Mason D, Powrie F. Control of immune pathology by regulatory T cells. Curr Opin Immunol 1998; 10:649-655.
43. Boitard C, Yasunami R, Dardenne M et al. T cell-mediated inhibition of the transfer of autoimmune diabetes in NOD mice. J Exp Med 1989; 169:1669-1680.
44. Fowell D, Mason D. Evidence that the T cell repertoire of normal rats contains cells with the potential to cause diabetes. Characterization of the CD4+ T cell subset that inhibits this autoimmune potential. J Exp Med 1993; 177:627-636.
45. Saoudi A, Seddon B, Heath V et al. The physiological role of regulatory T cells in the prevention of autoimmunity: The function of the thymus in the generation of the regulatory T cell subset. Immunol Rev 1996; 149:195-216.
46. Seddon B, Saoudi A, Nicholson M et al. CD4+CD8- thymocytes that express L-selectin protect rats from diabetes upon adoptive transfer. Eur J Immunol 1996; 26:2702-2708.
47. Han H-S, Jun H-S, Utsugi T et al. A new type of CD4+ suppressor T cell completely prevents spontaneous autoimmune diabetes and recurrent diabetes in syngeneic islet-transplanted NOD mice. J Autoimmun 1996; 9:331-339.
48. Hammond KJL, Poulton LD, Palmisano LJ et al. α/β-T cell receptor $(TCR)^+CD4^-CD8^-$ (NKT) thymocytes prevent insulin-dependent diabetes mellitus in nonoberse diabetic (NOD)/Lt mice by the influence of interleukin (IL)-4 and/or IL-10. J Exp Med 1998; 187:1047-1056.
49. Salomon B, Lenschow DJ, Rhee L et al. B7/CD28 costimulation is essential for the homeostasis of the CD4+CD25+ immunoregulatory T cells that control autoimmune diabetes. Immunity 2000; 12:431-440.
50. Kurrer MO, Pakala SV, Hanson HL et al. Beta cell apoptosis in T cell-mediated autoimmune diabetes. Proc Natl Acad Sci USA 1997; 94:213-218.
51. O'Brien BA, Harmon BV, Cameron DP et al. Apoptosis is the mode of beta-cell death responsible for the development of IDDM in the nonobese diabetic (NOD) mouse. Diabetes 1997; 46:750-757.
52. O'Brien BA, Harmon BV, Cameron DP et al. Beta-cell apoptosis is responsible for the development of IDDM in the multiple low-dose streptozotocin model. J Pathol 1996; 178:176-181.

53. Scaglia L, Smith FE, Bonner-Weir S. Apoptosis contributes to the involution of beta cell mass in the post partum rat pancreas. Endocrinology 1995; 136:5461-5468.
54. Chervonsky AV, Wang Y, Wong FS et al. The role of Fas in autoimmune diabetes. Cell 1997; 89:17-24.
55. Itoh N, Imagawa A, Hanafusa T et al. Requirement of Fas for the development of autoimmune diabetes in nonobese diabetic mice. J Exp Med 1997; 186:613-618.
56. Kagi D, Odermatt B, Seiler P et al. Reduced incidence and delayed onset of diabetes in perforin-deficient nonobese diabetic mice. J Exp Med 1997; 186:989-997.
57. Amrani A, Verdaguer J, Anderson B et al. Perforin-independent β-cell destruction by diabetogenic $CD8^+$ T lymphocytes in transgenic nonobese diabetic mice. J Clin Invest 1999; 103:1201-1209.
58. Pakala SV, Chivetta M, Kelly CB et al. In autoimmune diabetes the transition from benign to pernicious insulitis requires an islet cell response to tumor necrosis factor alpha. J Exp Med 1999; 189:1053-1062.
59. Shultz LD, Schweitzer PA, Christianson SW et al. Multiple defects in innate and adaptive immunologic function in NOD/LtSz-scid mice. J Immunol 1995; 154:180-191.
60. Suri A, Katz JD. Dissecting the role of CD4+ T cells in autoimmune diabetes through the use of TCR transgenic mice. Immunol Rev 1999; 169:55-65.
60a. Lieberman SM, Evans AM, Han B et al. Identification of the βcw antigen targeted by a prevalent population of pathogenic CD8+ T cells in autoimmune diabetes. Proc Natl Acad Sci USA 2003; 8384-8388.
61. Smith KM, Olson DC, Hirose R et al. Pancreatic gene expression in rare cells of thymic medulla: Evidence for functional contribution to T cell tolerance. Int Immunol 1997; 9:1355-1365.
62. Vafiadis P, Bennett ST, Todd JA et al. Insulin expression in human thymus is modulated by INS VNTR alleles at the IDDM2 locus. Nat Genet 1997; 15:289-292.
63. Pugliese A, Zeller M, Fernandez A et al. The insulin gene is transcribed in the human thymus and transcription levels correlate with allelic variation at the INS VNTR-IDDM2 susceptibility locus for type I diabetes. Nat Genet 1997; 15:293-297.
64. Heath VL, Moore NC, Parnell SM et al. Intrathymic expression of genes involved in organ specific autoimmune disease. J Autoimmun 1998; 11:309-318.
65. Throsby M, Homo-Delarche F, Chevenne D et al. Pancreatic hormone expression in the murine thymus: Localization in dendritic cells and macrophages. Endocrinology 1998; 139:2399-2406.
66. Sospedra M, Ferrer-Francesch X, Dominguez O et al. Transcription of a broad range of self-antigens in human thymus suggests a role for central mechanisms in tolerance toward peripheral antigens. J Immunol 1998; 161:5918-5929.
67. Daniel D, Gill RG, Schloot N et al. Epitope specificity, cytokine production profile and diabetogenic activity of insulin-specific T cell clones isolated from NOD mice. Eur J Immunol 1995; 25:1056-1062.
68. Daniel D, Wegmann DR. Protection of nonobese diabetic mice from diabetes by intranasal or subcutaneous administration of insulin peptide B-(9-23). Proc Natl Acad Sci USA 1996; 93:956-960.
69. Wong FS, Karttunen J, Dumont C et al. Identification of an MHC class I-restricted autoantigen in type 1 diabetes by screening an organ-specific cDNA library. Nat Med 1999; 5:1026-1031.
70. Zhang ZJ, Davidson L, Eisenbarth G et al. Suppression of diabetes in nonobese diabetic mice by oral administration of porcine insulin. Proc Natl Acad Sci USA 1991; 88:10252-10256.
71. Hutchings P, Cooke A. Protection from insulin-dependent diabetes mellitus affected by insulin antigens in incomplete Freund's adjuvant depends on route of administration. J Autoimmun 1998; 11:127-130.
72. Harrison LC, Dempsey-Collier M, Kramer DR et al. Aerosol insulin induces regulatory CD8 gamma delta T cells that prevent murine insulin-dependent diabetes. J Exp Med 1996; 184:2167-2174.
72a. Moriyama H, Abiru N, Paronen J et al. Evidence for a primary islet autoantigen (preproinsulin) for insulitis and diabetes in the NOD mouse. Proc Natl Acad Sci USA 2003; 100(18):10376-81.
73. Karounos DG, Bryson JS, Colhen DA. Metabolically inactive insulin analog prevents type I diabetes in prediabetic NOD mice. J Clin Invest 1997; 100:1344-1348.
74. Zekzer D, Wong FS, Wen L et al. Inhibition of diabetes by an insulin-reactive CD4 T-cell clone in the nonobese diabetic mouse. Diabetes 1997; 46:1124-1132.
75. Kim J, Richter W, Aanstoot HJ et al. Differential expression of GAD65 and GAD67 in human, rat and mouse pancreatic islets. Diabetes 1993; 42:1799-1808.
76. Hagopian WA, Karlsen AE, Gottsater A et al. Quantitative assay using recombinant human islet glutamic acid decarboxylase (GAD65) shows that 64K autoantibody positivity at onset predicts diabetes type. J Clin Invest 1993; 91:368-374.
77. Kaufman DL, ClareSalzler M, Tian J et al. Spontaneous loss of T-cell tolerance to glutamic acid decarboxylase in murine insulin-dependent diabetes [see comments]. Nature 1993; 366:69-72.

78. Tisch R, Yang X-D, Singer SM et al. Immune response to glutamic acid decarboxylase correlates with insulitis in nonobese diabetic mice. Nature 1993; 366:72-75.
79. Cetkovic-Cvrlje M, Gerling IC, Muir A et al. Retardation or acceleration of diabetes in NOD/Lt mice mediated by intrathymic administration of candidate β-cell antigens. Diabetes 1997; 46:1975-1982.
80. Tian J, Atkinson M, ClareSalzer M et al. Nasal administration of glutamate decarboxylase (GAD65) peptides induces Th2 responses and prevents murine insulin dependent diabetes. J Exp Med 1996; 183:1561-1567.
81. Zekzer D, Wong FS, Ayalon O et al. GAD-reactive CD4+ Th1 cells induce diabetes in NOD/SCID mice. J Clin Invest 1998; 101:68-73.
82. Panina-Bordignon P, Lang R, Van Endert PM et al. Cytotoxic T cells specific for glutamic acid decarboxylase in autoimmune diabetes. J Exp Med 1995; 181:1923-1927.
83. Geng L, Solimena M, Flavell RA et al. Widespread expression of an autoantigen-GAD65 transgene does not tolerize nonobese diabetic mice and can exacerbate disease. Proc Natl Acad Sci USA 1998; 95:10055-10060.
84. French MB, Allison J, Cram DS et al. Transgenic expression of mouse proinsulin II prevents diabetes in nonobese diabetic mice. Diabetes 1996; 46:34-39.
85. Bridgett M, Cetkovic-Cvrlje M, O'Rourke R et al. Differential protection in two transgenic lines of NOD/Lt mice hyperexpressing the autoantigen GAD65 in pancreatic beta-cells. Diabetes 1998; 47:1848-1856.
86. Yoon JW, Yoon CS, Lim HW et al. Control of autoimmune diabetes in NOD mice by GAD expression or suppression in beta cells. Science 1999; 284:1183-1187.
87. von Boehmer H, Sarukhan A. GAD, a single autoantigen for diabetes [comment]. Science 1999; 284:1135-1137.
88. Atkinson M, Serreze D, Tian J et al. Control of Autoimmune Diabetes. Science 1 A.D.; 287:191a.
89. Solimena M, De Camilli P. From Th1 to Th2: Diabetes immunotherapy shifts gears. Nat Med 1996; 2:1311-1312.
90. Wasmeier C, Hutton JC. Molecular cloning of phogrin, a protein-tyrosine phosphatase homologue localized to insulin secretory granule membranes. J Biol Chem 1996; 271:18161-18170.
91. Hawkes CJ, Wasmeier C, Christie MR et al. Identification of the 37-kDa antigen in IDDM as a tyrosine phosphatase-like protein (phogrin) related to IA-2. Diabetes 1996; 45:1187-1192.
92. Payton MA, Hawkes CJ, Christie MR. Relationship of the 37,000- and 40,000-M_r tryptic fragments of islet antigens in insulin-dependent diabetes to the protein tyrosine phosphatase-like molecule IA-2 (ICA512). J Clin Invest 1995; 96:1506-1511.
93. Rabin DU, Pleasic SM, Palmer-Crocker R et al. Cloning and expression of IDDM-specific human autoantigens. Diabetes 1992; 41:183-186.
94. Lan MS, Lu J, Goto Y et al. Molecular cloning and identification of a receptor-type protein tyrosine phosphatase, IA-2, from human insulinoma. DNA Cell Biol 1994; 13:505-514.
95. Lu J, Li Q, Xie H et al. Identification of a second transmembrane protein tyrosine phosphatase, IA-2β, as an autoantigen in insulin-dependent diabetes mellitus: Precursor of the 37-kDa tryptic fragment. Proc Natl Acad Sci USA 1996; 93:2307-2311.
96. Rabin DU, Pleasic SM, Shapiro JA et al. Islet cell antigen 512 is a diabetes-specific islet autoantigen related to protein tyrosine phosphatases. J Immunol 1994; 152:3183-3188.
97. Myers MA, Laks MR, Feeney SJ et al. Antibodies to ICA512/IA-2 in rodent models of IDDM. J Autoimmun 1998; 11:265-272.
98. Kelemen K, Crawford ML, Gill RG et al. Cellular immune response to phogrin in the NOD mouse: Cloned T-cells cause destruction of islet transplants. Diabetes 1999; 48:1529-1534.
99. Gepts W. Pathologic anatomy of the pancreas in juvenile diabetes mellitus. Diabetes 1965; 14:619-633.
100. Gepts W, LeCompte PM. The pancreatic islets in diabetes. Am J Med 1981; 70:105-115.
101. Bottazzo GF, Dean BM, McNally JM et al. In situ characterization of autoimmune phenomena and expression of HLA molecules in the pancreas in diabetic insulitis. N Engl J Med 1985; 313:353-360.
102. Van Vliet E, Roep BO, Meulenbroek L et al. Human T cell clones with specificity for insulinoma cell antigens. Eur J Immunol 1989; 19:213-216.
103. Naquet P, Ellis J, Tibensky D et al. T cell autoreactivity to insulin in diabetic and related nondiabetic individuals. J Immunol 1988; 140:2569-2578.
104. Roep BO, Arden SD, deVries RP et al. T-cell clones from a type-1 diabetes patient respond to insulin secretory granule proteins. Nature 1990; 345:632-634.
105. Roep BO, Kallan AA, Hazenbos WLW et al. T-cell reactivity to 38kD insulin-secretory-granule protein in patients with recent-onset type 1 diabetes. Lancet 1991; 337:1439-1441.

106. Atkinson MA, Kaufman DL, Campbell L et al. Response of peripheral-blood mononuclear cells to glutamate decarboxylase in insulin-dependent diabetes. Lancet 1992; 339:458-459.
107. Honeyman MC, Cram DS, Harrison LC. Glutamic acid decarboxylase 67-reactive T cells: A marker of insulin- dependent diabetes. J Exp Med 1993; 177:535-540.
108. Harrison LC, De Aizpurua HJ, Loudovaris T et al. Reactivity to human islets and fetal pig proislets by peripheral blood mononuclear cells from subjects with preclinical and clinical insulin-dependent diabetes. Diabetes 1991; 40:1128-1133.
109. Durinovic-Bello I, Hummel M, Ziegler AG. Cellular immune response to diverse islet cell antigens in IDDM. Diabetes 1996; 45:795-800.
110. Hawkes CJ, Schloot NC, Marks JB et al. T-Cell Lines Reactive to an Immunodominant Epitope of the Tyrosine Phosphatase-Like Autoantigen IA-2 in Type 1 Diabetes. Diabetes 2000; 49:356-366.
111. Klemetti P, Hyoty H, Roivainen M et al. Relation between T-cell responses to glutamate decarboxylase and coxsackievirus B4 in patients with insulin-dependent diabetes mellitus. J Clin Virol 1999; 14:95-105.
112. Ou D, Mitchell LA, Metzger DL et al. Cross-reactive rubella virus and glutamic acid decarboxylase (65 and 67) protein determinants recognised by T cells of patients with type I diabetes mellitus. Diabetologia 2000; 43:750-762.
113. Honeyman MC, Coulson BS, Stone NL et al. Association between rotavirus infection and pancreatic islet autoimmunity in children at risk of developing type 1 diabetes. Diabetes 2000; 49:1319-1324.
114. Zhang J, Markovic-Plese S, Lacet B et al. Increased frequency of interleukin 2-responsive T cells specific for myelin basic protein and proteolipid protein in peripheral blood and cerebrospinal fluid of patients with multiple sclerosis. J Exp Med 1994; 179:973-984.
115. Lohmann D, Krug J, Lampeter EF et al. Cell-mediated immune reactions against B cells and defect of suppressor cell activity in type 1 (insulin-dependent) diabetes mellitus. Diabetologia 1986; 29:421-425.
116. Quiniou-Debrie MC, Debray-Sachs M, Dardenne M et al. Anti-islet cellular and humoral immunity, T-cell subsets, and thymic function in type I diabetes. Diabetes 1985; 34:373-379.
117. Phipps RP, Stein SH, Roper RL. A new view of prostaglandin E regulation of the immune response. Immunol Today 1991; 12:349-352.
118. Peterson LD, van der KM, de Vries RR et al. Autoreactive and immunoregulatory T-cell subsets in insulin-dependent diabetes mellitus. Diabetologia 1999; 42:443-449.
119. Roep BO, Atkinson MA, Van Endert PM et al. Autoreactive T cell Responses in Insulin-dependent (Type 1) Diabetes Mellitus. Report of the first international workshop for standardization of T cell assays. J Autoimmun 1999; 13:267-282.
120. Nepom GT, Lippolis JD, White FM et al. Identification and modulation of a naturally processed T cell epitope from the diabetes-associated autoantigen human glutamic acid decarboxylase 65 (hGAD65). Proc Natl Acad Sci USA 2001; 98:1763-1768.
121. Amrani A, Verdaguer J, Serra P et al. Progression of autoimmune diabetes driven by avidity maturation of a T-cell population. Nature 2000; 406:739-742.
121a. Trudeau JD, Kelly-Smith C, Verchere CB et al. Prediction of spontaneous autoimmune diabetes in NOD mice by quantification of autoreactive T cells in peripheral blood. J Clin Invest 2003; 111:217-223.

CHAPTER 5

Type 1 Database Mellitus:
An Inflammatory Disease of the Islet

Regine Bergholdt, Peter Heding, Karin Nielsen, Runa Nolsøe, Thomas Sparre, Joachim Størling, Jørn Nerup, Flemming Pociot and Thomas Mandrup-Poulsen

Type 1 diabetes mellitus (T1DM) is generally considered to be a classical autoimmune disease, in which immune tolerance is broken by environmental factors in genetically predisposed individuals, leading to recognition of specific β-cell antigens by helper T-cells that subsequently activate effector T-cells, and these effector T-cells then directly mediate β-cell killing. However, as reviewed in the previous chapters, the cellular and molecular pathogenesis of Type 1 diabetes is more complex.

Although there seems to be broad agreement that autoimmunity is defined as termination of a natural unresponsive state to "self", exhaustive and unifying consensus on the definition and classification of autoimmune disease has been difficult to obtain. According to the classical papers by Witebsky,[1,2] definition of a disease as an autoimmune state depends on

1. detectable circulating or cell-bound antibodies reactive with an autoantigen,
2. identification of the autoantigen
3. presence of mononuclear cell inflammation in the target tissue and
4. ability to transfer disease with lymphoid cells or with serum

In type 1 diabetes, there is still uncertainty regarding the nature of the autoantigen(s), passive transfer of the disease by the use of autoantibodies, e.g., by transplacental transfer from diabetic mother to nondiabetic fetus has not been demonstrated, and the transfer of disease by lymphoid cells from man to experimental animal has not been reproducibly demonstrated.

As discussed in the above chapters, and reviewed in reference 3 there is still uncertainty as to the importance of classical T-cell effector mechanisms, such as the Fas/FasL, perforin/granzyme and membrane-bound TNF/TNF-related apoptosis inducing ligand (TRAIL). Further, recent studies in the NOD mouse have clearly demonstrated that although β2-microglubulin^{null} and thereby MHC class I and CD8+ T-cell-deficient mice do not develop diabetes,[4] later studies have shown that class I restricted T-cells are needed for the initiation but not the later effector phase.[5]

Taken together, these observations indicate that β-cell destruction in Type 1 diabetes cannot yet be definitively classified as an autoimmune disease, but certainly as an immune-mediated disease. This is most clearly demonstrated by the ability of immunosuppressants to prevent destruction of residual β-cell function in recent onset Type 1 diabetic patients.[6]

A common feature shared by immune-mediated diseases is the build-up of an inflammatory infiltrate in the target organ. There is increasing evidence from other immune-mediated diseases such as inflammatory bowel disease, multiple sclerosis and rheumatoid arthritis that many components of the inflammatory response, including macrophages, CD4+ and CD8+ T-cells, inflammatory mediators in the form of cytokines and free oxygen and nitrite oxide radicals, prostaglandins, complement factors, etc. all contribute to tissue destruction. Type 1 diabetes

Immunology of Type 1 Diabetes, Second Edition, edited by George S. Eisenbarth.

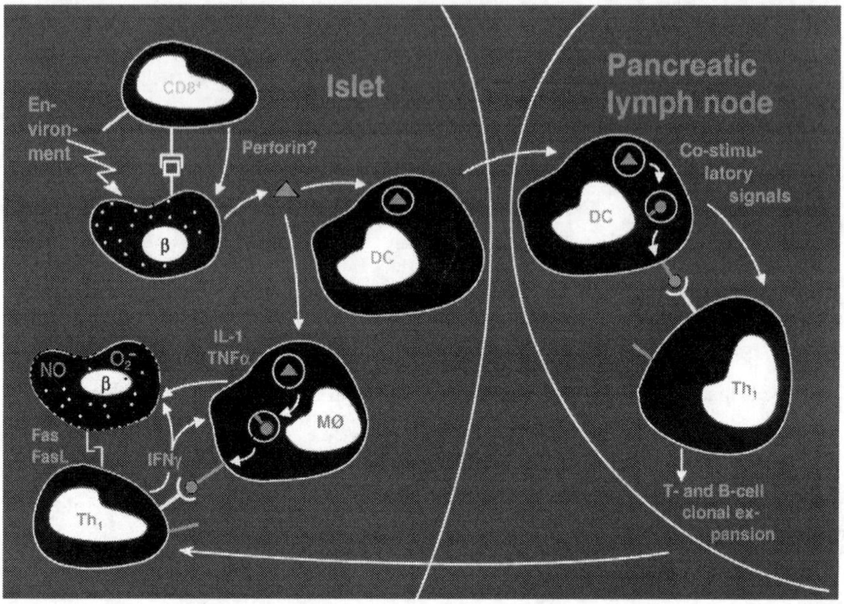

Figure 1. The inflammatory model for the pathogenesis of Type 1 diabetes.

does not seem to be exceptional in this regard. In the past 15 years, in vitro and animal studies have underlined the importance of inflammatory mediators.[3,7-9] Thus, combinations of the proinflammatory cytokines IL-1β, IFNγ, TNFα and IL-6 are synergistically cytotoxic to β-cells, in rodent islets by inducing a mixture of β-cell necrosis and apoptosis, in human islets mainly by inducing β-cell apoptosis. These proinflammatory cytokines are present early in the inflammatory infiltrate in animals models with Type 1 diabetes, and antagonists of proinflammatory cytokines prevent diabetes development in such models (reviewed in ref. 3). There is an extensive literature on the effect of free oxygen and nitric oxide radicals elaborated either by infiltrating immune cells or as a result of cytokine-induced β-cell specific expression of enzymes generating these radicals (inducible nitric oxide synthase).

Based on the above observation(s), we suggest an inflammatory model for the pathogenesis of Type 1 diabetes as depicted in Figure 1. The model suggests that environmental factors, most likely in the form of common viruses, induce a MHC Class I restricted presentation of β-cell antigen. This antigen is recognized by CD8+ T-cells that cause a limited MHC Class I restricted β-cell damage, either via cytotoxic cytokines as secreted IFNγ and/or secreted TNFα/membrane-bound TNFα/TRAIL or the perforin/granzyme system. Liberated β-cell components, such as insulin or GAD possibly in glycosylated immature forms not previously "seen" by the immune system are taken up by dendritic cells in the islets and transported to regional pancreatic lymph nodes, where the antigens are processed and presented to CD4+ T-cells. After clonal expansion the CD4+ T-cells will home to the islets, tracing antigen gradients, chemokines and endothelial adhesion molecules induced by the early CD8+ T-cell mediated inflammatory response. The activated CD4+ T-cells will recruit and activate specific as well as nonspecific inflammatory cells that then build up the inflammatory insulitis infiltrate. The effector phase of the β-cell destruction is mediated by cytokines via induction of proapoptotic signaling selectively in β-cells and/or by inducing β-cell expression of Fas, marking the β-cells for MHC Class II nonrestricted CD4+ T-cell mediated killing via interaction between the Fas ligand on CD4+ T-cells and Fas on the β-cells.

There are many unanswered questions in this model: How and why is tolerance broken to the β-cell? What is the nature of the antigens recognized by the CD8+ T-cell in the initiating phase? Are these antigens different or identical to those that cause activation of the CD4+ T-cell system, and how do these components become antigenic? What regulatory mechanisms normally suppress the CD4+ T-cell response and how is this suppression lost? Is the role of the CD4+ T-cell mainly to orchestrate the immunological response in the islet or does the Th_1-cell have specific effector functions? What is the relative importance of cytokines and the Fas/FasL system as effector mechanisms? What are the specific cellular signaling mechanisms leading to necrosis/apoptosis of the β-cell? What is the relative role of nitric oxide dependent and nitric oxide independent intracellular toxic effectors? To which extent can this model be extrapolated to human Type 1 diabetes? These questions are important to clarify for our future possibilities to provide rational pharmacological intervention.

The inflammatory model of Type 1 diabetes has been extensively reviewed.[3,7,9] The purpose of this chapter is therefore to focus on three areas of the inflammatory model depicted in Figure 1, where we believe particular advances have been produced in the past years:

1. studies of candidate genes of susceptibility that control individual components of the inflammatory process and/or β-cell response to mediators of this process
2. the use of expression profiling by DNA chip microarray or proteomic analysis to identify deleterious and protective genes induced in the β-cell in the inflammatory process and finally
3. studies on how inflammatory mediators signal β-cell necrosis/apoptosis.

Genetic Aspects of the Inflammatory Model for the Pathogenesis of Type 1 Diabetes

As detailed in Chapter 7 and reviewed in reference 10 Type 1 diabetes is considered a complex genetic disease where several genes together with unknown environmental factors determine disease risk.

What is the genetic evidence supporting the inflammatory model for the pathogenesis of Type 1 diabetes? As there is considerable familial clustering of Type 1 diabetes as evidenced by lambda – s (risk for siblings relative to population risk) of 15, it can be suspected that several of the events outlined in the model are under genetic control. From published genome scans in Type 1 diabetes[11-15] and as reviewed in Chapter 7, it is evident that the HLA locus, *IDDM1*, is the major genetic susceptibility component to Type 1 diabetes, though several minor loci have been identified. Here we will focus only on the contribution of polymorphisms in candidate genes presumed to directly influence the pathogenetic process as shown in the inflammatory model (Fig. 1). These candidate genes include cytokine genes, genes involved in T cell regulation and inflammation and genes determining protective and deleterious mechanisms in the β-cell. Several, but not all, are found in chromosomal regions linked to Type 1 diabetes. In Table 1, the various candidate gene polymorphisms are listed, as well as their chromosomal position, whether association to Type 1 diabetes has been demonstrated and at what level in the model of the pathogenesis they might be involved.

The *HLA* locus accounts for 40-50% of the genetic susceptibility to Type 1 diabetes.[16,17] The *HLA-DQ8* and *HLA-DQ2* alleles in the class II region are considered to be the major genetic risk factors and seem to affect class II mediated antigen presentation in the inflammation in the islets, as well as the role of HLA-DQ restriction in the development of central and peripheral tolerance.[18,19]

The cytokine genes are obvious candidate genes as they are effector molecules in the inflammatory model of the pathogenesis of Type 1 diabetes. In most of the analyzed cytokine genes polymorphisms have been demonstrated, some revealing Type 1 diabetes association, and for some of them it has even been shown that they correlate to the expression level of the protein, suggesting a functional significance, see Table 1.

Table 1. ***List of polymorphisms of candidate genes in the model for the pathogenesis of Type 1 diabetes and presumed level of action in the model. * TDT: transmission disequilibrium test***

Gene	Position	Polymorphism	Association Significance		Functional	Possible Target	References
			Case-control	**TDT***			
HLA	6p21	Alleles of DQA1, DQB1, DRB1, DPB1	Yes	Yes	Influences conformation of Ag-binding groove and thereby Ag-presentation.	Ag-presenting cell	Reviewed in Chapter 7, (10)
Cytokines:							
IL1B	2q12-q22	Exon 5, T to C	Yes	No	Allele-dosage effect on LPS stimulation	Effector molecule, acting on β-cells, costimulatory cytokine for T-cells, macrophages	(20)
		-511, C to T	No		on IL-1 secretion		
IL1RN	2q12-q22	Intron 2, 86bp repeat	Yes	No	2-allele: Increased expression 1/1: reduced expression		(21,34,35)
IL1RI	2q12-q22	PstI-a (5´UTR) PstI-b (5´UTR) AluI (5´UTR)	Yes	No No No			(22,23,132)
		HinfI (5´UTR)		Yes	Allele-dosage effect		(24)
IFNG	12q14	Intron 1, CA-repeat	Yes No	 No	2-allele: increased in vitro expression	Effector molecule and potentiation of IL-1β action on β-cells.	(133-136)

Table 1. Continued

Gene	Position	Polymorphism	Association Significance		Functional	Possible Target	References
			Case-control	**TDT***			
TNFA	6p21	Several promoter polymorphisms	Yes/No		One of the polymorphisms: different expression of different genotypes, but in linkage disequilibrium with HLA cl.II	Effector molecule and potentiation of IL-1β action on β-cells.	(137,138)
IL12B	5q31.1-q33.1	3´UTR (TaqI-RFLP)	Yes		1-allele: Increased express.	Influence on T-cell function	(36)
				No	No functional significance		(37)
IL6	7p21	-174, G/C	Yes	No	Allele dosage effect	Macrophages, β-cell.	(38)
T-cells:							
CTLA4	2q33	3´UTR, (AT)n		Yes	Allele dosage effect	Downregulation of T-cell function and regulation of immune responses	Reviewed in Chapter 7, (139)
		Exon 1, A49G		Yes	Thr to Ala in leader peptide		
				No	Allele dosage effect		
		(-318, C to T)		No	Allele dosage effect		
CD4	12p12	-1188(CTTTT) 4 –15 repeats	Yes	Yes	Allele dosage effect	Early phase of T-cell activation and clonal expansion	(140,141)

Table continued on next page

Table 1. Continued

Gene	Position	Polymorphism	Association Significance		Functional	Possible Target	References
			Case-control	TDT*			
Fas	10q24.1	15 mutations		No		T-cell interaction, regulation of immune responses	(41)
FasL	1q23	3 mutations		No		T-cell interaction	(42)
β-cell:							
INS	11p15.5	-596, 5´UTR VNTR	Yes	Yes	Different classes: different INS transcription in pancreas and thymus	Autoantigen / Shaping of T-cell repertoire in thymus	Reviewed in Chapter 7 (142)
NOS2	17q11.2	3 promoter-polymorphisms	No	No		NO-product ion in β-cells and macrophages	(143)
		10 in coding exons, 9 of them: Exon 16 C150T:	No	No Yes	Ser(608)Leu		(49)
NFKB	4q24	CA repeat in regulatory region	Yes			NF-κB production in β-cells	(52)
BCL2	18q21.3	Ala43Thr	Yes	No	Increased sensitivity to apoptosis	β-cells, T-cells	(53,54)
SOD2	6q21	TaqI RFLP	Yes			Free radical scavenger in β-cells	(45)

Specific alleles of the interleukin-1β gene, *IL1B*, the interleukin-1 receptor antagonist gene, *IL1RN* and the interleukin-1 receptor type 1 gene, *IL1RI*, polymorphisms are significantly associated with Type 1 diabetes,[20-24] and it has been examined whether disease-associated alleles have quantitative consequences for the synthesis and secretion of the peptides. Assessment of composite haplotypes of *IL1A, IL1B* and *IL1RI* seems to give the best prediction.[25-28] This approach has been used in studies of peridontitis[29-32] and gastric cancer.[33] For *IL1B* +3954C>T polymorphism a clear allele-dosage effect of the Type 1 diabetes associated allele has been demonstrated in LPS stimulated monocytes, with higher IL-1β secretion when this allele is present, and highest when present in duplicate.[20] Similarly, Type 1 diabetic patients homozygous for the disease-associated allele of the *IL1RN* gene had lower circulating levels of IL-1RA,[21] whereas increased IL-1RA levels have been found for the 2-allele.[34,35] In the *IL1RI* gene, a polymorphism in the 5′UTR showed linkage to Type 1 diabetes and significant differences in IL-1RI plasma-levels correlated to genotype.[24] These studies suggest genetic regulation of monokine antagonism.

Functional effects of some of the variants of the interferon-γ gene, *IFNG*, and the tumor necrosis factor-α gene, *TNFA*, have also been demonstrated, see Table 1, further supporting the genetic control of specific elements of the pathogenic model. The significance of another cytokine gene, the *IL12B* gene is not yet clear. Association has been shown in some populations[36] but not in other,[37] suggesting genetic heterogeneity. Apart from *IL12B*, other cytokine genes, such as *IL6*, can exert their effect at the level of the β-cell as well as at the level of T cell regulation. The G allele of the *IL6* promoter polymorphism at position -174 is associated with an increased risk to Type 1 diabetes,[38] and with an increased promoter activity as well as higher plasma IL-6 levels.[39]

The major role of CD4 and CTLA-4 lies in the activation of CD4+ T cells upon antigen stimulation and in the following clonal expansion of T lymphocytes. The G49 allele of exon 1 of *CTLA4*, which is significantly more frequently transmitted to diabetic offspring, leads to less cell surface CTLA-4 expression as well as reduced mRNA expression and is associated with reduced control of T cell proliferation.[40] From this, one can hypothesize that the combined effect of the diabetes associated polymorphisms in *CD4* and *CTLA4* might lead to a more pronounced T cell activation and clonal expansion.

FasL mediated apoptosis is important in maintaining peripheral self-tolerance. However, *FAS* and *FASL* polymorphisms and haplotypes are not associated with diabetes.[41,42] Fas and FasL may however still play roles in the pathogenesis of Type 1 diabetes, albeit this is not under genetic control.

The insulin gene, *INS*, which is expressed specifically in the β-cell and in the thymus, is an early detectable auto antigen in Type 1 diabetes. The class I alleles of the *INS VNTR*, which confers genetic risk to diabetes, leads to lower insulin expression in the thymus as well as higher insulin expression in the β-cell compared to the dominant protective class III alleles. This may attenuate the development of central tolerance to insulin whilst providing high antigen expression in the β-cell.[43] Certain class III alleles, which silence thymic *INS* expression, also confer genetic predisposition to Type 1 diabetes.[44] There is evidence for interaction between the *INS* and *HLA* loci in conferring susceptibility to Type 1 diabetes.[15]

Free oxygen radicals play a role in the immune mediated β-cell destruction. *SOD2* and *NOS2*, whose gene products are the O_2^- scavenger manganese superoxide dismutase (MnSOD) and inducible NO synthetase (iNOS) respectively, are important candidate genes in this regard. Polymorphisms in the MnSOD gene, *SOD2*, are associated with susceptibility to Type 1 diabetes.[45] Cytokine induced β-cell destruction correlates with MnSOD activity.[46] Variants of MnSOD with reduced activity have been described[47] and these might enhance the sensitivity to cytokine mediated beta cell destruction and thereby the susceptibility to Type 1 diabetes. The reduced activity of MnSOD might be under genetic control – although not shown. Furthermore, upregulation of MnSOD protects islet cells from IL-1 induced damage.[48]

A Ser608Leu polymorphism has bee identified in *NOS2*.[49] The Leu608-allele was preferentially transmitted to HLA DR3/DR4 positive diabetic offspring indicating an interaction between the *HLA* locus and *NOS2*.[49] The polymorphism is located in close proximity to an autoinhibitory domain of the molecule and it is possible that it might affect NO production. Furthermore the *A14* allele of a *NOS2* promoter polymorphism has in two studies been shown to exert protection against severe diabetic retinopathy[50] and diabetic nephropathy,[51] respectively. Interestingly, the protective allele has in a promoter activity assay been shown to exhibit increased promoter activity.[50]

Key regulatory elements in apoptotic as well as signal transduction pathways might be under genetic control in the pathogenesis of Type 1 diabetes. Polymorphisms in genes encoding central signaling molecules in the apoptotic pathways might play a role both on the level of the T-cell as well as on the level of the β-cell, while gene polymorphisms in signal transduction pathways are mainly thought to exert their effects in the β-cell itself possibly affecting the magnitude and direction of the response of the β-cell to cytokines. Examples as such disease associated polymorphisms are reported in nuclear factor kappa beta (NF-κB)[52] and in the antiapoptotic molecule Bcl2.[53,54]

As can be seen from Table 1 and summarized above, several of the polymorphisms genetically associated with Type 1 diabetes are either affecting cytokine/monokine production, the regulation of the central as well as the peripheral immune tolerance and the character of the immune response, or affecting β-cell defense mechanisms. The combined effect of many these polymorphisms supports the inflammatory model for the pathogenesis of Type 1 diabetes in that they lead to high cytokine/monokine production, a T cell response skewed towards a Th1 response and poor β-cell defense mechanisms against damage by free radicals.

It has been speculated that a balance between protective and deleterious mechanisms in and around the β-cells exist.[55] There is evidence that at least some of the mechanisms involved in this balance are under genetic control. In a mathematical model of the onset of Type 1 diabetes,[56] it has been proposed that the process consist of separated elements, each with a quantitative effect and that they can each be speed-limiting in the process, suggesting that Type 1 diabetes can be viewed as a dynamical process, in which an unstable condition has arisen. A combination of multiple events in different cascades and signaling pathways, all pushing in the same direction leading to Type 1 diabetes, is therefore likely (Fig. 2). The key regulatory elements of these events are not known in details and the genetic basis for them therefore also unknown. Many of the "key molecules" identified by transcriptome and proteome analyses, as discussed below, have yet to be analyzed in this context. The combination may differ between ethnic groups, between families or between individuals, hence specific candidate genes or regions are very difficult to identify. Even though association might be difficult to show, determination of a functional significance of a gene in such a cascade is still possible. Knowledge about the "functional genomics" of the genes in the cascades, that is the genes involved in β-cell destruction, is therefore still very important.

Expression Profiling Studies of Genes and Proteins Regulated by Inflammatory Mediators

Cytokine-mediated β-cell destruction requires de novo protein synthesis,[57-61] but the signals initiating gene transcription and protein synthesis and the patterns of these are incompletely understood. Over the last few years new technologies have made it possible to study expression profiles of both proteins and mRNA transcripts. By using microarrays and genechip arrays it is possible to display several thousands of mRNAs or Expressed Sequence Tags (EST) at the same time from cells and tissue and compare different conditions by analyzing the changes in the expression level. It is well known that not all mRNAs present in a cell are translated into proteins and that mRNA encodes only for unmodified preforms of the proteins. Therefore active proteins and posttranslational modified proteins are not necessarily detected by the array (transcriptome) analysis. In contrast, proteome analysis defined as the analysis of the entire

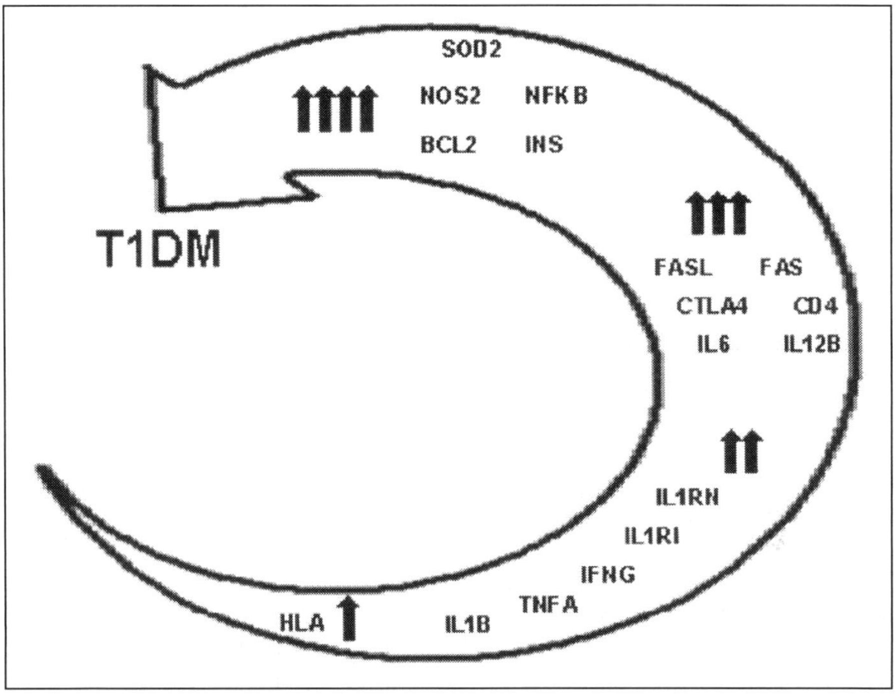

Figure 2. Schematical outline of possible cascade-like events in the dynamical process leading to Type 1 diabetes.

protein complement expressed by the genome of an organism, or in multicellular organisms, as the protein complement expressed by a tissue or a differentiated cell,[62] has been used to asses a global image of the protein expression pattern and changes therein. Analyses at the protein level are important, since the gene-transcripts, mRNAs are by themselves not able to catalyze cellular processes. They have to be translated into proteins, which then can work as building-stones of cellular organelles, enzymes, signal molecules or defense molecules etc. Some proteins work together in kinase cascades, where changes in the expression level of a protein may initiate changes in several other proteins.

Both transcriptome and proteome analyses have been used to describe the molecular events taking place in purified rat β-cells and islets exposed to cytokines involved in the development of T1DM. When β-cells and islets are exposed to cytokines multiple changes in the expression profiles of mRNA and proteins are seen.[55,63-68] These changes are illustrated by up- and down-regulation of the expression levels as well as de novo synthesis of both mRNA and proteins. We have previously demonstrated at the protein level that IL-1β induces perturbation in protein patterns (PIPP),[60,68] where PIPP are the total number of up- and down-regulation of protein expression level and/or gel-position shift resulting from post-translatory modifications. In rat insulinoma (RIN) cells exposed to IL-1β, 149 out of 2,660 detectable transcripts changed in expression level with (±) 3-fold.[65] Similar results were obtained in purified Wistar rat β-cells exposed to IL-1β alone or in combination with IFNγ where 162 out of 3,000 detectable transcripts changed expression level at least 3-fold.[66] At the protein level in Wistar Furth rat islets exposed to IL-1β, 105 out of 1,900 detectable proteins changed significantly expression levels ($p<0.01$).[68]

Cytokine receptor binding and activation of the cytokine-signaling pathways influence several different pathways demonstrating that the energy generation, insulin production/β-cell

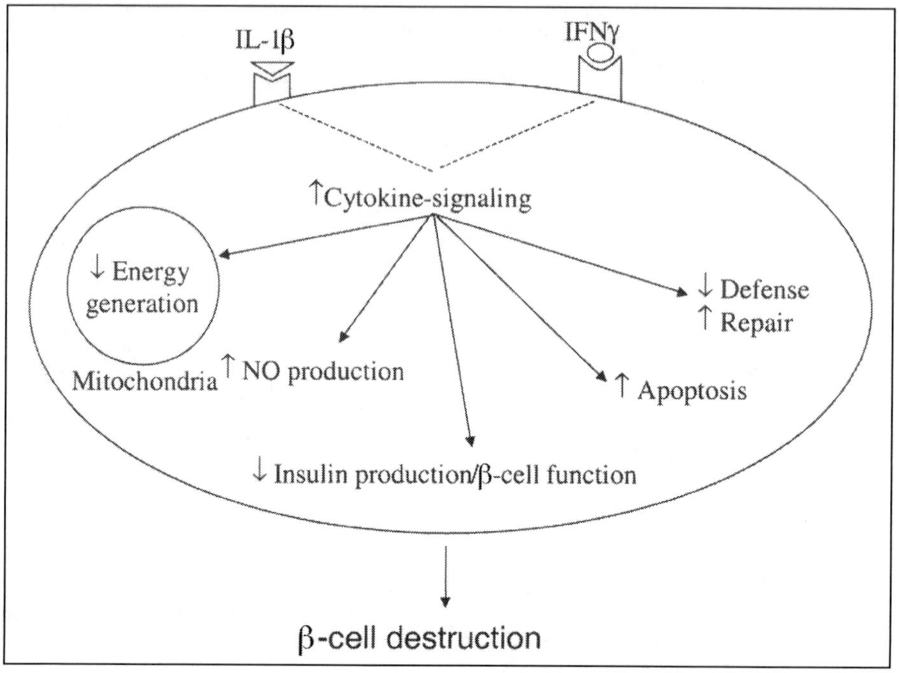

Figure 3. Important pathways changed in expression profiling studies after cytokine-exposure.

function and cellular defense are decreased, whereas the nitrite oxide (NO) production, cytokine-signaling, apoptosis and cellular repair are increased.

The changes seen in expression profiling studies involve many biochemical pathways (Fig. 3), and the overall picture is that mechanisms favoring β-cell destruction generally dominate. The identified transcripts and proteins may affect viability, function and recognition of the β-cell by the immune system. Although the new powerful and informative techniques give us a detailed picture of the genes expressed in the β-cells in response to cytokines, further studies of the function, interaction and relative significance are needed. Current methodologies describe changes observed at specific cytokine doses and time windows. Selecting other time-points or other cytokine concentrations or combinations of cytokines may result in other changes in the protein/gene expression levels. Furthermore selection of a certain time-point and specific concentrations of cytokines do not allow discrimination between "primary" and "secondary" effects of cytokines as well as dynamic changes over time. Finally, other criteria include selection of significance level or fold-variation, which obviously will affect the number of changes identified.

Below we have outlined some transcripts and proteins of special interest in Type 1 diabetes that suggest that a race between protective and destructive mechanisms take place in the cytokine-exposed β-cell.

Cytokine-Signaling

The effects of cytokines on cells are mediated through membrane-bound receptors. The binding of cytokines to receptors initiates a cascade of several different kinases and transcription factors (see below). After IL-1β exposure of RIN-cells, genes such as Janus kinase 2 and 3 (JAK2/3), extracellular signal-regulated kinase 3 (Erk-3) and NF-κB were up-regulated.[65] The same pattern was found after IL-1β or IL-1β/IFNγ exposure of purified rat β-cells, where genes encoding Jak-2, Erk-3, NF-κB and c-jun were up-regulated and mitogen activated phosphokinase

(MAPK) phosphatase were down-regulated.[66] Cytokine induced activation of ERK1/2, p38 and NF-κB leads to iNOS expression,[67,68] suggesting that these signaling pathways play a role in cytokine-induced NO-dependent β-cell destruction. In contrast, negative regulators of cytokine-signaling, such as genes encoding CL100 tyrosine protein phosphatase and inhibitor kappa B alfa (I-κBα),[66] were also up-regulated, but perhaps not sufficiently so to inhibit cytokine-induced NO production.

Energy Generation

IL-1β has previously been demonstrated to have an inhibitory effect on β-cell mitochondrial energy generation.[69] This is supported by proteome analysis, since proteins, such as ATP synthase α and β involved in the energy generation pathway, are down-regulated in rat islets after exposure to IL-1β.[68] The same pattern was found at the transcriptome level where genes such as ATPase and cytochrome b5 reductase in purified rat β-cells were down-regulated after exposure to IL-1β alone or in combination with IFNγ.[66] These results support a decreased β-cells energy generation in response to cytokine-exposure.

Insulin Production/β-Cell Function

Several studies have demonstrated that the insulin expression in the β-cells is decreased after exposure to cytokines.[7] In the proteome analysis the expression level of neuroendocrine convertase 2 percursor (NEC2, also called prohormone convertase 2 (PC2)) was down-regulated in rat islets after exposure to IL-1β.[68] PC1 and PC2 are involved in the processing of insulin. Furthermore, gene expression level of PC1 was down-regulated in purified rat β-cells after exposure to IL-1β and IFNγ.[66] It has been shown that cytokines decrease the expression of NEC2 resulting in delay in proinsulin conversion.[70] Other genes important for β-cell function such as genes related to glucose uptake (glucose transporter-2 (GLUT-2)) and insulin gene transcription/β-cell development (islet factor-1 (Isl-1)) were also down-regulated after cytokine-exposure of both RIN and purified rat β-cells.[65,66] This is supported in microarray analysis, where blocking of NF-κB activation, a transcription factor for iNOS, prevented cytokine-induced down-regulation of important genes related to β-cell function, including glucose uptake (GLUT-2), insulin processing (PC1), insulin release (phospholipase D_1 and cholecystokinin-A receptor) and insulin gene transcription/β-cell development (Isl-1 and pancreatic duodenum homeobox 1).[6]

NO Production

Cytokines (in particular IL-1β and IFNγ) have been shown to induce free radicals that are toxic for the β-cells like superoxide (O_2^-), hydrogen peroxide (H_2O_2) and NO in β-cells and islets of Langerhans.[7,55,71] The genes encoding iNOS and argininosuccinate synthetase were both up-regulated after exposure of purified rat β-cells to IL-1β or IL-1β/IFNγ.[66] iNOS and sepiapterin reductase, involved in the biosynthesis of tetrahydrobiopterin, a cofactor for iNOS[72], were also up-regulated in RIN-cells exposed to IL-1β.[65] These three genes are involved in the production of NO, suggesting that NO plays a role in the cytokine-mediated β-cell destruction.

Calcium

Ca^{2+} is involved in NO-mediated β-cell destruction by inducing endoplasmic reticulum (ER) Ca^{2+} release, which leads to ER stress and induction of apoptosis.[73] Expression of several different transcripts/proteins important for Ca^{2+} handling is changed by cytokines. In purified rat β-cells stimulated with IL-1β or IL-1β/IFNγ, Ca^{2+} ATPase type 2 was down-regulated, which may result in ER Ca^{2+} depletion and ER stress.[66] Furthermore transcripts encoding calcineurin A α, intracellular calcium-binding protein (MRP14), Ca^{2+}/Calmodulin-dependent protein kinase IV, protein p9Ka and C kinase substrate changed expression levels in RIN-cells after exposure to IL-1β.[65] These data suggest that cytokine exposure leads to change in the Ca^{2+} level, which may influence several intracellular functions such as ER stress, transcription and apoptosis.[73-75] The exact mechanism(s) of Ca^{2+} regulating genes/proteins remains to be further elucidated (see below).

Apoptosis

Programmed cell death (apoptosis) is induced in cytokine-exposed rat and human islets.[76,77] Apoptosis is characterized by cell shrinkage, chromatin condensation and DNA fragmentation (reviewed in ref. 78). Several proteins are involved in apoptosis and some of these have been identified in rat islets at the proteome level. These include lamins A and B and transforming growth factor β (TGF-β) receptor interacting protein, which were up-regulated following IL-1β exposure.[68] At the transcriptome level, caspase-1 was found up-regulated in IFNγ-exposed rat islets[77] and death protein-5 and Fas were up-regulated after IL-1β and IL-1β/IFNγ exposure of purified rat β-cells.[66,79] In RIN-cells exposed to IL-1β several genes involved in the apoptotic pathways were identified, such as up-regulation of interleukin-1β-converting enzyme-related protease CCP32.[65] These changes in protein and gene expression levels all support that apoptosis is initiated during cytokine-mediated β-cell destruction.

Defense/Repair Mechanisms

Activation or over-expression of heat shock proteins (HSPs) have been shown to have a protective role in cytokine- or NO-mediated β-cell destruction.[80,81] Several proteins, such as glutathione peroxidase (GP), glutathione-S-transferase (GST) and manganese superoxide dismutase (MnSOD) are involved in the scavenging of free radicals, such as O_2^-, H_2O_2 and NO. In the proteome analysis several HSPs were up-regulated after exposure of rat islets to IL-1β,[68] and in the transcriptome analysis genes encoding HSPs, metallothionein, MnSOD and GST (M5) were up-regulated after exposure to cytokines.[65,66] However, this increase in defense/repair mechanisms may be insufficient to prevent cytokine-mediated β-cell destruction, resulting in β-cell death. This may in part be explained by the fact that cytokines also down-regulate defense/repair genes, such as GP and Gas-6 growth arrest (protection against apoptosis) in purified rat β-cells exposed to IL-1β or IL-1β/IFNγ,[66] and GST (Yb), SOD B and selenium dependent GP in RIN-cells exposed to IL-1β.[65]

In conclusion, transcriptome and proteome analyses support the action that a race between protective and destructive mechanisms is induced in β-cells by cytokines. The expression level of many genes and proteins are changed in response to cytokines and important cellular pathways are affected.

Following receptorbinding of IL-1β, TNFα and IFNγ several, but distinct, signaling cascades are activated. IL-1β-induced signaling involves activation of the mitogen-activated protein kinase (MAPK) pathways constituting c-jun N-terminal kinase (JNK), p38 and extracellular signal-regulated kinase (ERK). These kinases are responsible for phosphorylation and subsequent activation of multiple transcription factors. Furthermore, IL-1β mediates activation of the transcription factor nuclear factor κB (NF-κB) pathway, a pathway commonly shared by TNFα. IFNγ signaling involves induction of Janus Kinases (JAKs) leading to activation of signal transducers and activators of transcription-1 (STAT-1). The network of activated transcription factors induces significant alterations in gene regulation and consequently protein expression, some critically involved in β-cell dysfunction.

Signal Transduction of Cytokine-Mediated β-Cell Death

The form of β-cell death induced by cytokines is mostly apoptotic, but also necrotic cell death is observed in rodent β-cells upon cytokine exposure.[3] The signaling pathways elicited by receptor binding of IL-1β, TNFα and IFNγ are complex and far from fully understood. Although certain signaling elements may be shared to some extent, these cytokines utilize discrete signal transduction pathways (Fig. 4).

IL-1β, TNFα and IFNγ bind to their respective receptors on the cell surface leading to induction of several, but distinct, signaling cascades resulting in activation of various transcription factors, thereby leading to altered gene-expression. Because IL-1β is the most potent β-cell

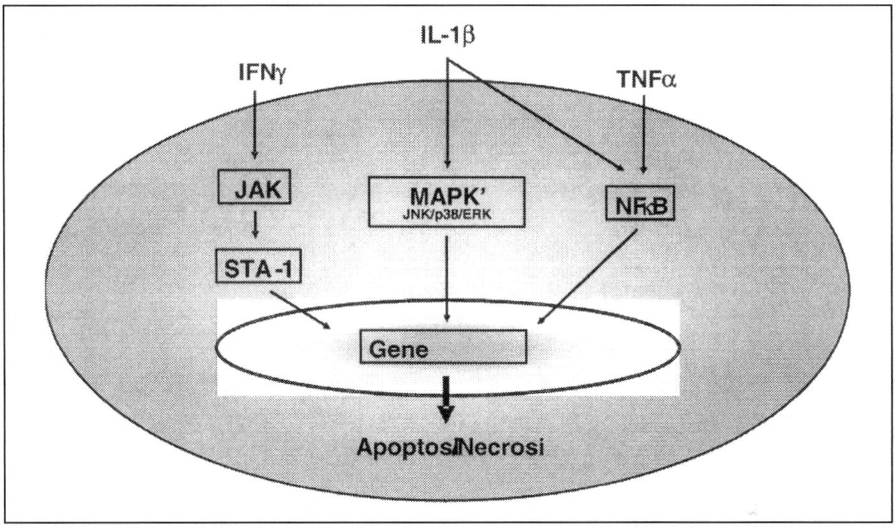

Figure 4. Schematic presentation of IL-1β, TNFα and IFNγ signaling.

cytotoxic cytokine, signal transduction elicited by IL-1β has been intensively studied in β-cells. At present, two IL-1β-induced signaling events or pathways that impact on β-cell fate, i.e., death or survival, have been established. These are activation of mitogen-activated protein kinases (MAPK)[82,83] and activation of the transcription factor NF-κB).[84,85] The latter pathway is in addition commonly utilized by TNFα.[86] IFNγ signaling involves first and foremost induction of JAKs leading to activation of signal transducers and activators of transcription-1 (STAT-1).[87] As illustrated in Figure 4, NF-κB, STAT-1 and other cytokine induced transcription factors, induce alterations in the regulation of several genes and consequently protein expression. These expression patterns (up-/down-regulation and/or posttranslational modifications) represent pathways of protective and deleterious nature.[60,68,88]

Possibly, as a result of its degree of specialization, the differentiated β-cell is highly susceptible to deleterious signaling by IL-1β, TNFα and IFNγ. As will be discussed below, the action or impact of the individual signaling elements and resulting response have been dissected to some extent and have provided valuable information about the complex mechanisms underlying cytokine mediated β-cell destruction, i.e., aspects of Type 1 diabetes at the molecular level.

Mitogens and stress stimuli (including pro-inflammatory cytokines) induce the activation of several but distinct signaling pathways. This involves the G-protein Ras as the main early activator of the classical ERK pathway, and TRAF and G-protein Rho are early mediators of the p38 and JNK pathway. Collectively Ras, TRAF and Rho interact with and activate selective members of MAP kinase kinase kinases (MAPKKK), which in turn activate MAP kinase kinases (MAPKK) by dual phosphorylation. Finally, the MAPKK phophorylate and activate the MAPKs ERK, p38 and JNK. Once activated, the MAPKs activate a range of transcription factors (including Elk-1, ATF-2 and c-jun), which regulates the transcription of multiple genes.

The MAPK Pathways

IL-1β activates mitogen-activated protein kinases (MAPKs) in rodent β-cells. Classically, MAPKs transmit signals from the cytosol to the nucleus by phosphorylating and thereby activating various transcription factors leading to altered gene expression. MAPKs are activated by dual phosphorylation on one threonine and one tyrosine residue in a three-module kinase cascade (Fig. 5).[89,90] Three major MAPK family members have been described, namely c-jun N-terminal kinase (JNK), p38 and extracellular signal-regulated kinase (ERK), all of which

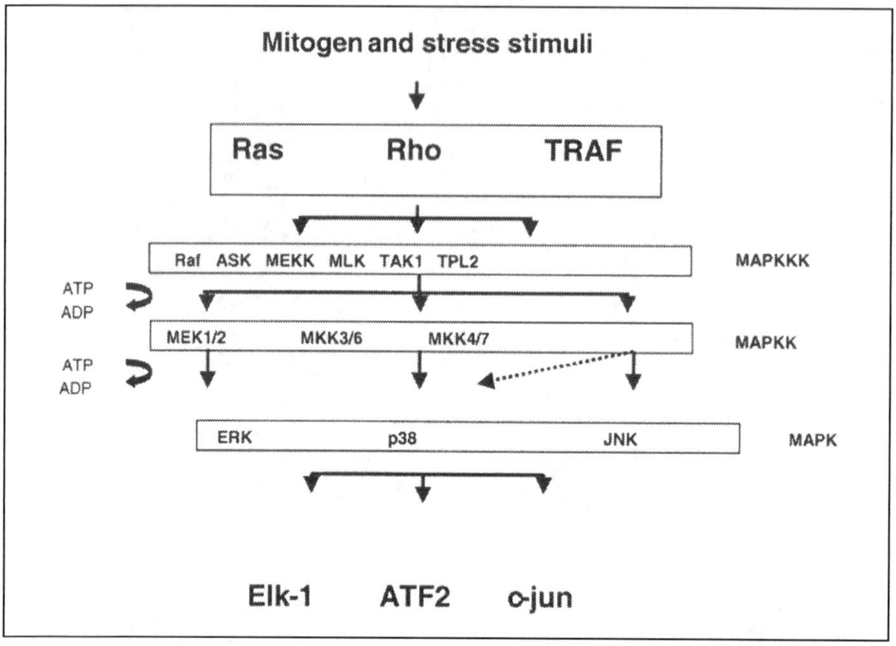

Figure 5. Schematic presentation of the MAPK pathways.

have been shown to be activated by IL-1β in β-cells.[91,92] IL-1β-induced activity of MAPKs is in addition, to some extent, potentiated by IFNγ and TNFα,[93] suggesting cross talk at this level of signaling between cytokine-induced pathways. The main MAPK mediator of cytokine-induced β-cell apoptosis appears to be JNK. Whereas inhibition of the p38 and/or the ERK pathways only moderately reduce cytokine-induced primary β-cell apoptosis,[94,95] inhibition of JNK activity confers nearly full protection against apoptosis caused by IL-1β in clonal β-cells.[82,83,96] Thus, although the involvement of JNK in cytokine-induced apoptosis of primary β-cells needs to be confirmed, JNK seems to be a central transducer of cytokine-induced signaling events associated with β-cell apoptosis.

The NF-κB Pathway

NF-κB is a ubiquitously expressed transcription factor that is activated in response to a diverse set of stress stimuli including pro-inflammatory cytokines such as IL-1β and TNFα. NF-κB potentially regulates the expression of several NF-κB responsive genes. However, the effect NF-κB activation is not uniform - it orchestrates the expression of a subset of genes in a manner that is a highly stimulus- and cell type-specific.[86] The signaling components involved in IL-1β-induced NF-κB activation in the β-cell is not known in detail, but as defined in other cell systems, it probably utilizes a common pathway initiated by IL-1β (Fig. 6). With regards to cytokine mediated β-cell destruction, cytokine dependent NF-κB activation is a critical point, since inhibition of its activity reduces rodent as well as human β-cell apoptosis and necrosis.[84,85]

TNFα binding to a p60 receptor induces recptor trimerization. TNF receptor-associated death domain (TRADD) is recruited to the receptor complex, followed by Fas-associated death domian (FADD) binding to TRADD. FADD in turn recruits TNF receptor associated factor 2 (TRAF2), which associate/activate NF-κB inducing kinase (NIK). Binding of IL-1 to the IL-1β receptor type I (IL-1RI) leads to assembly of the IL-1β receptor complex, which include IL-1 receptor associated protein (IL-1RAcP), MyD88, IL-1β receptor associated kinase 1 and 2 (IRAK 1/2) and TRAF 6 which in turn associate/activate to NIK.

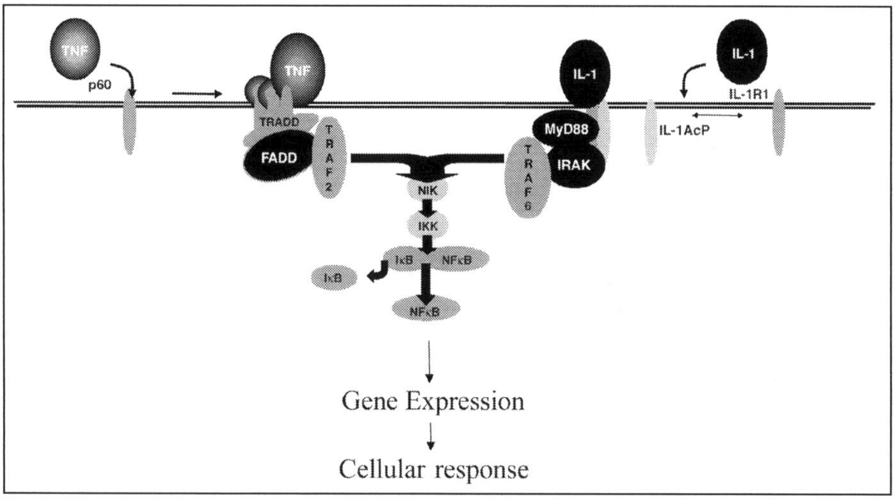

Figure 6. The NF-κB pathway.

Downstream events of NIK is shared among IL-1β and TNFα signaling and involves NIK mediated phosphorylation and activation of a protein kinase complex constituting dimers of IκB kinase (IKK). In the resting cell, NF-κB is found in the cytoplasm as a two subunit (p50 and p65) protein complex bound to the inhibitory κB (IκB) proteins. IKK phosphorylates IκB leading to proteasome-dependent degradation of IκB and subsequently liberation of NF-κB in its transcriptionally active form. Liberated NF-κB translocates to the nucleus where it regulates the transcription of selected genes.

JAK/STAT and β-Cell Destruction

IFNγ-mediated signaling involves JAKs, which becomes phosphorylated as a result of association with the intracellular portion of the IFNγ receptor (Fig. 7). Once activated, JAKs recruit and in turn activate STAT-1 monomers by phosphorylation. STAT-1 then dimerizes and translocates to the nucleus where it regulates STAT-1 responsive genes,[87] some of which apparently are responsible for initiation of β-cell apoptosis.[77]

IFNγ signaling involves association of homodimeric IFNγ with two type I integral membrane proteins IFNγR1 (IFNγ receptor subunit 1) each connected to a Janus kinase 1 (JAK1). This leads to recruitment of two IFNγR2 (IFNγ receptor subunit 2) each associated with a JAK2, and resulting in activation of the JAKs by auto- and transphosphorylation. The signal transducer and activator of transcription 1 (STAT1) is then recruited to two binding sites in the receptor complex. The receptor-associated STATs are subsequently phosphorylated by the receptor-bound JAKs leading to dissociation and dimerisation of the STATs. The STAT1-homodimers translocates to the nucleus where they bind to specific sequences to promote induction of selective target genes.

In most cell systems, cytokines induce a rapid endogen up-regulation and/or activation of proteins that in a negative feedback loop terminates the induced signal transduction.[97] Such feedback mechanisms are important in many regards, but foremost they inhibit the possible deleterious effects due to uncontrolled gene transcription. One group of such inhibitory proteins is the suppressors of cytokine signaling (SOCS), which constitute a family of proteins that acts in a classically negative feedback loop to suppress constitutive signaling initiated by multiple cytokines.[97] In clonal rat β-cells, over-expression of SOCS-3 inhibits the toxic effect of both IL-1β and IFNγ.[98] However, endogen up-regulation of SOCS-3, following IL-1β stimulation, is dramatically delayed compared to other tissues.[98] This suggests that β-cells are highly

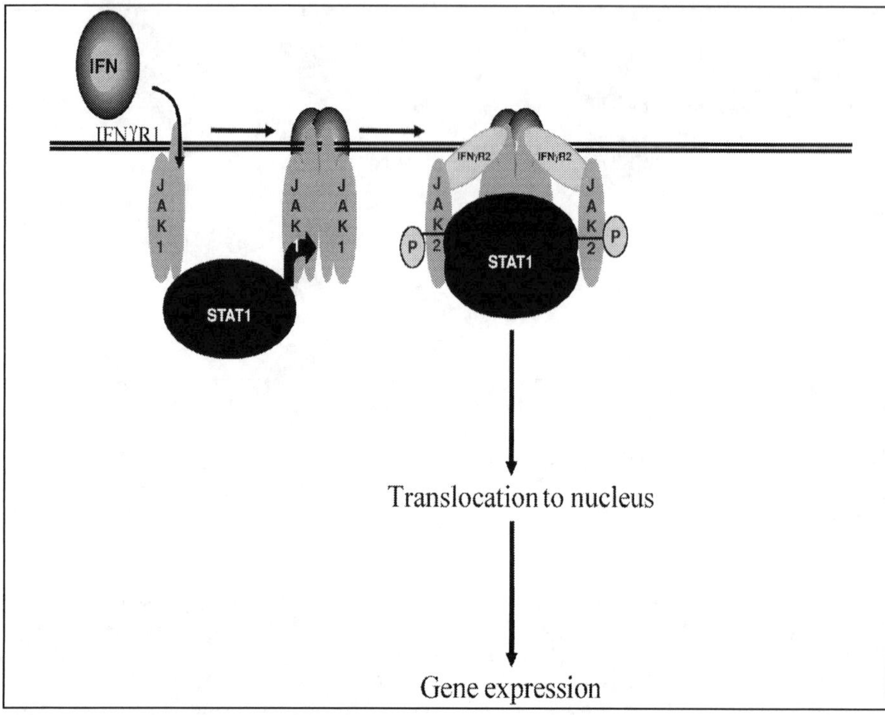

Figure 7. The JAK/STAT pathway.

susceptible to IL-1β and IFNγ-mediated cytotoxicity due to an inadequate ability to rapidly up-regulate molecules, i.e., SOCS, which terminate cytokine signaling.

Effector Mechanisms of Cytokine Mediated β-Cell Destruction

Reactive Oxygen Species (ROS)

Cytokine mediated β-cell destruction involves the toxicity of intracellular generated reactive oxygen species (ROS), which have the capacity to oxidize and thereby damage critical cellular components.[99] Being a diverse group of molecules with variable reactivity, the most pronounced ROS in the IL-1β-stimulated β-cell is NO, which is formed during an enzymatic conversion of L-arginine to L-citrulline. IL-1β, potentiated by IFNγ and TNFα, induces selective β-cell expression of iNOS[3] in a manner that is essentially dependent on NF-κB activation[85,100] but also regulated by p38 and ERK.[92] NO is capable of modifying the function/action of numerous proteins, including transcription factors, ion channels and proteins involved in signal transduction.[101] Using synthetic substrate analogues of L-arginine as competitive inhibitors of iNOS, it was suggested that NO plays an important role in mediating cytokine-induced inhibitory effects on rodent β-cell function. However, despite induction of iNOS and subsequent NO production, cytokine-mediated death of human islets occurs independently of iNOS-generated NO.[76,102,103] Recent efforts using iNOS$^{-/-}$ transgenic mice have suggested that NO production is predominantly involved in promoting necrosis, because purified β-cells from these animals are rescued from cytokine-induced necrotic, but not apoptotic cell death.[104] Thus, the exact role and contribution of NO in cytokine signaling and cell death of particularly human β-cells are still not clear. Most likely both NO-dependent and -independent mechanisms operate, and probably the relative contribution of these signaling mechanisms differs in rodent versus human β-cells. The role of NO may predominantly be to amplify and prolong cytokine signaling via JNK and p38, as iNOS inhibition attenuates IL-1β-induced JNK and p38 activity in isolated rat islets.[105]

In addition to NO generation, exposure of human islets to cytokine combinations induces production of other ROS such as the superoxide anion (O_2^-),[106] which if reacting with NO, may generate the highly reactive radical peroxynitrite ($ONOO^-$).[99] Apparently, β-cells may be particularly susceptible to oxidative stress, due to their relative low antioxidant enzyme expression levels.[107,108] In accordance with this hypothesis, anti-oxidant administration or over-expression of antioxidant enzymes reduce cytokine-induced β-cell death in both rodent and human β-cells.[106,109-111]

NO and ROS may cause β-cell toxicity via different mechanisms: (1) NO inactivates the mitochondrial Krebs cycle enzyme aconitase by nitrosylation of Fe-S groups thereby preventing glucose oxidation and ATP generation;[112] (2) ROS damages cellular DNA by causing DNA strand breaks thereby activating DNA repair mechanisms including the enzyme poly(ADP ribose)polymerase (PARP) which may lead to cell death through depletion of cellular nicotinamide adenine dinucleotide (NAD^+) and[113] (3) the ROS-induced DNA strand breaks may itself be sufficient to induce apoptosis, possibly through activation of the tumor suppressor protein p53.[114] In addition to these possibilities, novel findings indicate that NO may induce β-cell death by causing ER stress by depletion of ER Ca^{2+}.[73] In line with this, exposure of primary rat β-cells to IL-1β + IFNγ leads to NO-dependent downregulation of expression of smooth ER Ca^{2+} ATPase 2 (SERCA2),[66] which possibly results in ER Ca^{2+} depletion and ER stress. Because ER stress has been shown to be coupled to activation of JNK in nonβ-cells,[115] NO-mediated β-cell ER stress may be coupled to the pro-apoptotic JNK pathway.

JNK

The mechanisms by which JNK regulates apoptosis in β-cells are presently not known. However, JNK is capable of activating various transcription factors including ATF2 and c-jun,[90] which by regulation of critical apoptotic genes may contribute to induction of apoptosis. Another potential target of pro-apoptotic signaling by JNK is p53. JNK is capable of phosphorylating p53 in vitro and expression of a constitutively active form of MEKK1, an upstream kinase of JNK, in mouse fibroblasts leads to accumulation of p53 and subsequent apoptosis.[116] Furthermore, other reports have suggested that JNK may induce transcription-independent apoptosis by direct phosphorylation and thereby inhibition of anti-apoptotic proteins of the Bcl-2 family of apoptosis-regulating proteins.[117-119]

NF-κB

In primary rat β-cells, analysis of IL-1β and IL-1β + IFNγ mediated NF-κB activation has revealed a range of genes regulated by NF-κB. As defined for iNOS expression,[85,100] NF-κB regulates the expression of an array of genes, and an up-regulation of several pro-apoptotic genes and a down-regulation of several β-cell specific genes has been observed.[67] Collectively, these findings suggest that IL-1β + IFNγ-mediated NF-κB activity causes dedifferentiation and impaired function of the β-cell and induces deleterious/pro-apoptotic β-cell genes. In agreement with these observations, inhibition of NF-κB activation by over-expression of a nondegradable IκB mutant reduces IL-1β + IFNγ-mediated apoptosis in primary rat β-cells[85] and blocks IL-1β mediated Fas expression and subsequent Fas-antigen mediated apoptosis in human islets.[84]

Calcium

Entry of Ca^{2+} via low voltage-activated (LVA) or high voltage-activated (HVA) Ca^{2+} channels has been implicated in β-cell apoptosis, since blockers of these Ca^{2+} channels can prevent cytokine-induced apoptosis.[120,121] This may appear as a paradox because glucose-induced Ca^{2+} entry is an important determinant of insulin granule exocytosis. Thus, at the same time, Ca^{2+} influx may regulate basal β-cell functions like insulin secreting, but also lethally contributes to signaling evoked by cytokines. As blockade of HVA channels have been found to abrogate IL-1β-stimulated JNK activity in β-cells,[122] one role of Ca^{2+} influx through voltage-activated channels in cytokine-induced β-cell apoptosis may be to contribute to pro-apoptotic signaling via JNK.

Caspases

Members of the caspase family of cysteine proteases are key executioners of apoptosis being responsible for the cleavage of many cellular proteins resulting in dismantling of the cell. Caspases are present as inactive precursors, which are activated by proteolytic processing. Different stimuli-specific routes lead to activation of proximal caspases (caspases 2, 8, 9 and 10), which in turn cleave and activate effector caspases (caspases 3, 6 and 7) causing cellular disassembly.[123] In rodent and human β-cells, IFNγ has been shown to up-regulate caspase 1 (also known as interleukin-1 converting enzyme (ICE)),[77] suggesting involvement of caspases in cytokine-mediated β-cell death.

In summary, cytokine-mediated β-cell destruction involves a complex scenario of different signaling pathways and it is obvious that several distinct, both NO-dependent as well as -independent, β-cell destructive pathways exist. However, several key signaling components have been identified and in particular JNK and NF-κB are essential regulators of cytokine signaling leading to β-cell impairment and death.

Conclusions and Perspectives

There have been remarkable advances in our understanding of the molecular pathogenesis of Type 1 diabetes over the last decade. The Human Genome Project and new technologies are generating an amount of data unprecedented in biology. Science now has the tools to characterize the pathophysiological nuances and inherited variations that interact over time and lead to common diseases as Type 1 diabetes.

The multiple approaches to understand cytokine-mediated β-cell cytotoxicity presented here may be useful in identification of relevant mechanisms for β-cell death after cytokine exposure. Data obtained by the outlined methodologies may lead to the hitherto most detailed and complex picture of the molecular processes leading to β-cell destruction in vitro. Although the picture is complicated and far from complete we are looking at the ailing beta cell through a new window and the challenge now is to learn to fully understand what we see.

Genetic, expression and signal transduction studies have revealed that a race between protective and deleterious mechanisms takes place in β-cells when exposed to cytokines, and that the balance between these mechanisms is crucial, (Fig. 8). The overall picture in the cytokine-exposed β-cell is: Decrease in energy generation, insulin production and β-cell function, increase in NO production, activation of the MAPK and NF-κB cytokine-signaling pathways, increases in pro-apoptotic proteins and genes, and decreases in cellular defense proteins.

In β-cells the destructive processes prevail. Here illustrated by a decrease in energy generation, insulin production/β-cell function and cellular defense and an increase in NO production, cytokine-signaling, apoptosis and cellular repair.

The proposed model may also provide valuable to identify therapeutical intervention strategies in Type 1 diabetes and serve as a tool for testing such strategies. By using transgenic or knock-out mice of identified susceptibility genes, by detailing expression profiling e.g., of primary and secondary effects, and by modifying signal transduction new targets for intervention in the diseases process are likely to be identified. Most obvious strategies, as suggested by the model, include immunosuppression. Agents such as cyclosporine[6] are able to arrest the development of the disease, but carry too many demonstrable and theoretical side effects to be useful as lifelong therapy.

Manipulation of the costimulatory signals, e.g., by CTLA-4 activating immunoglobulins, has also been envisioned as a potential strategy for achieving therapeutically useful immunosuppression or tolerance. CTLA4-Ig has been used successfully in other immune-mediated diseases as psoriasis,[124,125] treatment of graft-versus-host disease in allogenic bone marrow trans-

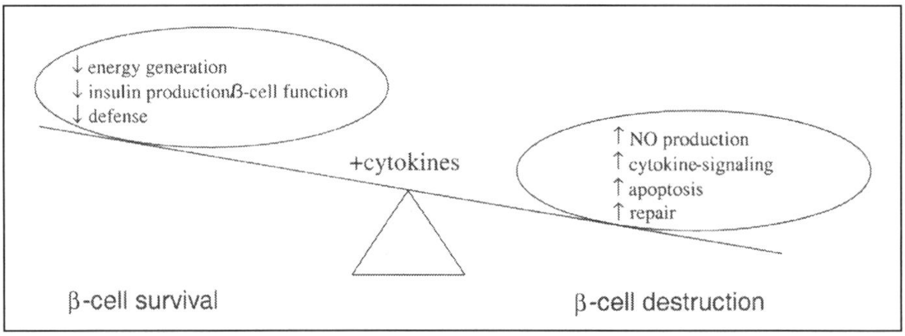

Figure 8. The functional effect on cytokine-induced pathways changes the balance between the destructive and the protective mechanisms.

plantation,[126] in mouse models of systemic lupus erythematosus (SLE),[127] and the effect of CTLA4-Ig in experimental models of RA is also promising.[128] Clinical randomized trials on the use of CTLA4-Ig in Type 1 diabetes are however missing.

The demonstrated effectiveness of superoxide dismutase and other oxygen free radical scavengers in preventing the development of disease recurrence in transplanted pancreatic tissue[129-131] and from in vitro studies (reviewed in ref. 3) suggests that agents directed at blocking inflammatory mediators, such as interleukins and oxygen radicals, may be effective in preventing damage to the susceptible pancreatic β-cells. Interestingly, interferon β for multiple sclerosis and tumor necrosis factor α antagonists for rheumatoid arthritis and Crohn's disease are amongst the first new treatments for autoimmunity approved by the Food and Drug Administration in 20 years.

Further genetics, expression profiling and signal transduction analyses within Type 1 diabetes research may eventually complete the picture that the first analyses have just begun to make visible. The perspective of this is the development of new and specific intervention modalities in β-cell destruction in Type 1 diabetes. Making the β-cell more resistant to mediators of the immune system may increase the survival time of transplanted islets or engineered β-cells, and potentially prevent the ongoing β-cell destruction in predisposed individuals.

References

1. Witebsky E. Historical roots of present concepts of immunopathology. In: Grabar P, Miescher P, eds. Immunopathology 1st International Symposium B. Schwabe & Co, 1959:1-13.
2. Milgrom F, Witebsky E. Autoantibodies and autoimmune diseases. JAMA 1962; 181:706-716.
3. Eizirik DL, Mandrup-Poulsen T. A choice of death – the signal-transduction of immune-mediated beta-cell apoptosis. Diabetologia 2001; 44:2115-2133.
4. Serreze DV, Chapman HD, Varnum Ds et al. Initiation of autoimmune diabetes in NOD/Lt mice is MHC class I-dependent. J Immunol 1997; 158:3978-3986.
5. DiLorenzo TP, Graser RT, Ono T et al. Major histocompatibility complex class I-restricted T cells are required for all but the end stages of diabetes development in nonobese diabetic mice and use a prevalent T cell receptor a chain gene rearrangement. Proc Natl Acad Sci USA 1998; 95:12538-12543.
6. Group C-ERCT: Cyclosporin-induced remission of IDDM after early intervention: Association with enhanced insulin secretion. Diabetes 1988; 37:1574-1582.
7. Mandrup-Poulsen T. The role of interleukin-1 in the pathogenesis of IDDM. Diabetologia 1996; 39:1005-1029.
8. Eizirik DL, Hoorens A. β-cell dysfunction and death. Adv Mol Cell Biol 1999; 29:47-73.
9. Rabinovitch A. An update on cytokines in the pathogenesis of insulin-dependent diabetes mellitus. Diabetes Metab Rev 1998; 14:129-151.
10. Pociot F, McDermott MF. Genetics of Type 1 Diabetes Mellitus. Genes Immun 2002; 3(5):235-49.
11. Davies JL, Kawaguchi Y, Bennett St et al. A genome-wide search for human type-1 diabetes susceptibility genes. Nature 1994; 371:130-136.

12. Concannon P, Gogolin-Ewens KJ, Hinds DA et al. A second-generation screen of the human genome for susceptibility to insulin-dependent diabetes mellitus. Nat Genet 1998; 19:292-296.
13. Mein CA, Esposito L, Dunn Mg et al. A search for type 1 diabetes susceptibility genes in families from the United Kingdom. Nat Genet 1998; 19:297-300.
14. Cox NJ, Wapelhorst B, Morrison VA et al. Seven regions of the genome show evidence of linkage to type 1 diabetes in a consensus analysis of 767 multiplex families. Am J Hum Genet 2001; 69:820-830.
15. Nerup J, Pociot F. For the European Consortium for IDDM genome studies. A genomewide scan for type 1-diabetes susceptibility in Scandinavian families: Identification of new loci with evidence of interactions. Am J Hum Genet 2001; 69:1301-1313.
16. Risch N. Assessing the role of HLA-linked and unlinked determinants of disease. Am J Hum Genet 1987; 40:1-14.
17. Noble J, Valdes A, Cook M et al. The role of hla class-ii genes in insulin-dependent diabetes-mellitus - molecular analysis of 180 caucasian, multiplex families. Am J Hum Genet 1996; 59:1134-1148.
18. Lee KH, Wucherpfennig KW, Wiley DC. Structure of a human insulin peptide-HLA-DQ8 complex and susceptibility to type I diabetes. Nat Immunol 2001; 2:501-507.
19. Todd JA, Wicker LS. Genetic protection from the inflammatory disease type 1 diabetes in humans and animal models. Immunity 2001; 15:387-395.
20. Pociot F, Mølvig J, Wogensen L et al. A TaqI polymorphism in the human interleukin-1β (IL-1β) gene correlates with IL-1β secretion in vitro. Eur J Clin Invest 1992; 22:396-402.
21. Mandrup-Poulsen T, Pociot F, Mølvig J et al. Monokine antagonism is reduced in patients with IDDM. Diabetes 1994; 43:1242-1247.
22. Bergholdt R, Karlsen AE, Johannesen J et al. The Danish study group of diabetes in childhood: Characterization of polymorphisms of an interleukin-1 receptor type 1 gene (IL1RI) promoter region (P2) and their relation to insulin-dependent diabetes mellitus (IDDM). Cytokine 1995; 7:727-732.
23. Metcalfe KA, Hitman GA, Pociot F et al. An association between type-1 diabetes and the interleukin-1 receptor-type-1 gene. Human Immunol 1996; 51:41-48.
24. Bergholdt R, Larsen ZM, Andersen NA et al. Characterization of new polymorphisms in the 5' UTR of the human interleukin-1 receptor type 1 (IL1R1) gene: Linkage to type 1 diabetes and correlation to IL-1RI plasma level. Genes Immun 2000; 1:495-500.
25. Katila H, Hanninen K, Hurme M. Polymorphisms of the interleukin-1 gene complex in schizophrenia. Molecular Psychiatr 1999; 4:179-181.
26. Joos L, McIntyre L, Ruan J et al. Association of IL-1 beta and IL-1 receptor antagonist haplotypes with rate of decline in lung function in smokers. Thorax 2001; 56:863-866.
27. Boiardi L, Salvarani C, Timms J et al. Interleukin-1 cluster and tumor necrosis factor-alpha gene polymorphisms in polymyalgia rheumatica. Clin Exp Rheumatol 2000; 18:675-681.
28. Feakes R, Sawcer S, Broadley S et al. Interleukin 1 receptor antagonist (IL-1ra) in multiple sclerosis. J Neuroimmunol 2000; 105:96-101.
29. Diehl S, Wang Y, Brooks C et al. Linkage disequilibrium of interleukin-1 genetic polymorphisms with early-onset periodontitis. J Periodont 1999; 70:418-430.
30. Lang N, Tonetti M, Suter J et al. Effect of interleukin-1 gene polymorphisms on gingival inflammation assessed by bleeding on probing in a periodontal maintenance population. J Periodont Res 2000; 35:102-107.
31. McDevitt M, Wang H, Knobelman C et al. Interleukin-1 genetic association with periodontitis in clinical practice. J Periodont 2000; 71:156-163.
32. Papapanou P, Neiderud A, Sandros J et al. Interleukin-1 gene polymorphism and periodontal status - A case-control study. J Clin Periodontol 2001; 28:389-396.
33. El-Omar E, Carrington M, Chow W et al. Interleukin-1 polymorphisms associated with increased risk of gastric cancer. Nature 2000; 404:398-402.
34. Danis VA, Millington M, Hyland VJ et al. Cytokine production by normal human monocytes: inter-subject variation and relation to an IL-1 receptor antagonist (IL-1Ra) gene polymorphism. Clin Exp Immunol 1995; 99:303-310.
35. Hurme M, Santtila S. IL-1 receptor antagonist (IL-1ra) plasma-levels are coordinately regulated by both IL-1ra and IL-1-beta genes. Eur J Immunol 1998; 28:2598-2602.
36. Morahan G, Huang DX, Ymer SI et al. Linkage disequilibrium of a type 1 diabetes susceptibility locus with a regulatory IL12B allele. Nat Genet 2001; 27:218-221.
37. Johansson S, Lie BA, Thorsby E et al. The polymorphism in the 3 ' untranslated region of IL12B has a negligible effect on the susceptibility to develop type 1 diabetes in Norway. Immunogenetics 2001; 53:603-605.

38. Jahromi MM, Millward BA, Demaine AG. A polymorphism in the promoter region of the gene for interleukin-6 is associated with susceptibility to type 1 diabetes mellitus. J Interferon Cytokine Res 2000; 20:885-888.
39. Fishman D, Faulds G, Jeffery R et al. The effect of novel polymorphisms in the interleukin-6 (il-6) gene on il-6 transcription and plasma il-6 levels, and an association with systemic-onset juvenile chronic arthritis. J Clin Invest 1998; 102:1369-1376.
40. Ligers A, Teleshova N, Masterman T et al. CTLA-4 gene expression is influenced by promoter and exon 1 polymorphisms. Genes Immun 2001; 2:145-152.
41. Nolsoe RL, Kristiansen OP, Sangthongpitag K et al. Complete molecular scanning of the human Fas gene: mutational analysis and linkage studies in families with Type I diabetes mellitus. Diabetologia 2000; 43:800-808.
42. Nolsoe RL, Kristiansen OP, Larsen ZM et al. Complete mutation scan of the human Fas ligand gene: Linkage studies in Type I diabetes mellitus families. Diabetologia 2002; 45:134-139.
43. Pugliese A, Zeller M, Fernandez A et al. The insulin gene is transcribed in the human thymus and transcription levels correlate with allelic variation at the ins vntr-iddm2 susceptibility locus for type-1 diabetes. Nat Genet 1997; 15:293-297.
44. Vafiadis P, Ounissi-Benkalha H, Palumbo M et al. Class III alleles of the variable number of tandem repeat insulin polymorphism associated with silencing of thymic insulin predispose to type 1 diabetes. J Clin Endocrinol Metab 2001; 86:3705-3710.
45. Pociot F, Lorenzen T, Nerup J. A manganese superoxide dismutase (SOD2) gene polymorphism in insulin-dependent diabetes mellitus. Dis Markers 1993; 11:267-274.
46. Andrade J, Conde M, Ramirez R et al. Protection from nicotinamide inhibition of interleukin-1-beta-induced rin cell nitric-oxide formation is associated with induction of mnsod enzyme-activity. Endocrinology 1996; 137:4806-4810.
47. Borgstahl GE, Parge HE, Hickey MJ et al. The structure of human mitochondrial manganese superoxide dismutase reveals a novel tetrameric interface of two 4-helix bundles. Cell 1992; 71:107-118.
48. Hirose K, Longo DL, Oppenheim JJ et al. Overexpression of mitochondrial manganese superoxide dismutase promotes the survival of tumor cells exposed to interleukin-1, tumor necrosis factor, selected anticancer drugs, and ionizing radiation. Faseb J 1993; 7:361-368.
49. Johannesen J, Pie A, Pociot F et al. Linkage of the human inducible nitric oxide synthase gene to type 1 diabetes. J Clin Endocrinol Metab 2001; 86:2792-2796.
50. Warpeha K, Xu W, Liu L et al. Genotyping and functional analysis of a polymorphic (CCTTT)(n) repeat of NOS2A in diabetic retinopathy. Faseb J 1999; 13:1825-1832.
51. Johannesen J, Tarnow L, Parving H et al. CCTTT-repeat polymorphism in the human NOS2-promoter confers low risk of diabetic nephropathy in type 1 diabetic patients. Diabetes Care 2000; 23:560-562.
52. Hegazy D, O'Reilly D, Yang B et al. NFkappaB polymorphisms and susceptibility to type 1 diabetes. Genes Immun 2001; 2:304-308.
53. Komaki S, Kohno M, Matsuura N et al. The polymorphic 43Thr bcl-2 protein confers relative resistance to autoimmunity: an analytical evaluation. Human Genet 1998; 103:435-440.
54. Heding P, Karlsen A, Veijola R et al. No evidence of a functionally significant polymorphism of the BCL2 gene in Danish, Finnish and Basque type 1 diabetes families. Genes Immun 2001; 2:398-400.
55. Nerup J, Mandrup-Poulsen T, Helquist S et al. On the pathogenesis of IDDM. Diabetologia 1994; 37:82-89.
56. De-Blasio BF, Bak P, Pociot F et al. Onset of type I diabetes - A dynamical instability. Diabetes 1999; 48:1677-1685.
57. Spinas GA, Hansen BS, Linde S et al. Interleukin-1 dose-dependently effects the biosynthsis of (pro) insulin in isolated rat islets of Lanferhans. Diabetologia 1987; 30:474-480.
58. Helqvist S, Polla BS, Johannesen J et al. Heat shock protein induction in rat pancreatic islets by recombinant human interleukin 1 beta. Diabetologia 1991; 34:150-156.
59. Eizirik DL, Björklund A, Welsh N. Interleukin-1-induced expression of nitric oxide synthase in insulin producing cells is preceded by c-fos induction and depends on gene transcription and protein synthesis. FEBS Lett 1993; 317:62-66.
60. Andersen HU, Fey SJ, Larsen PM et al. Interleukin-1β induced changes in the protein expression of rat islets: A computerized database. Electrophoresis 1997; 18:2091-2103.
61. Chen MC, Schuit F, Pipeleers DG et al. IL-1beta induces serine protease inhibitor 3 (SPI-3) gene expression in rat pancreatic beta-cells. Detection by differential display of messenger RNA. Cytokine 1999; 11:856-862.

62. Wilkins MR, Pasquali C, Appel RD et al. From proteins to proteomes: Large scale protein identification by two-dimensional electrophoresis and amino acid analysis. Biotechnology (NY) 1996; 14:61-65.
63. Andersen HU, Mose Larsen P, Fey SJ et al. Two-dimensional gel electrophoresis of rat islet proteins. Interleukin-1β-induced changes in protein expression are reduced by L-argenine depletion and nicotinamide. Diabetes 1995; 44:400-407.
64. Christensen UB, Larsen PM, Fey SJ et al. Islet protein expression changes during diabetes development in islet syngrafts in BB-DP rats and during rejection of BB- DP islet allografts. Autoimmunity 2000; 32:1-15.
65. Rieneck K, Bovin LF, Josefsen K et al. Massive parallel gene expression profiling of RINm5F pancreatic islet β-cells stimulated with interleukin-1β. APMIS 2000; 108:855-872.
66. Cardozo AK, Kruhøffer M, Leeman R et al. Identification of novel cytokine induced genes in pancreatic β-cells by high density oligonucleotide arrays. Diabetes 2001; 50:909-920.
67. Cardozo AK, Heimberg H, Heremans Y et al. A Comprehensive analysis of cytokine-induced and NFκB-dependent genes in primary rat pancreatic β-cells. J Biol Chem 2001; 276:48879-48886.
68. Larsen PM, Fey SJ, Larsen MR et al. Proteome analysis of interleukin-1beta-induced changes in protein expression in rat islets of Langerhans. Diabetes 2001; 50:1056-1063.
69. Sandler S, Bendtzen K, Eizirik DL et al. Metabolism and β-cell function of rat pancreatic islets exposed to human interleukin-1β in the presence of high glucose concentration. Immunol Lett 1990; 26:245-252.
70. Hostens K, Pavlovic D, Zambre Y et al. Exposure of human islets to cytokines can result in disproportionately elevated proinsulin release. J Clin Invest 1999; 104:67-72.
71. Tiedge M, Lortz S, Munday R et al. Protection against the cooperative toxicity of nitric oxide and free radicals by overexpression of antioxidant enzymes in bioengineered insulin-producing RINm5F cells. Diabetologia 1999; 42:849-855.
72. Nagatsu T, Ichinose H. Regulation of pteridine-requiring enzymes by the cofactor tetrahydrobiopterin. Mol Neurobiol 1999; 19:76-96.
73. Oyadomari S, Takeda K, Takiguchi M et al. Nitric oxide-induced apoptosis in pancreatic beta cells is mediated by the endoplasmic reticulum stress pathway. Proc Natl Acad Sci USA 2001; 98:10845-10850.
74. Sheng M, Thompson MA, Greenberg ME. CREB: A Ca(2+)-regulated transcription factor phosphorylated by calmodulin-dependent kinases. Science 1991; 252:1427-1430.
75. Asai A, Qiu JH, Narita Y et al. High level calcineurin activity predisposes neuronal cells to apoptosis. J Biol Chem 1999; 274:34450-34458.
76. Delaney CA, Pavlovic D, Hoorens A et al. Cytokines induce deoxyribonucleic acid strand breaks and apoptosis in human pancreatic islet cells. Endocrinology 1997; 138:2610-2614.
77. Karlsen AE, Pavlovic D, Nielsen K et al. Interferon-gamma induces interleukin-1 converting enzyme expression in pancreatic islets by an interferon regulatory factor-1-dependent mechanism. J Clin Endocrinol Metab 2000; 85:830-836.
78. Schulze-Osthoff K, Ferrari D, Los M et al. Apoptosis signaling by death receptors. Eur J Biochem 1998; 254:439-459.
79. Eizirik DL, Darville MI. β-cell apoptosis and defense mechanisms: Lessons from type 1 diabetes. Diabetes 2001; 50(Suppl 1):S64-S69.
80. Scarim AL, Heitmeier MR, Corbett JA. Heat shock inhibits cytokine-induced nitric oxide synthase expression by rat and human islets. Endocrinology 1998; 139:5050-5057.
81. Burkart V, Liu H, Bellmann K et al. Natural resistance of human beta cells toward nitric oxide is mediated by heat shock protein 70. J Biol Chem 2000; 275:19521-19528.
82. Ammendrup A, Maillard A, Nielsen K et al. The c-Jun amino-terminal kinase pathway is preferentially activated by interleukin-1 and controls apoptosis in differentiating pancreatic beta-cells. Diabetes 2000; 49:1468-1476.
83. Bonny C, Oberson A, Steinmann M et al. IB1 reduces cytokine-induced apoptosis of insulin-secreting cells. J Biol Chem 2000; 275:16466-16472.
84. Giannoukakis N, Rudert WA, Trucco M et al. Protection of human islets from the effects of interleukin-1β by adenoviral gene transfer of an IκB repressor. J Biol Chem 2000; 275:36509-36513.
85. Heimberg H, Heremans Y, Jobin C et al. Inhibition of cytokine-induced NF-κB activation by adenovirus- mediated expression of a NF-κB super-repressor prevents β-cell apoptosis. Diabetes 2001; 50:2219-2224.
86. Karin M, Ben Neriah Y. Phosphorylation meets ubiquitination: The control of NF-[kappa]B activity. Annu Rev Immunol 2000; 18:621-663.
87. Schindler C. Cytokines and JAK-STAT signaling. Exp Cell Res 1999; 253:7-14.

88. John NE, Andersen HU, Fey SJ et al. Cytokine- or chemically derived nitric oxide alters the expression of proteins detected by two-dimensional gel electrophoresis in neonatal rat islets of Langerhans. Diabetes 2000; 49:1819-1829.
89. Chang L, Karin M. Mammalian MAP kinase signalling cascades. Nature 2001; 410:37-40.
90. Davis RJ. Signal transduction by the JNK group of MAP kinases. Cell 2000; 103:239-252.
91. Welsh N. Interleukin-1β-induced ceramide and diacylglycerol generation may lead to activation of the c-Jun NH2-terminal kinase and the transcription factor ATF-2 in the insulin-producing cell line RINm5F. J Biol Chem 1996; 271:8307-8312.
92. Larsen CM, Wadt KA, Juhl LF et al. Interleukin-1β-induced rat pancreatic islet nitric oxide synthesis requires both the p38 and extracellular signal-regulated kinase 1/2 mitogen-activated protein kinases. J Biol Chem 1998; 273:15294-15300.
93. Andersen NA, Larsen CM, Mandrup-Poulsen T. TNFα and IFNγ potentiate IL-1β induced mitogen activated protein kinase activity in rat pancreatic islets of Langerhans. Diabetologia 2000; 43:1389-1396.
94. Pavlovic D, Andersen NA, Mandrup-Poulsen T et al. Activation of extracellular signal-regulated kinase (ERK)1/2 contributes to cytokine-induced apoptosis in purified rat pancreatic β-cells. Eur Cytokine Netw 2000; 11:267-274.
95. Saldeen J, Lee JC, Welsh N. Role of p38 mitogen-activated protein kinase (p38 MAPK) in cytokine-induced rat islet cell apoptosis. Biochem Pharmacol 2001; 61:1561-1569.
96. Bonny C, Oberson A, Negri S et al. Cell-permeable peptide inhibitors of JNK: Novel blockers of beta-cell death. Diabetes 2001; 50:77-82.
97. Yasukawa H, Sasaki A, Yoshimura A. Negative regulation of cytokine signaling pathways. Annu Rev Immunol 2000; 18:143-164.
98. Karlsen AE, Ronn SG, Lindberg K et al. Suppressor of cytokine signaling 3 (SOCS-3) protects beta -cells against interleukin-1beta - and interferon-gamma -mediated toxicity. Proc Natl Acad Sci USA 2001; 98:12191-12196.
99. Chapple IL. Reactive oxygen species and antioxidants in inflammatory diseases. J Clin Periodontol 1997; 24:287-296.
100. Kwon G, Corbett JA, Rodi CP et al. Interleukin-1 beta-induced nitric oxide synthase expression by rat pancreatic beta-cells: evidence for the involvement of nuclear factor kappa B in the signaling mechanism. Endocrinology 1995; 136:4790-4795.
101. Stamler JS. Redox signaling: nitrosylation and related target interactions of nitric oxide. Cell 1994; 78:931-936.
102. Eizirik DL, Sandler S, Welsh N et al. Cytokines suppress human islet function irrespective of their effects on nitric oxide generation. J Clin Invest 1994; 93:1968-1974.
103. Rabinovitch A, Suarez-Pinzon WL, Strynadka K et al. Human pancreatic islet β-cell destruction by cytokines is independent of nitric oxide production. J Clin Endocrinol Metab 1994; 79:1058-1062.
104. Liu D, Pavlovic D, Chen MC et al. Cytokines induce apoptosis in β-cells isolated from mice lacking the inducible isoform of nitric oxide synthase ($iNOS^{-/-}$). Diabetes 2000; 49:1116-1122.
105. Binzer J, Andersen NA, Størling J et al. Interleukin-1β induced p38 and c-jun-N-terminal kinase activities in isolated rat pancreatic islets is nitric oxide dependent. Diabetes Metab 2001; 17:A39.
106. Rabinovitch A, Suarez-Pinzon WL, Strynadka K et al. Human pancreatic islet beta-cell destruction by cytokines involves oxygen free radicals and aldehyde production. J Clin Endocrinol Metab 1996; 81:3197-3202.
107. Lenzen S, Drinkgern J, Tiedge M. Low antioxidant enzyme gene expression in pancreatic islets compared with various other mouse tissues. Free Radic Biol Med 1996; 20:463-466.
108. Tiedge M, Lortz S, Drinkgern J et al. Relation between antioxidant enzyme gene expression and antioxidative defense status of insulin-producing cells. Diabetes 1997; 46:1733-1742.
109. Sumoski W, Baquerizo H, Rabinovitch A. Oxygen free radical scavengers protect rat islet cells from damage by cytokines. Diabetologia 1989; 32:792-796.
110. Rabinovitch A, Suarez-Pinzon WL, Sorensen O et al. Inducible Nitric Oxide Synthase (iNOS) in Pancreatic Islet of Nonobese Diabetic Mice: Identification of iNOS-Expression Cells and Relationships to cytokines Expressed in Islets. Endocrinology 1996; 137:2093-2099.
111. Lortz S, Tiedge M, Nachtwey T et al. Protection of insulin-producing RINm5F cells against cytokine-mediated toxicity through overexpression of antioxidant enzymes. Diabetes 2000; 49:1123-1130.
112. Welsh N, Eizirik DL, Bendtzen K et al. Interleukin-1β-induced nitric oxide production in isolated rat pancreatic islets requires gene transcription and may lead to inhibition of the Krebs cycle enzyme aconitase. Endocrinology 1991; 129:3167-3173.

113. Radons J, Heller B, Burkle A et al. Nitric oxide toxicity in islet cells involves poly(ADP-ribose) polymerase activation and concomitant NAD+ depletion. Biochem Biophys Res Commun 1994; 199:1270-1277.
114. Messmer UK, Ankarcrona M, Nicotera P et al. p53 expression in nitric oxide-induced apoptosis. FEBS Lett 1994; 355:23-26.
115. Urano F, Wang X, Bertolotti A et al. Coupling of stress in the ER to activation of JNK protein kinases by transmembrane protein kinase IRE1. Science 2000; 287:664-666.
116. Fuchs SY, Adler V, Pincus MR et al. MEKK1/JNK signaling stabilizes and activates p53. Proc Natl Acad Sci USA 1998; 95:10541-10546
117. Maundrell K, Antonsson B, Magnenat E et al. Bcl-2 undergoes phosphorylation by c-Jun N-terminal kinase/stress- activated protein kinases in the presence of the constitutively active GTP-binding protein Rac1. J Biol Chem 1997; 272:25238-25242.
118. Fan M, Goodwin M, Vu T et al. Vinblastine-induced phosphorylation of Bcl-2 and Bcl-XL is mediated by JNK and occurs in parallel with inactivation of the Raf-1/MEK/ERK cascade. J Biol Chem 2000; 275:29980-29985.
119. Kharbanda S, Saxena S, Yoshida K et al. Translocation of SAPK/JNK to mitochondria and interaction with Bcl-x(L) in response to DNA damage. J Biol Chem 2000; 275:322-327.
120. Wang L, Bhattacharjee A, Zuo Z et al. A low voltage-activated $Ca2^{+}$ current mediates cytokine-induced pancreatic beta-cell death. Endocrinology 1999; 140:1200-1204.
121. Zaitsev SV, Appelskog IB, Kapelioukh IL et al. Imidazoline compounds protect against interleukin 1beta-induced beta- cell apoptosis. Diabetes 2001; 50 Suppl 1:S70-S76.
122. Størling J, Zaitsev SV, Andersen NA et al. Calcium signalling via L-type calcium channel mediates IL-1-induced activation of C-jun N-terminal kinase and p38 in pancrearic beta-cells. Diabetologia 2001; 44:A 129.
123. Thornberry NA, Lazebnik Y. Caspases: Enemies within. Science 1998; 281:1312-1316.
124. Abrams JR, Lebwohl MG, Guzzo CA et al. CTLA4Ig-mediated blockade of T-cell costimulation in patients with psoriasis vulgaris. J Clin Invest 1999; 103:1243-1252.
125. Abrams JR, Kelley SL, Hayes E et al. Blockade of T lymphocyte costimulation with cytotoxic T lymphocyte-associated antigen 4-immunoglobulin (CTLA4Ig) reverses the cellular pathology of psoriatic plaques, including the activation of keratinocytes, dendritic cells, and endothelial cells. J Exp Med 2000; 192:681-693.
126. Guinan EC, Boussiotis VA, Neuberg D et al. Transplantation of anergic histoincompatible bone marrow allografts. N Engl J Med 1999; 340:1704-1714.
127. Finck BK, Linsley PS, Wofsy D. Treatment of murine lupus with ctla4ig. Science 1994; 265:1225-1227.
128. Sfikakis PP, Stamatis-Nick L. CTLA-4 in rheumatoid arthritis. In: Pociot F, ed. CTLA-4 in Autoimmune Diseases. Georgetown, TX: Landes Bioscience, 2004; In press.
129. Nomikos IN, Prowse SJ, Carotenuto P et al. Combined treatment with nicotinamide and desferrioxamine prevents islet allograft destruction in NOD mice. Diabetes 1986; 35:1302-1304.
130. Nomikos IN, Wang Y, Lafferty KJ. Involvement of O2 radicals in 'autoimmune' diabetes. Immunol Cell Biol 1989; 67(Pt 1):85-87.
131. Pileggi A, Molano RD, Berney T et al. Heme oxygenase-1 induction in islet cells results in protection from apoptosis and improved in vivo function after transplantation. Diabetes 2001; 50:1983-1991.
132. Kristiansen OP, Pociot F, Johannesen J et al. Linkage disequilibrium testing of four interleukin-1 gene- cluster polymorphisms in Danish multiplex families with insulin-dependent diabetes mellitus. Cytokine 2000; 12:171-175.
133. Awata T, Matsumoto C, Urakami T et al. Association of polymorphism in the interferon g gene with IDDM. Diabetologia 1994; 37:1159-1162.
134. Pociot F, Veijola R, Johannesen J et al. Analysis of an interferon-gamma gene (ifng) polymorphism in danish and finnish insulin-dependent diabetes-mellitus (iddm) patients and control subjects. J Interferon Cytokine Res 1997; 17:87-93.
135. Pravica V, Asderakis A, Perrey C et al. In vitro production of IFN-gamma correlates with CA repeat polymorphism in the human IFN-gamma gene. Eur J Immunogenet 1999; 26:1-3.
136. Jahromi M, Millward A, Demaine A. A CA repeat polymorphism of the IFN-gamma gene is associated with susceptibility to type 1 diabetes. J Interferon Cytokine Res 2000; 20:187-190.
137. Wilson AG, Devries N, Pociot F et al. An allelic polymorphism within the human tumor-necrosis-factor-alpha promoter region is strongly associated with HLA- A1, HLA-B8, and HLA-DR3 alleles. J Exp Med 1993; 177:557-560.
138. Pociot F. Insulin-dependent diabetes mellitus - a polygenic disorder ? Dan Med Bull (Thesis) 1996; 43:216-248.

139. Kristiansen OP, Larsen ZM, Pociot F. CTLA-4 in autoimmune diseases-a general susceptibility gene to autoimmunity? Genes Immun 2000; 1:170-184.
140. Ghabanbasani MZ, Buyse I, Legius E et al. Possible association of CD3 and CD4 polymorphisms with inslin-dependent diabetes mellitus (IDDM). Clin Exp Immunol 1994; 97:517-521.
141. Kristiansen OP, Zamani M, Johannesen J et al. Linkage and association between a CD4 gene polymorphism and iddm in danish iddm patients. Diabetes 1998; 47:281-283.
142. Bennett ST, Lucassen AM, Gough SCL et al. Susceptibility to human type 1 diabetes at *IDDM2* is determined by tandem repeat variation at the insulin gene minisatellite locus. Nat Genet 1995; 9:284-291.
143. Johannesen J, Pociot F, Kristiansen OP et al. No evidence for linkage in the promoter region of the inducible nitric oxide synthase gene (NOS2) in a Danish type 1 diabetes population. Genes Immun 2000; 1:362-366.

CHAPTER 6

The Immunobiology of Pancreatic Islet Transplantation

Marilyne Coulombe and Ronald G. Gill

Insulin-dependent diabetes mellitus (IDDM, Type 1A diabetes) is the result of autoimmune destruction of pancreatic islet beta cells.[1] As such, the only cure for this irreversible loss of islet tissue is through the replacement of insulin-producing capacity. Presently, this can be accomplished by the transplantation of either the whole pancreas organ or isolated pancreatic islets. The rationale for pancreas or islet grafting is to prevent the debilitating complications associated with the disease process. Indeed, both pancreas[2-5] and islet[6-8] transplantation has been shown to prevent or arrest diabetic complications. There has been controversy over the relative utility of whole pancreas versus islet transplantation in insulin-dependent diabetes. In clinical practice, pancreas grafting has had an advantage over islet transplantation in terms of the proportion of recipients achieving a significant period of insulin-independence.[9] However, islet transplantation could provide important advantages unavailable with regard to pancreas transplantation. These include the potential for modifying tissue immunogenicity (immunomodulation) through in vitro culture or gene therapy approaches, the application of various forms of tissue encapsulation to avoid immune recognition (immuno-isolation), the establishment of cryopreserved tissue banks and the potential for grafting in immune-privileged sites. Islet grafting also allows for the use of alternative tissue sources, potentially the use of tissue from different species (xenografts) thus overcoming the shortage of available allogeneic tissue. Unfortunately, until recently, islet transplantation has shown only minimal evidence of clinical success in Type 1A (autoimmune) diabetic recipients.[10-14]

Type 1A diabetes represents a unique problem for the field of transplantation. Unlike other conditions of end-stage organ failure where transplantation is required to sustain life, diabetes is not immediately life-threatening due to the availability of exogenous insulin therapy. Thus, for type 1 diabetes, the risks associated with current regimens of immunosuppression used in transplantation, such as neoplasia, opportunistic infections and nephrotoxicity, certainly rival the consequences of disease-associated complications. Therefore, the criteria for islet transplantation in diabetic recipients prior to the manifestations of disease complications must be much more stringent than that for patients with other forms of organ failure. Due to this restriction, islet replacement was initially performed only in diabetic patients receiving simultaneous solid organ grafts (usually kidney) with concomitant immunosuppression.[14] Since 1990, only 8.2% of 267 IDDM patients have been insulin-independent for more than one year.[14] Recently, however, the feasibility of islet transplantation for Type 1 diabetic patients was demonstrated by a group in Edmonton, Canada[15] when they achieved 7 consecutive cases of insulin independence in such recipients. This required two sequential transplants of large numbers of high quality islets, induction therapy with antibody to the IL-2 receptor and a steroid-free immunosuppressive protocol.[15] Although this is a remarkable advance for the field of islet transplantation, two allogeneic donors were required to achieve freedom from exogenous insulin therapy and these recipients are bound to life-long immunosuppressive therapy. The future goal of

Immunology of Type 1 Diabetes, Second Edition, edited by George S. Eisenbarth.

transplantation, however, must be the grafting of islet tissue in the diabetic recipient prior to the onset of disease complications, using minimal host immunosuppression. As such, the purpose of this chapter is to examine the immunological obstacles hampering the clinical application of islet transplantation.

There are three main immunological reactions that pose barriers to the large-scale application of clinical islet transplantation for Type 1 diabetes:

1. Allograft rejection: the immune response to tissue derived from histoincompatible members of the same species,
2. Xenograft rejection: the immune response to tissue derived from a different species and
3. Disease recurrence: the re-enactment of tissue-specific autoimmunity in the transplanted tissue leading to graft destruction.

First, allograft rejection must be prevented by therapies that would reduce or eliminate the requirement for long-term use of immunosuppressive drugs. Secondly, the shortage of available allogeneic cadaveric donors could be resolved by consideration of tissue from other species as a potential source of donor material (xenografts) but this will require a greater understanding of the xenogeneic response. Third, there is evidence that islet transplants can succumb to disease recurrence, the reoccurrence of the underlying autoimmune pathogenesis that can destroy transplanted islet tissue both in animals models of autoimmune diabetes (the nonobese diabetic (NOD) mouse[16-18] and the Bio Breeding (BB) rat[19]) and in humans.[20-22] The nature of disease pathogenesis in islet transplants is a focus of another contribution to this volume and so will not be primarily addressed here. This chapter will focus on these types of T cell dependent reactions encountered in islet transplantation and, in particular, the role of the CD4+ T cell in each case. Understanding of the processes resulting in islet allograft or xenograft destruction in nonautoimmune as well as autoimmune environments will be crucial for developing future strategies to circumvent these responses.

Pancreatic Islet Allografting: The Allograft Paradox

The Two-Signal Hypothesis for T Cell Activation

According to the classic view of allograft immunity, major histocompatibility complex (MHC) antigens are inherently strong immunogens; that is, binding of the MHC molecule by the antigen-specific receptor directly triggered T cell activation.[23] Although allograft rejection clearly represents a response to transplantation antigens, especially MHC antigens, there has been a persistent problem with this classic view. Early attempts to isolate and characterize transplantation antigens found them to be quite weak immunogens.[24] Later studies by Batchelor and coworkers[25] demonstrated that MHC antigens, either class I or class II, isolated from the cell surface were poor immunogens. These observations lead us to a transplantation paradox: allograft immunity represents a violent response to antigens that are inherently weak immunogens.

The solution to this paradox was provided by the proposition of Lafferty and Cunningham that two signals were required for T cell activation,[26] a concept derived from the original two-signal model for self-nonself discrimination developed by Bretscher and Cohn.[27] According to the Lafferty/ Cunningham model, signal one (antigen) was provided by binding of the T cell receptor (TcR) and signal two was provided by an inductive molecule, or costimulator (CoS), produced by a metabolically active antigen-presenting cell (APC) (Fig. 1). Although previous studies show that cytokines can demonstrate CoS activity in vitro,[28] recent work has brought attention to the role of accessory molecule interactions between T cells and APCs involved in providing costimulatory signals required for T cell activation. Key cell surface molecules involved in this process are CD28,[29] CTLA-4[30] and CD154[31] expressed by T cells interacting with the B7/BB1 and CD40 molecules expressed by APCs and activated B cells.[32] Other accessory molecules expressed by APCs, such as ICAM-1 and LFA[33-36] may also have important costimulatory functions. Regardless of the exact nature of costimulatory signals, cells capable of exhibiting CoS activity will have a stimulatory phenotype, whereas other cells,

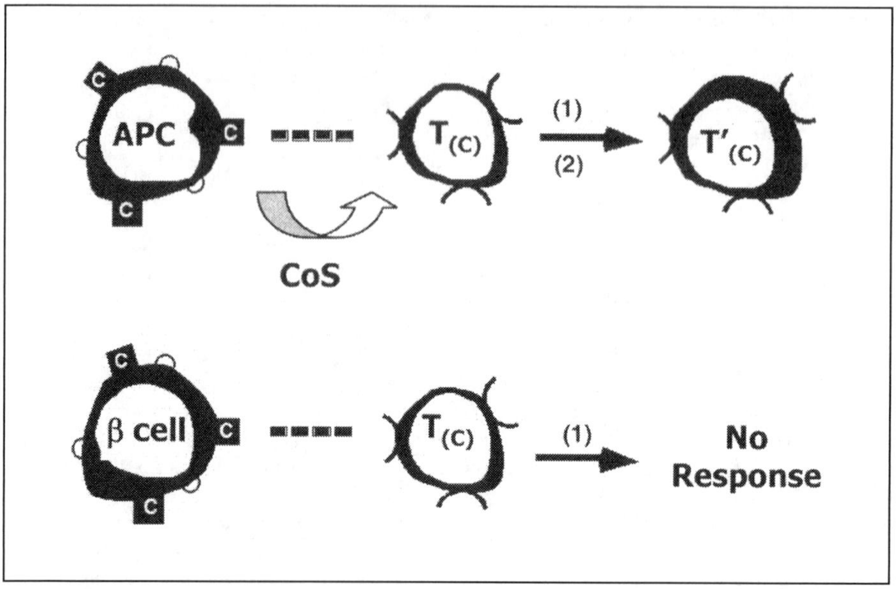

Figure 1. The stimulator cell model. Alloantigens presented by a metabolically active APC will provide a source of both signals (antigen and costimulator) required for T cell activation (T cell specificity indicated by the parentheses). Alloantigens presented on the surface of a parenchymal cell, such as a β cell, will not be directly immunogenic because these cells cannot provide a source of the second signal. According to this model, costimulator activity is triggered by engagement of a control structure ('C') on the surface of the APC.

such as tissue parenchymal cells, which express the same antigens but do not produce CoS activity will have a nonstimulatory phenotype.

The role of MHC in this process is that of a control structure; that is, CoS activity is triggered by engagement of MHC on the surface of a stimulatory APC. Experimental evidence supporting the role of MHC as a control molecule comes from a study by Nabavi and coworkers that indicates that class II MHC signaling triggers the surface expression of the costimulatory molecule B7.[37] This two-signal model provides a solution to the transplantation paradox. The immunogenicity of MHC antigen is not due to its inherent antigenic properties, but rather to its functional role as a control structure for the expression of CoS activity. This model predicts that MHC antigens will only be immunogenic while on the surface of a metabolically active APC; isolated MHC antigens or MHC antigens on the surface of parenchymal cells will not be immunogenic. The significant contribution of this two-signal hypothesis to transplantation biology is the concept that alloantigen alone is not the barrier to allografting. Rather, tissue immunogenicity will be a function of APCs, or "passenger leukocytes", resident within the graft.[38]

Alteration of Tissue Immunogenicity

The notion that alloantigen presentation between stimulatory APCs and parenchymal cells within the graft represent qualitatively different forms of antigen presentation led back to an early concept that graft immunogenicity can be reduced by appropriate tissue pretreatment.[24] That is, selective depletion of stimulatory APCs within the graft should lead to prolonged graft acceptance. Tissue pretreatment has been strikingly successful for facilitating allograft acceptance of endocrine tissues. Early work indicated that pretreatment of thyroid tissues in 95% O_2 culture would facilitate long-term graft acceptance in nonimmunosuppressed, allogeneic recipient mice.[39] This technique was applied to the modification of pancreatic islet allograft immunogenicity.[40] Grafting experimentally induced diabetic mice with nontreated islet allografts leads to euglycemia followed by a rapid return to hyperglycemia due to graft rejection. Following organ culture in 95% O_2 for 7 days, islet allografts function indefinitely in recipient mice.[40,41] Application of 95% O_2 culture has also allowed successful islet allografting in rats[42] and in outbred mice[43] though short-term immunosuppression with cyclosporin A was necessary for long-term graft survival in this latter study. A variety of other tissue pretreatment techniques have likewise led to long-term islet allograft acceptance. Graft pretreatment with low temperature culture,[44] anti-Ia or dendritic cell antibodies[45-47] ultraviolet irradiation[48] and low pH media[49] have all facilitated islet allograft acceptance. Hegre and coworkers[50] have shown that perinatal rat islets isolated by in vitro cultivation are accepted in nonimmunosuppressed, allogeneic recipients. Further, appropriate graft pretreatment techniques have led to prolonged graft acceptance of a variety of tissues including thyroid and parathyroid,[39,51,52] ovarian tissue,[53] and keratinocytes.[54] Taken together, a wide range of techniques with a variety of donor-recipient strain combinations indicate that allograft acceptance can be facilitated by appropriate pretreatment of the graft.

Antigenicity Versus Immunogenicity of MHC Alloantigens

A key question centers on the mechanism whereby pretreated tissues survive in immunocompetent, allogeneic recipients. Is the immunogenicity of such pretreated grafts altered due to the modulation of graft antigens, or by the elimination/inactivation of passenger leukocytes? This issue was important in light of the finding that hyperbaric oxygen culture of thyroid tissue led to reduced antigen expression of the graft.[52,55] Studies by Hullet and coworkers[52] show that thyroid grafts cultured in hyperbaric oxygen survive even in recipient mice primed against donor antigens. These results indicated that the regulation of graft antigen expression might play a role in subsequent graft survival. Although antigen modulation may occur during some forms of organ culture, a number of studies indicate that this modulation is not necessary for allograft acceptance. That is, loss of immunogenicity does not require a loss of antigenicity in the graft.

Varied studies have indicated that neither class I nor class II MHC molecules are the primary barrier to allografting. In the late 1970's, Batchelor's laboratory addressed this issue of graft immunogenicity versus antigenicity by examining the role of donor antigen expression in long-term kidney allograft survival. These investigators demonstrated that kidney grafts would survive in allogeneic recipients that had been tolerized by a regimen of donor-specific antibody administration.[56] When such long-surviving kidney allografts were transplanted from the primary recipient to a secondary recipient of the same haplotype, the kidney did not elicit a rejection response in the untreated, secondary recipient.[56] The failure of long-surviving kidney grafts to activate a rejection response could not be attributed to a lack of either class I or class II antigen expression in the graft.[57] This led these investigators to conclude that the effect resulted from a gradual replacement of donor-derived passenger leukocytes in the graft with host-derived APC. Importantly, they showed that immunization with donor strain dendritic cells (a source of stimulatory APCs) in very low numbers would trigger rejection of the established kidney graft.[57] Clearly, the antigens of the grafted kidney could not elicit alloimmunity but could serve as a target of immune destruction once the animal was primed.

Similarly, islet allografts cultured in 95% O_2 are vulnerable to rejection by active immunization of the recipient.[58] Challenge of the recipient animal with donor-type APCs during the immediate post-transplant period triggers acute rejection of the established islet graft. Active immunization of the host also triggers the rejection of established cultured thyroid[59] and keratinocyte[54] allografts. These results illustrate the distinction between graft antigenicity and immunogenicity; the cultured graft cannot trigger allograft immunity but can serve as a target for immune destruction. The issue of class II MHC immunogenicity was elegantly examined by Markmann and coworkers.[60] These investigators studied the response to fetal pancreas grafts from transgenic mice expressing a class II MHC alloantigen (I-E) under control of the insulin promoter (ins-I-E), leading to tissue-specific, ectopic expression of the I-E molecule on the surface of pancreatic islet β cells. Importantly, the I-E class II MHC molecule was expressed on islet parenchymal cells and not on hematopoietic cells.[61] Despite a high level of expression of the transgene product on the β cell surface, such fetal pancreas grafts did not trigger alloimmunity in I-E-parental strain mice. However, immunizing recipient mice with I-E+ spleen cells (as a source of APCs) triggered a prompt rejection response to the ins-I-E graft, indicating that the ins-I-E fetal pancreas could serve as a target for immune destruction. What these experiments clearly demonstrated was that the class II MHC antigen was not inherently immunogenic and so was not sufficient to trigger allograft immunity.

LaRosa and Talmage[55] later tested the relationship between donor antigen expression and the survival of cultured thyroid allografts. In these experiments thyroid grafts were cultured in 95% O_2 to reduce tissue immunogenicity. These cultured thyroid grafts were then grafted directly or were pretreated with interferon gamma (IFN-γ) to induce high levels of class I MHC antigen expression on the graft. Whereas fresh (noncultured) thyroid grafts were uniformly rejected, the acceptance of cultured grafts was not altered with IFN-γ treatment of the graft. That is, acceptance of the cultured graft was not due simply to the alteration of antigen expression. Our own experiments using IFN-γ treated oxygen-cultured islet allografts also indicate that the enhanced expression of class I MHC molecules on islet tissue does not increase the rejection response.[62] Conversely, we have demonstrated that transgenic expression of the CoS molecule B7-1 (CD80) on islet β cells led to acute allograft rejection even when donor-derived APCs were putatively eliminated by pretransplant culture.[62] Taken together, these studies indicate that donor-derived costimulatory activity, not MHC alloantigens per se, plays the dominant role in stimulating allorejection. Given that the goal of islet transplantation is to eliminate or reduce the immunosuppression of the recipient, tissue pretreatment aimed at eliminating APCs from the graft may, at least in part, be a useful tool in future clinical islet allografting.

Pancreatic Islet Xenografting: The Xenograft Paradox

A major dilemma facing clinical organ and tissue grafting is the shortage of allogeneic donor tissue for transplantation. This problem is amplified in islet transplantation where multiple donor pancreata are required to provide sufficient islet mass to normalize blood glucose levels in Type I diabetic recipients.[15] Therefore, the consideration of tissues from other species (xenografts) as a potential source of donor material is necessary. Unfortunately, the nature of cellular and humoral responses to xenogeneic tissues appears to differ substantially from allograft immunity.[63]

The classification by Calne[64] denoted xenografts as discordant or concordant based on the presence or absence, respectively, of preformed or natural antibodies against the donor. The presence of such natural antibodies roughly correlates with the phylogenetic disparity between species and are thought to be responsible for the initiation of hyperacute rejection which occurs between discordant species combinations.[65-68] A major target of such natural antibodies appears to be carbohydrate moieties expressed on the xenogeneic vascular endothelium.[69] While hyperacute rejection presents a major barrier to the survival of anastomosed solid-organ xenografts, neo-vascularized xenografts — such as skin and cellular transplants — generally do not appear sensitive to hyperacute rejection but rather involve cellular immunity.[65,66,70] While natural antibodies can recognize cellular tissue grafts such as islets,[71,72] our own studies indicate that vigorous islet xenograft rejection can occur independently of xenoreactive antibodies.[73] The relative lack of vulnerability of cellular xenografts to hyperacute rejection is likely due to the fact that the graft vasculature is of host rather than donor origin, thus eliminating the vascular target of antibody-mediated hyperacute rejection.[65,67,70]

Understanding of the cellular immune response to xenogeneic tissues is less clear than is understanding of the allograft response as illustrated by an important paradox concerning xenoreactivity: While cellular xenograft immunity in vivo — apart from hyperacute rejection

— is generally more vigorous than allograft rejection,[65,66,70] the intensity of the T cell-dependent response to xenogeneic APCs in vitro tends to decrease as the phylogenetic disparity between stimulating and responding species increases.[63,66,74-76] This hyporeactivity of the inter-species T cell-APC interaction in vitro could be attributed to three primary defects in the response.

Low Precursor Frequency

The view that the TcR repertoire is positively selected in the thymus using the MHC of the species as a template[77] would leave the possibility that the frequency of xenoreactive T cells might be quite low relative to the high precursor frequency of alloreactive T cells. This possibility would account for the low xenoreactive precursor frequency on a quantitative basis.

Species-Specific Accessory / Coreceptor Molecule Interactions

A number of cell surface molecules, in addition to the antigen-specific TcR, facilitate T cell activation. For example, the CD4 and CD8 coreceptor molecules, which characterize T cell subsets, interact with conserved regions of class I and class II MHC molecules, respectively.[78] Importantly, the interaction of these accessory/coreceptor molecules with their respective ligands can display species-specificity.[76,79] The interaction of CD4 and CD8 with MHC molecules from different species can be quite weak. While human CD4 is capable of functionally associating with mouse class II,[80] the reciprocal binding of mouse CD4 to human class II is quite weak.[81,82] Further, the association of CD8 with the $\alpha 3$ domain of class I MHC also displays species-specificity between human and mouse.[83,84]

Species-Specificity of Cytokines Required for T Cell Activation

A number of cytokines exhibit species specificity/preference such that cytokines necessary for T cell activation may not function across particular species barriers.[74,76,85,86] Studies addressing these issues have found that reconstitution of xenoreactivity in vitro can occur by restoring appropriate accessory/coreceptor molecule interactions[82-84,87] or by the addition of the relevant responder-type cytokines.[74,75,85] Such studies indicate that the potential frequency of xeno-specific T cells may be fairly high with the lack of reactivity being due to species-specific interactions, which are secondary to the TcR repertoire.

'Direct' Versus 'Indirect' Antigen Presentation

In light of these deficiencies in the xenogeneic T cell-APC interaction, a key question is "How does a low reactivity in vitro account for the violent reactivity in vivo?" One solution to this paradox can be inferred by considering the consequence of two pathways of host sensitization by the graft: direct (donor APC-dependent) and indirect (host APC-dependent) presentation (Fig. 2). Direct antigen presentation would be the result of host T cell recognition and activation by donor-derived APCs. Both CD4 and CD8 T cells can be activated by this direct pathway, however, because islets are devoid of class II MHC antigen, class I MHC-restricted CD8 T cells would be predominantly activated in a response to an islet transplant. The dependence of islet allograft immunity on donor APCs implies that rejection of islet allografts occurs via the direct pathway of antigen presentation. It is also this direct pathway of activation that is expected to be sensitive to the species-specific interactions required for xenogeneic T cell activation mentioned above. The indirect pathway would involve the processing and presentation of graft antigens by host APCs. While this indirect (responder APC-dependent) pathway can occur in allorecognition,[88-91] empirically it does not have an adverse effect on APC-depleted islet allografts and may, in fact, be involved in the induction of donor-specific tolerance.[92-94] Some xenogeneic combinations, however, may depend on this indirect pathway for T cell activation. The operation of the indirect pathway in xenoreactivity is shown by the observation that the mouse anti-human[95] and human-anti mouse[96] responses in vitro require responder-type APCs. Allogeneic responses, or some xenogeneic combinations such as mouse anti-rat, do not require responder-type APCs.[6]

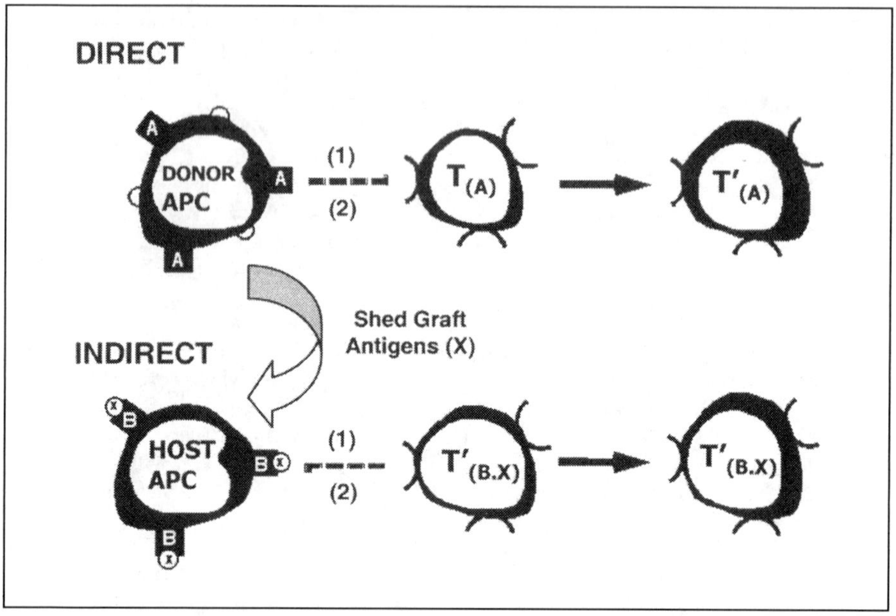

Figure 2. Direct versus indirect antigen presentation. Direct: Donor-type APCs present antigen ('A') to T cells along with an appropriate second signal (costimulator) required for T cell activation (T cell specificity indicated by the parentheses). Such a T cell will be specific for graft antigens directly on the surface of donor cells. Indirect: Graft antigens, generically termed 'x', will be processed by host-type APCs of genotype 'B' and presented in association with class II MHC. Responsive T cells will not be directly graft specific but rather specific for graft antigens in association with host class II MHC (specificity 'B.X').

The function of the indirect (host APC-dependent) pathway may solve the xenograft paradox: Although the xenogeneic APC cannot activate responding T cells via the direct pathway, the antigenic disparity of the xenograft relative to the host might favor a strong response to antigens processed by host APC. This form of xeno-presentation would be expected to activate host CD4 T cells specific for xenogeneic antigens in the context of self class II MHC. The function of this indirect pathway may explain why xenograft rejection is more dependent on CD4 T cells than is allograft rejection in some cases.[97-100]

Role of T Cell Subsets in Islet Graft Rejection: A Central Role for CD4 T Cells

Numerous studies into the mechanism of immune responses to islet transplants lead to the central thesis that CD4+ T cells play a major role in allograft and xenograft rejection as well as autoimmune disease recurrence. For example, anti-CD4 monoclonal antibody therapy can prevent or delay each of these types of responses. Transient in vivo depletion of CD4 T cells prevents the acute rejection of both islet allografts[101-103] and xenografts[98,99,104,105] in nonautoimmune prone strains of mice. In addition, the same therapy will prevent the recurrence of autoimmune diabetes in islet isografts transplanted into spontaneously diabetic NOD mice.[17] Paradoxically, the induction of donor-specific tolerance in several allograft models, including recipients of APC-depleted islets,[106] also appears to be CD4 T cell dependent. However, while each of these processes show a dependence on CD4+ T cells, this does not mean that the role of the CD4 cell is the same in each case. Three possible functions for the CD4+ T cell can be envisioned in the rejection process (Fig. 3): (1) A helper cell for the activation of CD8 T cells which in turn mediate graft destruction, (2) A helper cell for antibody-producing

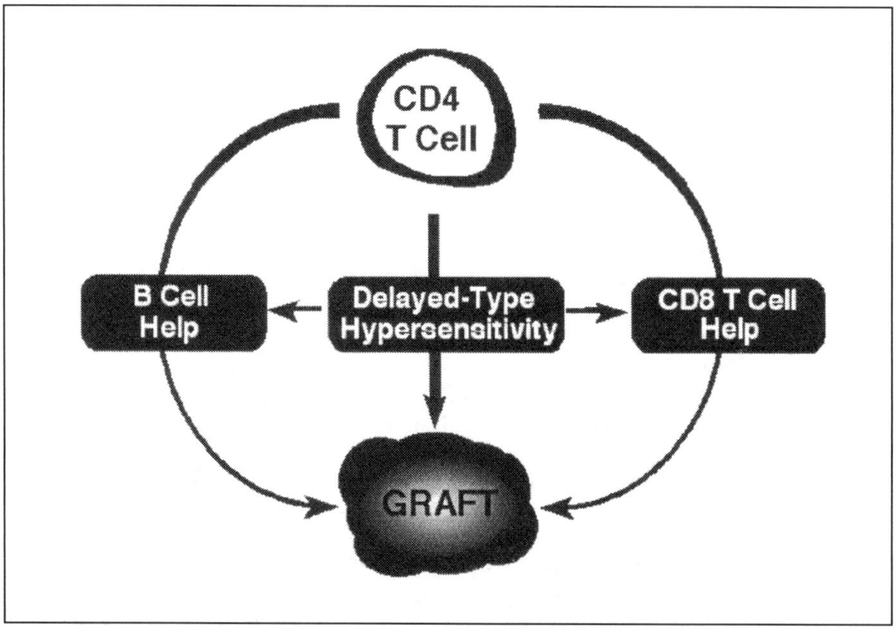

Figure 3. Possible roles for class II MHC-restricted CD4 T cells in the rejection response.

B cells, leading to a graft-specific antibody response, or (3) An effector cell mediating islet graft destruction by direct interaction with the graft and/or through inflammatory tissue damage.[70]

In vivo depletion of either CD4 or CD8 T cells leads to prolonged survival of islet grafts in allogeneic recipients indicating that both subsets are required for allograft rejection.[101-103,107,108] In adoptive transfer studies to severe-combined-immune-deficient (SCID) mice, islet allograft immunity could be reconstituted with unfractionated lymphocytes but not with lymphocytes depleted of either CD4 or CD8 T cells, indicating that both subsets are required.[73] Other studies have demonstrated that primed CD8 T cells are both necessary and sufficient for the rejection of islet allografts.[107] Thus, the primary role of CD4 T cells in islet allograft immunity appears to be to facilitate the function of CD8 T cells that in turn mediate graft rejection.

A difference in the role of CD4 T cells in islet xenograft immunity is demonstrated by examining the requirement for other lymphocyte sub-populations in these responses. Unlike islet allograft immunity, xenograft immunity does not appear to require class I MHC-restricted CD8 T cells.[73,99,108] Our cell transfer studies into severe-combined-immune-deficient C.B-17scid mice show that while islet allograft immunity requires both CD4 and CD8 T cells, CD4 T cells are both necessary and sufficient to reconstitute these animals for islet xenograft immunity.[73,108] B cells were not required for either response. Furthermore, studies show that islet allografts from transgenic mice which are deficient in class I MHC expression enjoy prolonged survival as allografts[109,110] but are acutely rejected as xenografts.[111] In addition, relative to islet allografts, rat islet xenografts show prolonged survival in MHC class II-deficient recipients and purified CD4 T cells fail to adoptively transfer xenograft rejection in immune-deficient mice that are also devoid of host MHC class II antigen.[70] Taken together, these results indicate that neither CD8 T cells nor their corresponding class I MHC target antigens are required for islet xenograft rejection. From these findings we form the hypothesis that a major pathway for islet xenograft immunity is through indirect antigen presentation of xenoantigens in association with host APCs which then elicits a CD4-dependent inflammatory response. Interestingly, such a response has been suggested as a model for CD4-dependent inflammatory islet damage in spontaneous diabetes.[112]

The Role of Donor APCs in Allograft Versus Xenograft Immunity

The difference in the nature of allogeneic versus xenogeneic responses has important implications concerning the role of donor-type APCs (passenger leukocytes) in triggering rejection responses. In the discussion of the allograft response above, the case was made for improving graft survival by eliminating/inactivating stimulatory APCs within the graft which were responsible for triggering host immunity. That islet allograft immunity is dependent on donor APCs implies that allograft rejection occurs predominantly via the direct pathway of antigen presentation. Implicit in this approach is the assumption that the indirect pathway, the presentation of graft antigens by host-type APC, is not a major pathway for eliciting a graft-destructive response (Fig. 2). The observation that APC-depleted islet allografts survive in nonimmunosuppressed recipients is consistent with this concept. The xenograft, however, represents a different situation. If the xenograft response is primarily a function of the indirect pathway of antigen presentation (Fig. 2), then the host response should be dependent of host-type, and not donor-type APCs. This predicts that removal of APCs from islet xenografts should have little influence on the rejection response. This issue remains controversial in that our studies suggest that APC-depleted xenografts are readily rejected[108] while other studies have shown a benefit to islet xenograft survival through tissue pretreatment.[104,105] Thus, the potential role for pretreating xenogeneic tissues to reduce immunogenicity remains to be clarified.

Islet Xenografting in Autoimmune Diabetes

An important goal for islet xenotransplantation will be grafting in spontaneous, autoimmune diabetes. Previous studies show that the recurrence of disease in islet grafts in NOD mice is CD4 T cell dependent and does not require donor-type APCs [17] nor does it require donor class I MHC expression.[109] As such, the pathogenesis of autoimmune destruction to islet transplants appears to be similar to islet xenograft immunity and may be a result of antigen presentation via the indirect pathway. Combined with the finding that islet xenograft immunity is CD4 T cell dependent, anti-CD4 therapy should facilitate xenograft survival in diabetic autoimmune NOD mice. This has been shown to be the case for fetal pig pancreas xenografts.[98,113] An important aspect of these results was that xenografts actually showed superior survival relative to allografts in anti-CD4 treated NOD mice. We have generated similar results whereby rat islet xenografts show prolonged function relative to BALB/c islet allografts in anti-CD4-treated diabetic NOD mice[114] (Table 1). On one hand, such results may reflect the fact that xenograft immunity is more exquisitely CD4-dependent than is allograft immunity. Indeed, CD8 T cells may play a greater role in allograft rejection than in xenograft rejection.[97,99]

However, another hypothesis is that the autoimmune disease process displays a degree of species-specificity; that is, islet-reactive T cells from NOD mice may preferentially recognize murine islet-associated antigens. This concept is supported by the findings by Haskins et al that a panel of islet-reactive T cell clones derived from NOD mice show little cross-reactivity with islet antigens from differing species.[115] The key rationale for considering islet xenotransplantation has generally been to fill the shortage of available allogeneic tissues. However, if autoimmune diabetes indeed is species-specific, then islet xenografts may have an additional benefit by being relatively resistant to the disease process. A clear definition of the degree of cross-species recognition of autoimmune T cells will require identifying the relevant autoantigens which are targeted by the disease process.

Future Direction: Donor-Specific Tolerance Induction

Wide spread application of pancreatic islet grafting in Type I diabetes requires minimal recipient manipulation. Therefore, the ultimate goal of islet grafting, as in transplantation generally, is the induction of donor-specific tolerance via some form of transient induction

Table 1. Allograft versus xenograft survival in anti-CD4-treated NOD mice

Donor	Anti-CD4 Treatment	*n*	Graft Function (days)	MST ± SEM[a]
BALB/c mouse	-	12	0,0,0,0,0,2,5,6, 7,8,9,11	4.0 ± 1.2
BALB/c mouse	+	12	0,0,2,2,5,7,14, 21,30,33,37,70	18.4 ± 6.1
DA rat	-	11	2,2,3,4,4,5,6,6, 6,7,7	4.7 ± 0.6
DA rat	+	15	0,12,21,30,31,33, 40,42,43,45,46, 47,54,66,69	38.6 ± 4.7

Spontaneously diabetic nonobese diabetic (NOD) mice were treated with the anti-CD4 monoclonal antibody GK1.5(10 mg/kg/day) days –7, -1, 0, 1, 2, 7, 14, 21, and 28 relative to transplantation on day 0. Recipients were grafted with 450 untreated pancreatic allografts (BALB/c mouse) or xenografts (DA rat) beneath the kidney capsule. Graft function was indicated by restoration of euglycemia; graft failure was indicated by the first of consecutive blood glucose values above the normal range. Rat xenograft survival was significantly longer (P<0.05) than BALB/c allograft survival.

[a] Mean survival time; SEM=standard error of the mean

regimen. Understanding the basic mechanism(s) of tolerance may lead to clinical graft survival with a brief course of immunomodulatory agents, with or without transient immunosuppression. This has been accomplished by short-term drug treatment,[42,116] donor bone marrow infusion,[117-119] intrathymic islet grafting,[120] and by various monoclonal antibodies that putatively block T cell coreceptors, or costimulatory or adhesion molecules. These tolerizing antibodies bind to CD4 and/or CD8 T cell coreceptors,[101,102,121,122] CD28 / CTLA4[123-125] or CD154[126-129] costimulatory molecules, adhesion molecules such as LFA-1[130,131] or molecules involved in signal transduction.[132] Protocols combining such therapies[133-136] have also been successful and several of these transient therapies have been applied to xenogeneic tissues.[104,137-142] Differences in the nature of the rejection responses to allografts versus xenografts raises the possibility that the nature of tolerance induction to these sources of tissue may differ as well. The induction of donor hematopoietic chimerism has been shown to be an effective tolerogenic protocol for both allografting and xenografting.[118] Also, as in islet allografting, anti-CD4 monoclonal antibody treatment has resulted in long-term acceptance and even tolerance induction to islet xenografts.[104,105,108,113] Interestingly, other forms of recipient treatment, such as administration of the hybrid molecule CTLA4Ig (which interferes with the CTLA-4/B7 interaction),[137] antibodies blocking CD28 and CD40 pathways,[141] cytokine intervention therapies[143-145] and donor-specific antibody treatment[146] have all resulted in marked prolongation of islet xenografts. In most cases, the mechanism(s) of graft prolongation or tolerance induction obtained by these varied approaches and whether similar approaches will be applicable for the induction of tolerance in an autoimmune environment[147-152] remains to be clarified. However, the potential for tissue pretreatment along with an increasing variety of tolerogenic regimens, and the prospect of immuno-isolation technology[153] and gene therapy approaches,[154] provide the foundation for clinically-relevant approaches to islet transplantation in the future.

References

1. Eisenbarth GS. Type I diabetes mellitus: A chronic autoimmune disease. N Engl J Med 1986; 314:1360-1368.
2. Kennedy WR, Navarro X, Goetz FC et al. Effects of pancreatic transplantation on diabetic neuropathy. N Engl J Med 1990; 322:1030-7.
3. Fioretto PF, Steffes MW, Sutherland DER et al. Reversal of lesions of diabetic neuropathy after pancreas transplantation. N Engl J Med 1998; 339:69-75.
4. Gross CR, Limwattananon C, Matthees BJ. Quality of life after pancreas transplantation: A review. Clin Transplantation 1998; 12:351-361.
5. Pearce IA, Ilango B, Sells RA et al. Stabilisation of diabetic retinopathy following simultaneous pancreas andkidney transplant. Br J Opthalmol 2000; 84:736-740.
6. Gray BN, Watkins E. Prevention of vascular complications of diabetes by pancreatic islet transplantation. Arch Surg 1976; 111:254-257.
7. Federlin K, Bretzel RG. The effect of islet transplantation on complications in experimental diabetes of the rat. WorldJ Surg 1984; 8:16-178.
8. Woehrle W, Spitzer D, Linn T et al. The effect of early islet transplantation on prevention ofnephropathy in the spontaneously diabetic BB rat. Transplant Proc 1990; 22:819-920.
9. Robertson RP, Davis C, Larsen J et al. Pancreas and islet transplantation for patients with diabetes. Diabetes Care 2000; 23:112-116.
10. Scharp DW, Lacy PE, Santiago JV et al. Results of our first nine intraportal islet allografts in type 1, insulin-dependent diabetic patients. Transplantation 1991; 51:76-85.
11. Ricordi C, Tzakis AG, Carroll PB et al. Human islet isolation and allotransplantation in 22 consecutive cases. Transplantation 1992; 53:407-414.
12. Warnock GL, Kneteman NM, Ryan EA et al. Long-term follow-up after transplantation ofinsulin-producing pancreatic islets into patients with Type 1 (insulin-dependent) diabetes mellitus. Diabetologia 1992; 35:89-95.
13. Gores PF, Najarian JS, Stephanian E et al. Insulin independence in type Idiabetes after transplantation of unpurified islets from single donor with 15-deoxyspergualin. Lancet 1993; 341:19-21.
14. Hering BJ. International islet transplant registry. Newsletter #8. University of Minnesota 1999.
15. Shapiro AMJ, Lakey JRT, Ryan EA et al. Islet transplantation in seven patients with type I diabetes mellitus using a glucocorticoid-free immunosuppressive regimen. N Engl J Med 2000; 343:230-8.
16. Terada M, Lennartz K, Mullen Y. Allogeneic and syngeneic pancreas transplantation in nonobese diabetic mice. Transplant Proc 1987; 19:960.
17. Wang Y, Pontesili O, Gill RG et al. The role of CD4+ and CD8+ T cells in the destruction of islet grafts by spontaneously diabetic mice. Proc Natl Acad Sci USA 1991; 88:527-531.
18. Wang T, Singh B, Warnock GL et al. Prevention of recurrence of IDDM in islet-transplanted diabetic NOD mice by adjuvant immunotherapy. Diabetes 1992; 41:114-117.
19. Naji A, Silvers W, Kimura H et al. Influence of islet and bone marrow transplantation on thediabetes and immunodeficiency of BB rats. Metabolism 32 Suppl 1983; 1:62-68.
20. Sutherland DER, Sibley R, Xu XZ et al. Twin-to-twin pancreas transplantation: Reversal and reenactment of the pathogenesis of type I diabetes. Transplant Assoc Am Physicians 1984; 97:80-87.
21. Tyden G, Reinholt FP, Sundkvist G et al. Recurrence of autoimmune diabetes mellitus in recipients of cadaveric pancreatic grafts. N Engl J Med 1996; 335:860-863.
22. Stegall MD, Lafferty KJ, Kam I et al. Evidence of recurrent autoimmunity in human allogeneic islet transplantation. Transplantation 1996; 61(8):1272-1274.
23. Medawar PB. The Uniqueness of the Individual. New York: Basic Books, 1957.
24. Snell GD. The homograft reaction. Annu Rev Microbiol 1957; 11:439-458.
25. Batchelor J, Welsh K, Burgos H. Transplantation antigens per se are poor immunogens within a species. Nature 1978; 273:54-56.
26. Lafferty KL, Cunningham AJ. A new analysis of allogeneic interactions. Aust J Exp Biol Med Sci 1975; 53:27-42.
27. Bretscher P, Cohn M. A theory of self-nonself discrimination. Science 1970; 169:1042-1049.
28. Talmage D, Woolnough J, Hemmingsen H et al. Activation of cytotoxic T cells by nonstimulating tumor cells and spleen cell factor(s). Proc Natl Acad Sci USA 1977; 74:1610-1614.
29. June CH, Ledbetter JA, Linsley PS et al. Role of the CD28 receptor in T-cell activation. Immunol Today 1990; 11:211-216.
30. Linsley PS, Brady W, Urnes M et al. CTLA-4 is a second receptor for the B cell activation antigen B7. J Exp Med 1991; 174:561-69.
31. Grewal IS, Foellmer HG, Grewal KD et al. Requirement for CD40 ligand in costimulation induction, T cell activation, and experimental allergic encephalomyelitis. Science 1996; 273:1864-1867.

32. Schwartz RH. Costimulation of T lymphocytes: The role of CD28, CTLA-4, and B7/BB1 in interleukin-2 production and immunotherapy. Cell 1992; 71:1065-1068.
33. Damle NK, Klussman K, Linsley PS et al. Differential costimulatory effects of adhesion molecules B7, ICAM-1, LFA-3, and VCAM-1 on resting and antigen-primed CD4+ T lymphocytes. J Immunol 1992; 148:1985-1992.
34. Van Seventer GA, Shimizu Y, Horgan KJ et al. The LFA-1 ligand ICAM-1 provides an important costimulatory signal for T cell receptor-mediated activation of resting T cells. J Immunol 1990; 144:4579-4586.
35. Bachmann MF, McKall-Faienza K, Schmits R et al. Distinct roles for LFA-1 and CD28 during activation of naive T cells: Adhesion versus costimulation. Immunity 1997; 7:549-557.
36. Ni HT, Deeths MJ, Li W et al. Signaling pathways activated by leukocyte function-associated Ag-1- dependent costimulation. J Immunol 1999; 162(9):5183-9.
37. Nabavi H, Freeman GJ, Gault A et al. Signalling through the MHC class II cytoplasmic domain is required for antigen presentation and induces B7 expression. Nature 1992; 360:266-268.
38. Lafferty KJ, Prowse SJ, Simeonovic CJ. Immunobiology of tissue transplantation: A return to the passenger leukocyte concept. Annu Rev Immunol 1983; 1:143-173.
39. Lafferty K, Bootes A, Dart G et al. Effect of organ culture on the survival of thyroid allografts in mice. Transplantation 1976; 22:138-149.
40. Bowen K, Andrus L, Lafferty K Successful allotransplantation of mouse pancreatic islets to nonimmunosuppressed recipients. Diabetes 1980; 29(Suppl. 1):98-104.
41. Coulombe M, Gill RG. Tolerance induction to cultured islet allografts. I Characterization of the tolerant state. Transplantation 1994; 57(8):1195-1200.
42. Haug CE, Gill RG, Babcock SK et al. Cyclosporine-induced tolerance requires antigens capable of initiating an immune response. J Immunol 1987; 139:2947-2949.
43. Simeonovic CJ, Prowse SJ, Lafferty KJ. Survival and function of islet allografts in outbred mice. Transpl Proceed 1986; 18:316-318.
44. Lacy P, Davie J, Finke E. Prolongation of islet allograft survival following in vitro culture (24° C) and a single injection of ALS. Science 1979; 204:312-313.
45. Faustman D, Hauptfeld V, Lacy P et al. Prolongation of murine islet allograft survival by pretreatment of islets with antibody directed to Ia determinants. Proc Natl Acad Sci USA 1981; 78:5156-5159.
46. Morrow C, Sutherland D, Steffes M et al. Lack of donor-specific tolerance in mice with established anti-Ia-treated islet allografts. Transplantation 1983; 36:691-694.
47. Faustman D, Steinman R, Gebel H et al. Prevention of rejection of murine islet allografts by pretreatment with anti-dendritic cell antibody. Proc Natl Acad Sci USA 1984; 81:3864-3868.
48. Lau H, Reemtsma K, Hardy M. Prolongation of rat islet allograft survival by direct ultraviolet irradiation of the graft. Science 1984; 223:607-608.
49. La Rosa F. Abrogation of mouse pancreatic islet allograft rejection by a four-day culture. Transplantation 1988; 46(1):330-333.
50. Hegre O, Hickey G, Marshall S et al. Modification of allograft immunogenicity in perinatal islets isolated and purified in vitro. Transplantation 1984; 37(3):227-233.
51. La Rosa F, Talmage D. The abrogation of thyroid allograft rejection by culture in acid medium. Transplantation 1987; 44:592.
52. Hullett DA, Landry AS, Leonard DK et al. Enhancement of thyroid allograft survival following organ culture. Transplantation 1989; 47:24-27.
53. Jacobs B. Ovarian allograft survival. Transplantation 1974; 18:454.
54. Ramrakha P, Sharp R, Yeoman H et al. The influence of MHC-compatible and MHC-incompatible antigen-presenting cells on the survival of MHC-compatible cultured murine keratinocyte allografts. Transplantation 1989; 48:676-680.
55. La Rosa FG, Talmage DW. Major histocompatibility complex antigen expression on parenchymal cells of thyroid allografts is not by itself sufficient to induce rejection. Transplantation 1990; 49:605-609.
56. Batchelor JR, Welsh KI, Maynard A et al. Failure of long surviving, passively enhanced kidney allografts to provoke T-dependent alloimmunity. I. Retransplantation of (AS x AUG) F1 kidneys into secondary AS recipients. J Exp Med 1979; 150:455-464.
57. Lechler RI, Batchelor JR. Restoration of immunogenicity to passenger cell-depleted kidney allografts by the addition of donor strain dendritic cells. J Exp Med 1982; 155:31-41.
58. Bowen K, Prowse S, Lafferty K. Reversal of diabetes by islet transplantation: Vulnerability of the established allograft. Science 1981; 213:1261-1262.
59. Vesole D, Dart G, Talmage D. Rejection of stable cultured allografts by active or passive (adoptive) immunization. Proc Natl Acad Sci USA 1982; 79:1626-1628.

60. Markmann J, Lo D, Naji A et al. Antigen presenting function of class II MHC expressing pancreatic beta cells. Nature 1988; 336:476-479.
61. Lo D, Burkly L, Widera G et al. Diabetes and tolerance in transgenic mice expressing class II MHC molecules in pancreatic beta cells. Cell 1988; 53:159-168.
62. Coulombe M, Yang H, Guerder S et al. Tissue Immunogenicity. The role of MHC antigen and the lymphocyte costimulator B7-1. J Immunol 1996; 157:4790-4795.
63. Gill RG, Wolf L. Immunobiology of cellular transplantation. Cell Transplant 1995; 4(4):361-370.
64. Calne RY. Organ transplantation between widely disparate species. Transplant Proc 1970; 2:550.
65. Auchincloss HJ. Xenografting: A review. Transplant Rev 1990; 4(1):14-27.
66. Sachs DH, Bach FH. Immunology of xenograft rejection. Human Immunol 1990; 28:245-251.
67. Platt JL, Bach FH. The barrier to xenotransplantation. Transplantation 1991; 52(6):937-947.
68. Logan JS. Prospects for xenotransplantation. Curr Opin in Immunol 2000; 12:563-568.
69. Galili U. Interaction of the natural anti-Gal antibody with a-galactsyl epitopes: A major obstacle for xenotransplantation in humans. Immunol Today 1993; 14:480-482.
70. Gill RG. Direct and indirect pathways of immunity to pancreatic islet transplants. Transplant Rev 1998; 12:85-95.
71. Hamelmann W, Gray DWR, Cairns TDJ et al. Immediate destruction of xenogeneic islets in a primate model. Transplantation 1994; 58:1109-1114.
72. Galili U, Tibell A, Samuelsson B et al. Increased anti-Gal activity in diabetic patients transplanted with fetal porcine islet cell clusters. Transplantation 1995; 59(11):1549-1556.
73. Gill R, Wolf L, Daniel D et al. CD4 T cells are both necessary and sufficient for islet xenograft rejection. Transplant Proc 1994; 26:1203.
74. Lafferty KJ, Warren HS, Woolnough JA et al. Immunological induction of T lymphocytes: Role of antigen and the costimulator. Blood Cells 1978; 4:395-404.
75. Alter BJ, Bach FH. Cellular basis of the proliferative response of human T cells to mouse xenoantigens. J Exp Med 1990; 171:333-338.
76. Moses RD, Winn HJ, Auchincloss HJ. Evidence that multiple defects in cell-surface molecule interactions across species differences are responsible for diminished xenogeneic T cell responses. Transplantation 1992; 53(1):203-209.
77. Sprent J. The thymus and T cell tolerance. Today's Life Sciences 1991; 3(12):14-20.
78. Swain SL. T cell subsets and the recognition of MHC class. Immunol Rev 1983; 74:129-142.
79. Simon AR, Warrens AN, Sykes M. Efficacy of adhesive interactions in pig-to human xenotransplantation. Immunol Today 1999; 20:323-330.
80. von Hoegen P, Miceli MC, Tourvieille B et al. Equivalence of human and mouse CD4 in enhancing antigen responses by a mouse class II-restricted T cell hybridoma. J Exp Med 1989; 170:1879-1886.
81. Vignali DAA, Moreno J, Schiller D et al. Species-specific binding of CD4 to the b2 domain of major histocompatibility complex class II molecules. J Exp Med 1992; 175:925-932.
82. Barzaga-Gilbert E, Grass D, Lawrence S et al. Species specificity and augmentation of responses to class II major histocompatibility complex molecules in human CD4 transgenic mice. J Exp Med 1992; 175:1707-1715.
83. Irwin MJ, Heath WR, Sherman LA. Species-restricted interactions between CD8 and the a3 domain of class I influence the magnitude of the xenogeneic response. J Exp Med 1989; 170:1091-1101.
84. Kalinke U, Arnold B, Hammerling GJ. Strong xenogeneic HLA response in transgenic mice after introducing an a3 domain into HLA B27. Nature 1990; 348:642-644.
85. Alter BJ, Bach FH. Lymphocyte reactivity in vitro 1. Cellular reconstitution of purified lymphocyte response. Cell Immunol 1970; 1:207-18.
86. Collins MKL. Species specificity of interleukin 2 binding to individual receptor components. Eur J Immunol 1989; 19:1517-1520.
87. Greenstein JL, Foran JA, Gorga JC et al. The role of T cell accessory molecules in the generation of class II-specific xenogeneic cytolytic T cells. J Immunol 1986; 136(7):2358-2363.
88. Golding H, Singer A. Role of accessory cell processing and presentation of shed H-2 alloantigens in allospecific cytotoxic T lymphocyte responses. J Immunol 1984; 133:597.
89. Stock PG, Ascher NL, Chen S et al. Evidence for direct and indirect pathways in the generation of the alloimmune response against pancreatic islets. Transplantation 1991; 52:704-709.
90. Benichou G, Takizawa PA, Oslon CA et al. Donor major histocompatibility complex (MHC) peptides are presented by recipient MHC molecules during graft rejection. J Exp Med 1992; 175:305-308.
91. Fangmann J, Dalchau R, Fabre J. Rejection of skin allografts by indirect allorecognition of donor class I MHC peptides. J Exp Med 1992; 175:1521.

92. Coulombe M, Gill RG. Tolerance induction to cultured islet allografts. II. Status of anti-donor reactivity in tolerant animals. Transplantation 1994; 57(8):1201-1207.
93. Shoskes DA, Wood KJ. Indirect presentation of MHC antigens in transplantation. Immunol Today 1994; 15(1):32-38.
94. Gill RG, Coulombe M, Lafferty KJ. Pancreatic islet allograft immunity and tolerance: The two-signal hypothesis revisited. Immunol Rev 1996; 149:75-96.
95. Yoshizawa K, Yano A. Mouse T lymphocytes proliferative responses specific for human MHC products in mouse anti-human xenogeneic MLR. J Immunol 1984; 132(6):2820-2829.
96. Brunswick M, Lake P. Functional interactions of human and murine lymphoid cells. Cell Immunol 1986; 103:441.
97. Pierson III RN, Winn HJ, Russell PS et al. Xenogeneic skin graft rejection is especially dependent on CD4+ T cells. J Exp Med 1989; 170:991-996.
98. Mandel TE, Koulmanda M. The survival of xeno-, allo- and isografts in NOD mice and xenografts in other strains, after immunosuppression with anti-CD4 monoclonal antibody. Diab Nutr Metab 1992; 5 (Suppl. 1):91-96.
99. Gill RG, Coulombe M. Rejection of pancreatic islet xenografts does not require CD8+ T-lymphocytes. Transplant Proc 1992; 24(6):2877-2878.
100. Gould DS, Auchincloss Jr H. Direct and indirect recognition: The role of MHC antigens in graft rejection. Immunol Today 1999; 20(2):77-82.
101. Shizuru JA, Gregory AK, Chao CTB et al. Islet allograft survival after a single course of treatment of recipient with antibody to L3T4. Science 1987; 237:278-280.
102. Hao L, Wang Y, Gill RG et al. Role of the L3T4+ T cell in allograft rejection. J Immunol 1987; 139:4022-4026.
103. Gill RG, Rosenberg AS, Lafferty KJ et al. Characterization of primary T cell subsets mediating rejection of pancreatic islet grafts. J Immunol 1989; 143(7):2176-2178.
104. Ricordi C, Lacy PE, Sterbenz K et al. Low-temperature culture of human islets or in vivo treatment with L3T4 antibody produces a marked prolongation of islet human-to-mouse xenograft survival. Proc Natl Acad Sci USA 1987; 84:8080-8084.
105. Lacy PE, Ricordi C, Finke EH. Effect of transplantation site and aL3T4 treatment on survival of rat, hamster, and rabbit islet xenografts in mice. Transplantation 1989; 47(5):761-766.
106. Coulombe M, Yang H, Wolf LA et al. Tolerance to antigen-presenting cell-depleted islet allografts is CD4 T cell dependent. J Immunol 1999; 162:2503-2510.
107. Prowse SJ, Warren HS, Agostino M et al. Transfer of sensitised Lyt2+ cells triggers acute rejection of pancreatic islet allografts. Aust J Exp Biol Med Sci 1983; 61(2):181-185.
108. Wolf LA, Coulombe M, Gill RG. Donor antigen-presenting cell-independent rejection of islet xenografts. Transplantation 1995; 60(10):1164-1170.
109. Markmann JF, Bassiri H, Desai NM et al. Indefinite survival of MHC class I-deficient murine pancreatic islet allografts. Transplantation 1992; 54:1085-1089.
110. Osorio RW, Ascher NL, Jaenisch R et al. Major histocompatibility complex class I deficiency prolongs islet allograft survival. Diabetes 1993; 42:1520-1527.
111. Desai NM, Bassiri H, Kim J et al. Islet allograft, islet xenograft, and skin allograft survival in CD8+ T lymphocyte-deficient mice. Transplantation 1993; 55:718-722.
112. Gill RG, Lafferty KJ. The role of islet transplantation in the treatment of insulin-dependent diabetes mellitus. Imm Aller Clin No Amer 1989; 9(1):165-186.
113. Simeonovic CJ, Wilson JD. CD4+ T-cell depletion in mice facilitates induction of tolerance to pig proislet xenografts: A comparison of NOD and CBA/H recipient models. Diab Nutr Metab 1992; 5(Suppl. 1):133-138.
114. Gill R, Coulombe M. Islet xenografting in autoimmune diabetes. Transplant Proc 1994; 26:1140.
115. Haskins K, Portas M, Bergman B et al. Pancreatic islet-specific T-cell clones from nonobese diabetic mice. Proc Natl Acad Sci USA 1989; 86:8000-8004.
116. Hao L, Calcinaro F, Gill R et al. Facilitation of specific tolerance induction in adult mice by RS-61443. Transplantation 1992; 53:590-595.
117. Li H, Colson YL, Ildstad ST. Mixed allogeneic chimerism achieved by lethal and nonlethal conditioning approaches induces donor-specific tolerance to simultaneous islet allografts. Transplantation 1995; 60:523-529.
118. Wekerle T, Sykes M. Mixed chimerism as an approach for the induction of transplantation tolerance. Transplantation 1999; 68:459-467.
119. Fuchimoto Y, Huang CA, Yamada K et al. Mixed chimerism and tolerance without whole body irradiation in a large animal model. J Clin Invest 2000; 105:1779-1789.
120. Posselt AM, Barker CF, Tomaszewski JE et al. Induction of donor-specific unresponsiveness by intrathymic islet transplantation. Science 1990; 249:1293-1295.

121. Pearson TC, Madsen JC, Larsen CP et al. Induction of transplantation tolerance in adults using donor antigen and anti-CD4 monoclonal antibody. Transplantation 1992; 54:475-483.
122. Cobbold SP, Adams E, Marshall SE et al. Mechanisms of peripheral tolerance and suppression induced by monoclonal antibodies to CD4 and CD8. Immunol Rev 1996; 149:5-33.
123. Pearson TC, Alexander DZ, Winn KJ et al. Transplantation tolerance induced by CTLA4-Ig. Transplantation 1994; 57(12):1701-1706.
124. Tran HM, Nickerson PW, Restifo AC et al. 1997. Distinct mechanisms for the induction and maintenance of allograft tolerance with CTLA4-Fc treatment. J Immunol 159:2232-2239.
125. Lakkis FG, Konieczny BT, Saleem S et al. Blocking the CD28-B7 T cell costimulation pathway induces long term cardiac allograft acceptance in the absence of IL-4. J Immunol 1997; 158:2443-2448.
126. Parker DC, Greiner DL, Phillips NE et al. Survival of mouse pancreatic islet allografts in recipients treated with allogeneic small lymphocytes and antibody to CD40 ligand. Proc Natl Acad Sci USA 1995; 92:9560-9564.
127. Honey K, Cobbold SP, Waldmann H. CD40 ligand blockade induces CD4+ T cell tolerance and linked suppression. J Immunol 1999; 163:4805-4810.
128. Kenyon NS, Fernandez LA, Lehmann R et al. Long-term survival and function of intrahepatic islet allografts in baboons treated with humanized anti-CD154. Diabetes 1999; 48:1473-1481.
129. Kenyon NS, Chatzipetrou M, Masetti M et al. Long-term survival and function of intrahepatic islet allografts in rhesus monkeys treated with humanized anti-CD154. Proc Natl Acad Sci USA 1999; 96:8132-8137.
130. Nishihara M, Gotoh M, Ohzato H et al. Awareness of donor alloantigens in antiadhesion therapy induces antigen-specific unresponsiveness to islet allografts. Transplantation 1997; 64:965-970.
131. Nicolls MR, Colulombe M, Yang H et al. Anti-LFA-1 therapy induces long term islet allograft acceptance in the absence of IFN-g or IL-4. J Immunol 2000; 164:3627-3634.
132. Auersvald LA, Rothstein DM, Oliveira SC et al. Indefinite islet allograft survival in mice after a short course of treatment with anti-CD45 monoclonal antibodies. Transplantation 1997; 63:1355-1369.
133. Larsen CP, Elwood ET, Alexander DZ et al. Long-term acceptance of skin and cardiac allografts after blocking CD40 and CD28 pathways. Nature 1996; 381:434-438.
134. Kirk AD, Harlan DM, Armstrong NN et al. CTLA4-Ig and anti-CD40 ligand prevent renal allograft rejection in primates. Proc Natl Acad Sci USA 1997; 94:8789-8794.
135. Pearson TC, Alexander DZ, Hendrix R et al. CTLA4-Ig plus bone marrow induces long-term allograft survival and donor-specific unresponsiveness in the murine model. Evidence for hematopoieticchimerism. Transplantation 1996; 61(7):997-1004.
136. Durham MM, Bingaman AW, Adams AB et al. Cutting edge: Administration of anti-CD40 ligand and donor bone marrow leads to henopoietic chimerism and donor-specific tolerance without cytoreductive conditioning. J Immunol 2000; 165:1-4.
137. Lenschow DJ, Zeng Y, Thistlethwaite JR et al. Long-term survival of xenogeneic pancreatic islet grafts induced by CTLA4Ig. Science 1992; 257(7):789-792.
138. Zeng Y, Bluestone JA, Ildstad ST et al. Long-term functional xenograft tolerance after intrathymic islet transplantation (Lewis rat —> B6 mouse). Transplant Proc 1993; 25:438-439.
139. Mayo GL, Posselt AM, Barker CF et al. Prolongation of survival of donor-strain islet xenografts (rat-mouse) by intrathymic inoculation of xenogeneic islet and bone marrow cells. Transplantation 1994; 58(1):107-109.
140. Li H, El-Ezz A, Gambrell BE et al. Mixed xenogeneic chimerism and tolerance III. A nonlethal approach to induce donor-specific transplantation tolerance to simultaneous islet xenografts. Transplantation 1995; 60:59-65.
141. Elwood ET, Larsen CP, Cho HR et al. Prolonged acceptance of concordant and discordant xenografts with combined CD40 and CD28 pathway blockade. Transplantation 1998; 65:1422-1428.
142. Sablinski T, Emery DW, Monroy R et al. Long-term discordant xenogeneic (porcine-to-primate) bone marrow engraftment in a monkey treated with porcine-specific growth factors. Transplantation 1999; 67(7):972-7.
143. Carel J, Schreiber RD, Falqui L et al. Transforming growth factor b decreases the immunogenicity of rat islet xenografts (rat to mouse) and prevents rejection in association with treatment of the recipient with a monoclonal antibody to interferon g. Proc Natl Acad Sci USA 1990; 87:1591-1595.
144. Carel J-C, Sheehan KC, Schreiber RD et al. Prevention of rejection of transforming growth factor b-treated rat-to-mouse islet xenografts by monoclonal antibody to tumor necrosis factor. Transplantation 1992; 55(2):456-458.
145. Gerasimidi A, Sheehan KCF, Schreiber RD et al. Tumor necrosis factor-a prevents rejection of islet xenografts (rat to mouse). Diabetes 1993; 42:651-657.

146. Faustman D, Coe C. Prevention of xenograft rejection by masking of donor HLA class I antigens. Science 1991; 252:1700-1702.
147. Gerling IC, Serreze DV, Christianson SW et al. Intrathymic islet cell transplantation reduces beta-cell autoimmunity and prevents diabetes in NOD/Lt mice. Diabetes 1992; 41(12):1672-6.
148. Battan R, Mordes JP, Abreau S et al. Evidence that intrathymic islet transplantation does not prevent diabetes or subsequent islet graft destruction in RT6-depleted, diabetes-resistant biobreeding/ Worcester rats. Transplantation 1994; 57(5):731-736.
149. Li H, Kaufman CL, Boggs SS et al. Mixed allogeneic chimerism induced by a sublethal approach prevents autoimmune diabetes and reverses insulitis in nonobese diabetic (NOD) mice. J Immunol 1996; 156:380-388.
150. Weber CJ, Hagler MK, Chryssochoos JT et al. CTLA4-Ig prolongs survival of microencapsulated neonatal porcine islet xenografts in diabetic NOD mice. Cell Transplant 1997; 6(5):505-508.
151. Markees TG, Serreze DV, Phillips NE et al. NOD mice have a generalized defect in their response to transplantation tolerance induction. Diabetes 1999; 48:967-974.
152. Smith RM, Mandel TE. Pancreatic islet xenotransplantation: The potential for tolerance induction. Immunol Today 2000; 21:42-8.
153. Lanza RP, Chick WL. Immunoisolation: At a turning point. Immunol Today 1997; 18:135-139.
154. Giannoukakis N, Rudert WA, Robbins PD et al. Targeting autoimmune diabetes with gene therapy. Diabetes 1999; 48:2107-2121.

Chapter 7

Type 1 Diabetes Mellitus of Man:
Genetic Susceptibility and Resistance

Alberto Pugliese and George S. Eisenbarth

Introduction

Insulin-dependent diabetes mellitus (IDDM), or type 1 diabetes, is usually a chronic disease characterized by the autoimmune destruction (Type 1A) of pancreatic β-cells and complete insulin deficiency.[1] Type 1B diabetes refers to insulin dependent diabetes not of immune etiology and is not the subject of this chapter. Type 1A diabetes is quite heterogeneous in its clinical expression,[2] and it can be often confused with type 2 diabetes in those patients who develop diabetes at a later age. A large body of evidence indicates that inherited genetic factors influence both susceptibility to and resistance to the disease. Although a significant proportion of patients with type 1 diabetes lack a positive family history for the disease, there is significant familial clustering with an average prevalence of 6% in siblings compared to 0.4% in the US Caucasian population. The familial clustering (λs) can be calculated as the ratio of the risk to siblings over the disease prevalence in the general population, and thus $\lambda s = 6/0.4 = 15$.[3,4]

One's genetic susceptibility is also dependent on the degree of genetic identity with the proband, and in fact the risk of diabetes in families has a nonlinear correlation with the number of alleles shared with the proband. The highest risk is naturally observed in monozygotic twins (100% sharing) followed by first, second, and third degree relatives (50%, 25%, 12.5% sharing, respectively). Based on such estimates of observed risk, it has been suggested that diabetes susceptibility may be linked to a major locus and that several other minor loci may contribute to diabetes risk in an epistatic way. This model generates the risk curve that best parallels the risk curve obtained from observed risk estimates.[5] The moderate disease concordance observed even amongst identical twins (usually 30-50%, 70% in studies with longest follow-up) implies that inherited genes provide increased susceptibility.[6-13] The increasing availability of microsatellite markers, single nucleotide polymorphims (SNPs), automated typing technology,[14] and large collections of families with affected sib-pairs, including those in the Human Biological Data Interchange (HBDI) and British Diabetic Association (BDA)-Warren repositories,[15] have enabled the design of large scale, genome-wide scans to map IDDM susceptibility loci. Many loci have now been linked to diabetes and there is evidence for epistatic interactions among different loci. The prevailing opinion is that type 1A diabetes is a polygenic disorder. However, these findings may also be noted if the disease was genetically heterogeneous with different major loci determining disease risk in different families. There is evidence that complex parental effects and other mechanisms regulating gene expression may influence susceptibility by modifying the transmission and expression of inherited genes. It is also an intriguing possibility that additional genetic factors or their expression may be acquired after birth, perhaps through environmental exposures. Such factors may play a key role in triggering β-cell autoimmunity in susceptible individuals. This chapter will review the current knowledge about the genetics of type 1A diabetes in humans.

Immunology of Type 1 Diabetes, Second Edition, edited by George S. Eisenbarth.

Table 1. Susceptibility loci for type 1 diabetes

Locus	Chromosome	Candidate Genes	Markers	LOD*
IDDM1	6p21.3	*HLA DR/DQ*	-	65.8
IDDM2	11p15.5	INSULIN VNTR	-	4.28
	16q22-24		D16S3098	3.93
IDDM3	15q26		D15S107	
IDDM4	11q13.3	*MDU1, ZFM1, RT6, ICE, LRP5, FADD, CD3*	FGF3, D11S1917	
IDDM5	6q25	*MnSOD*	ESR, a046Xa9	1.96
IDDM6	18q12-q21	*JK (Kidd), ZNF236*	D18S487, D18S64	
IDDM7	2q31-33	*NEUROD, IGRP*	D2S152, D251391	2.62
IDDM8	6q25-27		D6S281, D6S264, D6S446	1.81
IDDM9	3q21-25		D3S1303, D10S193	
IDDM10	10p11-q11		D10S565	2.8
IDDM11	14q24.3-q31	*ENSA, SEL-1L*	D14S67	
IDDM12	2q33	*CTLA-4*	(AT)n 3' UTR, A/G Exon 1	See *IDDM7*
IDDM13	2q34	*IGFBP2, IGFBP5, NEUROD, HOXD8*	D2S137, D2S164, D2S1471	See *IDDM7*
IDDM15	6q21		D6S283, D6S434, D6S1580	
IDDM17	10q25		D10S1750, D10S1773	1.56
IDDM18	5Q31.1-33.1	IL12B	IL12B	
	1q42			2.2
	16p11-13			1.74
	17q25			1.81
	19q11			1.8

*If no LOD score indicated, < 1.56 for region in Cox genome screen (suggestive LOD > 2.2, $p < 7.4\times10^{-4}$, significant LOD > 3.6, $p < 2.2\times10^{-5}$) Rows without a LOD score gave no positive signal or relatively low signals. Candidate gene abbreviations are explained in the text. The *IDDM14* and *IDDM16* denominations have not been assigned to any locus.

Inherited Susceptibility Loci

Both association studies and linkage analysis using various analytical methods have been used to identify susceptibility loci, and these are conventionally noted using the abbreviation IDDM and a number, e.g., *IDDM1*, *IDDM2*, etc., with the number usually reflecting the order in which such loci were reported. Using the candidate gene approach, association studies dating now to about two decades ago provided evidence for the first two susceptibility loci, the HLA region (*IDDM1*) and the insulin gene (*INS*) locus (*IDDM2*). These two loci only contribute a portion of the disease familial clustering, suggesting the existence of additional loci. Genome scans conducted by several groups confirmed linkage with *IDDM1* and *IDDM2* and provided evidence for the existence of at least 15 additional loci. A comprehensive list of the known susceptibility loci is shown in Table 1 with LOD scores from the combined genome screen of Cox and coworkers analyzing 767 families (HBDI and BDA repository).[16]

The Major Susceptibility Locus *IDDM1*

The major susceptibility locus lies within the HLA (human leukocyte antigen) region on the short arm of chromosome 6[17] and it provides up to 40-50% of the inheritable diabetes risk.[18] The HLA complex was first linked to diabetes when associations with several HLA class I antigens (HLA-B8, -B18, and -B15) were discovered by serological typing and affected sib-pairs

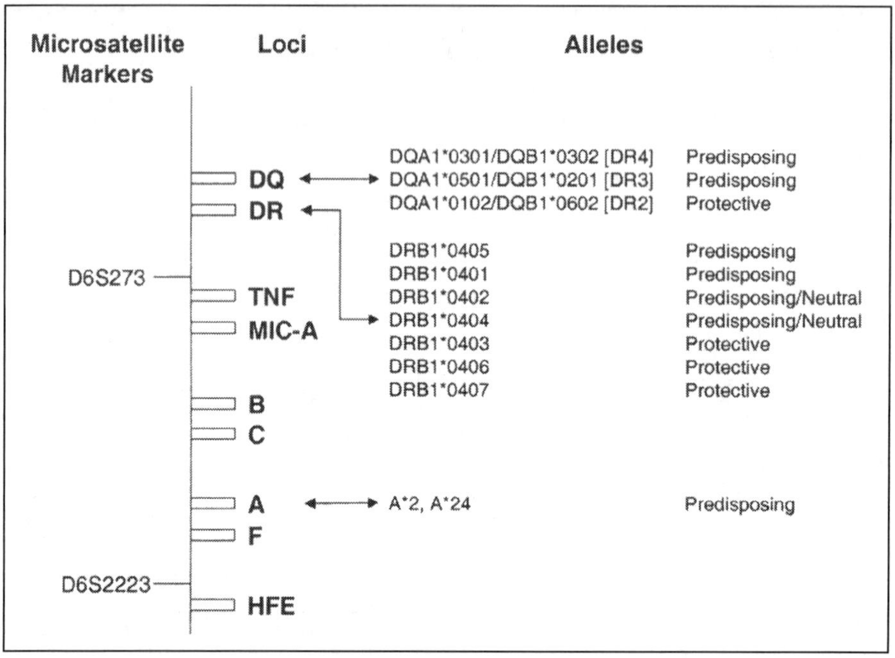

Figure 1. The HLA Region and IDDM Susceptibility. Schematic representation of the HLA region showing microsatellite markers, loci, and alleles associated with IDDM susceptibility. Distances between loci are grossly approximated.

showed evidence of linkage.[19-21] With the development of novel typing reagents, HLA class II genes were shown to be even more strongly associated with the disease.[21-23] However, other loci within or near the HLA complex appear to modulate diabetes risk, and add further complexity to the analysis of *IDDM1*-encoded susceptibility. Alleles, modes of inheritance and putative mechanisms of susceptibility encoded for at the *IDDM1* locus are discussed below. A schematic representation of the HLA region and its association with IDDM is shown in Figure 1.

The HLA Class II Region

The great majority of patients carry the HLA-DR3 or -DR4 class II antigens and approximately 30% of patients are DR3/DR4 heterozygotes. The DR3/DR4 genotype confers the highest diabetes risk with a synergistic mode of action, followed by DR4 and DR3 homozygosity, respectively.[24] Following the development of DNA-based sequencing and typing technology, the HLA-DQ locus was found to be the most strongly associated with diabetes susceptibility. This locus encodes for several variants of the HLA-DQ molecule, a heterodimer consisting of two glycoproteic chains (α and β) involved in immune recognition and antigen presentation to CD4 T cells. In Caucasians, the HLA-DQ heterodimers (the α-chains are labeled DQA1 and the β-chains DQB1) encoded by the DQA1*0301, DQB1*0302 and DQA1*0501, DQB1*0201 alleles have the strongest association with diabetes. These alleles are in linkage disequilibrium with the HLA-DR4 and -DR3 alleles (Table 2), respectively,[25] and linkage disequilibrium often extends centromeric and telomeric of the class II region.[26-28]

Allelic variation at the DQB1 locus differentiates IDDM susceptibility among the two most common HLA-DR4 haplotypes found in Caucasians based on the presence of the DQB1*0302 or DQB1*0301 allele. Most patients with DR4 carry the DQB1*0302 allele, while the DQB1*0301 and *0302 alleles are more evenly distributed in the general population. An independent effect has not been demonstrated for DQB1*0201 because of the strong link-

Table 2. HLA class II DR-DQ linkage patterns and IDDM susceptibility in whites

HLA-DR	*DQA1*	*DQB1*	*DRB1*	*Susceptibility*
DR2	0102	0602	1501	Protective
DR2	0102	0502 (AZH)	1601	Predisposing
DR2	0103	0601	1502	Neutral
DR3	0501	0201	0301	High Risk
DR4	0301	0302	0401	High Risk
DR4	0301	0302	0402	Predisposing
DR4	0301	0302	0403	Neutral
DR4	0301	0302	0404	Predisposing
DR4	0301	0302	0405	High Risk
DR4	0301	0301	0401	Neutral
DR4	0301	0303	0401	Neutral
DR7	0201	0303	0701	Protective
DR6	0101	0503	1401	Protective

age disequilibrium between DQB1*0201 and DRB1*0301 on Caucasian DR3 haplotypes. However, DQB1*0201 does not confer increased susceptibility in association with DRB1*0701 on DR7 haplotypes. The different risk conferred by DQB1*0201 when on a chromosome with DR3 or DR7 may be explained by the different DQA1 alleles associated with DQB1*0201 on such haplotypes (*0501 on DR3, *0201 on DR7).[29,30] Other susceptibility loci may also be in linkage disequilibrium with DQB1*0201 on DR3 haplotypes.[18] Trans-complementation of DQ α- and β-chains from opposite haplotypes has been recently demonstrated, and this significantly increases the diversity of class II antigens participating in the immune response and the potential for HLA-DQ contribution to IDDM susceptibility. A trans-complementing DQ β heterodimer would be unique to a heterozygous individual and usually it would not be expressed in his parents. Thus, this phenomenon has been proposed as an explanation for the increased IDDM risk observed in DR4, DQA1*0301,DQB1*0302/DR3, DQA1*0501,DQB1*0201 heterozygotes.[30,31]

It has also been noted that DQB1*0302 differs from DQB1*0301 at position 57, where it lacks an aspartic acid residue, similar to the I-A molecule of the nonobese diabetic (NOD) mouse (reviewed in ref. 32). The DQB1*0201 allele also lacks aspartic acid at position 57, and it has been proposed that this residue may be involved in the molecular mechanism underlying *IDDM1*-encoded susceptibility.[25,28] In fact, the amino acid residue at position 57 of the DQ-β chain appears to be critical for peptide binding and recognition.[33] Other residues of the DQ-β chain may influence peptide binding and diabetes susceptibility, and in particular the combined variation of residues at positions 57 and 70 seem to more strongly correlate with diabetes risk.[34,35] An arginine residue at position 52 of the DQ-α chain also correlates with diabetes susceptibility.[29] The importance of the residue at position 57 has been disputed by trans-racial studies showing that DQB1 alleles found with increased frequency in Japanese patients carry instead of lack an aspartic acid residue at this position.[36] Patients carrying similar alleles are also found among Caucasians.[37-46] Moreover, certain low risk DQB1 genotypes also lack aspartic acid at position 57, including DQB1*0302/DQB1*0201 (DR7), and DQB1*0201 (DR3)/DQB1*0201 (DR7). By using X-ray crystallography, investigators have recently determined the three-dimensional structure of the DQ8 molecule (encoded by DQA1*0301/DQB1*0302) complexed with an immunodominant peptide of the insulin molecule.[47] The DQ8 structure suggests that the residue at position 57 contributes to the shaping of the P9 pocket, which together with the P1 and P4 pockets appear relevant to diabetes susceptibility. Moreover, the binding pockets of DQ8 were similar to those of DQ2 (encoded by DQA1*0501/DQB1*0201) and to those of the I-Ag7 molecule (corresponding to human DQ) from NOD mice. This finding suggests that diabetes may depend on an antigen-presentation event(s) that may be similar in humans and NOD mice.

Regardless of the debated contribution of the single residue at position 57, which may be very much dependent on genetic heterogeneity among populations, it is clear that *IDDM1*-encoded susceptibility is mostly conferred by alleles of the HLA-DQ locus in the class II region. The above conclusion is also supported by the fact that the DQA1*0102, DQB1*0602 heterodimer found on HLA-DR2 haplotypes confers dominant protection from the development of type 1 diabetes (reviewed in ref. 48). Among four common DR2 haplotypes observed in Caucasians, the DQA1*0102, DQB1*0602, DRB1*1501 haplotype is negatively associated with type 1 diabetes and is reported in less than 1% of patients in most populations studied, including those of Caucasian (both European and North-American),[25,29,42,43,49-53] Asian,[36,42,44,45] African-American,[41,42] and Mexican-American origin.[46,54] The DQB1*0602 allele in particular is the only class II allele exclusively found on protective DR2 haplotypes while all the other alleles (DQA1*0102, DQA1*0103, DQB1*0601, DQB1*0502, DRB1*1501, DRB1*1502, DRB1*1601) can be found on neutral or moderately predisposing DR2 haplotypes. Moreover, a few rare patients with type 1 diabetes have been described carrying mutated DQB1*0602 alleles or unusual DQA1/DQB1 alleles in cis with the usual DRB1*1501 allele.[35,55,56] Thus, the available evidence suggest that the diabetes-protective effect associated with DR2 haplotypes may be mostly mapped within the DQ locus and in particular to the DQB1*0602 allele. Although a number of patients with DQB1*0602 have been identified,[57] the overall number is very small. Protection appears to be dominant since DQB1*0602 protects from diabetes even in the presence of high risk HLA alleles.[49,58]

A recent study of transgenic mice has provided direct evidence that the DQ locus, and the DQB1*0302 allele in particular, can engender an immune response leading to the development of diabetes.[59] However, the mechanism by which the HLA-DQ locus influence IDDM susceptibility is the subject of intense speculation. Since HLA-DQ molecules are known to play a role in antigen presentation, it is a fair speculation that allelic variation at this locus may affect the binding and functional properties of DQ heterodimers and in turn the presentation of islet cell antigen-derived peptides to immunocompetent cells. Protective HLA molecules may have higher affinity for one or several peptides than predisposing molecules. This is as also suggested by the observation that the NOD mouse I-A^{g7} molecule, homologous to the human DQ8 molecule encoded by DQA1*0301/DQB1*0302, is generally a very poor peptide binder.[60] Therefore, it is plausible that predisposing HLA molecules may be ineffective at binding and presenting peptides derived from islet cell antigens. Indeed, the HLA-DQ molecules encoded by the protective DQA1*0102/DQB1*0602 and predisposing DQA1*0301/DQB1*0302 alleles appear to differ in their affinity and specificity for peptides derived from the insulin, glutamic acid decarboxylase (GAD), and tyrosine phosphatase IA-2 autoantigens.[61,62] Similar findings were recently reported for DR molecules, with protective DR2 (DRB1*1501) molecules displaying stronger affinity for (pro) insulin peptides than susceptible DR3 (DRB1*0301) molecules.[63] It is unclear whether genetically determined differences in peptide binding and presentation affects the shaping of the T-cell repertoire in the thymus or modulates immune responses in the extra-thymic periphery. A poor presentation in the thymus could impair mechanisms of negative selection allowing autoreactive T cells to escape deletion. In contrast, a protective HLA-DQ molecule could promote tolerance to β-cell molecules by eliciting more efficient antigen presentation and negative selection in the thymus. Although several studies involving the transgenic expression of MHC molecules in mice did not support this hypothesis,[64-66] a more recent study provided novel evidence for thymic deletion as a mechanism of protection associated with MHC genes in transgenic mice.[67] Moreover, the recent demonstration that insulin and other islet cell antigens are ectopically expressed in human thymus[68,69] indirectly supports the hypothesis that thymic self-antigen presentation and deletional mechanisms may be affected by the affinity and binding properties of HLA-DQ and HLA-DR molecules. An alternative hypothesis is that DQB1*0602-associated protection may be mediated outside the thymus through the stimulation of regulatory immune responses associated with peripheral tolerance. Such responses may control the balance between Th1 and Th2 lymphocytes. The predominance of TH2 responses is usually associated with lack of progression to overt diabetes (reviewed in ref. 70). There is indeed evidence that a nondiabetogenic immune response, mostly limited to the production of autoantibodies against the GAD autoantigen, may occur in first degree relatives with DQB1*0602,[58] and in whom the presence of DQB1*0602 and DQA1*0102 has been confirmed by direct sequencing of the second exon.[71] A similar response has been reported in patients with type 1 autoimmune polyendocrine syndrome who do not invariably progress to overt diabetes.[72,73] Moreover, a protective effect has also been reported in first degree relatives participating in the ongoing Diabetes Prevention Trial (DPT-1), although different degrees of protection may occur in different ethnicities.[74] The presence of GAD autoantibodies, often at high titers, may reflect the predominance of Th2 responses in relatives with DQB1*0602. Finally, the two hypotheses are not mutually exclusive and DQB1*0602-associated protection could be mediated both in the thymus and the periphery. Further studies will help define the molecular mechanisms and antigenic peptides regulating diabetes susceptibility and resistance associated with allelic variation at the *IDDM1* locus.

Two additional strongly protective haplotypes are DRB1*1401, DQA1*0101, DQB1*0503 and DRB1*0701, DQA1*0201, DQB1*0303. The DRB1*1401 haplotype is particularly interesting in that it is a DR allele with an apparent lack of transmission to affected children as dramatic as for DQB1*0602 (both DRB1*1401 and DQA1*0201/DQB1*0303 are relatively infrequent but strongly protective).[75] Other loci in the class II region have been associated with IDDM susceptibility besides HLA-DQ. Several studies indicate that DRB1 alleles significantly contribute and modulate diabetes susceptibility.[27,53,76-82] The DRB1*0405 and *0401 alleles have been reported as predisposing, *0402 and *0404 as mostly neutral, while *0403, *0406, and *0407 appear to be protective. Moreover, an independent association has been observed in Mexican American[83] and Caucasians with DPB1*0301.[84] The frequency of DPB1*0101 is increased in patients (almost exclusively found on DR3 haplotypes). The maternal transmissions of DRB1*0301-DPB1*0101 haplotypes to affected children occur twice as frequently as do paternal transmissions.[18] Transmissions of DR3 haplotypes carrying other DPB1 alleles occur at approximately equal maternal and paternal frequencies. It is controversial whether loci (TAP1, TAP2) encoding for peptide transporter genes associated with antigen processing and localized centromeric to the DQ loci may also affect IDDM susceptibility.[85,86] Homozygosity for the TAP2*0101 allele was associated with increased IDDM risk independent of HLA-DQ susceptibility in a French study,[87] but other studies have failed to show such independent effect and suggested linkage disequilibrium between the HLA-DQ and TAP2 loci.[88,89] Of note, a mutation at the same locus has been implicated as the cause of the class I deficiency associated with IDDM in studies in humans and NOD mice.[90-95]

The HLA Class I Region

A number of observations indicate that class II genes cannot explain all of the HLA association with IDDM. A role for HLA complex genes other than the DR-DQ or other class II genes was first demonstrated by Robinson et al.[96] They examined affected sib pairs with parents homozygous for the DR3 haplotype and used the HLA class I B locus to distinguish between the 2 DR3 haplotypes of the homozygous parent. Under the null hypothesis that no HLA region variation additional to that defined by the DR3 haplotype is involved in IDDM, the affected sib pairs should share the two parental DR3 haplotypes at the same frequency. Significant deviation from 50% sharing was observed. Since the DR3 haplotypes examined in this study could be assumed to be homogeneous for their DR-DQ alleles at the molecular level (DRB1*0301 DQA1*0501 DQB1*0201), this test implicated other HLA loci in IDDM susceptibility. Several other reports suggest that HLA class I genes may also influence susceptibility as well as age of onset[97,98] and the rate of β cell destruction.[48,99-103] More recently, an

IDDM-associated locus has been found in the class I region, telomeric to HLA-F.[104] By considering the transmission ratios of microsatellite variation from parents homozygous for the HLA class II DR-DQ genes (using the Homozygous Parent Transmission Disequilibrium Test), the possible confounding effect of linkage disequilibrium was removed. Evidence for a second IDDM locus in this region was demonstrated, near the HFE (hemochromatosis) gene and 8.5 Mb distal to the HLA class II loci. Analyses from three independent family data sets from Norway (100 families), Denmark (51 families), and UK (74 families) suggested the presence of additional type 1 diabetes gene(s). Allele 3 of marker D6S2223, 5.5 Mb telomeric of the class II region, was associated with type 1 diabetes when the haplotype was fixed for HLA-DRB1*03-DQA1*0501-DQB1*0201. In a case-control study allele 3 at D6S2223 was found to be reduced among DRB1*03-DQA1*0501-DQB1*0201 homozygous patients with type 1 diabetes compared to DR-DQ matched controls, thus corroborating the results of the family analysis.[105] The protective effects seem to be inherited as a recessive trait. Further studies are necessary to confirm these observations in multiple ethnic groups and to better characterize these putative additional HLA region susceptibility loci.

The HLA Class III Region

Moghaddam et al[106] analyzed 11 markers in the HLA region in IDDM patients and controls fully matched for the highest risk DQA1*0501-DQB1*0201/HLA-DQA1*0301-DQB1*0302 (DR3/DR4) genotype. Their study provided strong evidence that another critical region for IDDM susceptibility, approximately 200 kb in size, lies around the microsatellite locus D6S273 which is located between the *TNF* and *HSP70* genes. Another study has independently confirmed linkage with marker D6S273 showing evidence for nonrandom transmission from DRB1*03-DQA1*0501-DQB1*0201 homozygous parents.[104] Linkage disequilibrium analysis suggested that "diabetogenic haplotypes" might have resulted from a recombination telomeric of D6S1014 in the region of D6S273 and *TNF*. The *TNF* gene is a strong candidate since polymorphisms of this gene may affect the production of TNF-α (Tumor Necrosis Factor), a potent cytokine, and in turn the magnitude of immune responses. It has also been reported that *TNFa* polymorphisms are associated with age-of onset may and influence the inflammatory process leading to the destruction of pancreatic β-cell and the development of IDDM.[107] In addition, the class I chain-related *MIC-A* and *MIC-B* genes, located between the HLA-B and the *TNFa* genes, may also affect IDDM susceptibility. In a recent case-control study of Italian patients, the frequency of the *MIC-A5* allele was increased in patients while none of the *TNFa* alleles were statistically significantly associated with the disease. In this study, the *MIC-A5* allele was associated with IDDM independently of class II alleles, suggesting an independent contribution of this locus to diabetes risk.[108] Finally, *HSP70-2* and *HSP70-Hom* genes are also located in the class III region although there is no evidence for an independent association with IDDM from studies that could not circumvent linkage disequilibrium.[109-114]

Clinical Heterogeneity of Type 1 Diabetes in Relation to the IDDM1 *Locus*

Age dependent HLA heterogeneity has been observed in Caucasian IDDM patients, indicating that high risk HLA genotypes occur at a higher frequency among the younger age onset groups,[42,115-117] whereas older age at diagnosis is associated with an increased heterogeneity of DRB1 and a decreased heterogeneity of DPB1.[78] Caillat-Zucman and coworkers have found a decreased frequency of DR3 and DR4 haplotypes and of DR3/DR4 heterozygosity amongst patients who had developed diabetes after age 15.[116] In our analysis of 297 patients (Pugliese A, personal observation), we have observed a significantly reduced frequency of DR3/DR4 heterozygotes only in patients with IDDM onset after age 30. In the same group we have found an increased frequency of DR3 homozygosity, which seems to correlate with a slower rate of the disease process. The immunogenetic analysis of islet cell antibody (ICA) positive first-degree relatives from our family study has confirmed that DR3 (in the absence of DR4/DQB1*0302) is associated with a slower rate of progression to diabetes. This slower rate may be in part explained by a recessive lack of humoral response to insulin during the prediabetic period that we have recently noticed in a subset of first degree relatives at increased IDDM risk. This lack of humoral anti-insulin autoimmunity is mostly associated with DR3 homozygosity, but we could observe it also in relatives with DR5 or DR8 haplotypes. All these haplotypes carry DQA1alleles from the evolutionary lineage 4[118,119] sharing glutamic acid and phenylalanine amino acid residues at position 40 and 51 of the second exon. Thus, lack of or reduced humoral responsiveness to insulin during the prediabetic period may be associated with this particular subset of DQA1 alleles rather than with DR3 itself.[120] In our study, first-degree relatives homozygous for such DQA1 alleles were characterized by lack of or very low levels of insulin autoantibodies in contrast to relatives with DQA1 alleles from evolutionary lineages 1-3 (carrying glutamic acid and phenylalanine at positions 40 and 51).

The Insulin Gene-VNTR Locus *IDDM2*

The *IDDM2* locus has been mapped to a variable number of tandem repeats (VNTR) located ~0.5 kb upstream of the insulin gene (*INS*)[121-123](Fig. 2). This polymorphic repeat, also known as the insulin gene minisatellite or ILPR (insulin-linked polymorphic region), consists of a 14-15 bp unit consensus sequence (ACAGGGGTCTGGGG) with slight variations of the repeat sequence. Any number from 30 to several hundred repeats has been observed, but allele frequencies tend to cluster in the 30-60 repeat range (class I alleles) or at 120-170 repeats (class III alleles). The intermediate class II alleles are rare in Caucasians, and less rare in individuals of African descent.[124,125] The sequence of the VNTR is particularly G-rich, and it tends to form unusual DNA structures in vitro and in vivo, presumably through the formation of G-quartets.[126,127]

Shortly after its discovery,[121] the insulin VNTR was found to be associated with type 1 diabetes.[128] However, the 4.1 Kb region containing *INS* and its flanking regions contain several other polymorphisms in linkage disequilibrium.[129] All the polymorphisms identified within this region lie outside coding sequences, confirming that diabetes susceptibility must derive from genetic influences on the expression of the insulin gene. Extensive studies involving cross-match haplotype analysis and multiple DNA variant association analysis involving polymorphisms in the neighboring *HUMTHO1* (tyrosine hydroxylase) and *IGF2* genes provided strong evidence that the VNTR is the main susceptibility determinant in this region[130-134] (Fig. 2), although this has not been demonstrated beyond all doubt.[135,136]

The *INS* VNTR is now referred to as the *IDDM2* susceptibility locus. Homozygosity for the short class VNTR I alleles is found in ~75-85% of the patients compared to a frequency of 50-60% in the general population, suggesting that it predisposes to type 1 diabetes. In contrast, homozygosity for the longer class III VNTR alleles is rarely seen in patients and the class III VNTR is believed to confer a dominant protective effect.[128,137,138] The relative risk ratio of the I/I genotype vs. I/III or III/III has been reported to be moderate (in the 3-5 range) and it accounts for about 10% of the familial clustering of type 1 diabetes.[139] Moreover, by measuring the *HphI* polymorphism (in tight linkage disequilibrium with the VNTR),[129] Metcalfe et al[140] showed that homozygosity for the predisposing *INS* genotype increases the likelihood that identical twins will be concordant for type 1 diabetes.

Although VNTR alleles cluster in two main classes with divergent associations with type 1 diabetes, there is evidence that VNTR alleles are quite heterogeneous and may differ in their ability to modulate disease susceptibility. Further classification of VNTR alleles is indeed possible according to size differences, and at least 21 class I and 15 class III VNTR alleles were described by fluorescence-based DNA fragment sizing technology[133] (Fig. 2). Bennett et al grouped the 15 class III VNTR alleles identified according to two main modes of transmission based on the linkage disequilibrium pattern with alleles at the *HUMTHO1* locus on chromosome 11p15. Thus, by taking both size and flanking haplotypes into account class III VNTR alleles linked to the *HUMTH01* Z-8 allele were found more protective (very protective haplotype or VPH) than those linked to the *HUMTH01* Z allele (protective haplotype or PH)

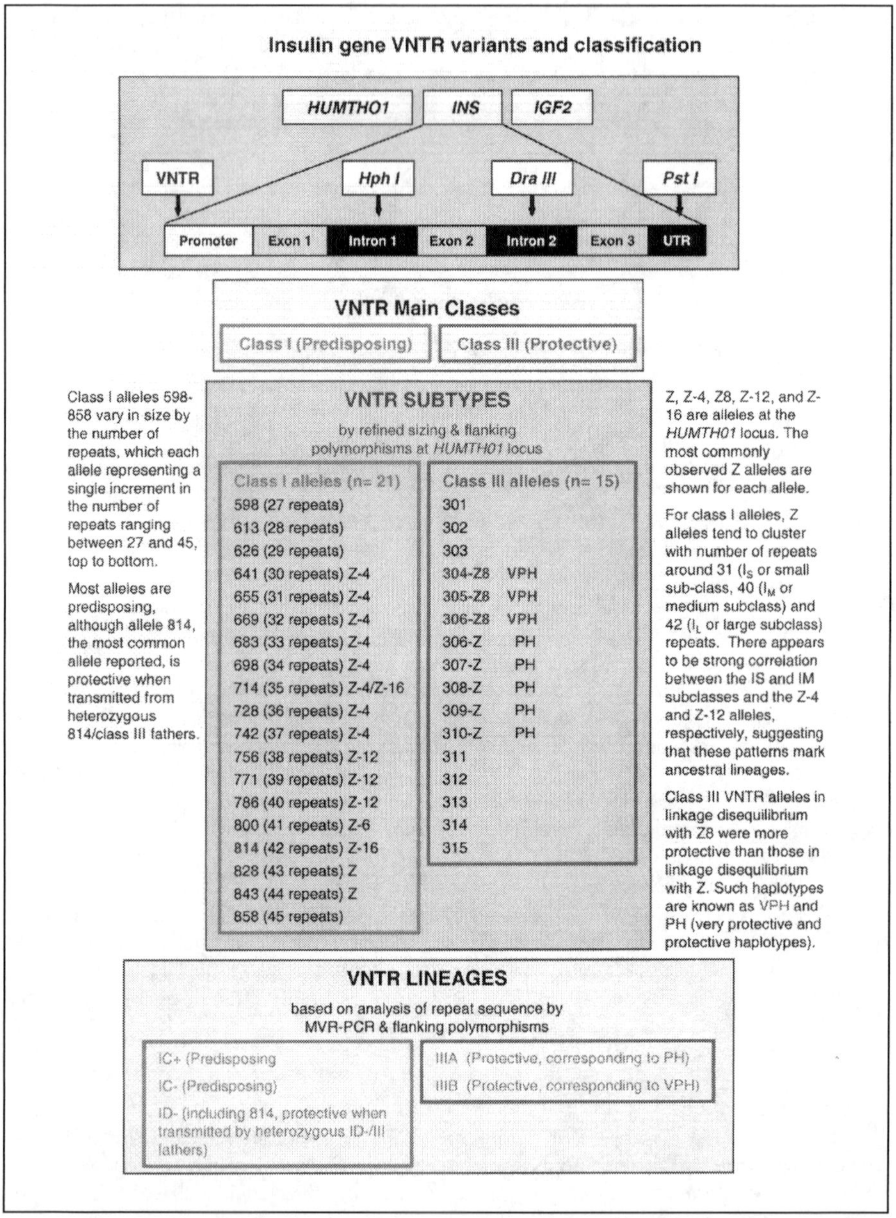

Figure 2. The *IDDM2* susceptibility locus. Top to bottom, the figure shows the *HUMTH01*, *INS* and IGF2 loci, as well as a schematic structure of the insulin gene with the approximate location of some of the most characterized polymorphic loci (VNTR, *HphI*, *DraIII*, *PstI*). Also shown are a schematic representation of the two main VNTR classes, their association with IDDM, as well as the VNTR alleles and allele lineages that have been defined with the variety of approaches described in the figure and in the papers referenced in the main text.

(Fig. 2).[133,141] However, certain VNTR alleles can be found in linkage disequilibrium with either the Z or Z-8 alleles. The variable degree of protection observed for these alleles may also be influenced by sequence heterogeneity and its effects on the VNTR physical state and transcriptional activity.[126,127,133,139,142] Sequencing studies have indeed identified several variants of the commonest VNTR repeat sequence that characterize yet another level of heterogeneity.[121,124,125,139,143]

More recent studies have analyzed the variant repeat distribution within the VNTR using minisatellite variant repeat mapping by PCR (MVR-PCR).[144] Some of the variation within the repetitive sequence most probably arises by mitotic replication slippage at an estimated frequency of 10^{-3} per gamete. However, sperm DNA analysis revealed a second class of mutation occurring at a frequency of approximately 2×10^{-5} that involved highly complex intra- and inter-allelic rearrangements which are probably meiotic in origin.[144] These events may help explain the heterogeneity of the VNTR locus. The combined analysis of the variant repeat distribution and of the haplotypes flanking the VNTR has allowed defining five new ancestral allele lineages.[145] By this approach, class III VNTR alleles can be divided into two diverging lineages, IIIA and IIIB (Fig. 2). These two lineages correspond to the PH and VPH haplotypes previously defined by Bennett et al.[133] Class I alleles can also be divided into three newly defined lineages, IC+, ID+ and ID-. The lineage denomination reflect the class of alleles, noted by the letter "I", while the letters "C" or "D" identify two different lineages defined by the very different distribution of variant repeats noted by multi-dimensional scaling.[145] The notation "+" or "-" refers to the presence (+) or absence (-) of a *MspI* restriction site at position +3,850, so that depending on this haplotypic analysis lineages could either be IC+, ID+ or ID-[145] IC+ and ID+ alleles are all predisposing to type 1 diabetes. In contrast, ID- alleles are protective when transmitted from ID-/III heterozygous fathers. Similar findings had been previously reported for the class I allele termed 814 (42 repeats), which is included in the ID- lineage (see next paragraph). The analysis of class ID- alleles into those of 42 repeats and those of other sizes suggested that the protective effect was a feature of all ID- alleles, irrespective of size. However, ID- alleles are clearly distinguished from all other alleles by a *MspI* variant within the *IGF2* gene. This suggests that at least for class I ID- alleles the susceptibility conferred by the VNTR may be modified by nearby sequences, and that in this case *IDDM2* susceptibility may have a multi-locus origin.[145] All together, the above studies suggest that the VNTR locus is extremely polymorphic, and that not only size of the VNTR but also sequence variation may play a significant role in modulating *INS* transcription and diabetes susceptibility.

Several studies have evaluated the effects of the VNTR on *INS* transcription. Transfection of rodent β-cell lines with reporter constructs representing the *INS* promoter flanked by class I or class III alleles resulted in three-fold differences going in opposite directions in reports from different laboratories.[122,123] These discrepant results may be due to species-specificity, differences among specific alleles within each class, or the absence of the genomic context necessary for the VNTR to have its physiologic effects. Studies on the transcriptional effects of the VNTR in vivo produced more meaningful results. In fetal pancreas RNA, the *INS* transcript in *cis* with the class III VNTR was expressed at lower levels (15-20%) than the class I transcript, a small but statistically significant difference.[145] Bennett et al[133] found a somewhat larger difference in adult pancreas. Moreover, it has been recently shown that single nucleotide differences in the VNTR sequence can affect *INS* transcription and correlate with the ability to form unusual DNA structures, both at the inter- and intra-molecular levels.[142] These findings have led to the hypothesis that VNTR variants may differ in their ability to stimulate transcription as a function of the binding of inter- and intra-molecular quartets with the transcription factor Pur1. However, the transcriptional activity of these variants observed in vitro may not always correspond to that in vivo, where overall transcription may depend on the interaction with other proteins involved in the transcriptional machinery and on differences among the various cell types that actively transcribe the insulin gene.

The evidence discussed thus far strongly supports the concept that the *INS* VNTR is the main factor conferring susceptibility at the *IDDM2* locus. However, there is a possibility that at least in some instances both *INS* and *IGF2* may be involved in modulating *IDDM2*-encoded susceptibility. The *INS* VNTR or other polymorphisms in the region do not alter the coding sequences in either *INS* or *IGF2*. Therefore, the biological effect must be mediated through effects on transcription, and such effects could influence the expression of both insulin and insulin growth factor-II (IGF-II). The studies described above report only marginal differences in pancreatic *INS* transcription, and the lower transcription associated with diabetes-protective class III VNTR alleles does not fit well with their dominant protective effect. It seems unlikely that such minor differences may influence susceptibility to a form of diabetes resulting from the autoimmune destruction of pancreatic β-cells. There has been much progress in the characterization of the autoimmune responses leading to diabetes, and there is strong evidence that the immune system reacts primarily against three molecules expressed by β-cells: insulin, glutamic acid decarboxylase (GAD), and a tyrosine-phosphatase like protein known as IA-2 or ICA512. Although insulin is one of several β-cell molecules targeted as an autoantigen, insulin is the only autoantigen in humans for which expression is specifically restricted to β-cells within the pancreatic islet. Autoantibodies against insulin have been reported in individuals at increased diabetes risk or patients at onset.[147-150] Although T cell responses have been difficult to detect consistently in peripheral blood[151-153] autoreactive T cells against the insulin peptide B9-23 were recently demonstrated in humans.[154] Moreover, T cells against the B9-B23 peptide represent approximately 40% of the lymphocytic infiltrate observed in the endocrine pancreas of NOD mice before diabetes onset,[155] and both CD4 and CD8 T cells recognizing insulin as their target have been reported in NOD mice.[156,157] Thus, it seems logical that the functional effects of *IDDM2* should be related to the function of the immune system and may specifically affect immune responsiveness to insulin.

A mechanistic hypothesis consistent with this assumption was formulated following the discovery that *INS* is actively transcribed in the thymus. Thymic *INS* transcription has now been reported in mouse,[158] rat,[159] and humans.[68,69] The human thymus was found to express low levels of *INS* message throughout fetal development and childhood but also during adulthood.[160] *INS* transcription in human thymus has been reported also by others[161] and similarly glucagon, somatostatin and GAD transcripts were reported in the mouse thymus.[158] Transcripts for the β-cell autoantigens GAD and IA-2 were also detected in human thymus.[68] Overall, genes encoding for several self-molecules have been found to be expressed in the thymus, including pancreatic and thyroid hormones, neuroendocrine molecules and other peripheral proteins.[162] Increasing evidence suggests that thymic expression of self-antigens may be crucial for the development of self-tolerance during the maturation of the immune system. This emerging concept is replacing the traditional beliefs that proteins with tissue-restricted or peripheral expression are not available for presentation in the thymus and that tolerance to such proteins can only be achieved through mechanisms of peripheral tolerance. Functional studies in transgenic mice and fetal organ thymic cultures have provided both in vivo and in vitro data showing that thymic expression of self-antigens and their levels of expression can dramatically affect the development of self-tolerance.[163-167]

The fact that negative selection of autoreactive thymocytes is dose-dependent suggested the hypothesis that different VNTR alleles may modulate tolerance to insulin by affecting insulin expression levels in the thymus. Consistent with this hypothesis, *INS* mRNA levels in the thymus were found to correlate with VNTR alleles in opposite fashion to that observed in the pancreas.[133] *INS* transcripts in *cis* with class III VNTR alleles are transcribed at much higher levels (on average 2-3 fold) than those in *cis* with class I VNTR alleles.[68,69] The increased transcription levels detected in thymus fit well with the dominant protective effect associated with class III VNTR alleles, as higher insulin levels in the thymus may more efficiently induce negative selection of insulin-specific T-lymphocytes (or improved selection of regulatory T cells). In contrast, homozygosity for diabetes-associated class I VNTR alleles determines lower insulin levels that may be associated with a less efficient deletion of insulin-specific autoreactive T-cells (or impaired selection of regulatory T cells). Although most of the thymus samples studied expressed both *INS* alleles, five out of 22 heterozygous samples expressed only one allele.[68,69] In all instances, the silenced allele was the one in *cis* with a class III VNTR. Such monoallelic expression resulting from the silencing of class III VNTR transcripts in the thymus may prevent the protective effect associated with the class III VNTR and explain the parent-of-origin effects discussed above. Vafiadis et al recently studied in more detail the class III alleles that were silenced in the thymus.[172] They developed a DNA fingerprinting method for identifying the type of alleles corresponding to the class III VNTR alleles that were found silenced in two thymus samples (S1, S2), and then analyzed the parental transmission of these type of class III alleles in a set of 287 diabetic children. Twelve of 18 possible transmissions of alleles matching the fingerprint of the S1 or S2 alleles were transmitted to the diabetic offspring, at a frequency of 0.67, which is significantly higher than the frequency of 0.38 seen in the remaining 142 class III alleles. These findings suggest that certain class III alleles may be predisposing instead of protective, and presumably these alleles are silenced in the thymus with obvious effects on the development of tolerance to insulin. Moreover, we have recently demonstrated monoallelic *INS* expression in the spleen of an 18 year-old Caucasian male, again preventing the expression of the *INS* transcript in *cis* with the class III VNTR allele.[173] Assuming that monoallelic expression in this subject was mediated by imprinting (parents were unavailable to determine the parental origin of the silenced allele), this finding suggests that the imprint status may be maintained beyond development and perhaps throughout life.

Direct support for the hypothesis that levels of *INS* expression in thymus and lymphoid organs can influence type 1 diabetes susceptibility is provided by the finding that transgenic expression of increased levels of proinsulin in the thymus (in MHC class II-positive cells) of NOD mice prevents insulitis and diabetes.[174] Proinsulin expression may be sufficient to obtain tolerance to insulin since most of the known immunodominant epitopes identified as targets of the insulin autoimmune responses in type 1 diabetes are shared by both insulin and proinsulin.[154,175-177] Moreover, there is evidence that proinsulin epitopes are targeted by autoimmune responses as well.[178] Further evidence for a critical role of insulin expression in the thymus comes from knockout mice. Mice have two insulin genes (*Ins1* and *Ins2*). *Ins2* is virtually the only insulin gene expressed in the thymus, and *Ins2* knockout mice are prone to develop insulin-specific autoreactive T cells.[178a] Moreover, when backcrossed into NOD mice, *Ins2* knockout mice develop spontaneous diabetes faster and the incidence in male mice increases to levels similar to those seen in female mice.[178b] *INS* transcription in human thymus also correlates with protein production in the thymus.[68,69] Proinsulin appears to be the main product of the insulin gene in the thymus.[160] This is not surprising since thymus cells expressing proinsulin are not likely to possess the refined machinery necessary to process proinsulin to mature insulin. Proinsulin is expressed in the thymus by a subset of antigen-presenting cells that appear to mediate a tolerogenic signal, both in mouse[169,170,179] and humans.[160] Moreover, similar cells and *INS* transcription have also been demonstrated in peripheral lymphoid organs, suggesting that insulin expression in lymphoid organs may also play a role in maintaining peripheral self-tolerance throughout life.[160] Halminen et al[180] reported that *IDDM2*-encoded susceptibility is associated with reduced insulin secretory capacity found in autoantibody-positive first-degree relatives (siblings) from the Childhood Diabetes in Finland study. This finding may be explained with more aggressive autoimmunity against insulin in subjects with the high-risk genotype and is consistent with the hypothesis that *IDDM2* may modulate immune responsiveness to insulin.

It is also possible that other loci in the 11p15 region may also contribute to *IDDM2*-encoded susceptibility, as recently suggested for certain class I VNTR allele lineages (ID-).[145] The four promoters of the *IGF2* gene are situated only 5-20 kb downstream of the *INS* VNTR, a distance that would allow enhancer effects. IGF-II is a growth factor with ubiquitous expression and has been implicated in biological functions that could be relevant to autoimmune diabetes.

These include inhibition of apoptosis[181] and the stimulation of β-cells proliferation,[182] functions that are of potential importance in resistance to immune injury and regeneration, respectively. IGF-II is also produced by T-lymphocytes[183] and can influence an activation-induced autocrine loop that would amplify clonal expansion of autoreactive T cells. However, the demonstration that the *INS*-VNTR does not influence *IGF2* transcription in the human thymus, pancreas, and leukocytes argues against a role for *IGF2* as a major contributor to *IDDM2*-encoded susceptibility.[184] It has also been suggested that IGF-II expression in the thymus may be important for the development of self-tolerance to IGF-II and other proteins of the IGF/insulin family, including insulin.[185-187] There also appears to be a thymus defect in IGF-II expression in the thymus of diabetes-prone BB rats, and this has led to the hypothesis that defective IGF-II expression in the thymus of BB rats may impair tolerance to insulin and favor diabetes development in BB rats.[188] However, it is controversial whether insulin is an autoantigen in the autoimmune diabetes of BB rats.[189-191] An alternative hypothesis is that the thymic IGF-II deficiency reported in this model may perhaps determine some more generic defects of the immune system and not necessarily affect tolerance to insulin.

Overall, the studies reviewed here suggest that the *IDDM2* is a quantitative trait resulting from allelic variation and, as discussed in a later paragraph, from complex parental and epigenetic effects at the VNTR locus. *IDDM2*-associated susceptibility and resistance may derive from quantitative differences in *INS* transcription in the specialized antigen presenting cells found in thymus and peripheral lymphoid tissues, where production of self-antigens such as proinsulin may be crucial for the shaping and maintaining of a self-tolerant T cell repertoire.[160] Such mechanisms may influence the probability of developing autoimmune responses to insulin, a key autoantigen in type 1 diabetes.

IDDM3

The initial evidence for linkage with marker D15S107 on chromosome 15q26 was initially reported in 250 Caucasian families from the U.K., U.S.A., and Canada. Families lacking the typical HLA predisposition provided most of the evidence for linkage, with sibling pair disease concordance or discordance being strongly affected by allele sharing at the D15S107 locus.[192] Linkage was confirmed in additional studies of 104 U.S.A. Caucasian families[193] and 81 Danish families.[194] In the Danish data set, the D15S107*130 allele provided increased susceptibility with a relative risk of 3.55 and the D15S107 locus was found to contribute up to 16% of the familial clustering of type 1 diabetes.[194] However, linkage was not confirmed in another data set of 265 Caucasian families.[195] At present there is no known candidate gene in the 15q26 region.

IDDM4

Several groups have reported evidence for the existence of the *IDDM4* susceptibility locus. This locus is tightly linked to the FGF3 marker on chromosome 11q13.[193,195-199] In a recent report,[200] researchers found evidence for decreased transmission (46.4%) to the affected offspring of a 15 Kb stretch of DNA containing two tightly linked alleles (D11S1917*03 and H0570polyA*02). In contrast, the D11S1917*03-H0570polyA*02 haplotype showed increased transmission (56.6%) to unaffected siblings. These results suggest that *IDDM4* susceptibility may derive from a gene very close to the D11S1917 marker. Moreover, similar to that discussed for *IDDM1* and *IDDM2*, these findings show that analysis of both predisposing and nonpredisposing alleles may be of value when mapping genes for common polygenic diseases.[200] A subsequent study provided further evidence for linkage with a peak LOD score of 3.4 at the D11S913 marker.[201] Moreover, the extended transmission disequilibrium test ETDT reveal significant association/linkage with the marker D11S987 (P=0.0004) within an interval of approximately 6 cM between D11S4205 and *GALN*.

Several candidate genes can be found in this chromosomal region. *MDU1*, encoding a cell-surface cell protein regulating intracellular calcium, and *ZFM1*, a nuclear protein, are both expressed in the pancreas. The RT6 gene lies also in this region, coding for a T-cell protein that is deficiently expressed in the BB rat animal model of diabetes.[197] The interleukin-converting enzyme (*ICE*) and *CD3* genes were also proposed as candidates to explain *IDDM4* susceptibility. However, the *CD3* gene has been excluded by both association and linkage analysis.[202] A previous report of an association of the *CD3* gene with IDDM could have been due to population stratification.[203] The gene coding for the low-density lipoprotein receptor related protein 5 (*LRP5*) has recently been mapped within the boundaries of the *IDDM4* locus and proposed as yet another candidate. However, its potential role in the pathogenesis of IDDM remains unclear.[204] Three additional genes have been identified in the LRP5 region: the CGI-85 gene and two novel genes, *C11orf24* and *C11orf23*. The *C11orf24* gene has no known similarity to other genes, and its function is unknown. *C11orf23* has similarity which genes involved in regulation of the cell cycle.[205] Finally, the gene coding for the Fas-associated death domain protein *FADD/MORT1* has been mapped to chromosome 11q13.3.[206] Both its chromosomal localization and function in apoptosis, a mechanism of cell death recently implicated in the autoimmune destruction of β-cells,[207,208] make it another plausible candidate for a susceptibility gene at the *IDDM4* locus. However, polymorphisms in the *FADD* and *GALN* genes were not found to be associated or in linkage with diabetes.[201]

IDDM5

Several but not all studies have shown linkage with the a046xa9 and *ESR* markers on chromosome 6q25.[195,196,199,209-211] The Mn-superoxide dismutase (*MnSOD*) is a likely candidate gene that could explain susceptibility at the *IDDM5* locus. Polymorphisms affecting the function of *MnSOD* could render β-cells more susceptible to free oxygen radicals damage. However, this region may contain a susceptibility gene that is common to more autoimmune diseases.[212]

IDDM6

Linkage with the 18q12-q21 region, and in particular with the Kidd Blood group locus (JK), was suggested almost 20 years ago[213] and linkage to the JK-D18S64 marker was recently confirmed by the first genome-wide scan based on the utilization of microsatellite.[196] The Transmission Disequilibrium Test (TDT) provided evidence for increased transmission of allele 4 of marker D18S487 to affected children in a total of 1,067 families from four different countries. Analysis using the TDT also provided evidence for genetic heterogeneity, which can often play as a confounding factor when mapping susceptibility genes in complex diseases.[214] Additional evidence for the existence of the *IDDM6* susceptibility locus near D18S487 was recently provided by another large study of 1,708 families from seven different countries.[215] There is evidence that this region may predispose to several autoimmune diseases.[212] A candidate gene has been recently identified in this region, *ZNF236*, a gene coding for a Kruppel-like zinc-finger protein. *ZNF236* is ubiquitously expressed in all human tissues tested. Its expression levels are highest in skeletal muscle and brain, intermediate in heart, pancreas, and placenta, and lowest in kidney, liver, and lung. Two alternative spliced forms of the *ZNF236* transcript have been found to be up-regulated in human mesangial cells in response to elevated levels of glucose, suggesting that *ZNF236* may be a candidate gene for diabetic nephropathy.[216]

IDDM7

Linkage on chromosome 2q near the marker D2S326 was initially reported in U.K. families[196] and was later observed for the D2S152 marker (2q31-q33) in 348 affected sibling pairs and 107 simplex families from three different populations.[217] Further analysis and expansion of the above data sets has failed to reproduce convincing evidence for a susceptibility gene in this region,[218] similar to studies of families from the U.S.A. and China.[195,219] However, linkage has been recently replicated in 241 Danish families.[220] The available evidence places *IDDM7* within two centiMorgans of D2S152, a chromosomal region that is syntenic with the NOD mouse chromosome 1 region containing the Idd5 susceptibility gene.[199,217] The *HOXD8* gene has been proposed as a possible susceptibility gene at the *IDDM7* locus.[221] Another potential candidate gene in this chromosomal region is *NEUROD*, another transcription factor regulating the expression of the insulin gene and playing an important role in the development of pancreatic β-cells. The *NEUROD* gene has been recently mapped to the long arm of human chromosome 2 (2q32).[222] A polymorphisms consisting of a nucleotide G-to-A transition results in the substitution of alanine to threonine at codon 45 (Ala45Thr). The analysis of this polymorphism in Japanese and Danish patients suggested an association with type 1 but not type 2 diabetes,[223,224] but a case-control study in France did not find a similar association.[225] Another candidate gene in this region is *GALNT3*, which encodes a polypeptide N-acetyl-galactosaminyltransferase-T3 (GalNAc-T3) and was recently mapped to a region 5-25 cM from D2S152.[226] GalNAc transferases may influence autoimmunity by glycosylating autoantigens. However, both a marker corresponding to *GALNT3* (D2S2363) and the T284A polymorphism in the *GALNT3* 3'UTR were not found to be linked with diabetes in Danish families.[220]

Most recently, the islet-specific glucose-6-phosphatase catalytic subunit-related protein (IGRP) has been identified as an autoantigen in the NOD mouse.[220a] The gene coding for IGRP falls within the region of *IDDM7*. While there is yet no direct evidence that IGRP is an autoantigen in humans, IGRP is probably a strong candidate gene to explain susceptibility at the *IDDM7* locus.

IDDM8

Several groups reported evidence for a susceptibility gene on chromosome 6q25-q27 linked to markers D6S264, D6S446,[193,196] and D6S281.[195,209,210] At present there is no known candidate gene in the 6q25-q27 region. Owerbach has recently defined a linkage disequilibrium map of nearly 1 Mbin the 6q27 region and identified multiple haplotypes associated with *IDDM8*, suggesting localization of this type 1 diabetes susceptibility gene to the terminal 200 kb of chromosome 6.[227] The *IDDM8* locus may also be subject to parental effects[228] and may confer susceptibility to rheumatoid arthritis as well.[229]

IDDM9

Limited evidence suggests a susceptibility locus on chromosome 3q21-q25 in linkage with marker D3S1303.[196] There is no known candidate gene in the 3q21-q25 region. *IDDM9* appears to be distinct from a susceptibility locus for Rheumatoid Arthritis reported on chromosome 3q.[230]

IDDM10

Another susceptibility locus may exist on chromosome 10p11-q11 (marker D10S193), and has been termed *IDDM10*.[196] Additional support for the existence of *IDDM10* was provided by the TDT analysis of 1,159 families with at least one affected child from the U.K., the U.S.A., Norway, Sardinia, and Italy.[231] At present there is no available candidate gene in the 10p11-q11 region. The *IDDM10* locus may be subject to parental effects.[228] *IDDM10* may play a stronger role in younger patients.[232]

IDDM11

IDDM11 appears to lie on chromosome 14q24.3-q31 and was linked to the microsatellite D14S67 using both maximum likelihood methods and affected sib pair methods. This represents the strongest evidence for linkage to any locus outside the HLA region. Similar to *IDDM3*, the strongest linkage (with the D14S67 marker) was obtained in a subset of families lacking increased HLA sharing among the affected offspring, suggesting that *IDDM11* may be an important susceptibility locus in families lacking strong HLA region predisposition.[233] Two candidate genes have been recently mapped to this chromosomal region. The *ENSA* gene encodes alpha-endosulfine, an endogenous regulator of β-cell K(ATP) channels.[234] The recombinant alpha-endosulfine has been shown to inhibit sulfanylurea binding to β-cell membranes, to reduce cloned K(ATP) channel currents, and to stimulate insulin secretion from β-cells. The *SEL-1L* gene codes for a negative regulator of the *NOTCH*, *LIN-12*, and *GLP-1* receptors, which are required for differentiation and maturation of cells as well as cell-cell interactions during development.[235] *SEL-1L* is abundantly expressed only in the pancreas, and appears to be involved in the down-regulation of mammalian Notch signaling, recently shown to be critical for the development of the pancreas and β-cells.[236] However, a study of families from Denmark and Sardinia found no evidence that *SEL-1L* is directly linked to diabetes.[237]

IDDM12

Linkage with markers on chromosome 2q33 was initially reported in a group of Italian families.[238] This chromosomal region contains the *CTLA-4* (cytotoxic T lymphocyte associated-4) and *CD28* genes, which encode two molecules that are intimately involved in the regulation of T-cell activation and proliferation. Differential regulation of these molecules could easily affect T-cell function and hence the regulation of immune responses. The *CTLA-4* gene is a strong candidate gene for autoimmune diseases since it encodes for a molecule that functions as a key negative regulator of T-cell activation, and the linked markers encompass a region containing an (AT)n microsatellite located in the 3' UTR of the *CTLA-4* gene.[238a] Moreover, the analysis of an A-G transition in the first exon of the *CTLA-4* gene, coding for a Thr/Ala substitution in the leader peptide, also showed preferential transmission to the affected siblings.[238] Although linkage was not observed in families from Sardinia, U.K., and U.S.A., preferential transmission was observed considering all of the above families together (n= 818).

Further confirmation of linkage with the *IDDM12-CTLA-4* locus came through linkage disequilibrium (association) analysis using a multi-ethnic collection of families with one or more affected children, which included families from Spain, France, China, Korea, and Mexican-Americans. In this study, the TDT revealed a highly significant deviation for transmission of alleles at the (AT)n microsatellite marker in the 3' untranslated region as well as the A/G polymorphism in the first exon of the *CTLA-4* gene. The overall evidence for transmission deviation of the *CTLA-4* A/G alleles remained highly significant even combining data sets (669 multiplex and 357 simplex families) from this study and the above families from Italy, U.K., U.S.A., Spain and Sardinia. However, significant heterogeneity was observed: U.K., Sardinian, and Chinese families did not show any deviation for the A/G polymorphism, while the U.S.A. families showed a weak transmission deviation. In contrast, strong deviation for transmission was seen in the three Mediterranean-European populations (Italian, Spanish and French), in Mexican-Americans, and in Koreans.[239] Linkage has also been reported in Japanese patients with IDDM and patients with autoimmune thyroid disease,[240,241] although not in all studies of Japanese patients.[242] Further evidence for linkage in the *CTLA-4/CD28* region was recently obtained in a large study of 960 families from Italy, Sardinia, U.K., and U.S.A.,[218] but not in the Danish population.[243] The *CTLA4* predisposing variant was increased diabetes risk in synergy with HLA-DR3 but not with HLA-DR4 in a German population.[244]

In a recent study, a multiethnic (U.S. Caucasian, Mexican-American, French, Spanish, Korean, and Chinese) collection of 178 simplex and 350 multiplex families was typed for 10 polymorphic markers within a genomic interval of approximately 300 kb, which contains the candidate genes *CTLA4* and *CD28*. The transmission disequilibrium test revealed significant association/linkage with three markers within *CTLA4* and two immediate flanking markers (D2S72 and D2S105) on each side of *CTLA4* but not with more distant markers including the candidate gene *CD28*. Because these markers are contained within a phagemid artificial chromosome clone of 100 kb, the *IDDM12* locus is likely to correspond to either *CTLA4* itself or to or an unknown gene in very close proximity.[245] Moreover, the Idd5 susceptibilty locus in NOD mice overlaps with *CTLA-4*.[246]

Further evidence identifying *IDDM12* with the *CTLA-4* gene comes form the combined analysis of 3,671 families from the U.S., U.K., Finland and other European populations. Linkage was found with both IDDM and autoimmune thyroid disease (AITD).[238a] Two main haplotypes were identified, a protective haplotype (+49A-CT60A) and a predisposing one (+49G-CT60G; O.R. 1.15). In addition, expression studies in peripheral blood cells demonstrated the existence of an alternatively spliced form of CTLA-4, which lacks exon 3 and is soluble. Of note, the protective haplotype was associated with higher levels of soluble CTLA-4 mRNA (50% greater). These findings suggest that variation in the leels of CTLA-4, in pariticular its soluable form, may affect the magnitude of the immune response and the risk of developing autoimmunity.

IDDM13

The *IDDM13* susceptibility locus appear to lie in the 2q34 region and is linked to the D2S164 marker in Caucasian families from Australia and the U.K. (BDA Repository).[247] Similarly to *IDDM3* and *IDDM11*, *IDDM13* may be of particular interest since it was detected in nonHLA identical siblings, suggesting that yet another locus may be an important susceptibility factor in subjects lacking the typical HLA predisposition. It has also been suggested that *IDDM13* may be active early during the evolution of diabetes since linkage was found also in prediabetic subjects. Moreover, it appears that *IDDM13* may favor diabetes development predominantly in males.[248] Several biological explanations are possible for these findings, including X and Y linkage, effects of sex hormones on gene expression, and quasi-linkage between sex chromosomes and autosomes. Another study in Japanese families has confirmed linkage to the D2S137 microsatellite in siblings lacking HLA predisposition.[249] In contrast, a recent study found no or little evidence for *IDDM13* in a data set including 352 U.K. families and 94 U.S.A. families.[218] Some evidence for linkage and association of the *IDDM13*/D2S137-D2S1471 region (approximately 3.5 cM) was found in Danish families.[243] It is of interest that *IDDM13*, *IDDM7*, and *IDDM12*, are all located on chromosome 2q31-35. This region may correspond to the syntenic mouse region on chromosome 1 containing Idd5, possibly a multigenic susceptibility locus in the NOD mouse. The *IDDM13* locus may correspond to the insulin growth factor-2 and -5 binding proteins (*IGFBP2*, *IGFBP5*), that are expressed at decreased levels in patients with type 1 diabetes.[250] A number of polymorphisms of *IGFBP2*, *IGFBP5* and other genes in the region (including *NEUROD*, *HOXD8*, and *CTLA4*) were not associated with diabetes in a case-control study.[251]

IDDM15

Linkage with the microsatellite D6S283 on chromosome 6q21 has been reported in families from France, Denmark, and the U.S.A. This locus is linked to HLA in males but not in females.[210] *IDDM15* is the third locus localized on 6q together with *IDDM5* and *IDDM8*, but there is no evidence that these loci interact or are linked to *IDDM1* on chromosome 6p. Multilocus analysis shows that linkage (measured by Lod scores) decreases with increasing distance from *IDDM1* (Lod Score *IDDM15*>*IDDM5*>*IDDM8*).[210] As discussed later in this chapter, parental effects may influence susceptibility at the *IDDM15* locus, and it has been suggested that the susceptibility gene at this locus may correspond to an imprinted gene associated with transient neonatal diabetes mellitus.[210,252,253]

IDDM17

Unlike all the preceding susceptibility loci, which have been mostly pinpointed studying large collection of families with affected sibling pairs, evidence for the *IDDM17* locus has been found studying a large Bedouin Arab family with 19 affected individuals.[254] *IDDM17* maps to the long arm of chromosome 10 (10q25). Recombination events occurring on this haplotype place *IDDM17* within an 8-cM interval between markers D10S1750 and D10S1773. Two other markers, D10S592 and D10S554, showed evidence of linkage disequilibrium with the disease locus. Remarkably, one chromosome 10 haplotype, the B haplotype, was transmitted from a heterozygous parent to 13 of 13 affected offspring compared to 10 of 23 unaffected siblings. A 273-bp allele at D10S592 was transmitted to 8 of 10 affected offspring compared to 3 of 14 unaffected siblings, and a 151-bp allele at D10S554 was transmitted to 15 of 15 affected offspring compared with 10 of 24 unaffected siblings. Moreover, all of the affected members in this family carry one or two high-risk HLA-DR3 haplotypes that are rarely found in other family members. Thus, the study of this family suggests the alternative hypothesis that type 1 diabetes may be a oligogenic disease rather than a polygenic one, and that perhaps just two or three genes may suffice to explain all of the inherited susceptibility in a given family.

IDDM18

Morahan and coworkers reported linkage dysequilibrium between a single base pair change in the 3' UTR (untranslated region) of the *IL-12B* gene and type 1 diabetes in two Australian cohorts.[255] The *IL-12* gene is an important candidate in terms of immune function. Unfortunately multiple studies from other populations, including British and U.S. have failed to find evidence of association with the reported polymorphism (see Table 1).[256] The functional influence of the 3' UTR also remains unconfirmed.

Other Susceptibility Loci

Linkage has been reported with a few other loci that have not received an official denomination. The glucokinase gene (*GCK*) on chromosome 7 was linked to IDDM in 339 affected sib-pair families, but this finding has not been reproduced in other studies.[257] Linkage was also reported for the D1S1617 marker on chromosome 1q (D1S1617),[258] and yet another locus may lie on chromosome X linked to markers DX6678 and DXS1068. This locus may influence the male-female bias in HLA-DR3-positive patients.[259] Note that the *IDDM14* and *IDDM16* denominations have not been officially assigned to any locus. Of importance, the combined analysis of multiple data sets showed the most dramatic linkage (LOD=3.83) after *IDDM1* and *IDDM2* with a region on chromosome 16q22-q24 in association with D16S3098. This was the only "significant" LOD score (outside of *IDDM1* and *IDDM2*) in this study of 767 mulitplex families (Table 1).[256]

Parental Effects on Inherited Genes Expression

The genetics of type 1 diabetes is further complicated by the possible existence of parental effects acting on the transmission and expression of inherited genes. Several studies have shown that diabetes risk differs in the offspring of diabetic mothers and fathers, although the results of different studies have been discrepant.[260-262] It is also controversial whether parent of origin effects influence the transmission of *IDDM1* alleles to the diabetic offspring.[263-265] Recent evidence also suggests that parental influences may be operative at the *IDDM8*, *IDDM10*, and *IDDM15* loci.[210,228]

Parental effects also influence the transmission of the VNTR alleles at the *IDDM2* locus, and probably this is the most studied locus in this regard. The first report of linkage at the *IDDM2* locus found evidence, in a small subset of families that were informative for parental origin, that the excess allele sharing was exclusively paternal.[130] Most of the subsequent studies of intra-familial association demonstrated a statistically significant difference only for paternally transmitted alleles.[137,266-268] These observations may be explained by imprinting, a mechanism that regulates gene expression by silencing either the maternal or the paternal allele. The silencing effect results from the epigenetic modification (probably mediated by methylation) of the DNA during the passage from the male or the female germline. This modification of the DNA marks the genetic material as maternal or paternal (parental imprint). Of note, the insulin gene is located in a region of the human genome that is known to be subject to parental imprinting.[268] The *IGF2* gene, which is adjacent to the insulin gene on chromosome 11p15, was the first human gene found to be imprinted and it is expressed exclusively from the paternal chromosome.[269] Several other genes in the region are expressed from the paternal or maternal chromosomes only, at least in some tissues or developmental stages.[270] *INS* is expressed from both copies in the pancreas of mice,[271] human fetuses of 7-20 weeks gestation[146] and adult humans.[141,173] However, we recently observed monoallelic *INS* expression in the pancreas of a 40-week old female fetus.[271] *INS* is also expressed monoallelically, and specifically from the paternal chromosome in the mouse yolk sack.[133] In addition, evidence has been recently presented for the imprinted paternal expression of *INS* in the human yolk sac.[272] Thus, imprinted expression can depend on the tissue and possibly the developmental stage.[273] The effects of imprinting may influence insulin expression during development and susceptibility to insulin/growth-related diseases in later life.

There is also evidence for even more complex mechanisms regulating *INS* transcription. Bennett et al[274] studied more than 1,300 triads (two parents and affected child) and showed that the most common class I VNTR allele among Caucasians, termed 814 in arbitrary electrophoretic mobility units, has a protective effect similar to that of class III VNTR alleles. A protective effect of the 814 allele was independently confirmed in Basque families.[275] Bennett et al[274] observed this protective effect only when the 814 allele was inherited from fathers with an 814/III VNTR genotype. In contrast, fathers with a 814/I VNTR genotype transmitted both the 814 and other class I VNTR alleles to their diabetic children at similar frequency. This unusual transmission pattern suggests that the 814 allele may behave differently in the offspring depending on the father's nontransmitted allele. This phenomenon could be explained by paramutation, a mechanism initially described in plants that requires some kind of physical interaction between homologous chromosomal domains in the premeiotic nucleus of the male germline.[276] According to this mechanism, the function of the 814 class I VNTR allele can be modified by some interactions with the paternal class III allele that is not transmitted to the offspring. This hypothesis finds additional support in the recent finding that both cis- and trans-allelic interactions influence imprinting at the *Ins2* locus in the mouse.[277] Morphological evidence for a similar interaction has been presented for another imprinted locus (the Prader-Willi/Angelman syndrome locus on chromosome 15) in somatic cells.[278] The data presented here suggest that besides allelic variation, parent-of-origin effects and complex epigenetic phenomena can dramatically influence *INS* transcription. It is important to notice that these phenomena can only be studied by evaluating *INS* transcription in selected tissues and correlating these data to the VNTR genotype. Thus, expression studies and genotyping must be combined to fully dissect the contribution of the insulin gene to diabetes susceptibility.

Other Nonmendelian Regulatory Mechanisms

Besides parent of origin effects and other epigenetic phenomena, there is also evidence that alternative splicing can affect gene expression in a tissue specific manner and predispose to certain conditions.[279] These include type 1 diabetes, multiple sclerosis, and other neurological diseases.[280,281]

In the case of type 1 diabetes, alternative splicing may affect the probability that one would mount autoimmune responses to the autoantigen IA-2. IA-2 is a tyrosine-phosphatase-like protein enriched in the secretory granules of islet and neuroendocrine cells,[282] and consists of a single transmembrane (TM) region (residues 577-600) and extra- and intra-cellular domains.[283,285-287] An alternatively spliced variant of the IA-2 transcript has been discovered through the sequencing of a clone (ICA512.bdc) derived from a human pancreas library that is routinely used as a source of antigen in a specific assay for the detection IA-2 autoantibodies.[284] This alternatively spliced transcript lacks exon 13 (Δexon 13), which codes for 73 amino acids (aa 557-629) encompassing the TM and juxta-membrane domains. Because several epitopes recognized by autoimmune responses in IDDM overlap with these domains, we hypothesized that splicing of the *IA-2* gene could be differentially regulated in pancreas compared to thymus and peripheral lymphoid tissues, which we reported to also express IA-2.[68,160,280]

We studied IA-2 expression in islets, thymus and spleen from nondiabetic human tissue donors (no tissues were available from patients) of different ages (from 21 weeks of gestation to 52 years of age) and from diverse racial groups. IA-2 mRNA expression and splicing were evaluated by RT-PCR using primers that amplify both full-length and Δexon 13 transcripts, and the identity of the products was verified by sequencing. All pancreatic samples expressed full-length IA-2 transcript and approximately 50% also expressed the Δexon 13 transcript. IA-2 mRNA was also detected in approximately 70% of the lymphoid organs studied, which included thymus and spleen specimens. Of the lymphoid tissues found to express IA-2 transcripts, all thymus and spleen specimens exclusively expressed the Δexon 13 transcript, including cases for which we could simultaneously evaluate IA-2 expression in pancreas, thymus, and spleen. This indicated that the observed differences reflect tissue specific splicing and not individual variation. We also detected another alternatively spliced IA-2 transcript in which 129 bp of exon 14 are spliced out, resulting in the deletion of 43 amino acids (aa 653-695) in the intracellular domain. This transcript was detected in about 50% of the pancreatic samples studied but essentially in none of the thymus and spleen specimens. Immunostaining of frozen thymus, spleen, and lymph node sections with two rabbit antibodies recognizing either intra-cellular (aa 601-979) or extra-cellular portions (aa 389-576) of IA-2[282] demonstrated that IA-2 mRNA is translated into protein in lymphoid tissues.[160,280] Because the Δexon 13 transcript is the only IA-2 transcript detected in lymphoid organs, the staining observed can only derive from the translation of the Δexon 13 transcript into protein.

Thus, we demonstrated that alternative splicing causes differential IA-2 mRNA and protein expression in pancreas compared to lymphoid organs. Such differences may affect immune responsiveness to specific epitopes and help explain why IA-2 and not many other islet proteins become targets of autoimmunity in IDDM. Tolerance to linear or conformational epitopes typical of the full-length protein or of the Δexon 14 variant may not be achieved if these epitopes are expressed in islets but not in lymphoid organs. The specific lack of expression of the TM/Juxta-membrane domains (exon 13) in lymphoid organs helps in explaining why epitopes from these domains are often targeted by autoimmune responses in IDDM. Autoantibodies against IA-2 epitopes encoded by exons 13 and 14 have been reported in patients and can precede the appearance of autoantibodies against other intra-cellular epitopes (epitope spreading). Only a few T-cell epitopes have been identified due to the limitations of current T-cell assays[288-292] and T-cell reactivity against IA-2 epitopes encoded by exons 13 and 14 or against epitopes generated by alternative splicing of these exons has not yet been directly analyzed. However, there is evidence that the HLA-DR4 restricted, naturally processed 654-674 epitope (exon 14) is recognized by autoreactive T-cells.[293]

It has to be noted that all of the subjects studied were nondiabetic tissue donors that belonged to diverse races and age groups. The absolute concordance of the data in such a heterogeneous albeit small sample group suggests that differential IA-2 splicing may occur commonly in the population. The fact that only a minority of individuals develops spontaneous autoimmune responses against IA-2 may have several explanations. First, IA-2 autoantibodies are usually detected during the later stages of the prodromic period preceding diabetes symptoms, suggesting that IA-2 autoimmunity may be secondary to inflammatory damage taking place during the earlier stages of the disease process. This may also be one of the reasons why IDDM patients do not apparently react against other tissues expressing IA-2 (i.e., adrenal), since TM and intra-cellular IA-2 epitopes may be unavailable to the immune system without inflammation and tissue damage. Second, other genetic factors (genes coding for HLA antigens, cytokines, etc.) and environmental exposures may be necessary to trigger autoimmunity. Thus, alternative splicing can cause differential expression of the IA-2 autoantigen and act as a permissive factor for the development of autoimmune responses to this molecule. Similar to the parent-of-origin effects affecting insulin gene expression in thymus[68,69] and peripheral lymphoid organs,[173] differential IA-2 splicing appears to function as mechanism regulating gene expression independent of inherited alleles at the insulin and IA-2 loci. Although recent investigations had excluded linkage with *IA-2* polymorphisms,[218] these findings suggest that expression studies for selected candidate genes in tissues relevant to the disease process can help dissect the complex genetics of a multi-factorial disease such as type 1 diabetes.

"Acquired" Genetic Polymorphisms

The modest disease concordance rates observed in monozygotic twins suggest that factors other than inherited genes must play a role in determining progression to overt disease in those individuals carrying predisposing genes. Environmental factors (viruses, diet) are suspected to be such determinants (reviewed in ref. 294). Viruses could trigger specific autoimmune responses through mechanisms of molecular mimicry or by mediating a direct insult to β-cells. It is also an intriguing and yet unproven possibility that novel genes may be acquired through, or their expression stimulated by, environmental factors (viruses or diet) after birth. Unlike more common viruses, retroviruses can integrate in the human genome. Retroviral genes can be either inherited or acquired after birth, and common viral infections and/or sex hormone changes associated with puberty may activate quiescent retroviruses. Such acquired expression may trigger the development of diabetes in genetically predisposed individuals either via cross-reactivity or immunity against novel viral antigens previously unknown to the immune system. This could drive immunity against the tissue that is expressing the novel gene, or to any tissue expressing molecules with significant cross-reactivity. Thus, environmental factors may provide or activate genes that could act as "disease genes". This hypothesis was recently supported by the finding that a human endogenous retrovirus, termed IDDMK1, 222, is apparently expressed and released from leukocytes in patients with type 1 diabetes but not in controls.[295] It is unclear whether this or similar retroviruses could be expressed in the endocrine pancreas, but it was suggested that IDDMK1, 222 could drive the same T-cell receptor restriction observed in T-lymphocytes infiltrating the endocrine pancreas of two children who died at the onset of diabetes.[296] However, the role of IDDMK1, 222 has been seriously questioned by later studies. In fact, IDDMK1, 222 was found expressed at similar frequencies in patients and controls in several studies and no evidence for auto-reactivity against this virus has been reported.[297,299-305]

Concluding Remarks

A large body of evidence indicates that genetic factors influence both susceptibility to and resistance to type 1 diabetes. Several chromosomal regions have been linked with the disease, suggesting that this is a polygenic disorder and there is evidence that certain alleles provide genetic protection from diabetes, as clearly shown for the *IDDM1* (HLA-DQB1*0602) and *IDDM2* (*INS*-VNTR class III) and *IDDM12(CTLA-4)* loci. A protective allele may also exist at the *IDDM4* locus. Nine of the susceptibility loci reported lie in just three chromosomes (6, 2 and 11). Chromosome 6 contains *IDDM1* on 6p21.3, *IDDM5* on 6q25, *IDDM8* on 6q25-27, and *IDDM15* on 6q21. Chromosome 2 contains *IDDM7* on 2q31-q33, *IDDM12* on 2q33, and *IDDM13* on 2q34. Chromosome 11 contains *IDDM2* on 11p15.5 and *IDDM4* on 11q13.3. Moreover, susceptibility loci on chromosomes 2 and 6 seem to cluster within fairly small distances. This may facilitate coinheritance of susceptibility and epistatic interactions. Most of the loci identified through linkage analysis, however, provide only modest contribution to the familial clustering of diabetes, and no other major susceptibility loci have been reported besides *IDDM1*.

Significant genetic heterogeneity has been consistently noted among and within different data sets, and this could decrease the probability that other major susceptibility genes, if heterogeneous and occurring in a minority of families, could be identified with the current analytical power and study design. The identification of *IDDM17* in a large pedigree may reflect this phenomenon, and also suggests the alternative hypothesis that type 1 diabetes may be oligogenic rather than polygenic. Moreover, not all studies have consistently reproduced evidence for linkage, and a second-generation genome-wide scan performed in a large data set of US and UK families combined using multi-point linkage methods has failed to confirm linkage for most of the previously identified loci.[258] In contrast, another study of 263 UK pedigrees found confirmation for many of the same loci.[298] It is conceivable that genetic heterogeneity among populations and differences in the statistical methods used may have a major impact on the outcome of linkage studies. Thus, a definitive confirmation of linkage may require the analysis of additional and larger data sets and, as recently suggested, more powerful analytical methods involving linkage disequilibrium-based and haplotype mapping approaches.[298] Consistent with the interpretation that discrepant results may often result from inadequate sample size, a third genome scan was performed using a new collection of 225 multiplex families and merging the data from all three major genome scans.[16] The combined sample of 831 affected sibling pairs provided 90% power to detect linkage for loci with lambda(s)= 1.3 at P= 7.4×10^{-4}. Three chromosomal regions showed significant evidence of linkage (LOD scores >4): 6p21 (*IDDM1*), 11p15 (*IDDM2*), and 16q22-q24. Four additional regions showed suggestive evidence (LOD scores $\geq$ 2.2): 10p11 (*IDDM10*), 2q31 (*IDDM7*, *IDDM12*, and *IDDM13*), 6q21 (*IDDM15*), and 1q42. Additional analyses that included the presence of specific high-risk HLA genotypes and age at disease onset provided evidence of linkage for additional loci including the putative *IDDM8* locus on chromosome 6q27. It is expected that large collaborations such as the 13th International HLA Working Group (http://www.ihwg.org) and the Type 1 Diabetes Genetic Consortium (http://www.t1dgc.org) will allow the study of much larger and diversified data sets and improve our ability to identify diabetes susceptibility genes. At present, however, most of the susceptibility genes lying in the chromosomal regions linked to diabetes remain unknown. A few candidate genes are available at some loci and it is expected that progress will be further facilitated by the availability of sequence data from the Human Genome Project.

Progress has also being made in dissecting the molecular mechanisms controlling diabetes susceptibility. It is a fair speculation that *IDDM1* may be involved in antigen presentation and control immune responsiveness to one or more islet cell antigens, either in the thymus or in the periphery, or both. As regards *IDDM2*, VNTR alleles appear to influence *INS* expression in the thymus, and thus may affect the development of tolerance to insulin. Thus, *IDDM1* and *IDDM2* may influence basic immunological functions such as antigen presentation and thymic selection, and appear to control the specificity of the immune response rather than mediate a generic predisposition to autoimmunity. It is important to note that the susceptibility alleles identified at the best characterized loci (*IDDM1*, *IDDM2*, and *IDDM12*) are commonly observed in the population and all appear to have normal sequence. This suggests that susceptibility may simply derive from polymorphisms affecting the function and expression of certain proteins. It is plausible that a number of loci may map to regulatory sequences (such as VNTRs) and act as quantitative traits. Complex parental effects and other genetic mechanisms such as alternative splicing may modify the transmission and expression of inherited genes, further complicate the genetics of diabetes. It is also an intriguing possibility that additional genetic factors or their expression may be acquired after birth, perhaps through environmental exposures. Such factors may play a key role in triggering the process leading to β-cell destruction. Future studies are expected to lead to the full identification and characterization of susceptibility and perhaps disease genes, and this will improve our ability to predict disease risk by genetic testing and deepen our understanding of the key mechanisms leading to type 1 diabetes. Such improved knowledge should be instrumental to the development of novel and effective strategies for prevention and cure.

References

1. Schranz DB, Lernmark A. Immunology in diabetes: An update. Diabetes Metab Rev 1998; 14:3-29.
2. Seissler J, de Sonnaville JJ, Morgenthaler NG et al. Immunological heterogeneity in type I diabetes: Presence of distinct autoantibody patterns in patients with acute onset and slowly progressive disease. diabetol 1998; 41:891-897.
3. Risch N, Ghosh S, Todd JA. Statistical evaluation of multiple-locus linkage data in experimental species and its relevance to human studies: Application to nonobese diabetic (NOD) mouse and human insulin- dependent diabetes mellitus (IDDM). Am J Hum Genet 1993; 53:702-714.
4. Todd JA, Farrall M. Panning for gold: Genome-wide scanning for linkage in type I diabetes. Hum Mol Genet 1996; 5:1443-1448.
5. Rich SS. Mapping genes in diabetes: Genetic epidemiological perspective. Diab 1990; 39:1315-1319.
6. Barnett AH, Eff C, Leslie RD et al. Diabetes in identical twins. A study of 200 pairs. diabetol 1981; 20:87-93.
7. Srikanta S, Ganda OP, Eisenbarth GS et al. Islet cell antibodies and beta cell function in monozygotic triplets and twins initially discordant for Type I diabetes mellitus. N Engl J Med 1983; 308:322-325.
8. Srikanta S, Ganda OP, Jackson RA et al. Type I diabetes mellitus in monozygotic twins: Chronic progressive beta cell dysfunction. Ann Intern Med 1983; 99:320-326.
9. Millward BA, Alviggi L, Hoskins PJ et al. Immune changes associated with insulin dependent diabetes may remit without causing the disease: A study in identical twins. Br Med J (Clin Res Ed) 1986; 292:793-796.
10. Kumar D, Gemayel NS, Deapen D et al. North-American twins with IDDM. Genetic, etiological, and clinical significance of disease concordance according to age, zygosity, and the interval after diagnosis in first twin. Diab 1993; 42:1351-1363.
11. Olmos P, A'Hern R, Heaton DA et al. The significance of the concordance rate for type 1 (insulin- dependent) diabetes in identical twins. diabetol 1988; 31:747-750.
12. Leslie RD, Pyke DA. Escaping insulin dependent diabetes. BMJ 1991; 302:1103-1104.
13. Verge CF, Gianani R, Yu L et al. Late progression to diabetes and evidence for chronic beta cell autoimmunity in identical twins of patients with type I diabetes (Abstract). J Investig Med 1995; 43:2.
14. Hearne CM, Ghosh S, Todd JA. Microsatellites for linkage analysis of genetic traits. TIG 1992; 8:288-294.
15. Lernmark A. Human cell lines from families available for diabetes research. Diabetol 1991; 34(1):61.
16. Cox NJ, Wapelhorst B, Morrison VA et al. Seven regions of the genome show evidence of linkage to type 1 diabetes in a consensus analysis of 767 multiplex families. Am J Hum Genet 2001; 69:820-830.
17. Johnson AH, Hurley CK, Hartzman RJ et al. HLA: The major histocompatibility complex of man. In: Henry JB, ed. Clinical Diagnosis & Management by Laboratory Methods. Philadelphia: WB Saunders, 1991.
18. Noble JA, Valdes AM, Cook M et al. The role of HLA class II genes in insulin-dependent diabetes mellitus: Molecular analysis of 180 Caucasian, multiplex families. Am J Hum Genet 1996; 59:1134-114.
19. Cudworth AG, Woodrow JC. Evidence for HL-A-linked genes in "juvenile" diabetes mellitus. Br Med J 1975; 3:133-135.
20. Cudworth AG, Woodrow JC. Genetic susceptibility in diabetes mellitus: Analysis of the HLA association. Br Med J 1976; 2:846-848.
21. Deschamps I, Lestradet H, Bonaiti C et al. HLA genotype studies in juvenile insulin-dependent diabetes. diabetol 1980; 19:189-193.
22. Rotter JI, Anderson CE, Rubin R et al. HLA genotypic study of insulin-dependent diabetes. The excess of DR3/DR4 heterozygotes allows rejection of the recessive hypothesis. Diab 1983; 32:169.
23. Wolf E, Spencer KM, Cudworth AG et al. The genetic susceptibility to Type 1 (insulin-dependent) diabetes: Analysis of the HLA-DR association. Diabetol 1983; 24:224-230.
24. Thomson G, Robinson WP, Kuhner MK et al. Genetic heterogeneity, modes of inheritance, and risk estimates for a joint study of Caucasians with insulin-dependent diabetes mellitus. Am J Hum Genet 1988; 43:799-816.
25. Todd JA, Bell JI, McDevitt HO et al. HLA-DQB gene contributes to susceptibility and resistance to insulin-dependent diabetes mellitus. Nature 1987; 329:599-604.
26. Horn GT, Bugawan TL, Long CM et al. Allelic sequence variation of the HLA-DQ loci: Relationship to serology and to insulin-dependent diabetes mellitus susceptibility. Proc Natl Acad Sci USA 1988; 85:6012-6016.

27. Sheehy MJ, Scharf SJ, Rowe JR et al. A diabetes-susceptible HLA haplotype is best defined by a combination of HLA-DR and -DQ alleles. J Clin Invest 1989; 83:830-835.
28. Morel PA, Dorman JS, Todd JA et al. Aspartic acid at position 57 of the HLA-DQ beta chain protects against type I diabetes: A family study. Proc Natl Acad Sci USA 1988; 85:8111-8115.
29. Khalil I, D'Auriol L, Gobet M et al. A combination of HLA-DQ? Asp57-negative and HLA DQ? Arg52 confers susceptibility to insulin-dependent diabetes mellitus. J Clin Invest 1990; 85:1315-1319.
30. Khalil I, Deschamps I, Lepage V et al. Dose effect of cis- and trans-encoded HLA-DQ alpha beta heterodimers in IDDM susceptibility. Diab 1992; 41:378-384.
31. Nepom BS, Schwarz D, Palmer JP et al. Transcomplementation of HLA genes in IDDM. Diab 1987; 36:114-117.
32. Bellgrau D, Pugliese A. NOD mouse and BB rat: Genetics and immunologic function. In: Eisenbarth GS, Lafferty KJ, eds. Type I Diabetes: Molecular, Cellular, and Clinical Immunology. New York: Oxford University Press, 1996:53-75.
33. Kwok WW, Domeier ME, Johnson ML et al. HLA-DQB1 codon 57 is critical for peptide binding and recognition. J Exp Med 1996; 183:1253-1258.
34. Sanjeevi CB, DeWeese C, Landin-Olsson M et al. Analysis of critical residues of HLA-DQ6 molecules in insulin-dependent diabetes mellitus. Tissue Antigens 1997; 50:61-65.
35. Hoover ML, Marta RT. Molecular Modelling of HLA-DQ suggests a mechanism of resistance in type I diabetes. Scand J Immunol 1997; 45:193-202.
36. Awata T, Kuzuya T, Matsuda A et al. High frequency of aspartic acid at position 57 of HLA-DQ B-chain in Japanese IDDM patients and nondiabetic subjects. Diab 1990; 39:266-269.
37. Ronningen KS, Iwe T, Halstensen TS et al. The amino acid at position 57 of the HLA-DQ beta chain and susceptibility to develop insulin-dependent diabetes mellitus. Hum Immunol 1989; 26:215-225.
38. Todd JA, Mijovic C, Fletcher J et al. Identification of susceptibility loci for insulin-dependent diabetes mellitus by trans-racial gene mapping. Nature 1989; 338:587-589.
39. Jenkins D, Mijovic C, Fletcher J et al. Identification of susceptibility loci for type I (insulin-dependent) diabetes by trans-racial gene mapping. Diabetol 1990; 33:387-395.
40. Mijovic CH, Barnett AH, Todd JA. Genetics of diabetes. Trans-racial gene mapping studies. Baillieres Clin Endocrinol Metab 1991; 5:321-340.
41. Mijovic CH, Jenkins D, Jacobs KH et al. HLA-DQA1 and -DQB1 alleles associated with genetic susceptibility to IDDM in a black population. Diab 1991; 40:748-753.
42. Ronningen KS, Spurkland A, Tait BD et al. HLA class II associations in insulin-dependent diabetes mellitus among blacks, caucasoids, and japanese. New York, NY: Oxford University Press, 1991:713-722.
43. Ronningen KS, Spurkland A, Iwe T et al. Distribution of HLA-DRB1, -DQA1 and -DQB1 alleles and DQA1-DQB1 genotypes among Norwegian patients with insulin-dependent diabetes mellitus. Tissue Antigens 1991; 37:105-111.
44. Penny MA, Jenkins D, Mijovic CH et al. Susceptibility to IDDM in a Chinese population. Role of HLA class II alleles. Diab 1992; 41:914-919.
45. Ikegami H, Kawaguchi Y, Yamato E et al. Analysis by the polymerase chain reaction of histocompatibility leucocyte antigen-DR9-linked susceptibility to insulin-dependent diabetes mellitus. J Clin Endocrinol Metab 1992; 75(5):1381-1385.
46. Sanjeevi CB, Zeidler A, Shaw S et al. Analysis of HLA-DQA1 and -DQB1 genes in Mexican Americans with insulin-dependent diabetes mellitus. Tissue Antigens 1993; 42:72-77.
47. Lee KH, Wucherpfennig KW, Wiley DC. Structure of a human insulin peptide/HLA-DQ8 complex and susceptibility to type 1 diabetes. Nat Immunol 2001; 2:501-507.
48. Pugliese A. Genetic Protection from insulin-dependent diabetes mellitus. Diabetes Nutr Metab 1997; 10:169-179.
49. Baisch JM, Weeks T, Giles R et al. Analysis of HLA-DQ genotypes and susceptibility in insulin-dependent diabetes mellitus. N Engl J Med 1990; 322:1836-1841.
50. Sorrentino R, DeGrazia U, Buzzetti R et al. An explanation for the neutral effect of DR2 on IDDM susceptibility in central Italy. Diab 1992; 41:904-908.
51. Kockum I, Wassmuth R, Holmberg E et al. HLA-DQ primarily confers protection and HLA-DR susceptibility in type I (insulin-dependent) diabetes studied in population-based affected families and controls. Am J Hum Genet 1993; 53:150-167.
52. Hagopian WA, Sanjeevi CB, Kockum I et al. Glutamate decarboxylase-, insulin-, and islet cell-antibodies and HLA typing to detect diabetes in a general population-based study of Swedish children. J Clin Invest 1995; 95:1505-1511.

53. Caillat-Zucman S, Djilali-Saiah I, Timsit J. Insulin dependent diabetes mellitus (IDDM) joint report. 12th International Histocompatibility Workshop Study. In: Charron D, ed. Genetic diversity of HLA. Functional and medical implications. Paris: EDK, 1997:389-398.
54. Erlich HA, Zeidler A, Chang J et al. HLA class II alleles and susceptibility and resistance to insulin dependent diabetes mellitus in Mexican-American families. Nat Genet 1993; 3:358-364.
55. Erlich HA, Griffith RL, Bugawan TL et al. Implication of specific DQB1 alleles in genetic susceptibility and resistance by identification of IDDM siblings with novel HLA- DQB1 allele and unusual DR2 and DR1 haplotypes. Diab 1991; 40:478-481.
56. Zeliszewski D, Tiercy J, Boitard C et al. Extensive study of DRB, DQA, and DQB gene polymorphism in 23 DR2-positive, insulin-dependent diabetes mellitus patients. Hum Immunol 1992; 33:140-147.
57. Pugliese A, Zeller M, Yu L et al. Sequence analysis of the DQB1*0602 allele in rare patients with type 1 diabetes (Abstract). Diab 1998; 47(Suppl 1).
58. Pugliese A, Gianani R, Moromisato R et al. HLA-DQB1*0602 is associated with dominant protection from diabetes even among islet cell antibody-positive first-degree relatives of patients with IDDM. Diab 1995; 44:608-613.
59. Wen L, Wong FS, Tang J et al. In vivo evidence for the contribution of human histocompatibility leukocyte antigen (HLA)-DQ molecules to the development of diabetes. J Exp Med 2000; 191:97-104.
60. CarrascoMarin E, Shimizu J, Kanagawa O et al. The class II MHC I-A^{g7} molecules from nonobese diabetic mice are poor peptide binders. J Immunol 1996; 156:450-458.
61. Raju R, Munn SR, David CS. T cell recognition of human preproinsulin peptides depends on the polymorphism at HLA DQ locus: A study using HLA DQ8 and DQ6 transgenic mice. Hum Immunol 1997; 58:21-29.
62. Ettinger RA, Kwok WW. A peptide binding motif for HLA-DQA1*0102/DQB1*0602, the class II MHC molecule associated with dominant protection in insulin-dependent diabetes mellitus. J Immunol 1998; 160:2365-2373.
63. Geluk A, van Meijgaarden KE, Schloot NC et al. HLA-DR binding analysis of peptides from islet antigens in IDDM. Diab 1998; 47:1594-1601.
64. Nishimoto H, Kikutani H, Yamamura K et al. Prevention of autoimmune insulitis by expression of I-E molecules in NOD mice. Nature 1987; 328:432-434.
65. Bohme J, Schuhbaur B, Kanagawa O et al. MHC-linked protection from diabetes dissociated from clonal deletion of T cells. Science 1990; 249:293-295.
66. Slattery RM, Kjer-Nielsen L, Allison J et al. Prevention of diabetes in nonobese diabetic I-Ak transgenic mice. Nature 1990; 345:724-726.
67. Schmidt D, Verdaguer J, Averill N et al. A mechanism for the major histocompatibility complex-linked resistance to autoimmunity. J Exp Med 1997; 186:1059-1075.
68. Pugliese A, Zeller M, Fernandez A et al. The insulin gene is transcribed in the human thymus and transcription levels correlate with allelic variation at the INS VNTR-IDDM2 susceptibility locus for type I diabetes. Nat Genet 1997; 15:293-297.
69. Vafiadis P, Bennett ST, Todd JA et al. Insulin expression in human thymus is modulated by INS VNTR alleles at the IDDM2 locus. Nat Genet 1997; 15:289-292.
70. Rabinovitch A. An update on cytokines in the pathogenesis of insulin-dependent diabetes mellitus. Diabetes Metab Rev 1998; 14:129-151.
71. Pugliese A, Kawasaki E, Zeller M et al. Sequence analysis of the diabetes-protective human leukocyte antigen-DQB1*0602 allele in unaffected, islet cell antibody-positive first degree relatives and in rare patients with type 1 diabetes. J Clin Endocrinol Metab 1999; 84:1722-1728.
72. Gianani R, Verge CF, Moromisato-Gianani RI et al. Limited loss of tolerance to islet autoantigens in ICA+ first degree relatives of patients with type I diabetes expressing the HLA DQB1*0602 allele. J Autoimmun 1996; 9:423-425.
73. Tuomi T, Björses P, Falorni A et al. Antibodies to glutamic acid decarboxylase and insulin-dependent diabetes in patients with autoimmune polyendocrine syndrome type I. J Clin Endocrinol Metab 1996; 81:1488-1494.
74. Greenbaum CJ, Cuthbertson D, Eisenbarth GS et al. Islet cell antibody positive relatives with HLA-DQA1*0102, DQB1*0602: Identification by the Diabetes Prevention Trial-1. J Clin Endocrinol Metab 2000; 85:1255-1260.
75. Redondo MJ, Kawasaki E, Mulgrew CL et al. DR and DQ associated protection from type 1 diabetes: Comparison of DRB1*1401 and DQA1*0102-DQB1*0602. J Clin Endocrinol Metab 2000; 85:3793-3797.
76. Erlich HA, Bugawan TL, Scharf S et al. HLA-DQ? sequence polymorphism and genetic susceptibility to IDDM. Diab 1990; 39:96-103.

77. Nepom GT, Erlich H. MHC class-II molecules and autoimmunity. Ann Rev Immunol 1991; 9:493-525.
78. Tait BD, Drummond BP, Varney MD et al. HLA-DRB1*0401 is associated with susceptibility to insulin-dependent diabetes mellitus independently of the DQB1 locus. Eur J Immunogenet 1995; 22:289-297.
79. Cucca F, Muntoni F, Lampis R et al. Combinations of specific DRB1, DQA1, DQB1 haplotypes are associated with insulin-dependent diabetes mellitus in Sardinia. Hum Immunol 1993; 37:85-94.
80. Yasunaga S, Kimura A, Hamaguchi K et al. Different contribution of HLA-DR and -DQ genes in susceptibility and resistance to insulin-dependent diabetes mellitus (IDDM). Tissue Antigens 1996; 47:37-48.
81. Sanjeevi CB, Hook P, Landin-Olsson M et al. DR4 subtypes and their molecular properties in a population- based study of Swedish childhood diabetes. Tissue Antigens 1996; 47:275-283.
82. Undlien DE, Friede T, Rammensee HG et al. HLA-encoded genetic predisposition in IDDM: DR4 subtypes may be associated with different degrees of protection. Diab 1997; 46:143-149.
83. Erlich HA, Rotter JI, Chang JD et al. Association of HLA-DPB1*0301 with IDDM in Mexican-Americans. Diab 1996; 45:610-614.
84. Noble JA, Valdes AM, Thomson G et al. The HLA class II locus DPB1 can influence susceptibility to type 1 diabetes. Diab 2000; 49:121-125.
85. Spies T, Bresnahan M, Bahram S et al. A gene in the human major histocompatibility complex class II region controlling the class I antigen presentation pathway. Nature 1990; 348:744-747.
86. Colonna M, Bresnahan M, Bahram S et al. Alleleic variants of the human putative peptide transporter involved in antigen processing. Proc Natl Acad Sci USA 1992; 89:3932-3936.
87. Caillat-Zucman S, Bertin E, Timsit J et al. Protection from insulin-dependent diabetes mellitus is linked to a peptide transporter gene. Eur J Immunol 1993; 23:1784-1788.
88. Ronningen KS, Undlien DE, Ploski R et al. Linkage disequilibrium between TAP2 variants and HLA class II alleles: No primary association between TAP2 variants and insulin- dependent diabetes mellitus. Eur J Immunol 1993; 23:1050-1056.
89. Van Endert PM, Liblau RS, Patel SD et al. Major histocompatibility complex-encoded antigen processing gene polymorphism in IDDM. Diab 1994; 43:110-117.
90. Faustman D, Li X, Lin HY et al. Linkage of faulty major histocompatibility complex class I to autoimmune diabetes. Science 1991; 254:1756-1761.
91. Faustman D, Xiangping L, Lin H et al. Abnormal MHC class I presentation linked to autoimmunity (Abstract). Diab 1992; 41(suppl).
92. Gaskins HR, Monaco JJ, Leiter EH. Expression of intra-MHC transporter (Ham) genes and class I antigens in diabetes-susceptible NOD mice. Science 1992; 256:1826-1828.
93. Yan G, Shi L, Faustman D. Novel splicing of the human MHC-encoded peptide transporter confers unique properties. J Immunol 1999; 162:852-859.
94. Fu Y, Yan G, Shi L et al. Antigen processing and autoimmunity. Evaluation of mRNA abundance and function of HLA-linked genes. Ann NY Acad Sci 1998; 842:138-155.
95. Yan G, Shi L, Fu Y et al. Screening of the TAP1 gene by denaturing gradient gel electrophoresis in insulin-dependent diabetes mellitus: Detection and comparison of new polymorphisms between patients and controls. Tissue Antigens 1997; 50:576-585.
96. RobinsonWP, Barbosa J, Rich SS et al. Homozygous parent affected sib pair method for detecting disease predisposing variants: Application to insulin-dependent diabetes mellitus. Genet Epidemiol 1993; 10:273-288.
97. Labarthe DR, Nichaman MZ, Harrist RB et al. Development of cardiovascular risk factors from ages 8 to 18 in project heartbeat: Study design and patterns of change in plasma total cholesterol concentration. Circulation 1997; 95:2636-2642.
98. Demaine AG, Hibberd ML, Mangles D et al. A new marker in the HLA class I region is associated with the age at onset of IDDM. Diabetol 1995; 38:623-628.
99. Fennessy M, Metcalfe K, Hitman GA et al. A gene in the HLA class I region contributes to susceptibility to IDDM in the Finnish population. Childhood Diabetes in Finland (DiMe) Study Group [see comments]. Diabetol 1994; 37:937-944.
100. Langholz B, Tuomilehto-Wolf E, Thomas D et al. Variation in HLA-associated risks of childhood insulin-dependent diabetes in the Finnish population: I. Allele effects at A, B, and DR loci. DiMe Study Group. Childhood Diabetes in Finland. Genet Epidemiol 1995; 12:441-453.
101. Nakanishi K, Kobayashi T, Murase T et al. Association of HLA-A24 with complete ?-cell destruction in IDDM. Diab 1993; 42:1086-1098.
102. HoneymanMC, Harrison LC, Drummond B et al. Analysis of families at risk for insulin-dependent diabetes mellitus reveals that HLA antigens influence progression to clinical disease. Mol Med 1995; 1:576-582.

103. Nejentsev S, Reijonen H, Adojaan B et al. The effect of HLA-B allele on the IDDM risk defined by DRB1*04 subtypes and DQB1*0302. Diab 1997; 46:1888-1892.
104. Lie BA, Todd JA, Pociot F et al. The predisposition to Type 1 Diabetes linked to the human leukocyte antigen complex includes at least one nonclass II gene. Am J Hum Genet 1999; 64:793-800.
105. LieBA, Sollid LM, AscherH et al. A gene telomeric of the HLA class I region is involved in predisposition to both type 1 diabetes and coeliac disease. Tissue Antigens 1999; 54:162-168.
106. Moghaddam PH, de Knijf P, Roep BO et al. Genetic structure of IDDM1: Two separate regions in the major histocompatibility complex contribute to susceptibility or protection. Diab 1998; 47:263-269.
107. Obayashi H, Nakamura N, Fukui M et al. Influence of TNF microsatellite polymorphisms (TNFa) on age-at-onset of insulin-dependent diabetes mellitus. Hum Immunol 1999; 60:974-978.
108. Gambelunghe G, Ghaderi M, Cosentino A et al. Association of MHC Class I chain-related A (MIC-A) gene polymorphism with Type 1 diabetes. Diabetol 2000; 43:507-514.
109. Caplen NJ, Patel A, Millward A et al. Complement C4 and heat shock protein 70 (HSP70) genotypes and type I diabetes mellitus. Immunogenetics 1990; 32:427-430.
110. Pugliese A, Awdeh Z, Galluzzo A et al. No independent association between HSP70 gene polymorphism and IDDM. Diab 1992; 41:788-791.
111. Cascino I, D'Alfonso S, Cappello N et al. Gametic association of HSP70-1 promoter region alleles and their inclusion in extended HLA haplotypes. Tissue Antigens 1993; 42:62-66.
112. Pociot F, Ronningen KS, Nerup J et al. Polymorphic analysis of the human MHC-linked heat shock protein 70 (HSP70-2) and HSP70-Hom genes in insulin-dependent diabetes mellitus (IDDM). Scand J Immunol 1993; 38:491-495.
113. Kawaguchi Y, Ikegami H, Fukuda M et al. Polymorphism of HSP70 gene is not associated with type 1 (insulin-dependent) diabetes mellitus in Japanese. Diabetes Res Clin Pract 1993; 21:103-107.
114. Chuang LM, Jou TS, Wu HP et al. A rapid method to study heat shock protein 70-2 gene polymorphism in insulin-dependent diabetes mellitus. Pancreas 1996; 13:268-272.
115. Karjalainen J, Salmela P, Ilonen J et al. A comparison of childhood and adult type I diabetes mellitus. N Engl J Med 1989; 320:881-886.
116. Caillat-Zucman S, Garchon H-J, Timsit J et al. Age-dependent HLA genetic heterogeneity of type 1 insulin- dependent diabetes mellitus. J Clin Invest 1992; 90:2242-2250.
117. Valdes AM, Thomson G, Erlich HA et al. Association between type 1 diabetes age of onset and HLA among sibling pairs. Diab 1999; 48:1658-1661.
118. Gyllensten UB, Erlich HA. Ancient roots for polymorphism at the HLA-DQ alpha locus in primates. Proc Natl Acad Sci USA 1989; 86:9986-9990.
119. Erlich HA, Gyllensten UB. The evolution of allelic diversity at the primate major histocompatibility complex class II loci. Hum Immunol 1991; 30:110-118.
120. Pugliese A, Bugawan T, Moromisato R et al. Two subsets of HLA-DQA1 alleles mark phenotypic variation in levels of insulin autoantibodies in first degree relatives at risk for insulin-dependent diabetes. J Clin Invest 1994; 93:2447-2452.
121. Bell GI, Selby MJ, Rutter WJ. The highly polymorphic region near the human insulin gene is composed of simple tandemly repeating sequences. Nature 1982; 295:31-35.
122. Kennedy GC, German MS, Rutter WJ. The minisatellite in the diabetes susceptibility locus IDDM2 regulates insulin transcription. Nat Genet 1995; 9:293-298.
123. Lucassen AM, Screaton GR, Julier C et al. Regulation of insulin gene expression by the IDDM associated, insulin locus haplotype. Hum Mol Genet 1995; 4:501-506.
124. Owerbach D, Aagaard L. Analysis of a 1963-bp polymorphic region flanking the human insulin gene. Gene 1984; 32:475-479.
125. Rotwein P, Yokoyama S, Didier DK et al. Genetic analysis of the hypervariable region flanking the human insulin gene. Am J Hum Genet 1986; 39:291-299.
126. Hammond-Kosack MC, Docherty K. A consensus repeat sequence from the human insulin gene linked polymorphic region adopts multiple quadriplex DNA structures in vitro. FEBS Lett 1992; 301:79-82.
127. Hammond-Kosack MC, Dobrinski B, Lurz R et al. The human insulin gene linked polymorphic region exhibits an altered DNA structure. Nucleic Acids Res 1992; 20:231-236.
128. Bell GI, Horita S, Karam JH et al. A polymorphic locus near the human insulin gene is associated with insulin-dependent diabetes mellitus. Diab 1984; 33:176-183.
129. Cox NJ, Bell GI, Xiang KS. Linkage disequilibrium in the human insulin/insulin-like growth factor II region of human chromosome II. Am J Hum Genet 1988; 43:495-501.
130. Julier C, Hyer RN, Davies RN et al. Insulin-IGF2 region on chromosome 11p encodes a gene implicated in HLA-DR4-dependent diabetes susceptibility. Nature 1991; 354:155-159.

131. Bain SC, Prins JB, Hearne CM et al. Insulin gene region-encoded susceptibility to type I diabetes is not restricted to HLA-DR4-positive individuals. Nat Genet 1992; 2:212-215.
132. Owerbach D, Gabbay KH. Localization of a Type I diabetes susceptibilty locus to the variable tandem repeat region flanking the insulin gene. Diab 1993; 42:1708-1714.
133. Bennett ST, Lucassen AM, Gough SCL et al. Susceptibility to human type I diabetes at IDDM2 is determined by tandem repeat variation at the insulin gene minisatellite locus. Nat Genet 1995; 9:284-292.
134. Undlien DE, Bennett ST, Todd JA et al. Insulin gene region-encoded susceptibility to IDDM maps upstream of the insulin gene. Diab 1995; 44:620-625.
135. Doria A, Yang Y, Malecki M et al. Phenotypic characteristics of early-onset autosomal-dominant type 2 diabetes unlinked to known maturity-onset diabetes of the young (MODY) genes. Diab care 1999; 22:253-261.
136. Mijovic CH, Penny MA, Jenkins D et al. The insulin gene region and susceptibility to insulin-dependent diabetes mellitus in four races; New insights from Afro-Caribbean race-specific haplotypes. Autoimmunity 1997; 26:11-22.
137. Pugliese A, Awdeh ZL, Alper CA et al. The paternally inherited insulin gene B allele (1,428 FokI site) confers protection from insulin-dependent diabetes in families. J Autoimmun 1994; 7:687-694.
138. Bennett ST, Barnett AH, Bain SC et al. Dissecting IDDM-2-VNTR-encoded predisposition to type I diabetes. Autoimmunity 1995; 21:16.
139. Bennett ST, Todd JA. Human type 1 diabetes and the insulin gene: Principles of mapping polygenes. Annu Rev Genet 1996; 30:343-370.
140. Metcalfe KA, Hitman GA, Rowe RE et al. Concordance for type 1 diabetes in identical twins is affected by insulin genotype. Diab care 2001; 24:838-842.
141. Bennett ST, Wilson AJ, Cucca F et al. IDDM2-VNTR-encoded susceptibility to type 1 diabetes: Dominant protection and parental transmission of alleles of the insulin gene-linked minisatellite locus. J Autoimmun 1996; 9:415-421.
142. Lew A, Rutter WJ, Kennedy GC. Unusual DNA structure of the diabetes susceptibility locus IDDM2 and its effect on transcription by the insulin promoter factor Pur-1/MAZ. Proc Natl Acad Sci USA 2000; 97:12508-12512.
143. Owerbach D, Gabbay KH. The search for IDDM susceptibility genes: The next generation. [Review]. Diab 1996; 45:544-551.
144. Stead JD, Jeffreys AJ. Allele diversity and germline mutation at the insulin minisatellite. Hum Mol Genet 2000; 9:713-723.
145. Stead JD, Buard J, Todd JA et al. Influence of allele lineage on the role of the insulin minisatellite in susceptibility to type 1 diabetes. Hum Mol Genet 2000; 9:2929-2935.
146. Vafiadis P, Bennett ST, Colle E et al. Imprinted and genotype-specific expression of genes at the IDDM2 locus in pancreas and leukocytes. J Autoimmun 1996; 9:397-403.
147. Palmer JP, Asplin CM, Clemons P et al. Insulin antibodies in insulin-dependent diabetics before insulin treatment. Science 1983; 222:1337-1339.
148. Vardi P, Ziegler AG, Matthews JH et al. Concentration of insulin autoantibodies at onset of type I diabetes. Inverse log-linear correlation with age. Diab care 1988; 11:736-739.
149. Vardi P, Dib SA, Tuttleman M et al. Competitive insulin autoantibody assay. Prospective evaluation of subjects at high risk for development of type I diabetes mellitus. Diab 1987; 36:1286-1291.
150. Yu L, Robles DT, Abiru N et al. Early expression of antiinsulin autoantibodies of humans and the NOD mouse: Evidence for early determination of subsequent diabetes. Proc Natl Acad Sci USA 2000; 97:1701-1706.
151. Naserke HE, Ziegler A-G. Humoral and cellular immune response to insulin. Diabetes Nutrition and Metabolism 2001; 9:208-214.
152. Gottlieb PA, Sandell T, Babu S et al. Characterization of human IDDM T cell clones specific for an insulin peptide (Abstract). Diab 2001; 46.
153. Ellis T, Jodoin E, Ottendorfer E et al. Cellular immune responses against proinsulin: No evidence for enhanced reactivity in individuals with IDDM. Diab 1999; 48:299-303.
154. Alleva DG, Crowe PD, Jin L et al. A disease-associated cellular immune response in type 1 diabetics to an immunodominant epitope of insulin. J Clin Invest 2001; 107:173-180.
155. Wegmann DR, Norbury-Glaser M, Daniel D. Insulin-specific T cells are a predominant component of islet infiltrates in prediabetic NOD mice. Eur J Immunol 1994; 24:1853-1857.
156. Haskins K, Wegmann D. Diabetogenic T-cell clones. Diab 1996; 45:1299-1305.
157. Wong FS, Karttunen J, Dumont C et al. Identification of an MHC class I-restricted autoantigen in type 1 diabetes by screening an organ-specific cDNA library. Nat Med 1999; 5:1026-1031.
158. Jolicoeur C, Hanahan D, Smith KM. T-cell tolerance toward a transgenic β-cell antigen and transcription of endogenous pancreatic genes in thymus. Proc Natl Acad Sci USA 1994; 91:6707-6711.

159. Heath VL, Moore NC, Parnell SM et al. Intrathymic expression of genes involved in organ specific autoimmune disease. J Autoimmun 1998; 11:309-318.
160. Pugliese A, Brown D, Garza D et al. Self-antigen presenting cells expressing islet cell molecules in human thymus and peripheral lymphoid organs: Phenotypic characterization and implications for immunological tolerance and type 1 diabetes. J Clin Invest 2001; 107:555-564.
161. Sospedra M, Ferrer-Francesch X, Dominguez O et al. Transcription of a broad range of self-antigens in human thymus suggests a role for central mechanisms in tolerance toward peripheral antigens. J Immunol 1998; 161:5918-5929.
162. Werdelin O, Cordes U, Jensen T. Aberrant expression of tissue-specific proteins in the thymus: A hypothesis for the development of central tolerance. Scand J Immunol 1998; 47:95-100.
163. Heath WR, Allison J, Hoffmann MW et al. Autoimmune diabetes as a consequence of locally produced interleukin-2. Nature 1992; 359:547-549.
164. Ashton-Rickardt PG, Bandeira A, Delaney JR et al. Evidence for a differential avidity model of T cell selection in the thymus. Cell 1994; 76:651-663.
165. Sebzda E, Wallace VA, Mayer J et al. Positive and negative thymocyte selection induced by different concentrations of a single peptide. Science 1994; 263:1615-1618.
166. Oehen SU, Ohashi PS, Burki K et al. Escape of thymocytes and mature T cells from clonal deletion due to limiting tolerogen expression levels. Cell Immunol 1994; 158:342-352.
167. Sprent J, Webb SR. Intrathymic and extrathymic clonal deletion of T cells. Curr Opin Immunol 1995; 7:196-205.
168. Liblau RS, Tisch R, Shokat K et al. Intravenous injection of soluble antigen induces thymic and peripheral T-cells apoptosis. Proc Natl Acad Sci USA 1996; 93:3031-3036.
169. Smith KM, Olson DC, Hirose R et al. Pancreatic gene expression in rare cells of thymic medulla: Evidence for functional contribution to T cell tolerance. Int Immunol 1997; 9:1355-1365.
170. Hanahan D. Peripheral-antigen-expressing cells in thymic medulla: Factors in self- tolerance and autoimmunity. Curr Opin Immunol 1998; 10:656-662.
171. Salmon AM, Bruand C, Cardona A et al. An acetylcholine receptor alpha subunit promoter confers intrathymic expression in transgenic mice. Implications for tolerance of a transgenic self-antigen and for autoreactivity in myasthenia gravis. J Clin Invest 1998; 101:2340-2350.
172. Vafiadis P, Ounissi-Benkalha H, Palumbo M et al. Class III alleles of the variable number of tandem repeat insulin polymorphism associated with silencing of thymic insulin predispose to type 1 diabetes. J Clin Endocrinol Metab 2001; 86:3705-3710.
173. Miceli D, Zeller D, Brown C et al. Insulin gene expression and imprinting in pancreas and lymphoid organs (Abstract). Diab 1999; 48(Suppl. 1).
174. French MB, Allison J, Cram DS et al. Transgenic expression of mouse proinsulin II prevents diabetes in nonobese diabetic mice. Diab 1997; 46:34-39.
175. Kuglin B, Gries FA, Kolb H. Evidence of IgG autoantibodies against human proinsulin in patients with IDDM before insulin treatment. Diab 1988; 37:130-132.
176. Castano L, Ziegler AG, Ziegler R et al. Characterization of insulin autoantibodies in relatives of patients with type 1 diabetes. Diab 1993; 42:1202-1209.
177. Keilacker H, Rjasanowski I, Besch W et al. Autoantibodies to insulin and to proinsulin in type 1 diabetic patients and in at-risk probands differentiate only little between both antigens. Horm Metab Res 1995; 27:90-94.
178. Chen W, Bergerot I, Elliott JF et al. Evidence that a peptide spanning the B-C junction of proinsulin is an early Autoantigen epitope in the pathogenesis of type 1 diabetes. J Immunol 2001; 167:4926-4935.
178a. Chentoufi AA, Poylchronakos C. Insulin expression levels in the thymus modulate insulin-specific autoreactive T-cell tolerance: the mechanisms by which the IDDM2 locus may predispose to diabetes. Diabetes 2002; 51:1383-1390.
178b. Thebault-Baumont K, Dubois-Laforgue D, Krief P et al. Acceleration of type 1 diabetes mellitus in proinsulin 2-deficient NOD mice. J Clin Invest 2003; 111:851-857.
179. Throsby M, Homo-Delarche F, Chevenne D et al. Pancreatic hormone expression in the murine thymus: Localization in dendritic cells and macrophages. Endocrinology 1998; 139:2399-2406.
180. Halminen M, Veijola R, Reijonen H et al. Effect of polymorphism in the insulin gene region on IDDM susceptibility and insulin secretion. The Childhood Diabetes in Finland (DiMe) Study Group. Eur J Clin Invest 1996; 26:847-852.
181. Christofori G, Naik P, Hanahan D. Deregulation of both imprinted and expressed alleles of the insulin-like growth factor 2 gene during β-cell tumorigenesis. Nat Genet 1995; 10:196-201.
182. Hill DJ, Hogg J. Growth factor control of pancreatic B cell hyperplasia. Baillieres Clin Endocrinol Metab 1991; 5:689-698.

183. Nyman T, Pekonen F. The expression of insulin-like growth factors and their binding proteins in normal human lymphocytes. Acta Endocrinol (Copenh) 1993; 128:168-172.
184. Vafiadis P, Grabs R, Goodyer CG et al. A functional analysis of the role of IGF2 in IDDM2-encoded susceptibility to type 1 diabetes. Diab 1998; 47:831-836.
185. Geenen V, Achour I, Robert F et al. Evidence that insulin-like growth factor 2 (IGF2) is the dominant thymic peptide of the insulin superfamily. Thymus 1993; 21:115-127.
186. Geenen V, Lefebvre PJ. The intrathymic expression of insulin-related genes: Implications for pathophysiology and prevention of Type 1 diabetes. Diabetes Metab Rev 1998; 14:95-103.
187. Geenen V, Martens H, Brilot F et al. Thymic neuroendocrine self-antigens. Role in T-cell development and central T-cell self-tolerance. Ann NY Acad Sci 2000; 917:710-723.
188. Kecha-Kamoun O, Achour I, Martens H et al. Thymic expression of insulin-related genes in an animal model of autoimmune type 1 diabetes. Diabetes Metab Res Rev 2001; 17:146-152.
189. Whalen BJ, Marounek J, Weiser P et al. BB rat thymocytes cultured in the presence of islets lose their ability to transfer autoimmune diabetes. Diab 2001; 50:972-979.
190. Bieg S, Hanlon C, Hampe CS et al. GAD65 and insulin B chain peptide (9-23) are not primary autoantigens in the type 1 diabetes syndrome of the BB rat. Autoimmunity 1999; 31:15-24.
191. Mordes JP, Schirf B, Roipko D et al. Oral insulin does not prevent insulin-dependent diabetes mellitus in BB rats. Ann NY Acad Sci 1996; 778:418-421.
192. Field LL, Tobias R, Magnus T. A locus on chromosome 15q26 (IDDM3) produces susceptibility to insulin-dependent diabetes mellitus. Nat Genet 1994; 8:189-194.
193. Luo D-F, Bui MM, Muir A et al. Affected-sib-pair mapping of a novel susceptibility gene to insulin-dependent diabetes mellitus (IDDM8) on chromosome 6q25-q27. Am J Hum Genet 1995; 57:911-919.
194. Zamani M, Pociot F, Raeymaekers P et al. Linkage of type I diabetes to 15q26 (IDDM3) in the Danish population. Hum Genet 1996; 98:491-496.
195. Luo D-F, Buzzetti R, Rotter JI et al. Confirmation of three susceptibility genes to insulin-dependent diabetes mellitus: IDDM4, IDDM5, and IDDM8. Hum Mol Genet 1996; 5:693-698.
196. Davies JL, Kawaguchi S, Bennett ST et al. A genome-wide search for human type 1 diabetes susceptibility genes. Nature 1994; 371:130-136.
197. Hasimoto L, Habita C, Beressi JP et al. Genetic mapping of a susceptibility locus for Insulin-dependent diabetes mellitus on chromosome 11q. Nature 1994; 371(8):161-16.
198. Cordell HJ, Todd JA, Bennett ST et al. Two-locus maximum lod score analysis of a multifactorial trait: Joint consideration of IDDM2 and IDDM4 with IDDM1 in type I diabetes. Am J Hum Genet 1995; 57:920-934.
199. Buhler J, Owerbach D, Schaffer AA et al. Linkage analyses in type I diabetes mellitus using CASPAR, a software and statistical program for conditional analysis of polygenic diseases. Hum Hered 1997; 47:211-222.
200. Nakagawa Y, Kawaguchi Y, Twells RCJ et al. Fine mapping of the diabetes-susceptibility locus, IDDM4, on chromosome 11q13. Am J Hum Genet 1998; 63:547-556.
201. Eckenrode S, Marron MP, Nicholls R et al. Fine-mapping of the type 1 diabetes locus (IDDM4) on chromosome 11q and evaluation of two candidate genes (FADD and GALN) by affected sibpair and linkage-disequilibrium analyses. Hum Genet 2000; 106:14-18.
202. Pritchard LE, Kawaguchi Y, Reed PW et al. Analysis of the CD3 gene region and type I diabetes: Application of fluorescence-based technology to linkage disequilibrium mapping. Hum Mol Genet 1995; 4:197-202.
203. Wong S, Moore S, Orisio S et al. Susceptibility to type I diabetes in women is associated with the CD3 epsilon locus on chromosome 11. Clin Exp Immunol 1991; 83:69-73.
204. Hey PJ, Twells RC, Phillips MS et al. Cloning of a novel member of the low-density lipoprotein receptor family. Gene 1998; 216:103-111.
205. Twells RC, Metzker ML, Brown SD et al. The sequence and gene characterization of a 400-kb candidate region for IDDM4 on chromosome 11q13. Genomics 2001; 72:231-242.
206. Kim PK, Dutra AS, Chandrasekharappa SC et al. Genomic structure and mapping of human FADD, an intracellular mediator of lymphocyte apoptosis. J Immunol 1996; 157:5461-5466.
207. Chervonsky AV, Wang Y, Wong FS et al. The role of Fas in autoimmune diabetes. Cell 1997; 89:17-24.
208. Stassi G, De Maria R, Trucco G et al. Nitric oxide primes pancreatic beta cells for Fas-mediated destruction in insulin-dependent diabetes mellitus. J Exp Med 1997; 186:1193-1200.
209. Davies JL, Cucca F, Goy JV et al. Saturation multipoint linkage mapping of chromosome 6q in type I diabetes. Hum Mol Genet 1996; 5:1071-1074.

210. Delepine M, Pociot F, Habita C et al. Evidence of a nonMHC susceptiblity locus in type I diabetes linked to HLA on chromosome 6. Am J Hum Genet 1997; 60:174-187.
211. Perez dN, Bilbao JR, Calvo B et al. Analysis of chromosome 6q in Basque families with type 1 diabetes. GEPV-N. Basque-Navarre Endocrinology and Paediatric Group. Autoimmunity 2000; 33:33-36.
212. Vaidya B, Imrie H, Perros P et al. Evidence for a new graves disease susceptibility locus at chromosome 18q21. Am J Hum Genet 2000; 66:1710-1714.
213. Hodge SE, Anderson CE, Neiswanger K et al. Close genetic linkage between diabetes mellitus and kidd blood group. Lancet 1981; 2:893-895.
214. Merriman T, Twells R, Merriman M et al. Evidence by allelic association-dependent methods for a type 1 diabetes polygene (IDDM6) on chromosome 18q21. Hum Mol Genet 1997; 6:1003-1010.
215. Merriman TR, Eaves IA, Twells RC et al. Transmission of haplotypes of microsatellite markers rather than single marker alleles in the mapping of a putative type 1 diabetes susceptibility gene (IDDM6). Hum Mol Genet 1998; 7:517-524.
216. Holmes DI, Wahab NA, Mason RM. Cloning and characterization of ZNF236, a glucose-regulated Kruppel-like zinc-finger gene mapping to human chromosome 18q22-q23. Genomics 1999; 60:105-109.
217. Copeman JB, Cucca F, Hearne CM et al. Linkage disequilibrium mapping of a type 1 diabetes susceptibility gene (IDDM7) to chromosome 2q31-q33. Nat Genet 1995; 9:80-85.
218. Esposito L, Hill NJ, Pritchard LE et al. Genetic analysis of chromosome 2 in type 1 diabetes: Analysis of putative loci IDDM7, IDDM12 and IDDM13 and candidate genes NRAMP1 and IA-2 and the interleukin-1 gene cluster. IM DIAB Group. Diab 1998; 47:1797-1799.
219. Luo D-F, Maclaren NK, Huang H-S et al. Intrafamilial and case-control association analysis of D2S152 in insulin-dependent diabetes. Autoimmunity 1995; 21:143-147.
220. Kristiansen OP, Pociot F, Bennett EP et al. IDDM7 links to insulin-dependent diabetes mellitus in danish multiplex families but linkage is not explained by novel polymorphisms in the candidate gene galnt3. The danish study group of diabetes in childhood and the danish iddm epidemiology and genetics group. Hum Mutat 2000; 15:295-296.
220a. Lieberman SM, Evans AM, Han B et al. Identification of the beta cell antigen targeted by a prevalent population of pathogenic CD8+ T cells in autoimmune diabetes. Proc Natl Acad Sci USA 2003; 100:8384-8388.
221. Owerbach D, Gabbay KH. The HOXD8 locus (2q31) is linked to type I diabetes: Interaction with chromosome 6 and 11 disease susceptibility genes. Diab 1995; 44:132-136.
222. Tamimi R, Steingrimsson E, Copeland NG et al. The NEUROD gene maps to human chromosome 2q32 and mouse chromosome 2. Genomics 1996; 34:418-421.
223. Iwata I, Nagafuchi S, Nakashima H et al. Association of polymorphism in the NeuroD/BETA2 gene with type 1 diabetes in the Japanese. Diab 1999; 48:416-419.
224. Hansen L, Jensen JN, Urioste S et al. NeuroD/BETA2 gene variability and diabetes: No associations to late-onset type 2 diabetes but an A45 allele may represent a susceptibility marker for type 1 diabetes among Danes. Danish study group of diabetes in childhood and the danish IDDM epidemiology and genetics group. Diab 2000; 49:876-878.
225. Dupont S, Dina C, Hani EH et al. Absence of replication in the French population of the association between beta 2/NEUROD-A45T polymorphism and type 1 diabetes. Diabetes Metab 1999; 25:516-517.
226. Bennett EP, Hassan H, Mandel U et al. Cloning and characterization of a close homologue of human UDP-N-acetyl-alpha-D-galactosamine:Polypeptide N-acetylgalactosaminyltransferase-T3, designated GalNAc-T6. Evidence for genetic but not functional redundancy. J Biol Chem 1999; 274:25362-25370.
227. Owerbach D. Physical and genetic mapping of IDDM8 on chromosome 6q27. Diab 2000; 49:508-512.
228. Paterson AD, Naimark DM, Petronis A. The analysis of parental origin of alleles may detect susceptibility loci for complex disorders. Hum Hered 1999; 49:197-204.
229. Myerscough A, John S, Barrett JH et al. Linkage of rheumatoid arthritis to insulin-dependent diabetes mellitus loci: evidence supporting a hypothesis for the existence of common autoimmune susceptibility loci. Arthritis Rheum 2000; 43:2771-2775.
230. Paterson AD, Rahman P, Petronis A. IDDM9 and a locus for rheumatoid arthritis on chromosome 3q appear to be distinct. Hum Immunol 1999; 60:883-885.
231. Reed P, Cucca F, Jenkins S et al. Evidence for a type 1 diabetes susceptibility locus (IDDM10) on human chromosome 10p11-q11. Hum Mol Genet 1997; 6:1011-1016.
232. Paterson AD, Petronis A. Age of diagnosis-based linkage analysis in type 1 diabetes. Eur J Hum Genet 2000; 8:145-148.

233. Field LL, Tobias R, Thomson G et al. Susceptibility to insulin-dependent diabetes mellitus maps to a locus (IDDM11) on human chromosome 14q24.3-q31. Genomics 1996; 33:1-8
234. Heron L, Virsolvy A, Apiou F et al. Isolation, characterization, and chromosomal localization of the human ENSA gene that encodes alpha-endosulfine, a regulator of beta-cell K(ATP) channels. Diab 1999; 48:1873-1876.
235. Harada Y, Ozaki K, Suzuki M et al. Complete cDNA sequence and genomic organization of a human pancreas-specific gene homologous to caenorhabditis elegans sel-1. J Hum Genet 1999; 44:330-336.
236. Apelqvist A, Li H, Sommer L et al. Notch signalling controls pancreatic cell differentiation. Nature 1999; 400:877-881.
237. Pociot F, Larsen ZM, Zavattari P et al. No evidence for SEL1L as a candidate gene for IDDM11-conferred susceptibility. Diabetes Metab Res Rev 2001; 17:292-295.
238. Nistico L, Buzzetti R, Pritchard LE et al. The CTLA-4 gene region of chromosome 2q33 is linked to and associated with, type 1 diabetes. Belgian diabetes registry. Hum Mol Genet 1996; 5:1075-1080.
238a. Ueda H, Howson JM, Esposito L et al. Association of the T-cell regulatory gene CTLA4 with susceptibility to autoimmune disease. Nature 2003; 423:506-511.
239. Marron MP, Raffel LJ, Garchon HJ et al. Insulin-dependent diabetes mellitus (IDDM) is associated with CTLA4 polymorphisms in multiple ethnic groups. Hum Mol Genet 1997; 6:1275-1282.
240. Awata T, Kurihara S, Iitaka M et al. Association of CTLA-4 gene A-G polymorphism (IDDM12 locus) with acute-onset and insulin-depleted IDDM as well as autoimmune thyroid disease (Graves' disease and Hashimoto's thyroiditis) in the Japanese population. Diab 1998; 47:128-129.
241. Hayashi H, Kusaka I, Nagasaka S et al. Association of CTLA-4 polymorphism with positive anti-GAD antibody in Japanese subjects with type 1 diabetes mellitus. Clin Endocrinol (Oxf) 1999; 51:793-799.
242. Yanagawa T, Maruyama T, Gomi K et al. Lack of association between CTLA-4 gene polymorphism and IDDM in Japanese subjects. Autoimmunity 1999; 29:53-56.
243. Larsen ZM, Kristiansen OP, Mato E et al. IDDM12 (CTLA4) on 2q33 and IDDM13 on 2q34 in genetic susceptibility to type 1 diabetes (insulin-dependent). Autoimmunity 1999; 31:35-42.
244. Badenhoop K, Donner H, Pani M et al. Genetic susceptibility to type 1 diabetes: Clinical and molecular heterogeneity of IDDM1 and IDDM12 in a german population. Exp Clin Endocrinol Diabetes 1999; 107 Suppl 3:S89-S92.
245. Marron MP, Zeidler A, Raffel LJ et al. Genetic and physical mapping of a type 1 diabetes susceptibility gene (IDDM12) to a 100-kb phagemid artificial chromosome clone containing D2S72-CTLA4-D2S105 on chromosome 2q33. Diab 2000; 49:492-499.
246. Hill NJ, Lyons PA, Armitage N et al. NOD Idd5 locus controls insulitis and diabetes and overlaps the orthologous CTLA4/IDDM12 and NRAMP1 loci in humans [In Process Citation]. Diab 2000; 49:1744-1747.
247. Morahan G, Huang D, Tait BD et al. Markers on distal chromosome 2q linked to insulin-dependent diabetes mellitus. Science 1996; 272:1811-1813.
248. Paterson AD, Petronis A. Sex of affected sibpairs and genetic linkage to type 1 diabetes. Am J Med Genet 1999; 84:15-19.
249. Fu J, Ikegami H, Kawaguchi Y et al. Association of distal chromosome 2q with IDDM in Japanese subjects. diabetol 1998; 41:228-232.
250. Giannini S, Mohan S, Kasuya J et al. Characterization of insulin-like growth factor-binding proteins produced by cultured fibroblasts from patients with noninsulin-dependent diabetes mellitus, insulin-dependent diabetes mellitus, or obesity. J Clin Endocrinol Metab 1994; 79:1824-1830.
251. Owerbach D, Naya FJ, Tsai MJ et al. Analysis of candidate genes for susceptibility to type I diabetes: A case-control and family-association study of genes on chromosome 2q31-35. Diab 1997; 46:1069-1074.
252. Temple IK, James RS, Crolla JA et al. An imprinted gene(s) for diabetes? Nat Genet 1995; 9:110-112.
253. Haig D. Is human insulin imprinted? Nat Genet 1994; 7:10.
254. Verge CF, Vardi P, Babu S et al. Evidence for oligogenic inheritance of type 1A diabetes in a large Bedouin Arab family. J Clin Invest 1998; 102:1569-1575.
255. Lin DY, Feuer EJ, Etzioni R et al. Estimating medical costs from incomplete follow-up data. Biometrics 1997; 53:419-434.
256. Cox NJ, Wapelhorst B, Morrison VA et al. Seven regions of the genome show evidence of linkage to type 1 diabetes in a consensus analysis of 767 multiplex families. Am J Hum Genet 2001; 69:820-830.

257. Rowe RE, Wapelhorst B, Bell GI et al. Linkage and association between insulin-dependent diabetes mellitus (IDDM) susceptibility and markers near the glucokinase gene on chromosome 7. Nat Genet 1995; 10:240-242.
258. Concannon P, Gogolin-Ewens KJ, Hinds DA et al. A second-generation screen of the human genome for susceptibility to insulin-dependent diabetes mellitus. Nat Genet 1998; 19:292-296.
259. Cucca F, Goy JV, Kawaguchi Y et al. A male-female bias in type 1 diabetes and linkage to chromosome Xp in MHC HLA-DR3-positive patients. Nat Genet 1998; 19:301-302.
260. Warram JH, Krolewski AS, Gottlieb MS et al. Differences in risk of insulin-dependent diabetes in offspring of diabetic mothers and diabetic fathers. N Engl J Med 1984; 311:149-152.
261. Warram JH, Martin BC, Krolewski AS. Risk of IDDM in children of diabetic mothers decreases with increasing maternal age at pregnancy. Diab 1991; 40:1679-1684.
262. Bleich D, Polak M, Eisenbarth GS et al. Decreased risk of Type 1 diabetes in offspring of mothers who acquire diabetes during adrenarchy. Diab 1993; 42:1433-1439.
263. Vadheim CM, Rotter JI, Maclaren NK et al. Preferential transmission of diabetic alleles within the HLA gene complex. N Engl J Med 1986; 315:1314-1318.
264. Deschamps I, Hors J, Clerget-Darpoux F et al. Excess of maternal HLA-DR3 antigens in HLA DR3,4 positive type 1 (insulin-dependent) diabetic patients. diabetol 1990; 33:425-430.
265. Bain SC, Rowe BR, Barnett AH et al. Parental origin of diabetes-associated HLA types in sibling pairs with type I diabetes. Diab 1994; 43:1462-1468.
266. Polychronakos C, Kukuvitis A, Giannoukakis N et al. Parental imprinting effect at the INS-IGF2 diabetes susceptibility locus. diabetol 1995; 38:715-719.
267. Bui MM, Luo D-F, She JY et al. Paternally transmitted IDDM2 influences diabetes susceptibility despite biallelic expression of the insulin gene in human pancreas. J Autoimmun 1996; 9:97-103.
268. Hoffman AR, Vu TH. Genomic imprinting. Scientific Amer 1996; 52-61.
269. Giannoukakis N, Deal C, Paquette J et al. Parental genomic imprinting of the human IGF2 gene. Nat Genet 1993; 4:98-10.
270. Feinberg AP. Imprinting of a genomic domain of 11p15 and loss of imprinting in cancer: An introduction. Cancer Res 1999; 59:1743s-1746s.
271. Giddings SJ, King CD, Harman KW et al. Allele specific inactivation of insulin 1 and 2, in the mouse yolk sac, indicates imprinting. Nat Genet 1994; 6:310-313.
272. Moore GE, Abu-Amero SN, Bell G et al. Evidence that insulin is imprinted in the human yolk sac. Diab 2001; 50:199-203.
273. Deltour L, Montagutelli X, Guenet JL et al. Tissue- and developmental stage-specific imprinting of the mouse proinsulin gene, Ins2. Dev Biol 1995; 168:686-688.
274. Bennett ST, Wilson AJ, Esposito L et al. Insulin VNTR allele-specific effect in type 1 diabetes depends on identity of untransmitted paternal allele. The IM DIAB Group. Nat Genet 1997; 17:350-352.
275. Urrutia I, Calvo B, Bilbao JR et al. Anomalous behaviour of the 5' insulin gene polymorphism allele 814: Lack of association with Type I diabetes in Basques. GEPV-N Group. Basque-Navarre Endocrinology and Paediatrics. Diabetol 1998; 41:1121-1123.
276. Hollick JB, Dorweiler JE, Chandler VL. Paramutation and related allelic interactions. Trends Genet 1997; 13:302-308.
277. Duvillie B, Bucchini D, Tang T et al. Imprinting at the mouse Ins2 locus: Evidence for cis- and trans-allelic interactions. Genomics 1998; 47:52-57.
278. LaSalle JM, Lalande M. Homologous association of oppositely imprinted chromosomal domains. Science 1996; 272:725-728.
279. Black DL. Protein diversity from alternative splicing: A challenge for bioinformatics and post-genome biology. Cell 2000; 103:367-370.
280. Diez J, Zeller M, Park Y et al. Differential splicing of the IA-2 nRNA in pancreas and lymphoid organs as a permissive genetic mechanism for autoimmunity against the IA-2 type 1 diabetes autoantigen. Diab 2001; 50:895-900.
281. Klein L, Klugmann M, Nave KA et al. Shaping of the autoreactive T-cell repertoire by a splice variant of self protein expressed in thymic epithelial cells. Nat Med 2000; 6:56-61.
282. Solimena M, Dirkx Jr R, Hermel JM et al. ICA 512, an autoantigen of type I diabetes, is an intrinsic membrane protein of neurosecretory granules. EMBO J 1996; 15:2102-2114.
283. Rabin DU, Pleasic SM, Shapiro JA et al. Islet cell antigen 512 is a diabetes-specific islet autoantigen related to protein tyrosine phosphatases. J Immunol 1994; 152:3183-3188.
284. Park YS, Kawasaki E, Kelemen K et al. Humoral autoreactivity to an alternatively spliced variant of ICA512/IA-2 in Type I diabetes. diabetol 2000; 43:1293-1301.

285. Payton MA, Hawkes CJ, Christie MR. Relationship of the 37,000- and 40,000-M_r tryptic fragments of islet antigens in insulin-dependent diabetes to the protein tyrosine phosphatase-like molecule IA-2 (ICA512). J Clin Invest 1995; 96:1506-1511.
286. Lan MS, LU J, Goto Y et al. Molecular cloning and identification of a receptor-type protein tyrosine phosphatase, IA-2, from human insulinoma. DNA Cell Biol 1994; 13:505-514.
287. Bonifacio E, Lampasona V, Genovese S et al. Identification of protein tyrosine phosphatase-like IA2 (islet call antigen 512) as the insulin-dependent diabetes-related 37/40K autoantigen and a target of islet-cell antibodies. J Immunol 1995; 155:5419-5426.
288. Hawkes CJ, Schloot NC, Marks JB et al. T-cell lines reactive to an immunodominant epitope of the tyrosine phosphatase-like autoantigen IA-2 in Type 1 diabetes. Diab 2000; 49:356-366.
289. Roep BO, Atkinson MA, Van Endert PM et al. Autoreactive T cell responses in insulin-dependent (Type 1) diabetes mellitus. Report of the first international workshop for standardization of T cell assays. J Autoimmun 1999; 13:267-282.
290. Honeyman MC, Stone NL, Harrison LC. T-cell epitopes in type 1 diabetes autoantigen tyrosine phosphatase IA-2: Potential for mimicry with rotavirus and other environmental agents. Mol Med 1998; 4:231-239.
291. Dotta F, Dionisi S, Viglietta V et al. T-cell mediated autoimmunity to the insulinoma-associated protein 2 islet tyrosine phosphatase in type 1 diabetes mellitus. Eur J Endocrinol 1999; 141:272-278.
292. Schulz RM, Hawa M, Leslie RD et al. Proliferative responses to selected peptides of IA-2 in identical twins discordant for type 1 diabetes. Diabetes Metab Res Rev 2000; 16:150-156.
293. Peakman M, Stevens EJ, LohmannT et al. Naturally processed and presented epitopes of the islet cell autoantigen IA-2 eluted from HLA-DR4 [see comments]. J Clin Invest 1999; 104:1449-1457.
294. Akerblom HK, Knip M. Putative environmental factors in type 1 diabetes. Diabetes Metab Rev 1998; 14:31-67.
295. Conrad B, Weissmahr RN, Böni J et al. A human endogenous retroviral superantigen as candidate autoimmune gene in type 1 diabetes. Cell 1997; 90:303-313.
296. Conrad B, Weidmann E, Trucco G et al. Evidence for superantigen involvement in insulin-dependent diabetes mellitus aetiology. Nature 1994; 371:351-355.
297. Knerr I, Repp R, Dotsch J et al. Quantitation of gene expression by real-time PCR disproves a "retroviral hypothesis" for childhood-onset diabetes mellitus. Pediatr Res 1999; 46:57-60.
298. Mein CA, Esposito L, Dunn MG et al. A search for type 1 diabetes susceptibility genes in families from the United Kingdom. Nat Genet 1998; 19:297-300.
299. Kim A, Jun HS, Wong L et al. Human endogenous retrovirus with a high genomic sequence homology with IDDMK(1,2)22 is not specific for Type I (insulin-dependent) diabetic patients but ubiquitous. diabetol 1999; 42:413-418.
300. Jaeckel E, Heringlake S, Berger D et al. No evidence for association between $IDDMK_{1,2}22$, a novel isolated retrovirus, and IDDM. Diab 1999; 48:209-214.
301. Murphy VJ, Harrison LC, Rudert WA et al. Retroviral superantigens and type 1 diabetes mellitus. Cell 1998; 95:9-11.
302. Muir A, Ruan QG, Marron MP et al. The IDDMK(1,2)22 retrovirus is not detectable in either mRNA or genomic DNA from patients with type 1 diabetes. Diab 1999; 48:219-222.
303. Badenhoop K, Donner H, Neumann J et al. IDDM patients neither show humoral reactivities against endogenous retroviral envelope protein nor do they differ in retroviral mRNA expression from healthy relatives or normal individuals. Diab 1999; 48:215-218.
304. Lan MS, Mason A, Coutant R et al. HERV-K10s and immune-mediated (type 1) diabetes. Cell 1998; 95:14-16.
305. Lower R, Tonjes RR, Boller K et al. Development of insulin-dependent diabetes mellitus does not depend on specific expression of the human endogenous retrovirus HERV-K. Cell 1998; 95:11-14.

Chapter 8

Autoimmune Polyendocrine Syndromes

George S. Eisenbarth

Polyendocrine autoimmune syndromes include:

1. The autoimmune polyendocrine syndrome type II (APS-II, Schmidt's syndrome)[1-3]
2. Autoimmune polyglandular syndrome type I[4] (APS-I, APECED: autoimmune polyendocrinopathy-candidiasis-ectodermal dystrophy, MIM number 240300)
3. XPID (X-Linked Polyendocrinopathy, Immune Dysfunction and Diarrhea, also termed XLAAD X-Linked Autoimmunity Allergic-Dysregulation)
4. Nonorgan-specific autoimmunity (e.g., lupus erythematosus) associated with anti-insulin receptor antibodies[5,6]
5. Thymic tumors with associated endocrinopathy[7-9]
6. Graves' disease associated with insulin autoimmune syndrome[10,11]
7. POEMS syndrome (Plamacytoma, endocrinopathy, monoclonal gammopathy, and skin changes).[12,13]

The two major autoimmune endocrine syndromes, APS-I and APS-II, both have Addison's disease as a prominent component. The major autoimmune polyendocrine syndromes have a strong genetic component with the type II syndrome occurring in multiple generations and the type I syndrome in siblings. Both syndromes are associated with type 1 A diabetes (Table 1).

The major illnesses associated with both APS-I and APS-II are listed in Table 2 and differences between the syndromes are outlined in Table 3. Knowledge of disease associations and inheritance pattern facilitates early diagnosis of component illnesses.

Patients with APS-I and APS-II develop multiple diseases over time and approximately one out of seven relatives of such patients have an undiagnosed autoimmune disorder (most often hypothyroidism for the type II syndrome).[45] The individual polyendocrine autoimmune syndromes, their immunogenetics, pathogenesis and selected aspects of therapy will be reviewed in this chapter.

Autoimmune Polyendocrine Syndrome Type II (APS-II, Schmidt's Syndrome)

The type II syndrome is the most common autoimmune polyendocrine syndrome.[46] In 1926, Schmidt described two subjects with thyroiditis and Addison's disease.[1] Other diseases of APS II include Graves' disease (thyrotoxicosis), primary hypothyroidism,[19] insulin-dependent or type 1 diabetes mellitus (IDDM), celiac disease,[33,47,48] vitiligo[20,22-24] serositis,[34] IgA deficiency,[24] primary hypogonadism,[38] stiff-man syndrome,[39] alopecia, pernicious anemia,[20,21] myasthenia gravis,[28] and Parkinson's disease. Organ-specific autoantibodies in the absence of overt disease is also frequently present in patients and their relatives.[49] Some authors divide the APS-II syndrome based upon the specific disease components reserving APS-II for Addison's disease plus another disorder (e.g., APS-III type 1 diabetes and thyroiditis). In that the additional divisions at present provide limited prognostic information (e.g., patient with diabetes and thyroiditis is at risk for Addison's disease) we will use APS-II as inclusive of multiple autoimmune disorders with one or more autoimmune endocrine diseases.

Immunology of Type 1 Diabetes, Second Edition, edited by George S. Eisenbarth.

Table 1. Type I diabetes and polyendocrine autoimmunity

- As many as 18% of APS-I patients become diabetic and approximately 50% of patients with APS-II
- "Restricted" or GAD-ICA is common among APS II patients
- APS-I and II: DQA1*0102/DQB1*0602 Protection from Diabetes
- APS-II: Later mean age of diabetes onset than typical type 1A patients
- APS-I: Anti-GAD autoantibodies react with GAD on Western blots and inhibit enzymatic activity [14]

Table 2. Autoimmune polyendocrine syndromes and disease associations

Autoimmune Polyendocrine Syndrome		Autoimmune Polyendocrine Syndrome	
Type 1	**Incidence**	**Type II**	**Incidence**
Mucocutaneous candidiasis[4,15]	C*	Addison's disease[16]	C
Addison's disease[4,15]	C		
Hypoparathyroidism[15]	C		
Chronic active hepatitis[4,15,17]	C		
Graves' disease	I	Graves' disease	I
Autoimmune thyroiditis[15,18]	I	Autoimmune throiditis[19]	I
Pernicious anemia	C	Pernicious anemia[20,21]	C
Vitiligo	C	Vitiligo[20,22-24]	C
Type 1 DM (18%) [4,15]	I	Type 1 DM (50%)	C
Alopecia[25-27]	I	Alopecia	I
		Myasthenia gravis[28,29]	I
Malabsorption syndrome[15,30]	I	Celiac disease, Dermatitis, herpetiformis[31-33]	C
IgA deficiency	I	IgA deficiency[24]	I
		Serositis[34]	R
Asplenism[35]	C		
Ectodermial dysplasia[4]	C		
Keratitis[36]	C		
Hypogonadism[15,37]	C	Hypogonadism[38]	I
		Stiff-man syndrome[39]	R
		Parkinson's disease	I
Dental enamel and nail dystrophy[4]	C		
		Idiopathic heart block[20]	R
Pure red cell aplasia[40]	R		
		Idiopathic thrombocytopenia[41,42]	I
		Hypophysitis[43,44]	R

*C= common within syndrome; R= rare; I= intermediate

Genetic abnormalities underlying disease susceptibility are partially defined and include human leukocyte (HLA) haplotypes with DR3/DQ2 and DR4/DQ8 [DQ2:DQA1*0501, DQB1*0201; DQ8:DQA1*0301, DQB1*0302] (and with DRB1*0404 particularly in Addison's disease families).[50-53] Identified major environmental factors initiating a component disease include the wheat protein gliadin for celiac disease,[54] sulfhydryl containing medications for the insulin autoimmune syndrome, penicillamine as a rare cause of myasthenia gravis[55] and congenital rubella as a rare cause of type 1 diabetes.

Table 3. Comparison of APS-I and APS-II

APS-I	APS-II
Onset infancy	Older onset
Siblings affected (autosomal recessive, Chromosome 21, AIRE gene)	Multiple generations
HLA class II associated diseases	DR3/DR4 associated
Immunodeficiency	No defined immunodeficiency
Asplenism	
Mucocutaneous candidiasis	

Many of the diseases of the type II syndrome are associated with HLA antigens HLA-DR3 or HLA-DR4[51,56,57] (Table 4). In that the B locus allele HLA B8 is in linkage disequilibrium with HLA DR3 (associated at a population level), B8 is also associated with APS II. Primary adrenal insufficiency in the type II, and the type I syndrome, is strongly associated with both HLA-DR3 and HLA-DR4.[57] The association of HLA markers with disease can correlate with an inheritance of a common HLA-haplotype within families, but haplotypes with DR3 are often introduced into the family by more than one relative.[57] Several diseases that manifest in patients with polyendocrine autoimmune syndromes are not associated with HLA-B8 or HLA-DR3. These disorders include vitiligo,[58] pernicious anemia,[59] and autoimmune goitrous thyroiditis.[60-62] Addison's disease patients also frequently have DR4/DQ8, DR2 have DQB1*0602, and DR5 associated haplotypes. Approximately 30% of patients with Addison's disease are DR3/4, DQ2/DQ8 heterozygous compared to 2.4% of the general U.S. population.[57] This occurs in the U.S. for patients with or without type 1 diabetes. Between 70 to 80% of Addison's disease patients with DR4 have the DRB1*0404 allele. Such genotypic associations are likely to vary by country depending on the frequency of specific HLA DR and DQ haplotypes. This suggests that multiple HLA haplotypes can result in similar manifestations of pluriglandular autoimmune syndromes. Other genes that may contribute to Addison's disease include polymorphisms of the regulatory molecule CTLA-4 (G nucleotide substitutes for A) that have been implicated in Graves' disease[63-65] and Addison's disease,[66] and the "5.1" allele of the atypical class I HLA molecule MIC-A.[67,68]

Families with the type II polyendocrine syndrome should be evaluated over time to detect the presence of organ-specific antibodies indicating the possibility of a future endocrine malfunction (Table 4). All such relatives should be advised of the early symptoms and signs of the principal component diseases. Even though signs and symptoms of disease may be absent, patients with multiple disorders should be screened every few years with measurement of anti-islet antibodies, 21-hydroxylase autoantibodies and transglutaminase autoantibodies, a sensitive thyrotropin assay, and measurement of serum B12 levels.[69] ACTH and cosyntropin (adrenocerticotropin) -stimulated cortisol determination is indicated if 21-hydroxylase autoantibodies are detected.[57,70] Assays of anti-islet cell antibodies,[71,72] anti-thyroid and anti-adrenal antibodies (21-hydroxylase)[16,73-76] and anti-ovarian antibodies[38,77] help identify subjects at increased disease risk. An excellent autoantibody assay for celiac disease is now also available (determination of transglutaminase autoantibodies).[47] More than 20 years may elapse between the onset on one endocrinopathy and the diagnosis of the next. As many as 40-50% of subjects with Addison's disease will develop an associated endocrinopathy. A distinction must be made for subjects with isolated thyroid disease (relatively frequent in the general population) who have no family history of polyglandular syndrome type II. Such individuals have a relatively low probability of developing additional autoimmune disorders in comparison with individuals with rare autoimmune disorders such as Addison's disease or myasthenia gravis.

Table 4. Polyendocrine type II syndrome diseases

Disease	HLA Association	Initiator	Predominant Mediator	Autoantigen
Graves' disease	DR3	Iodine (rare) anti-CD52 (rare)	Antibody	TSH receptor[47]
Myasthenia gravis	DR3,DR7[105,106]	Thymoma[9,107] (rare) Penicillamine[106] (rare)	Antibody	Acetylcholine receptor
Insulin Receptor Hypoglycemia or diabetes (rare)	?	?	Antibody[108,109]	Insulin receptor
Hypoparathyroidism "geriatric" (rare)	?	?	Antibody[78]	Cell surface inhibitor
Insulin autoimmune	DR4,DRB1*0406[11]	Methimazole[110]	Antibody	Insulin
Celiac disease	DR3/DR2/DQ8	Gluten (common)	T cell	Transglutaminase
Type 1A Diabetes	DR3/4, DQ2/8	Rubella[111] (rare)	T cell	Insulin, GAD, ICA512 (IA-2), GM2-1
Addison's disease	DR3,DR4[57]	?	T cell	21-Hydroxylase P450-5cc[57]
Thyroiditis	DQB1*0201, DQA1*0301[113-115]	Iodine[112] Interferon α	T cell	Thyroglobulin, Peroxidase[116]
Pernicious anemia	?	?	T cell	Intrinsic factor H^+/K^+ ATPase
Vitiligo		Melanoma Antigen Immunization[118]	?	Melanocyte[117]
Chromosome Dysgenesis (Trisomy21 Turners)	DQA1*0301[119,120]		?	Thyroid, Islet, Celiac[121,122]

Rarely, hypoparathyroidism, a specific endocrine disturbance present in the type I syndrome, is identified in a patient with type II syndrome. Hypoparathyroidism in such type II polyendocrine autoimmune patients may result from a "suppressive" autoantibody[78,79] rather than parathyroid destruction as in the type I syndrome. In a patient with the type II syndrome, celiac disease is a more frequent cause of hypocalcemia than hypoparathyroidism. Other HLA-B8 and DR3 associated illnesses include selective IgA deficiency[80,81] juvenile dermatomyositis, and dermatitis herpetiformis,[33] alopecia, scleroderma,[82] autoimmune thrombocytopenia purpura,[83] multiple endocrine neoplasia, hypophysitis,[44] metaphyseal ostenopenia,[84] and serositis.[34]

Pathogenesis

Both autoreactive T cells and autoantibodies are pathogenic, depending on the specific disease. In Graves' disease, anti-thyrotropin (TSH) autoantibodies lead to thyroid hyperfunction[85] and anti-insulin receptor autoantibodies can result in either hypoglycemia or insulin resistance with hyperglycemia.[11,86] Type 1 diabetes is a T cell mediated disorder and an interesting case report describes a child developing diabetes with a mutation eliminating B lymphocytes.[87]

Initiating factors for the type II syndrome and its component illnesses are not established except for celiac disease (wheat protein gliadin), the insulin autoimmune syndrome (e.g., methmizole), myasthenia gravis (rarely penicillamine), type 1A diabetes (rarely congenital rubella), Graves' disease (rarely anti-CD52 monoclonal treating patients with multiple sclerosis[88]) and hypothryoidism (rarely inteferon α).[89] Patients with celiac disease, which is characterized by atrophy of intestinal villi associated with lymphocytic infiltration, have autoantibodies reacting with transglutaminase (the endomysial antigen)[47] and with less specificity and sensitivity with the wheat proetin gliadin. Removal of gliadin from the diet restores intestinal villi to normal.[90-92] In a similar manner, controversial data suggest that ingestion of the milk protein bovine albumin in the first few months of life may be associated with type 1 diabetes[93] while other investigators implicate casein. A number of drugs are associated with induction of autoimmunity including interferon-α (thyroiditis).[94-96] Remarkably 1/3 of multiple sclerosis patients treated with an anti-CD52 monoclonal antibody developed Graves' disease.[88] Apparently nonmultiple sclerosis patients treated with the same monoclonal do not develop Graves' disease.

Several autoantibodies are both disease specific (e.g., anti-acetylcholine receptor antibodies in myasthenia gravis[97] and anti-TSH receptor antibodies in Graves' disease[98]) and causal. "Causal" autoantibodies are associated with transplacental disease transmission. Other autoantibodies (e.g., antithyroid autoantibodies including anti-thyroid peroxidase, formerly termed anti-microsomal, and anti-thyroglobulin) are so frequent among patients and relatives as to be of little predictive value. For example, a relative with anti-thyroid peroxidase autoantibodies has a low risk of hypothyroidism unless evidence of abnormal thyroid function is also present (e.g., elevated TSH). In a similar manner, many individuals may have antibodies to parietal cells, H^+/K^+ adenosine triphosphatase of the stomach[99,100] and intrinsic factor, but the autoantibodies do not correlate well with abnormal gastric acid secretion or development of pernicious anemia.[59] Studies of the neonatal presence of such autoantibodies will be importat to determine if they increase risk of later disease.

In the APS-II syndrome, many ICA-positive individuals do not progress to diabetes, and diabetes risk is much lower than for ICA-positive first-degree relatives of patients with type 1 diabetes.[72] These nonprogressing ICA-positive polyendocrine patients usually express what has been termed "selective" or restricted ICA.[71] Such ICA reacts only with islet B cells, not A cells within rat islets and fails to react with mouse islets. Restricted ICA consists of unusual high titer autoantibodies reacting with glutamic acid decarboxylase (GAD). This unusual form of ICA confers a lower risk of type 1 diabetes as compared with nonrestricted ICA (reacts with multiple islet molecules) for both polyendocrine patients and relatives of patients with type 1 diabetes.

Other autoantibodies associated with the type II syndrome include anti-melanocytic, anti-adrenal (in particular 21-hydroxylase)[101] and anti-gonadal autoantibodies.[74] Anti-adrenal cortical antibodies have been used to predict adrenal insufficiency in the type 1 syndrome. Tryptophan hydroxylase autoantibodies have been rported in APS-I patients with intestinal disease.[30] It is noteworthy that many of the polyendocrine autoantibodies react with intracellular enzymes, including thyroid peroxidase (Hashimoto's thyroditis), glutamic acid decarboxylase (type 1 diabetes and stiff-man syndrome), 21 hydroxylase (Addison's disease), and cytochrome P450 cholesterol side chain cleavage enzyme[102] (Addison's disease). In addition, antibodies to hormones can be present, including anti-insulin, anti-thyroxine, and anti-intrinsic factor antibodies (pernicious anemia).

Antibodies to specific receptors are characteristic of given disorders (anti-acetylcholine receptor antibodies of myasthenia gravis, anti-TSH receptor antibodies of Graves' disease or hypothyroidism,[103] and oocyte sperm receptor autoantibodies associated with oophoritis). The large variety of target molecules, (e.g., type 1 diabetes), presence of high affinity IgG autoantibodies, and the sequential appearance over years of specific antibodies or disorders suggest that the production of most autoantibodies is secondary to tissue destruction and are antigen "driven".[104]

T cell autoimmunity has been much more difficult to study and correlate with disease compared to autoantibodies. Recent advances in T cell immunobiology, studies of animal models, and transfer of autoreactive T lymphocytes or affected human organs into immunodeficient mice should lead to more rapid progress in understanding T cell autoimmunity.

For decades, experimental animal models of organ-specific autoimmunity have been studied. These were dependent upon the injection of putative autoantigens into animals in the presence of adjuvants that enhance inflammation.[42] Thus, thyroiditis can readily be induced in selective strains of mice following injection of thyroglobulin or thyroid peroxidase in Freund's adjuvant. Anti-insulin autoantibodies can be induced following the administration of insulin peptide B:9-23, and these autoantibodies react with intact insulin and are not absorbed by the immunizing peptide.[123] T cell clones reacting with these molecules, or selected peptides, can be generated, and such clones when transferred into naive animals induce disease. Of note, several forms of immunization with such autoreactive clones can be used to make animals refractory to disease induction.[124] These studies provide clear evidence that autoreactive T cells are present in normal animals and they can be rapidly activated, given "appropriate" stimulation.

In addition, a study by Tung and associates suggests one mechanism whereby properties of T cell recognition may lead to multiple autoimmune disorders.[42] In studying experimental autoimmune oophoritis, they identified a peptide of the oocyte sperm receptor (ZP3) that upon injection in adjuvant induced disease. They then identified which of nine amino acids of this peptide were essential to activate autoimmune T cell clones or induce disease. T cells recognize only short peptides presented in the groove of class I or class II major histocompatibility complex (MHC) molecules on antigen-presenting cells. As few as three properly spaced amino acids (of nine) interacting with a T cell receptor can be sufficient to trigger T cell responses. Noting that a peptide of the acetylcholine receptor had the appropriate T cell binding motif associated with experimental oophoritis, they demonstrated that this peptide stimulated an oophoritis-derived T cell clone in vitro and when administered in vivo induced oophoritis. The requirement for sharing of as few as three of nine amino acids of a linear sequence for activation of autoreactive T cells provides a mechanism whereby inflammation directed at one organ may spread to additional tissues by T cell cross reactions to distinct peptides of different tissues. Such a model may also help explain disease associations. For example, Graves' thyroid disease is frequently complicated by autoimmunity directed at extraocular muscles leading to Graves' ophthalmopathy.[125] If such a mechanism for the "spreading" of autoimmunity is operative, it implies that mechanisms to suppress autoreactivity are particularly important. The possibilities that "molecular mimicry" may induce autoimmunity are greatly increased if short minimally homologous sequences are sufficient to stimulate cross-reactive T cell clones.

Interruption of normal T cell development can also result in multiple autoimmune disorders. The BB rat develops type 1 diabetes and thyroditis in association with a severe T cell immunodeficiency.[126] Neonatal thymectomy induces autoimmunity apparently by removing regulatory T cells.[127,128] A series of different regulatory T cells (e.g., CD4+CD25+, NK T cells) are now the subject of intense investigation.[129]

Autoimmune disorders appear to share a number of "nonspecific" abnormalities of T cell function or enumeration including increased numbers of cells expressing class II molecules ("Ia" positive T cells),[130] IL2 receptors, depressed autologous mixed lymphocyte responses,[131] and lack of NK T cells.[132] These abnormalities appear not to be disease specific and may relate to fundamental abnormalities predisposing to autoimmunity or reflect disease activity.

Therapy

The therapy of the APS-II syndrome depends upon the specific disease manifestation with a few caveats.[133] Patients with suspected Addison's disease and hypothyroidism, should be evaluated and treated for adrenal insufficiency prior to replacement of thyroid hormone to avoid Addisonian crisis. There is one fascinating case report of a patient with 21-hydroxylase autoantibodies treated for Graves' eye disease with a 6 month course of glucocorticoids. In this patient

21-hydroxylase autoantibodies became negative and adrenal function was restored to normal. This remission was reported to have lasted for 100 months at last follow up.[134]

Autoimmune Polyendocrine Syndrome Type I (APS-I)

Major components of the APS-I autoimmune polyendocrine syndrome include hypoparathyroidism, mucocutaneous candidiasis, Addison's disease, and hepatitis[4] (Table 2). Other diseases associated with APS-I include primary hypothyroidism, a malabsorption syndrome,[30] vitiligo, pernicious anemia, type 1 diabetes, alopecia,[26] primary hypogonadism,[37] cutaneous abnormalities,[4] and pure red cell aplasia.[40,135] The onset of the disease is usually in infancy and there is no association with specific HLA class II alleles (Table 2). Of note, in a study by Friedman and coworkers[35] four of nine patients with APS-I were asplenic. Chronic mucocutaneous candidiasis is often the first disease detected, followed by the later development of adrenal insufficiency. Decades may elapse between the diagnosis of one disease and another in the same individual and the order of disease appearance is not invariant.

The syndrome is inherited in an autosomal recessive manner linked to a gene (The AIRE: Autoimmune Regulator gene) on chromosome 21.[136-137a] It is likely that the immunodeficiency underlying disease susceptibility is secondary to autosomal recessive mutations of this transcription factor.[101] Studies of autoimmune disorders including Addison's disease, but in patients without the APS-I syndrome, indicate that AIRE mutations are not involved in these more common diseases.[138] The etiology of the mucocutaneous candidiasis in the absence of systemic candidiasis in the APS-I syndrome is unknown. The orally active antifungal drug ketoconazole has significantly improved the treatment of the mucocutaneous candidiasis associated with this disorder. Aggressive treatment of oral candidiasis is essential as death from oral carcinoma for APS-I patients is thought to be rlated to chronic infection.

Particular patterns of anti-adrenal autoantibodies are associated with this syndrome.[102,139,140] The presence of anti-adrenal autoantibodies (e.g., 21-hydroxylase) is strongly associated with subsequent development of Addison's disease. In addition, anti-GAD autoantibodies of patients with the APS I syndrome differ from anti-GAD autoantibodies in terms of being able to detect GAD on Western blots and inhibit enzymatic activity. A minority (18%) of patients with APS I develop type 1 diabetes.[15] Many patients express ICA and anti-GAD autoantibodies ["restricted" ICA(41%)] with relatively little progression to diabetes. Patients expressing multiple anti-islet autoantibodies (e.g., GAD and ICA512(IA-1)) are at higher risk for progression to diabetes. APS-I patients express additional autoantiboides consistent with widespread loss of tolerance to multiple self antigens.[141]

XPID (X-Linked Polyendocrinopathy Immune Dysfunction) and Diarrhea

The XPID syndrome presents in neonates with fatal autoimmunity. The mutated gene is a transcription factor that is also mutated in a mouse model (Scurfy gene).[3] In that the mouse model is cured with bone marrow transplanatation, such therapy has recently been tested in a single child with the disorder. The child became chimeric after bone marrow transplantation and a two year remission was induced.[142]

Anti-Insulin Receptor Antibodies

The presence of anti-insulin receptor autoantibodies is characterized by marked insulin resistance, but paradoxically, patients can also have severe hypoglycemia.[6] Approximately one third of the subjects have other autoimmune disorders. Patients with anti-insulin receptor antibodies frequently have nonorgan specific autoimmunity (e.g., Lupus erythematosus).

Thymic Tumors

Thymomas and thymic hyperplasias are associated with series of autoimmune diseases.[7-9,143-145] The most common autoimmune diseases are myasthenia gravis and red cell aplasia. Graves' disease, type 1 diabetes, and Addison's disease may also be associated with thymic tumors. Unique anti-acetylcholine receptor autoantibodies may be present with thymoma[145] and disease may be initiated by transcription of molecules within the tumor related to acetylcholine receptors.[146] There is a 1997 report of treatment of a patient with thymoma and pure red cell aplasia with octreotide and prednisone.[143]

POEMS Syndrome

These patients usually present with a sensory motor polyneuropathy, diabetes mellitus (50%), primary gonadal failure (70%), and a plasma cell dyscrasia with sclerotic bony lesions.[147,148] Temporary remission may result following radiotherapy. The syndrome is assumed to be secondary to circulating immunoglobulins but patients have excess vascular endothelial growth factor[149,150] as well as elevated IL1-β, IL-6, and TNF-α.[148]

Insulin Autoimmune Syndrome (Hirata Syndrome)

The insulin autoimmune syndrome, associated with Graves' disease and methimazole therapy is of particular interest due to a remarkably strong association with a specific HLA haplotype.[11] Such patients with elevated titers of anti-insulin autoantibodies frequently present with hypoglycemia. The disease in Japan is essentially confined to DR4-positive individuals with DRB1*0406.[110]

Other Disorders

Other diseases with polyendocrine autoimmunity are Kearns-Sayre syndrome[151] diabetes and thyroditis associated with trisomy 21,[152] DIDMOAD syndrome (diabetes insipidus, diabetes mellitus, optic atrophy, and nerve deafness)[153,154] and congenital rubella[111,155,156] associated with thyroditis and/or diabetes. Karounos and coworkers have described a rubella protein with homology to an islet protein, suggesting that molecular mimicry may initiate disease.[157] An alternative hypothesis has been proposed by Rabinowe and coworkers.[156] They described long-term T cell subset abnormalities in patients following congenital rubella infection. The Lambert Eaton Myasthemic Syndrome (LEMS) results from anti-Ca channel autoantibodies. This syndrome is frequently associated with small cell lung cancer. A recent report describes a high percentage of patients (particularly those with small cell carcinoma) with LEMS having autoantibodies racting with GAD65 and ICA512. Relatively few patients have LEMS and type 1 diabetes.

Conclusions

Polyendocrine autoimmune syndromes have played an important role in understanding autoimmune disorders and in particular type 1 diabetes. The initial evidence that type 1 diabetes was an autoimmune disorder came from its association with spontaneous Addison's disease, and lack of association with tuberculous Addison's disease.[2] The first demonstration of cytoplasmic islet cell autoantibodies occurred in patients with polyendocrine autoimmunity.[158] It is now recognized that a major subset of such ICA in polyendocrine patients represent unusual anti-GAD autoantibodies.[159]

The existence of families of related autoimmune disorders not only is clinically important but also suggests that these diseases are pathogenically related. This relationship probably is a result of three distinct phenomena. The first is inherited abnormalities of immune function, predisposing to the loss of tolerance to a series of self antigens. For instance the gene mutated in Scurfy mice and neonates with the XPID syndrome appears to be essential for regulatory T lymphocytes.[159a] Given such a predisposition, "normal" alleles of HLA genes within the major histocompatibility complex may then lead to targeting of specific organs. Different diseases are often associated with the same or similar HLA genotypes (e.g., DR3/DR4, DQ2/DQ8: type 1A diabetes and Addison's disease) and thus would increase the co-occurrence of the disorders. For other disorders the HLA associations may be antagonistic. For example, the DR2 haplotype (DRB1*1501, DQB1*0602) protects from type 1A diabetes but is associated with multiple sclerosis. In this case it may be as common to have siblings each with different disorders as individuals with both diseases given genetic susceptibility to autoimmunity in the family. The third phenomenon that may link these disorders, especially for tightly associated diseases such as Graves' ophthalmopathy and Graves' thyroid disease, is creation of pathogenic T or B[160] cells reacting with components of more than one tissue. In that T cell clones can respond to peptides that share no identical amino acids,[161] depending upon their three dimensional structure, the potential for such T-cell cross-reactivity must be enormous.[160]

Finally, the relationships between these diverse disorders suggest that as disease pathogenesis is elucidated and antigen-specific therapies are developed, the improved understanding of pathogenesis and improvements in therapy will be applicable to many autoimmune diseases.

References

1. Schmidt MB. Eine biglandulare erkrankung (Nebennieren und Schilddruse) bei morbus addisonii. Verh Dtsch Ges Pathol 1926; 21:212-221.
2. Nerup J. Addison's disease—clinical studies: A report of 108 cases. Acta Endocrinol 1974; 76:127-141.
3. Patel DD. Escape from tolerance in the human X-linked autoimmunity-allergic disregulation syndrome and the Scurfy mouse. J Clin Invest 2001; 107:155-157.
4. Ahonen P, Myllarniemi S, Sipila I et al. Clinical variation of autoimmune polyendocrinopathy - candidiasis - ectodermal dystrophy (APECED) in a series of 68 patients. N Engl J Med 1990; 322:1829-1836.
5. Flier JS, Kahn CR, Roth J et al. Antibodies that impair insulin receptor binding in an unusual diabetic syndrome with severe insulin resistance. Science 1975; 190:63-65.
6. Flier JS, Bar RS, Muggeo M et al. The evolving clinical course of patients with insulin receptor autoantibodies: Spontaneous remission or receptor proliferation with hypoglycemia. J Clin Endocrinol Metab 1978; 47:985-995.
7. Souadjian JV, Enriquez P, Silverstein MN et al. The spectrum of diseases associated with thymoma. Arch Intern Med 1974; 134:374-379.
8. Combs RM. Malignant thymoma, hyperthyroidism and immune disorder. South Med J 1968; 61:337-341.
9. Maggi G, Casadio C, Cavallo A et al. Thymoma: Results of 241 operated cases. Ann Thorac Surg 1991; 51:152-156.
10. Hirata Y, Ishizu H, Ouchi N et al. Insulin autoimmunity in a case with spontaneous hypoglycaemia. Japan J Diabetes 1970; 13:312-319.
11. Uchigata Y, Hirata Y. Insulin Autoimmune Syndrome (IAS, Hirata Disease). 1999; 133-148.
12. Imawari M, Akatsuka N, Ishibashi M et al. Syndrome of plasma cell dyscrasia, polyneuropathy, and endocrine disturbances. Ann Intern Med 1974; 81:490-493.
13. Miralles GD, O'Fallon JR, Talley NJ. Plasma-cell dyscrasia with polyneuropathy: The spectrum of POEMS syndrome. N Engl J Med 1992; 327:1919-1923.
14. Bjork E, Velloso LA, Kampe O et al. GAD autoantibodies in IDDM, stiff-man syndrome, and autoimmune polyendocrine syndrome type I recognize different epitopes. Diabetes 1994; 43:161-165.
15. Perheentupa J, Miettinen A. Autoimmune Polyendocrinopathy-Candidiasis-Ectodermal Dystrophy. 1999; 19-40.
16. De Bellis A, Bizzarro A, Rossi R et al. Remission of subclinical adrenocortical failure in subjects with adrenal autoantibodies. J Clin Endocrinol Metab 1993; 76:1002-1007.
17. Clemente MG, Obermayer-Straub P, Meloni A et al. Cytochrome P450 1A2 is a hepatic autoantigen in autoimmune polyglandular syndrome type 1. J Clin Endocrinol Metab 1997; 82:1353-1361.
18. Radetti G, Paganini C, Gentili L et al. Frequency of Hashimoto's thyroiditis in children with type 1 diabetes mellitus. Acta Diabetol 1995; 32:121-124.
19. Landin-Olsson M, Karlsson FA, Lernmark Å et al. Islet cell and thyrogastric antibodies in 633 consecutive 15- to 34-yr-old patients in the diabetes incidence study in Sweden. Diabetes 1992; 41:1022-1027.
20. Fairfax AJ, Leatham A. Idiopathic heart block: Association with vitiligo, thyroid disease, pernicious anemia, and diabetes mellitus. BMJ 1975; 4:322-324.
21. Davis RE, VcCann VJ, Stanton KG. Type 1 diabetes and latent pernicious anaemia. Med J Aust 1992; 156:160-162.
22. Peserico A, Rigon F, Semsenzato G et al. Vitiligo and polyglandular autoimmune disease with autoantibodies to melanin-producing cells. A new syndrome? Arch Dermatol 1981; 117:751-752.

23. Betterle C, Caretto A, Pedini B et al. Complement-fixing activity to melanin-producing cells preceding the onset of vitiligo in a patient with type I polyglandular failure. Arch Dermatol 1995; 128:123-124.
24. Torrelo A, España A, Balsa J et al. Vitiligo and polyglandular autoimmune syndrome with selective IgA deficiency. Int J Dermatol 1992; 31:343-344.
25. Stankler L, Bewsher PD. Chronic mucocutaneous candidiasis endocrine deficiency and alopecia areata. Br J Dermatol 1972; 86:238-245.
26. Garty BZ, Kauli R. Alopecia universalis in autoimmune polyglandular syndrome type I. West J Med 1990; 152:76-77.
27. Hedstrand H, Perheentupa J, Ekwall O et al. Antibodies against hair follicles are associated with alopecia totalis in autoimmune polyendocrine syndrome type I. J Invest Dermatol 1999; 113:1054-1058.
28. Bosch EP, Reith PE, Granner DK. Myasthenia gravis and Schmidt syndrome. Neurology 1994; 27:1179-1180.
29. Marino M, Ricciardi R, Pinchera A et al. Mild clinical expression of myasthenia gravis associated with autoimmune thyroid diseases. J Clin Endocrinol Metab 1997; 82:438-443.
30. Ekwall O, Sjoberg K, Mirakian R et al. Tryptophan hydroxylase autoantibodies and intestinal disease in autoimmune polyendocrine syndrome type 1 [letter]. Lancet 1999; 354:568-
31. Collin P, Salmi J, Hällström O et al. Autoimmune thyroid disorders and coeliac disease. Eur J Endocrinol 1994; 130:137-140.
32. Reijonen H, Ilonen J, Knip M et al. Insulin-dependent diabetes mellitus associated with dermatitis herpetiformis: Evidence for heterogeneity of HLA-associated genes. Tissue Antigens 1991; 37:94-96.
33. Reunala T, Salmi J, Karvonen J. Dermatitis herpetiformis and celiac disease associated with Addison's disease. Arch Dermatol 1987; 123:930-932.
34. Tucker Jr WS, Niblack GD, McLean RH et al. Serositis with autoimmune endocrinopathy: Clinical and immunogenetic features. Medicine 1987; 66:138-147.
35. Friedman TC, Thomas PM, Fleisher TA et al. Frequent occurrence of asplenism and cholelithiasis in patients with autoimmune polyglandular disease type I. Am J Med 1991; 91:625-630.
36. Merenmies L, Tarkkanen A. Chronic bilateral keratitis in autoimmune polyendocrinopathy -candidiadisectodermal dystrophy (APECED). A long-term follow-up and visual prognosis. Acta Ophthalmol Scand 2000; 78:532-535.
37. Ahonen P, Miettinen A, Perheentupa J. Adrenal and steroidal cell antibodies in patients with autoimmune polyglandular disease type I and risk of adrenocortical and ovarian failure. J Clin Endocrinol Metab 1987; 64:494-500.
38. Turkington RW, Lebovitz HE. Extra-adrenal endocrine deficiencies in Addison's disease. Am J Med 1967; 43:499-507.
39. Solimena M, Folli F, Aparisi R et al. Autoantibodies to GABA-ergic neurons and pancreatic beta cells in stiff-man syndrome. N Engl J Med 1990; 322:1555-1560.
40. Mandel M, Etzioni A, Theodor R et al. Pure red cell hypoplasia associated with polyglandular autoimmune syndrome type I. Isr J Med Sci 1989; 25:138-141.
41. Selinger S, Tsai J, Pulini M et al. Autoimmune thrombocytopenia and primary billary cirrhosis with hypoglycemia and insulin receptor autoantibodies. Ann Intern Med 1987; 107:686-688.
42. Rhim SH, Millar SE, Robey F et al. Autoimmune disease of the ovary induced by a ZP3 peptide from the mouse zona pellucida. J Clin Invest 1992; 89:28-35.
43. Muir A, Maclaren NK. Autoimmune diseases of the adrenal glands, parathyroid glands, gonads, and hypothalamic-pituitary axis. Endocrinol Metab Clin North Am 1991; 20:619-644.
44. Barkan AL, Kelch RP, Marshall JC. Isolated gonadotrope failure in the polyglandular autoimmune syndrome. N Engl J Med 1985; 312:1535-1540.
45. Eisenbarth GS, Wilson P, Ward F et al. HLA type and occurrence of disease in familial polyglandular failure. N Engl J Med 1978; 298:92-94.
46. Betterle C, Volpato M, Greggio AN et al. Type 2 polyglandular autoimmune disease ISchmitd's syndrome). J Pediatr Endocrinol Metab 1996; 9:113-123.
47. Bao F, Yu L, Babu S et al. One third of HLA DQ2 homozygous patients with type 1 diabetes express celiac disease associated transglutaminase autoantibodies. J Autoimmunity 1999; 13:143-148.
48. Papadopoulos KI, Hornblad Y, Hallengren B. The occurrence of polyglandular autoimmune syndrome type III associated with coeliac disease in patients with sarcoidosis. J Intern Med 1994; 236:661-663.
49. Furmaniak J, Sanders J, Rees Smith B. Autoantigens in the Autoimmune Endocrinopathies. 1999; 183-216.
50. Eisenbarth GS, Wilson PW, Ward F et al. The polyglandular failure syndrome: Disease inheritance, HLA- type and immune function. Ann Intern Med 1979; 91:528-533.

51. Maclaren NK, Riley WJ. Inherited susceptibility to autoimmune Addison's disease is linked to human leukocyte antigens-DR3 and/or DR4, except when associated with type 1 autoimmune polyglandular syndrome. J Clin Endocrinol Metab 1986; 62(3):455-459.
52. Butler MG, Hodes ME, Conneally PM et al. Linkage analysis in a large kindred with autosomal dominant transmission of polyglandular autoimmune disease type II (Schmidt syndrome). Am J Med Genet 1984; 18:61-65.
53. Farid NR, Larsen B, Payne R et al. Polyglandular autoimmune disease and HLA. Tissue Antigens 1980; 16:23-29.
54. Friis SU. Coeliac disease, pathogenesis and clinical aspects. Acta Pathol Microbiol Immunol Scand 1996; 61:5-48.
55. Drosos AA, Christou L, Galanopoulou V et al. D-penicillamine induced myasthenia gravis: Clinical, serological and genetic findings. Clin Exp Rheumatol 1993; 11:387-391.
56. Valenta LJ, Bull RW, Hackel E et al. Correlation of the HLA-A1, B8 haplotypes with circulating autoantibodies in a family with increased incidence of autoimmune disease. Acta Endocrinol 1982; 100:143-149.
57. Yu L, Brewer KW, Gates S et al. DRB1*04 and DQ alleles: Expression of 21-hydroxylase autoantibodies and risk of progression to Addison's disease. J Clin Endocrinol Metab 1999; 84:328-335.
58. Bloch MH, Sowers JR. Vitiligo and polyglandular autoimmune endocrinopathy. Cutis 1985; 36:417-419.
59. Petite J, Rosset N, Chapuis B et al. Genetic factors predisposing to autoimmune diseases. Study of HLA antigens in a family with pernicious anemia and thyroid diseases. Schweiz Med Wochenschr 1987; 117:2032-2037.
60. Inoue D, Sato K, Sugawa H et al. Apparent genetic difference between hypothyroid patients with blocking-type thyrotropin receptor antibody and those without, as shown by restriction fragment length polymorphism analyses of HLA-DP loci. J Clin Endocrinol Metab 1993; 77:606-610.
61. Cho BY, Chung JH, Shong YK et al. A strong association between thyrotropin receptor-blocking antibody-positive atrophic autoimmune thyroiditis and HLAL-DR8 and HLA-DQB1*0302 in Koreans. J Clin Endocrinol Metab 1993; 77:611-615.
62. Santamaria P, Barbosa JJ, Lindstrom AL et al. HLA-DQB1-associated susceptibility that distinguishes Hashimoto's thyroiditis from Graves' disease in type I diabetic patients. J Clin Endocrinol Metab 1994; 78:878-883.
63. Vaidya B, Imrie H, Perros P et al. The cytotoxic T lymphocyte antigen-4 is a major Graves' disease locus. Hum Mol Genet 1999; 8:1195-1199.
64. Heward JM, Allahabadia A, Armitage M et al. The development of Graves' disease and the CTLA-4 gene on chromosome 2q33. J Clin Endocrinol Metab 1999; 84:2398-2401.
65. Donner H, Rau H, Walfish PG et al. CTLA4 alanine-17 confers genetic susceptibility to Graves' disease and to type-1 diabetes mellitus*. J Clin Endocrinol Metab 1997; 82:143-146.
66. Vaidya B, Imrie H, Geatch DR et al. Association analysis of the cytotoxic T lymphocyte antigen-4 (CTLA-4) and autoimmune regulator-1 (AIRE1) genes in sporadic autoimmune Addison's disease. J Clin Endocrinol Metab 2000; 85:688-691.
67. Gambelunghe G, Falorni A, Ghaderi M et al. Microsatellite polymorphism of the MHC class I chain-related (MIC-A and MIC-B) genes marks the risk for autoimmune addison's disease. J Clin Endocrinol Metab 1999; 84:3701-3707.
68. Kristiansen OP, Larsen ZM, Pociot F. CTLA-4 in autoimmune diseases—a general susceptibility gene to autoimmunity? Genes Immun 2000; 1:170-184.
69. Schilling RF. Who has vitamin B_{12} deficiency? Proc Assoc Am Physicians 1996; 108:68-70.
70. de Carmo S, Kater CE, Dib SA et al. Autoantibodies against recombinant human steroidogenic enzymes 21-hydroxylase, side-chain cleavage and 17alpha-hydroxylase in Addison's disease and autoimmune polyendocrine syndrome type III. Eur J Endocrinol 2000; 142:187-194.
71. Bosi E, Becker F, Bonifacio E et al. Progression to type I diabetes in autoimmune endocrine patients with islet cell antibodies. Diabetes 1991; 40:977-984.
72. Wagner R, Genovese S, Bosi E et al. Slow metabolic deterioration towards diabetes in islet cell antibody positive patients with autoimmune polyendocrine disease. Diabetologia 1994; 37:365-371.
73. Blizzard RM, Kyle M. Studies of the adrenal antigens and antibodies in Addison's disease. J Clin Invest 1963; 42:1653-1660.
74. Soderbergh A, Winqvist O, Norheim I et al. Adrenal autoantibodies and organ-specific autoimmunity in patients with Addison's disease. Clin Endocrinol (Oxf) 1996; 45:453-460.
75. Julien JF, Samama P, Mallet J. Rat brain glutamic acid deciarboxylase sequence deduced from a cloned cDNA. J Neurochem 1990; 54:703-705.

76. Betterle C, Volpato M, Rees Smith B et al. II. Adrenal cortex and steroid 21-hydroxylase autoantibodies in children with organ-specific autoimmune diseases: Markers of high progression to clinical Addison's disease. J Clin Endocrinol Metab 1997; 82:939-942.
77. Tung KSK, Taguchi O, Teuscher C. Testicular and Ovarian Autoimmune Diseases. 1993.
78. Posillico JT, Wortsman J, Srikanta S et al. Parathyroid cell surface autoantibodies that inhibit parathyroid hormone secretion from dispersed human parathyroid cells. Bone Miner Res 1986; 1:475-483.
79. Posillico JT, Srikanta S, Eisenbarth G et al. Binding of monoclonal antibody (4F2) to its cell surface antigen on dispersed adenomatous parathyroid cells raises cytosolic calcium and inhibits parathyroid secretion. J Clin Endocrinol Metab 1987; 64:43-50.
80. Liblau RS, Caillat-Zucman S, Fischer AM et al. The prevalence of selective IgA deficiency in type 1 diabetes mellitus. Acta Pathol Microbiol Immunol Scand 1992; 100:709-712.
81. Alper CA, Marcus-Bagley D, Awdeh Z et al. Prospective analysis suggests susceptibility genes for deficiencies of IgA and several other immunoglobulins on the [HLA-B8, SC01, DR3] conserved extended haplotype. Tissue Antigens 2000; 56:207-216.
82. Valenta LJ, Elias AN. Familial scleroderma in a kindred with high incidence of autoimmune disease: Correlation with HLA-A1/B8 haplotype. Arch Dermatol 1987; 123:1438-1440.
83. Candrina R, Giustina A. Development of type II autoimmune polyglandular syndrome in a patient with idiopathic thrombocytopenic purpura. Isr J Med Sci 1988; 24:57-58.
84. Vela BS, Dorin RI, Hartshorne MF. Case report 631: Neo-osseous porosis (metaphyseal osteopenia) in polyglandular autoimmune (Schmidt) syndrome. Skeletal Radiol 1990; 19:468-471.
85. Kita M, Ahmad L, Marians RC et al. Regulation and transfer of a murine model of thyrotropin receptor antibody mediated Graves' disease. Endocrinology 1999; 140:1392-1398.
86. Hirata Y, Ishizu H. Elevated insulin-binding capacity of serum proteins in a case of spontaneous hypoglycemia and mild diabetes not treated with insulin. Tohoku J Exp Med 1972; 107:277-286.
87. Martin S, Wolf-Eichbaum D, Duinkerken G et al. Development of type 1 diabetes despite severe hereditary B-lymphocyte deficiency. N Engl J Med 2001; 345:1036-1040.
88. Coles AJ, Wing M, Smith S et al. Pulsed monoclonal antibody treatment and autoimmune thyroid disease in multiple sclerosis. Lancet 1999; 354:1691-1695.
89. Imagawa A, Itoh N, Hanafusa T et al. Autoimmune endocrine disease induced by recombinant interferon-α therapy for chronic active type C hepatitis. J Clin Endocrinol Metab 1995; 80:922-926.
90. Hoffenberg EJ, Haas J, Drescher A et al. A trial of oats in newly diagnosed celiac disease. J Pediatr 2000; 137:361-366.
91. Ventura A, Neri E, Ughi C et al. Gluten-dependent diabetes-related and thyroid-related autoantibodies in patients with celiac disease. J Pediatr 2000; 137:263-265.
92. Hoffenberg EJ, Bao F, Eisenbarth GS et al. Silent celiac disease in American children at genetic risk. J Pediatr Gastroenterol Nutr 2000; 31:S65-
93. Atkinson MA, Bowman MA, Kao K et al. Lack of immune responsiveness to bovine serum albumin in insulin- dependent diabetes. N Engl J Med 1993; 329:1853-1858.
94. Burman P, Totterman TH, Oberg K et al. Thyroid autoimmunity in patients on long term therapy with leukocyte- derived interferon. J Clin Endocrinol Metab 1986; 63:1086-1090.
95. Papo T, Oksenhendler E, Izembart M et al. Antithyroid hormone antibodies induced by interferon-α. J Clin Endocrinol Metab 1992; 75:1484-1486.
96. Gisslinger H, Gilly B, Woloszczuk W et al. Thyroid autoimmunity and hypothyroidism during long-term treatment with recombinant interferon-alpha. Clin Exp Immunol 1992; 90:363-367.
97. Bach J-F, Yamamoto AM, Djabiri F et al. Etiopathogenesis of Myasthenia Gravis (MG). 1997.
98. Sanders J, Oda Y, Roberts S et al. The interaction of TSH receptor autoantibodies with 125I-labelled TSH receptor. J Clin Endocrinol Metab 1999; 84:3797-3802.
99. Karlsson FA, Burman P, Loof L et al. Major parietal cell antigen in autoimmune gastritis and pernicious anemia is the acid producing HK-ATPase of the stomach. J Clin Invest 1988; 81:475-479.
100. De Block CE, De LI, Van Gaal LF. High prevalence of manifestations of gastric autoimmunity in parietal cell antibody-positive type 1 (insulin-dependent) diabetic patients. The Belgian diabetes registry. J Clin Endocrinol Metab 1999; 84:4062-4067.
101. Myhre AG, Halonen M, Eskelin P et al. Autoimmune polyendocrine syndrome type 1 (APS I) in Norway. Clin Endocrinol (Oxf) 2001; 54:211-217.
102. Soderbergh A, Rorsman F, Halonen M et al. Autoantibodies against aromatic L-amino acid decarboxylase identifies a subgroup of patients with Addison's disease. J Clin Endocrinol Metab 2000; 85:460-463.
103. Takasu N, Yamada T, Takasu M et al. Disappearance of thyrotropin-blocking antibodies and spontaneous recovery from hypothyroidism in autoimmune thyroiditis. N Engl J Med 1992; 326:513-518.

104. Verge CF, Gianani R, Kawasaki E et al. Number of autoantibodies (against insulin, GAD or ICA512/ IA2) rather than particular autoantibody specificities determines risk of type I diabetes. J Autoimmun 1996; 9:379-383.
105. Matsuki K, Juji T, Tokunaga K et al. HLA antigens in japanese patients with myasthenia gravis. J Clin Invest 1990; 392:399.
106. Garlepp MI, Dawkins RL, Christiansen FT. HLA antigens and acetylcholine receptor antibodies in penicillamine induced myasthenia gravis. BMJ 1983; 286:1442-1443.
107. Nagvekar N, Moody AM, Moss P et al. A pathogenetic role for the thymoma in myasthenia gravis. Autosensitization of IL-4- producing T cell clones recognizing extracellular acetylcholine receptor epitopes presented by minority class II isotypes. J Clin Invest 1998; 101:2268-2277.
108. Taylor SI, Grunberger G, Marcus-Samuels B et al. Hypoglycemia associated with antibodies to the insulin receptor. N Engl J Med 1982; 307:1422-1426.
109. Dons RF, Havlik R, Taylor SI et al. Clinical disorders associated with autoantibodies to the insulin receptor. J Clin Invest 1992.
110. Uchigata Y, Kuwata S, Tsushima T et al. Patients with Graves' disease who developed insulin autoimmune syndrome (Hirata disease) possess HLA-Bw62/Cw4/DR4 carrying DRB1*0406. J Clin Endocrinol Metab 1993; 77:249-254.
111. Mcevoy RC, Fedun B, Cooper LZ et al. Children at high risk of diabetes mellitus: New York studies of families with diabetes and of children with congenital rubella syndrome. Adv Exp Med Biol 1988; 246:221-227.
112. Rabinowe SL, Larsen PR, Antman EM et al. Amiodarone therapy and autoimmune thyroid disease. Increase in a new monoclonal antibody-defined T cell subset. Am J Med 1986; 81:53-57.
113. Shi Y, Zou M, Robb D et al. Typing for major histocompatibility complex class II antigens in thyroid tissue blocks: Association of Hashimoto's thyroiditis with HLA-DQA0301 and DQB0201 alleles. J Clin Endocrinol Metab 1992; 75:943-946.
114. Tomer Y, Barbesino G, Greenberg DA et al. Mapping the major susceptibility loci for familial Graves' and Hashimoto's diseases: Evidence for genetic heterogeneity and gene interactions. J Clin Endocrinol Metab 1999; 84:4656-4664.
115. Badenhoop K, Walfish PG, Rau H et al. Susceptibility and resistance alleles of human leukocyte antigen (HLA) DQA1 and HLA DQB1 are shared in endocrine autoimmune disease. J Clin Endocrinol Metab 1995; 80:2112-2117.
116. De Block CE. [Diabetes mellitus type 1 and associated organ-specific autoimmunity]. Verh K Acad Geneeskd Belg 2000; 62:285-328.
117. Naughton GK, Eisenger M, Bystryn J-C. Detection of antibodies to melanocytes in vitiligo by specific immunoprecipitation. J Invest Dermatol 1983; 81:540-542.
118. Xie Z, Chen D, Jiao D et al. Vitiligo antibodies are not directed to tyrosinase. Arch Dermatol 1999; 135:417-422.
119. Nicholson LB, Wong FS, Ewins DL et al. Susceptibility to autoimmune thyroiditis in Down's syndrome is associated with the major histocompatibility complex class II DQA*0301 allele. Clin Endocrinol 1994; 41:381-383.
120. Larizza D, Bianchi MM, Lorini R et al. Autoimmunity, HLA, Gm and Km polymorphism in Turner's syndrome. Autoimmunity 1989; 4:69-78.
121. Failla P, Ruberto C, Pagano MC et al. Celiac disease in Down's syndrome with HLA serological and molecular studies. J Pediatr Gastroenterol Nutr 1996; 23:303-306.
122. Fleming S, Cowell C, Bailey J et al. Hashimoto's disease in Turner's syndrome. Clin Invest Med 1988; 11:243-246.
123. Abiru N, Maniatis AK, Yu L et al. Peptide and MHC specific breaking of humoral tolerance to native insulin with the B:9-23 peptide in diabetes prone and normal mice. Diabetes 2001; 50:1274-1281.
124. Cohen IR, Young DB. Autoimmunity, microbial immunity and the immunological homunculus. Immunol Today 1991; 105:105-110.
125. Zhang Z-G, Wall JR, Bernard NF. Tissue distribution and quantitation of a gene expressing a 64-kDa antigen associated with thyroid-associated ophthalmopathy. Clin Immunol Immunopathol 1996; 80:236-244.
126. Martin AM, Maxson MN, Leif J et al. Diabetes-prone and diabetes-resistant BB rats share a common major diabetes susceptibility locus, iddm4: Additional evidence for a "universal autoimmunity locus" on rat chromosome 4. Diabetes 1999; 48:2138-2144.
127. Schranz DB, Lernmark A. Immunology in diabetes: An update. Diabetes Metab Rev 1998; 14:3-29.
128. Petterson A, Jacob H, Lernmark Å. Lessons from the Animal Models: The BB Rat 1996; 181-201.
129. Maloy KJ, Powrie F. Regulatory T cells in the control of immune pathology. Nat Immunol 2001; 2:816-822.

130. Jackson RA, Haynes BF, Burch WM et al. Ia+ T cells in new onset Graves' disease. J Clin Endocrinol Metab 1984; 59:187-190.
131. Kaye WA, Adri MN, Soeldner JS et al. Acquired defect in interleukin-2 production in patients with type I diabetes mellitus. N Engl J Med 1986; 315:920-924.
132. Wilson SB, Kent SC, Patton KT et al. Extreme Th1 bias of invariant Vα24JαQ T cells in type 1 diabetes. Nature 1998; 391:177-181.
133. Oelkers W, Diederich S, Bahr V. Therapeutic strategies in adrenal insufficiency. Ann Endocrinol (Paris) 2001; 62:212-216.
134. De Bellis AA, Falorni A, Laureti S et al. Time course of 21-hydroxylase antibodies and long-term remission of subclinical autoimmune adrenalitis after corticosteroid therapy: Case report. J Clin Endocrinol Metab 2001; 86:675-678.
135. Hara T, Mizuno Y, Nagata M et al. Human gamma delta T-cell receptor-positive cell-mediated inhibition of erythropoiesis in vitro in a patient with type I autoimmune polyglandular syndrome and pure red blood cell aplasia. Blood 1990; 75:941-950.
136. Aaltonen J, Björses P, Sandkuijl L et al. An autosomal locus causing autoimmune disease: Autoimmune polyglandular disease type I assigned to chromosome 21. Nat Genet 1994; 8:83-87.
137. Aaltonen J, Björses P, Perheentupa J et al. An autoimmune disease, APECED, caused by mutations in a novel gene featuring two PHD-type zinc-finger domains. Nat Genet 1997; 17:399-403.
137a. Anderson MS, Venzani ES, Klein L et al. Projection of an immunological self-shadow within the thymus by the Aire protein. Science 2002; 298:1395-1401.
138. Meyer G, Donner H, Herwig J et al. Screening for an AIRE1 mutation in patients with Addison's disease, type 1 diabetes, Graves' disease and Hashimoto's thyroiditis as well as in APECED syndrome. Clin Endocrinol (Oxf) 2001; 54:335-338.
139. Gylling M, Tuomi T, Bjorses P et al. ss-Cell autoantibodies, human leukocyte antigen II alleles, and type 1 diabetes in autoimmune polyendocrinopathy-candidiasis-ectodermal dystrophy. J Clin Endocrinol Metab 2000; 85:4434-4440.
140. Ekwall O, Hedstrand H, Haavik J et al. Pteridin-dependent hydroxylases as autoantigens in autoimmune polyendocrine syndrome type I. J Clin Endocrinol Metab 2000; 85:2944-2950.
141. Perniola R, Falorni A, Clemente MG et al. Organ-specific and nonorgan-specific autoantibodies in children and young adults with autoimmune polyendocrinopathy-candidiasis-ectodermal dystrophy (APECED). Eur J Endocrinol 2000; 143:497-503.
142. Baud O, Goulet O, Canioni D et al. Treatment of the immune dysregulation, polyendocrinopathy, enteropathy, X-linked syndrome (IPEX) by allogeneic bone marrow transplantation. N Engl J Med 2001; 344:1758-1762.
143. Palmieri G, Lastoria S, Colao A et al. Successful treatment of a patient with a thymoma and pure red-cell aplasia with octreotide and prednisone. N Engl J Med 1997; 336:263-265.
144. Marx A, Wilisch A, Schultz A et al. Pathogenesis of myasthenia gravis. Virchows Arch 1997; 430:355-364.
145. Mygland A, Aarli JA, Matre R et al. Ryanodine receptor antibodies related to severity of thymoma associated myasthenia gravis. J Neurol Neurosurg Psychiatr 1995; 57:843-846.
146. Geuder KI, Marx A, Witzemann V et al. Genomic organization and lack of transcription of the nicotinic acetylcholine receptor subunit genes in myasthenia gravis-associated thymoma. Lab Invest 1992; 452:458-
147. Amiel LL, Machover D, Droz JP. Dyscrasie plasmocytaire avec arteriopathie, polyneuropathie, syndrome endocrinien. Ann Intern Med 1975; 745:749-
148. Kim DE, Kim HJ, Kim YA et al. Kaposi's sarcoma herpesvirus-associated Castleman's disease with POEMS syndrome. Muscle Nerve 2000; 23:436-439.
149. Watanabe O, Maruyama I, Arimura K et al. Overproduction of vascular endothelial growth factor/ vascular permeability factor is causative in Crow-Fukase (POEMS) syndrome. Muscle Nerve 1998; 21:1390-1397.
150. Soubrier M, Dubost JJ, Serre AF et al. Growth factors in POEMS syndrome: Evidence for a marked increase in circulating vascular endothelial growth factor. Arthritis Rheum 1997; 40:786-787.
151. Scully RE, Mark EJ, McNeely WF et al. Case records of the Massachusetts general hospital: Case 34-1987. N Engl J Med 1987; 317:493-501.
152. Rabinowe SL, Rubin IL, George KL et al. Trisomy 21 (Down's syndrome): Autoimmunity, aging and monoclonal-antibody defined T-cell abnormalities. J Autoimmun 1989; 2:25-30.
153. Inoue H, Tanizawa Y, Wasson J et al. A gene encoding a transmembrane protein is mutated in patients with diabetes mellitus and optic atrophy (Wolfram syndrome). Nat Genet 1998; 20:143-148.
154. Karasik A, O'Hara C, Srikanta S et al. Genetically programmed selective islet beta-cell loss in diabetic subjects with Wolfram's syndrome. Diabetes Care 1989; 12:135-138.

155. Ou D, Jonsen LA, Metzger DL et al. CD4+ and CD8+ T-cell clones from congenital rubella syndrome patients with IDDM recognize overlapping GAD65 protein epitopes. Implications for HLA class I and II allelic linkage to disease susceptibility. Hum Immunol 1999; 60:652-664.
156. Rabinowe SL, George KL, Loughlin R et al. Congenital rubella. Monoclonal antibody-defined T cell abnormalities in young adults. Am J Med 1986; 81:779-782.
157. Karounos DG, Wolinsky JS, Thomas JW. Monoclonal antibody to rubella virus capsid protein recognizes a β-cell antigen. J Immunol 1993; 150:3080-3085.
158. Bottazzo GF, Florin-Christensen A, Doniach D. Islet-cell antibodies in diabetes mellitus with autoimmune polyendocrine deficiencies. Lancet 1974; 2:1279-1283.
159. Gianani R, Pugliese A, Bonner-Weir S et al. Prognostically significant heterogeneity of cytoplasmic islet cell antibodies in relatives of patients with type I diabetes. Diabetes 1992; 41:347-353.
159a. Hori S, Nomura T, Sakaguchi S. Control of regulatory T cell development by the transcription factor Foxp3. Science 2003; 299:1057-1061.
160. Weetman AP. Determinants of autoimmune thyroid disease. Nat Immunol 2001; 2:769-770.
161. Benoist C, Mathis D. Autoimmunity provoked by infection: How good is the case for T cell epitope mimicry? Nat Immunol 2001; 2:797-801.

CHAPTER 9

Epidemiology of Type 1 Diabetes Mellitus

Marian Rewers, Jill Norris and Dana Dabelea

Type 1 Diabetes

Type 1 diabetes accounts for about 10% of all diabetes, affecting approximately 1.4 million people in the U.S., and 10-20 million globally.[1,2] About 40% of persons with type 1 diabetes are younger than 20 years of age, thus making it one of the most common severe chronic diseases of childhood. In the U.S., where 30,000 new cases occur each year, type 1 diabetes affects 1:300 children and as many as 1:100 adults during the life span.[3,4] Type 1 diabetes is the leading cause of end-stage renal disease, blindness, and amputation, and a major cause of cardiovascular disease and premature death in the general population.[5] This disease results in over $5 billion in medical care expenditures per year, with costs for patients over 10 times those for persons without diabetes.[6] Both the U.S. Public Health Service[7] and the American Diabetes Association[8] have identified needs in the area of prevention of type 1 diabetes. Population-based studies, as well as family studies and laboratory data, have provided new insights into the pathogenesis of type 1 diabetes, its associated risk factors and its natural history. They also allowed development of plans for primary and secondary prevention of diabetes in persons with ongoing β-cell autoimmunity.

This chapter considers the whole spectrum of type 1 diabetes - from genetic susceptibility, through β-cell autoimmunity without diabetes, to clinical type 1 diabetes. Special emphasis is given to epidemiological data relevant to the primary and secondary prevention.

Natural History of Type 1A Diabetes

The American Diabetes Association and the World Health Organization have revised the classification of diabetes mellitus.[9,10] The new classification of diabetes is now primarily based on pathogenesis rather than on the requirement for insulin therapy. Based on the new system, type 1 diabetes (previously defined as insulin-dependent or juvenile diabetes) is caused by β cell_destruction, often immune mediated, that leads to loss of insulin secretion and absolute insulin deficiency. The etiologic agents that cause the autoimmune process and β cell destruction are not well established. Type 1 diabetes also includes cases that are thought not to be immune mediated, but are characterized by absolute insulin deficiency (insulinopenia).

The current classification of diabetes mellitus distinguishes type 1a (autoimmune)[11] and type 1b (not immune-mediated) form that remains poorly defined.[12,13] Type 1a is the most common form of diabetes among children and adolescents of European origin, usually characterized by acute onset and dependence on exogenous insulin for survival. In adults, the disease is nearly as frequent as in children, but often less dramatic onset may lead to misclassification as type 2 and a delayed insulin treatment. About 60% of persons with type 1 diabetes are diagnosed as adults.

Immunology of Type 1 Diabetes, Second Edition, edited by George S. Eisenbarth.

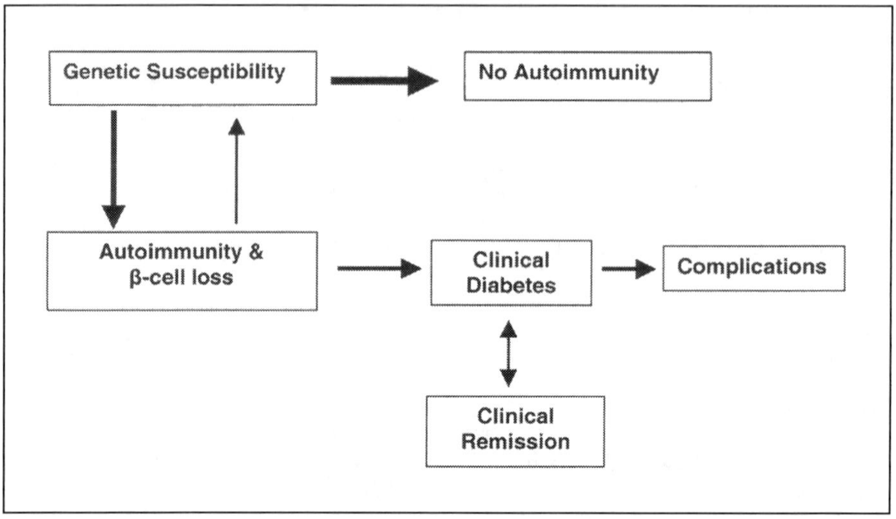

Figure 1.

The natural history of type 1a diabetes (Fig. 1) includes four distinct stages: (1) preclinical β-cell autoimmunity with progressive defect of insulin secretion; (2) onset of clinical diabetes; (3) transient remission; and (4) established diabetes associated with acute and chronic complications and premature death.

β-cell Autoimmunity

In most patients, the etiology of the autoimmune process and β-cell destruction is not known. The process is mediated by macrophages and T lymphocytes with detectable autoantibodies to various β-cell antigens. Currently, autoimmunity is defined by the presence of autoantibodies,[11] because their measurement is reliable and standardized across laboratories, in contrast to the cellular markers. The initial islet cell antibody (ICA) assay, using immunofluorescence and pancreatic tissue[14] has been notoriously difficult to standardize and has been replaced by a combination of specific β-cell autoantibodies to insulin (IAA),[15-17] glutamic acid decarboxylase (GAD),[18,19] and tyrosine phosphatase ICA512 (IA-2).[20] These tests have been shown to be quite sensitive and predictive in relatives of type 1 diabetes patients[21,22] and in the general population.[23]

Progression from β-Cell Autoimmunity to Clinical Diabetes

The duration of preclinical β-cell autoimmunity is variable and precedes the diagnosis of diabetes by up to 13 years.[24,25] In persons with persistent autoantibodies, there is an early loss of spontaneous pulsatile insulin secretion, progressive reduction in the acute insulin response to intravenous glucose load, followed by decreased response to other β-cell secretagogues, impaired oral glucose tolerance and fasting hyperglycemia.[26] However, a nonprogressive β-cell defect has been shown to exist for many years in monozygotic twins and other relatives of type 1 diabetes persons.

Studies in the first-degree relatives of type 1 diabetes patients[25,27,28] and in school children with no family history of type 1 diabetes[29-33] have reported ICA "remission" rates between 10-78%. Newer data suggest that while individual islet-specific autoantibodies may fluctuate in titers, it is very unusual to observe remission after two or more of such autoantibodies were present for even a few months.[34] Some people may loose β-cell autoantibodies or remain positive but do not progress to diabetes due to incomplete penetrance of susceptibility genes or insufficient exposure to the causative environmental agent(s). It is possible that β-cell autoimmunity remits spontaneously in genetically resistant persons or when the offending factor is removed, similar to celiac disease. Age also plays a role, because children younger than 10 years have a 3-fold increased risk of progressing from autoimmunity to type 1 diabetes, compared to older relatives.[35]

β-cell autoimmunity may remit and reappear in the course of viral infections or variable exposure to dietary causal factors. The cumulative β-cell damage and increases in insulin resistance with obesity and physical inactivity, may eventually cause diabetes at a later age. Those people in whom the disease process is slow may present with type 1 diabetes as adults, develop diabetes that does not require insulin treatment, or may even escape diabetes altogether. Markers of autoimmunity can be detected in 14-33% of diabetes patients classified on clinical grounds as "Type 2"[36,37] and are associated with early failure of oral hypoglycemic drug therapy and insulin dependence in these patients. A term "latent autoimmune diabetes of adults" (LADA)[38] has been coined for this slowly progressing form of type 1 diabetes.

Clinical Onset of Type 1 Diabetes

Most epidemiological data on type 1 diabetes are based on a clinical, standard definition that has been widely used for a long time: (1) physician diagnosis of diabetes, (2) daily insulin injections instituted at the time of diagnosis, and (3) residence by the patient in the area of registration at the time of first insulin administration. This practical definition is not taking into account the nuances of etiological classification into Type 1a, 1b, and other forms of diabetes.

In industrialized countries, 20-40% of type 1 diabetes patients younger than 20 years presents in diabetic ketoacidosis (DKA).[39-42] Younger age, female gender, HLA-DR4 allele,[43] lower socioeconomical status and lack of family history of diabetes have been associated with more severe presentation. Younger children present with more severe symptoms at diagnosis, because children younger than 7 have lost on average 80% of the islets, compared to 60% in those 7-14 year old and 40% in those older than 14.[44] Case fatality in industrialized countries ranges between 0.4-0.9%.[45] While brain edema is believed to be the major cause of onset death, the risk factors are poorly understood and heterogeneous.

Both DKA and onset deaths are largely preventable, because most of the patients have typical symptoms of polyuria, polydipsia, and weight loss for 2-4 weeks prior to diagnosis. The diagnosis is straightforward in almost all cases, based on the symptoms, random blood glucose over 200 mg/dl and/or HbA_{1c} >7%. Oral glucose tolerance test (OGGT) is rarely needed for diagnosis. In questionable cases, negative OGGT allows to rule out current diabetes and, in combination with negative β-cell autoantibodies, offers reassurance that diabetes is not around the corner. Traditionally, nearly all children with newly diagnosed type 1 diabetes were hospitalized. More recently, an increasing proportion of these children have been managed on an outpatient basis, especially in urban centers with specialized diabetes education and treatment facilities. In Colorado, U.S.A., the proportion of children receiving only outpatient care at onset increased from 6% in 1978 to 35% in 1988[46] and 75% in 2000. Hospitalization at onset does not improve short-term outcomes such as readmission for DKA or severe hypoglycemia,[39,42] if adequate family education and follow-up is available on outpatient basis. Onset hospitalizations and subsequenty acute complications have similar predictors: Biological (younger age, lower endogenous insulin secretion) and psychosocial (lower socioeconomic status, limited access to health care, dysfunctional family).

Prevalence of Type 1 Diabetes

Initial examination of a disease generally begins with cross-sectional data on its prevalence (i.e., the number of people in the population who have the disease at a given point in time). Selected estimates of type 1 diabetes prevalence in different populations with at least 90% ascertainment of cases are shown in Table 1. The prevalence of type 1 diabetes in children aged

Table 1. Prevalence of childhood type 1 diabetes, from selected studies

Location	Year of Study	Age (y)	Prevalence (per 100,000)
United States			
Rochester, MD	1970	0-14	57
Kentucky	1978-79	0-17	208
Erie County, PA	1961	0-15	61
Europe			
Denmark	1973	0-14	83
Finland	1979	0-14	191
Sweden	1977	0-14	300
Leicester	1984-85	0-14	99
Estonia	1988	0-14	60
Oceania			
Canterbury, NZ	1986	0-19	105
Tasmania, Australia	1984	0-14	57

less than 15 years ranges from 0.05 to 0.3% in most European and North American populations.[47] Comparisons of prevalence data may be subject to considerable bias, since prevalence is determined not only by disease incidence, but also by case survival, which may vary markedly across populations. Prevalence data however,are useful in determining the public health impact of type 1 diabetes. For example, it is estimated that there are approximately 123,000 individuals aged to 0-19 years in United States who currently have type 1 diabetes.[48]

Incidence of Type 1 Diabetes

Geographic location: One of the most striking characteristics of type 1 diabetes is the large geographic variability in the incidence[49-51] (Fig. 2). Scandinavia and the Mediterranean island of Sardinia have the highest incidence rates in the world while Oriental populations have the lowest rates. A child in Finland is 400 times more likely to develop diabetes than one in China. While there is a strong south-north gradient in the incidence, 'hot-spots' in warm climates have been reported (Sardinia, Puerto Rico, Kuwait). The geographic and ethnic variations in type 1 diabetes reflect the prevalence of susceptibility genes or that of causal environmental factors, or both.

In the general population, the prevalence of β-cell autoimmunity appears to be roughly proportional to the incidence of type 1 diabetes in the populations.[52,53] In contrast, the prevalence of β-cell autoimmunity in first-degree relatives of type 1 diabetic persons does not differ dramatically between high and low risk countries. The incidence of β-cell autoimmunity is higher in relatives younger than 5 years (3.7%/yr), compared to those 5-9 yr old (0.5%/yr).[54] During 5 years of follow-up, none of the relatives older than 10 has developed β-cell autoimmunity,[54] suggesting that β-cell autoimmunity develops primarily before the age of 5 and that it may remit in many cases.[32,33]

The clinical picture of the disease is similar in low- and high-risk areas,[55,56] making it unlikely that the inter-population difference are due to misclassification of different types of diabetes. However, in many population[57,58] type 2 diabetes is an increasing or already the predominant form of diabetes in children[59] making correct diagnosis and treatment increasingly difficult.

Age: Type 1 diabetes incidence peaks at the ages of 2, 4-6 and 10-14 years, perhaps due to alterations in the pattern of infections or increases in insulin resistance. The age-distribution of type 1 diabetes onset is similar across geographic areas and ethnic groups[56] (see also Fig. 3). There have been only a few studies that have examined the incidence of type 1 diabetes in

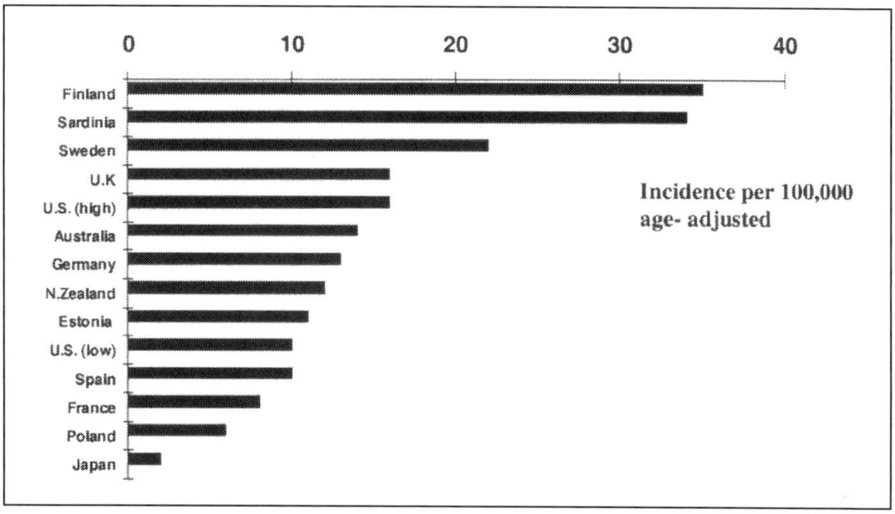

Figure 2. Worldwide incidence of type 1 diabetes in 0-14 year old children.[50,51,73]

adults, mainly because of the difficulty of distinguishing type 1 diabetes from insulin-requiring type 2 diabetes in older individuals. The incidence decreases in the third decade of life,[60-62] only to increase again in the fifth to seventh decades of life.[63,64] It has been speculated that the incidence of type 1 diabetes increases again in the fifth through seventh decades of life,[27,28] but there is no hard evidence of such increase, and it is not known whether there are etiologic differences between childhood- and adult-onset type 1 diabetes. The prevalence of ICA decreased with age in first-degree relatives participating in DPT-1 screening. It is not known whether the etiology differs between childhood- and adult-onset type 1diabetes, but it was not apparent among adult participants in the UKPDS. Over 30% of those aged 25-34 were positive for ICA and/or GAD autoantibodies, but the prevalence decreased with age, to less than 10% in those aged 55-65.[65] The presence of the autoantibodies and age of presentation of diabetes were strongly associated with the presence of the HLA-DRB1*03/DRB1*04-DQB1*0302 genotype.[66]

Race and ethnicity: There are striking racial differences in type 1 diabetes risk in multiracial populations, although not of the same magnitude as the geographic differences (Table 2). In the U.S., nonHispanic whites are about one and a half times as likely to develop type 1 diabetes as African Americans[67] or Hispanics[68] (Fig. 3). This is similar to the differences reported from Montreal, where children of British descent had about one and a half the risk of type 1 diabetes in children of French descent.[69]

Little data concerning β-cell autoimmunity is available for nonCaucasian populations. Among 86,000 relatives of type 1 diabetic patients screened for the DPT-1 trail, lower prevalence of ICA was observed in Asian Americans (2.6%) and in Hispanics (2.7%), compared to African Americans (3.3%) or nonHispanic whites (3.9%) (DPT-1, unpublished data 2001). Lower prevalence of GAD antibodies has also been reported in Oriental compared to Caucasian type 1 diabetes patients.[70]

Gender: In general, males and females have similar risk of type 1 diabetes,[71] with the pubertal peak of incidence in females preceding that in males by 1-2 years. In lower-risk populations, such as Japan or U.S. blacks, there is a female preponderance, while in high-risk groups, there is a slight male excess.[3,49,72] DPT-1 screening data (not adjusted for age, relation to proband of ethnicity) have suggested higher prevalence of ICA in male first-degree relatives (4.3%) than in females (3.3%). Interestingly, even within Europe, all populations with an incidence higher than 20/100,000 (Sardinia, U.K., Italy, Finland, Norway, etc) had male excess, whereas those

Table 2. Age adjusted incidence of type 1 diabetes per 100,000 per year in children 0-14 yrs of age of different ethnic origin

Location	Incidence	95% Confidence Interval
Allegheny County, PA		
White	16.2	14.1-18.4
Black	11.8	7.9-17.2
Colorado		
Non-Hispanic White	16.4	15.0-17.8
Hispanic	9.7	7.4-12.4
Montreal, Canada		
French	8.2	7.5 - 8.9
British	15.3	12.8-17.8
Italian	10.7	7.5-13.9
Jewish	17.2	10.4-24.0
Other	13.1	10.9-15.3

with a rate below 4.5.100,000 (the Baltic countries, Macedonia, Yugoslavia, Romania, etc) had female excess.[47,73,74]

Type 1 diabetes diagnosed in adulthood seems to be associated with male excess with a male: Female ratio between 1.3 and 2.15 in most populations of European origin,[75-79] but there is considerably less information concerning type 1 diabetes with onset in adult life. These findings contrast with data from animal models of type 1 diabetes (the NOD mouse), in which diabetes progression is almost twice as common in females.[80] Several pathways for these differ-

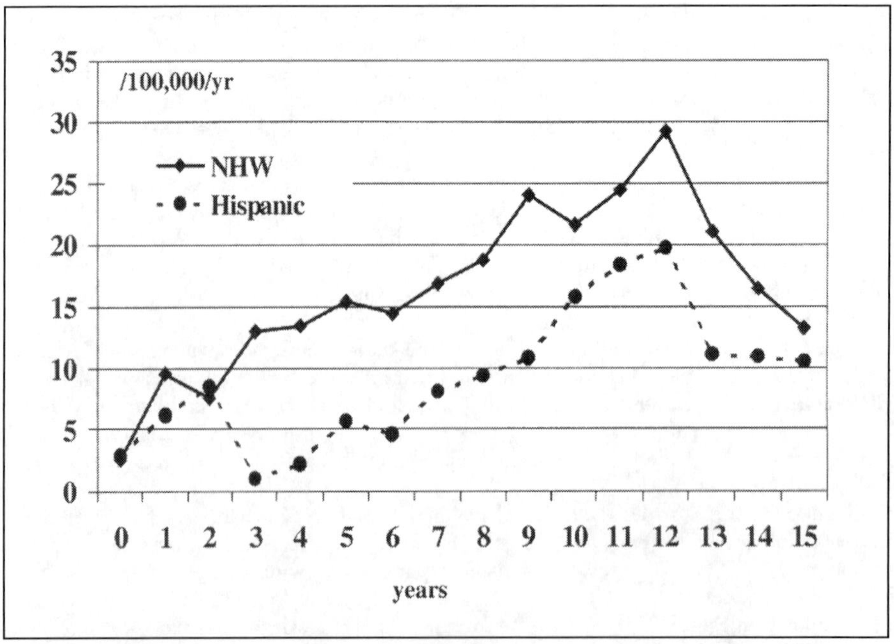

Figure 3. Age at diagnosis of type 1 diabetes, by sex. Data from the Colorado IDDM registry, 1978-1988.[68]

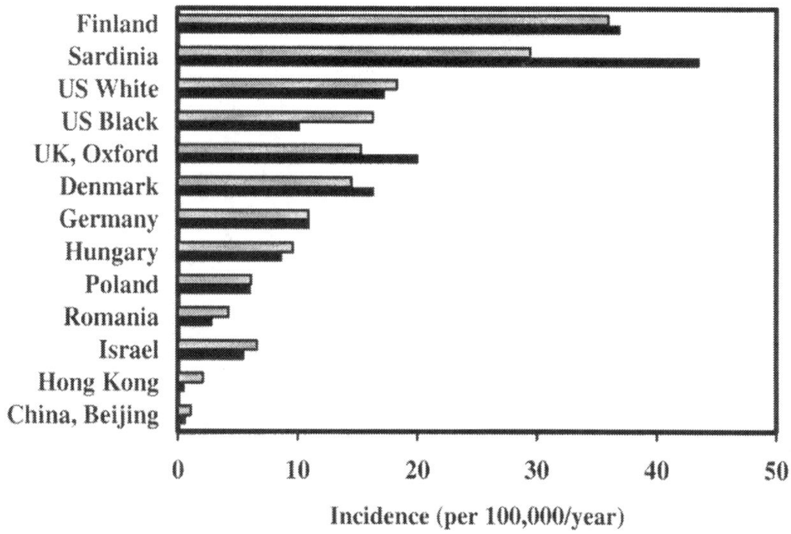

Figure 4. Gender differences in the incidence of type 1 diabetes (female= open bars).

ences have been explored, such as the effects of sex hormones, hormone substitution, and pregnancy on autoimmunity, the relationship between genetic risk factors for diabetes and sex,[81] but the reasons for these differences are yet not known.

Time: The incidence of type 1 diabetes varies markedly over time, both seasonally and annually. In the Northern Hemisphere, the incidence declines during the warm summer months; similarly in the Southern Hemisphere, the seasonal pattern exhibits a decline during the warm months of December and January, implicating a climatic factor.[82] This seasonal pattern appears to occur only in older children,[83,84] suggesting that factors triggering diabetes may be related to school attendance. The observed seasonality does not appear to be an artifact of health care seeking or access, but the seasonal patterns differ by the HLA-DR genotype.[85,86]

Most population-based registries have shown an increase in type 1 diabetes incidence over time.[87-92] Periodic outbreaks, sometimes of pandemic proportion, e.g., during 1984-86[88] appear to be superimposed on a steady secular increase in incidence. Figure 5 displays an example of parallel temporal incidence changes in populations that differ vastly in geographic location and the background risk of type 1 diabetes[56] (Allegheny, CO., U.S.A., and Wielkopolska, Poland. While the increase in type 1 diabetes incidence has affected all age groups, several studies reported particular increase among the youngest children.[93-97] No reliable information is available concerning potential seasonal or annual variation in the incidence of β-cell autoimmunity.

Genetic models are unable to explain the apparent temporal changes in the incidence.[98] Few studies have analyzed time trends using modeling procedures, looking specifically at the effects of age, calendar period, or birth cohort on the incidence of type 1 diabetes.[99-102] Figure 4 displays examples of two types of changes in the incidence of childhood onset type 1 diabetes that have been observed from late 1960s to mid 1980s: Periodic outbreaks superimposed on a steady secular increase (Finland), and outbreaks with little secular increase (Poland). These nonlinear calendar period effects were attributed to environmental factors causing epidemic peaks. The most recent study covering the period from 1984 to 1996 found only a strong linear effect of an increase of 2.3% per year, suggesting that the causal role of unknown environmental

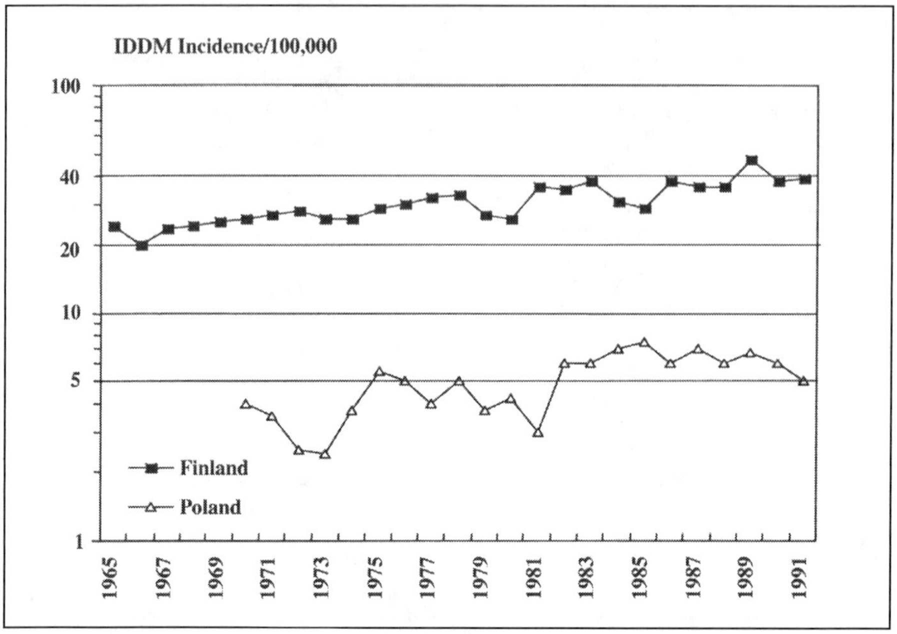

Figure 5. Temporal changes in the incidence of type 1 diabetes: A secular trend with superimposed outbreaks (Finland) and outbreaks with no clear secular trend (Poland).[99,100]

factors is diffusing over time.[102] Importantly, the increase in incidence was seen in the age group 0 to 14 years and also in the age group on 15-29 years. The incidence is increasing both in low and high incidence populations. A recent analysis of data on published incidence trends showed that the incidence of type 1 diabetes is globally increasing by 3.0% per year, and that the incidence of type 1 diabetes will be 40% higher in 2010 than in 1998.[103] The polio model, where autoimmune diabetes results from delayed exposure to infection that is benign when encounter in early childhood, could explain the recent increase in the incidence, but not the shift in diagnosis to earlier ages. Alternative explanation invokes the congenital rubella model where increased hygiene has led to a decline in herd immunity to common infections among women in child-bearing age. These women are more likely to develop viremia during pregnancy resulting in congenital persistent infection of β-cells and early onset type 1 diabetes in the offspring. This model could explain both the increasing incidence of diabetes and the decreasing age of disease onset.

Genetic Factors

Family history of type 1 diabetes: In moderate type 1 diabetes risk areas, such as the United States, the risk of type 1 diabetes by the age of 20 years is approximately 1:300 (Table 3). The risk is increased to about 1:50 in offspring of type 1 diabetes mothers and 1:15 in offspring of type 1 diabetes fathers (the reason for this parental gender difference is not known). The risk to siblings of type 1 diabetes probands ranges from 1:12 to 1:35[104,105] and is further increased, in HLA-identical siblings. It is estimated that by the age of 60 years approximately 10% of the first degree relatives develop type 1 diabetes.[106] Family history of type 1 diabetes is a surrogate measure of the combination of type 1 diabetes genes and environmental exposures shared by family members.

The risk of β-cell autoimmunity is higher, because not all autoantibody positive children develop diabetes by the age of 20. Prevalence estimates from cross-sectional studies shown in Table 1 are obviously less precise than cumulative incidence rates available for clinical diabetes.

Table 3. Risk, by the age of 20 years, of type 1 diabetes and beta-cell autoimmunity in the general population and family members of type 1 diabetic patients

Risk Group	Type 1 Diabetes	Pre-Diabetic Autoimmunity
General Population		
All HLA genotypes	1:300	1:30-1:100
HLA-DR3/4,DQB1*0302	1:15	1:10(?)
Family Members		
Maternal offspring	1:50	1:15
Paternal offspring	1:15	1:5
Siblings (all)	1:12 - 1:35	1:5
Monozygotic twins	1:3	1:1(?)
HLA-identical siblings	1:4	1:2(?)

'Familial' cases represent about 10% of type 1 diabetes and do not appear to be etiologically different from 'sporadic' cases in terms of the HLA-DR,DQ gene frequencies, seasonality of onset, prevalence of islet autoantibodies.[107] 'Familial' cases tend to have lower HbA1c and higher C-peptide levels than 'sporadic' cases, because relatives recognize diabetes symptoms earlier, however, these differences disappear soon after diagnosis.

Candidate genes. The primary loci of genetic susceptibility to type 1 diabetes have been mapped to the HLA-DR and DQ regions.[108-110] While 50 percent of nonHispanic whites in the United States have HLA-DR3 or DR4 allele, at least one of these alleles is present in 95 percent of patients with type 1 diabetes. The estimated risk for general population children who have the HLA-DR3/4,DQB1*0302 genotype is approximately 1:15.[111] Only 2.4% of the general population carries this genotype, compared to 30-40% of type 1 diabetes patients.

No particular HLA type seems to be associated with β-cell autoimmunity, although inconsistent associations between insulin or GAD autoantibodies and HLA-DR, DQ phenotypes have been reported. The HLA-DR2,DQB1*0602 haplotype, which almost completely protects from type 1 diabetes,[108] is found in about 15% of GAD and IAA positive young relatives of type 1 diabetes patients.[112-114] However, 90% of the first-degree relatives[115] and general population children who stay persistently autoantibody positive[29,116] express the HLA-DRB1*04,DQB1*0302. This may suggest that HLA genes are not involved in the initiation of β-cell autoimmunity, but rather determine progression to diabetes.

Candidate genes outside the HLA region are being identified.[117-123] It is increasingly apparent that the identification of true genetic associations in common multifactorial disease will require studies comprising thousands rather than the hundreds of individuals employed to date.[124] Perhaps even more subjects with β-cell autoimmunity need to be genotyped to precisely determine the role of HLA and additional type 1 diabetes candidate genes in the initiation of autoimmunity and progression to diabetes.

Environmental Factors

Twin[125] and family studies indicate that genetic factors alone cannot explain the etiology of type 1 diabetes. Seasonality, increasing incidence and epidemics of type 1 diabetes as well as numerous ecological, cross-sectional and retrospective studies suggest that certain viruses and components of early childhood diet may cause type 1 diabetes.

Viruses: Herpesviruses,[126,127] mumps,[128,129] rubella,[130,131] retroviruses,[132,133] and rotavirus[134] have been implicated. Viral infections appear to initiate autoimmunity rather than precipitate diabetes in subjects with autoimmunity. Two or more infections with similar viruses may be needed - mice persistently expressing a viral protein in the β-cells do not develop β-cell autoimmunity unless exposed to the same virus later in life.[135,136] ICA or IAA has been detected after mumps,[128] rubella, measles, chickenpox,[137] Coxsackie,[138] ECHO4,[139] and rotavirus[134] infections. Newborns and infants are particularly likely to develop a persistent infection and among patients with congenital rubella syndrome, 70% have ICA.[130]

An increased incidence of type 1 diabetes in patients with congenital rubella syndrome (CRS) is particularly interesting. While CRS is responsible for a minute proportion of type 1 diabetes and there is little evidence that postnatal rubella exposure to the wild strain[137] or to the MMR vaccine[140] causes type 1 diabetes, CRS provides an example of viral persistence leading to type 1 diabetes. The incubation period of type 1 diabetes in CRS patients is 5-20 years[131] and persistent rubella virus infection of the pancreas has been demonstrated in some cases. While CRS is not associated with particular HLA-DR alleles, the distribution of the HLA-DR3 and 4 alleles among patients with CRS and diabetes resembles that in nonCRS type 1 diabetes patients.[130] Finally, a molecular mimicry has been reported between a rubella virus protein and a 52 kD β-cell autoantigen.[141]

The evidence is strongest for picornaviruses, which include human (enteroviruses and rhinoviruses) and animal pathogens (e.g., mouse EMC virus and Theiler's virus). Enteroviruses have been most strongly linked to human type 1 diabetes, but convincing proof of causality remains elusive (for review see refs. 142,143). Case and autopsy reports,[144,145] epidemics of type 1 diabetes associated with concurrent epidemics of enteroviruses[146,147] and multiple cross-sectional seroepidemiological studies[142] have been suggestive, but not entirely convincing. At least 90% of type 1 diabetes patients demonstrate prolonged period of β-cell autoimmunity that is hardly compatible with an acute cytolytic enteroviral infection being a major cause. Enteroviral infection could, however, initiate β-cell autoimmunity through molecular mimicry between CBV P2-C protein and GAD[148]) or a persistent β-cell infection with impairment of insulin secretion and expression of self-antigens.

Cross-sectional studies of anti-Coxsackie antibodies in β-cell autoimmunity have been week and inconclusive[149] and have been recently replaced by studies based on detection of picornaviral RNA in bodily fluids using PCR. Prospective studies of nondiabetic relatives and general population children found a strong relation between enteroviral infections, defined by PCR, and development of islet autoantibodies in Finland[150-152] but not in the U.S (P. Graves, unpublished data 2000). Studies from Finland[150] and Sweden [153,154] have suggested in utero enteroviral infections can lead to type 1 diabetes in a significant proportion of the cases. Additional perinatal factors[154-160] and season of birth[161] have been associated with type 1 diabetes.

Possible protective effect of infections: In animal models, viral infection may protect the host from developing type 1 diabetes.[162-164] Evidence for such a protective effect in humans is speculative, at best.[165-167]

Routine childhood immunization. None of the routine childhood immunization have been shown to increase the risk of diabetes[140,167-170] or prediabetic autoimmunity.[171,172]

Dietary factors. Cow's milk or wheat introduced at weaning trigger insulitis and diabetes in animal models[173] perhaps through a molecular mimicry.[174] Human data are conflicting, but predominantly negative.[175-181] An ecological study suggested an association between decrease in breast-feeding and increase in type 1 diabetes incidence between 1940 and 1980.[182] Subsequent case-control studies have shown a negative,[183,184] positive[185] or no association.[186-188] Certain studies[183,189] but not others[185,187,188] suggested a dose-response relationship between the duration of breast-feeding and protection from type 1 diabetes. A meta-analysis found a 50% increase in type 1 diabetes risk associated with a breast-feeding duration of less than 3 months, and exposure to breast-milk substitutes prior to 3 months of age,[190] but a subsequent meta-analysis reported much lower risk estimates.[191] Breast-feeding may be viewed as a surrogate for the delay in the introduction of diabetogenic substances present in formula or early childhood diet. The reports that newly diagnosed diabetic children, compared with age-matched controls, have higher levels of serum antibodies against cow's milk and beta-lactoglobulin[192] as well as against bovine serum albumin[193] have been difficult to reproduce.[173] More recent cohort studies failed to find an association between infant diet exposures and beta-cell autoimmunity.[172,179,181] Interestingly, a study from Finland suggested that current cow's milk consumption was more closely linked to prediabetic autoimmunity and diabetes than infant exposure.[194] To resolve this controversy, a dietary intervention trial to prevent type diabetes by a short-term elimination of cow's milk from infant diet (TRIGR) is underway.[195]

Chemical compounds (streptozotocin[196] or dietary nitrates and nitrozamines[197]) induce β-cell autoimmunity in animal models. Circumstantial evidence suggest a connection between type 1 diabetes and consumption of foods and water containing nitrates, nitrites or nitrosamines.[198-200] Multiple hits of dietary β-cell toxins may render genetically resistant individuals susceptible to diabetogenic viruses leading to type 1 diabetes.[201]

Gene-Environment Interactions

Type 1 diabetes is likely caused by an interactive effect of genetic and environmental factors within a limited age-window. While both the susceptibility genes and the candidate environmental exposures appear to be quite common, the disease is still uncommon, raising a possibility of low penetrance.[202]

In mice, the host's genes restrict the diabetogenic effect of picornaviruses in a manner compatible with a recessive trait not related to the MHC. In humans, on the other hand, susceptibility to diabetogenic enteroviruses appears to be genetically restricted by HLA-DR and DQ alleles.[142] However, the allelic specificity is controversial[43,203-205] and may depend on the viral type and epidemicity. In general, the HLA-DR3 allele, present in most patients with type 1 diabetes, is associated with viral persistence.

Very few studies have examined a possibility of an interaction between the HLA genes and dietary exposured[206,194] (Fig. 6). The epidemiological data are limited, but suggest that an early exposure to cow's milk in relatives with HLA-DR3/4,DQB1*0302, DR3/3 or DRx/4,DQB1*0302 is not associated with development of β-cell autoantibodies.[207] It is unclear whether other genes are involved.

Remission ("Honeymoon Period")

Shortly after clinical onset, most of the patients experience a transient fall in insulin requirements due to improving β-cell function. Total and partial remissions have been reported in, respectively, 2-12% and 18-62% of young type 1 diabetes patients.[42,208,209] Older age and less severe initial presentation of diabetes[208-211] and low or absent ICA[211-213] or IA-2[214] have been consistently associated with deeper and longer remission. Evidence relating GAD autoantibodies,[211,214,215] nonCaucasian origin, HLA-DR3 allele, female gender and family history of type 1 diabetes to a less severe presentation, greater frequency of remission and slower deterioration of insulin secretion is inconclusive. Most studies,[208,213] but not all,[209,210] agree that preserved β-cell function is associated with better glycemic control (lower HbA_{1c}) and preserved α-cell glucagon response to hypoglycemia.[216] The prevalence of ICA (but not GAA) decreases from 87% at the time of type 1 diabetes diagnosis to 38-62% 2-3 years later,[212,217] faster in young boys, subjects lacking HLA-DR3 and 4, and those diagnosed between July and December.[217]

The natural remission is always temporary, ending with a gradual or abrupt increase in exogenous insulin requirements. Destruction of β-cells is complete within 3 years of diagnosis in most young children, especially those with the HLA DR3/4 phenotype.[218] It is much slower and often only partial in older patients,[219] 15% of whom have still some β-cell function preserved 10 years after diagnosis.[220]

Established Diabetes

Acute Complications

Acute complications of type 1 diabetes: DKA, hypoglycemia and infections are described in detail in other chapters. The risk of hospital admission for acute complication is 30/100 patient-years (p-yrs) in the first year of the disease and 20/100 p-yrs in the subsequent 3 years.[42] An estimated 26% of the patients have at least one episode of severe hypoglycemia within the initial 4 years of diagnosis, with little relation to the demographic or socioeconomical factors.

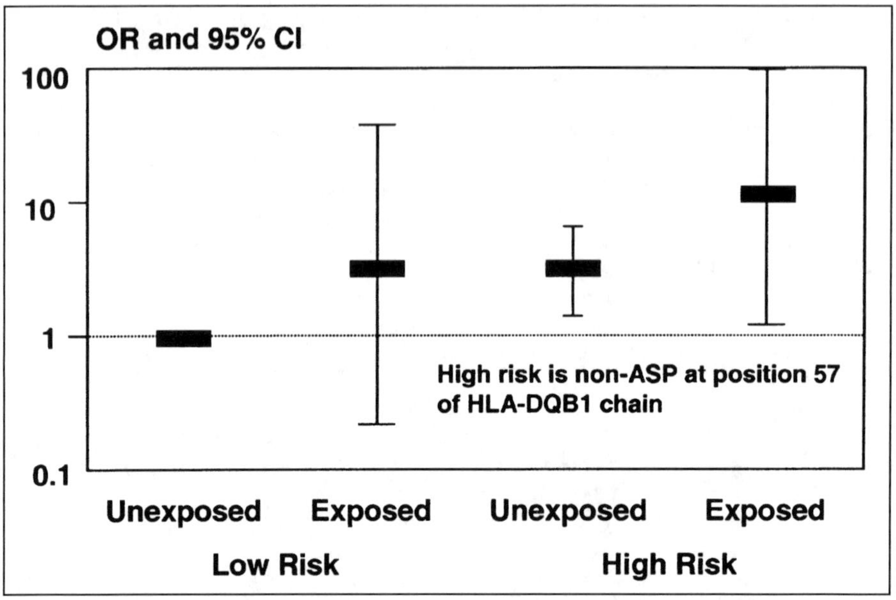

Figure 6. Odds ratios for type 1 diabetes by exposure to whole cow's milk prior to 3 months of age in low- and high- risk individuals, as determined by a genetic marker on the HLA-DQB1 chain.[206]

The incidence of severe hypoglycemic episodes varies between 6 and 20 per 100 person-years, depending on age, geographic location, and intensity of insulin treatment.[42]

Mortality

Insulin treatment dramatically prolongs survival but it does not cure diabetes. Although the absolute mortality at onset and within the first 20 years of type 1 diabetes is low (3-6%), it is 5 times higher for diabetic males and 12 times higher for diabetic females, compared to the general population.[221] This excess mortality is lowest in Scandinavia, intermediate in the U.S. and highest in countries where type 1 diabetes is rare, e.g., Japan,[222,223] probably due to a combination of the quality of care and access. Even in Finland, at least a half of the death is due to currently preventable causes such as acute complications, infections and suicide.[224] On the other hand, 40% of the patients survive over 40 years and a half of these have no major complications. Survival and avoidance of complications has been related to better metabolic control,[225] however, genetic factors also appear to be involved. Inconsistent associations have been reported between diabetic nephropathy and HLA-DR4[226] and several genes involved in blood pressure regulation.[225,227-233] Polymorphisms of paraoxonase[234,235] and A-IV[236] appear to play an important role in development of coronary artery disease in type 2 diabetes patients, but have not been extensively studied in persons with type 1 diabetes.

Microvascular Complications

Insulin treatment dramatically prolongs survival but it does not cure diabetes. The high incidence, associated severe morbidity,[5,237] mortality,[238] and enormous health care expenditures[239-241] make T1DM a prime target for interventions.

The Diabetes Control and Complications Trial demonstrated the efficacy of tight glycemic control in reducing the risk of late complications of type 1 diabetes.[242-244] Unfortunately, increased risk of hypoglycemia associated with tight control[245,246] and compliance problems have hampered full implementation of the trial recommendations, especially in children. Importantly, studies including ours have shown that factors other than hyperglycemia contribute significantly to development of microvascular complications.

Diabetic retinopathy: Diabetes is the most common cause of new cases of blindness in the U.S. adults. After 15+ years of T1DM, 80% of the patients have diabetic retinopathy and 25% have vision-threatening proliferative diabetic retinopathy (PDR).[247] Fundus photography is an effective method for detecting; however, this technique is used to detect disease associated with anatomic changes in the retina. Fluorescein angiography may detect earlier changes, such as breakdown of the blood-retinal barrier manifested by leakage of fluorescein, however, it is costly and associated with complications and is generally used to evaluate more severe diabetic retinopathy, e.g., macular edema, prior to laser photocoagulation. Novel sensitive and specific markers are needed to detect meaningful early changes predictive of incidence and progression of diabetic retinopathy and subsequent visual loss. Such markers could also potentially shorten the duration of clinical trials or allow smaller sample sizes by allowing more precise detection of regression or progression of diabetic retinopathy.

Alterations in retinal blood flow and loss of retinal pericytes[248] precede the earliest clinical stages of diabetic retinopathy. In contrast to the loss of retinal pericytes, changes in retinal arterial blood flow can be detected noninvasively using laser Doppler velocimetry,[249] video fluorescein angiography[250] or pulsatile ocular blood flow.[251] T1 DM patients with no retinopathy or mild NPDR show dilated major retinal arteries with reduced blood flow velocity, compared to that in nondiabetic controls.[250,252] With increasing duration of diabetes and progression of retinopathy, retinal blood flow shows a bimodal pattern of a transition from reduced to increased.[249,253]

Diabetic nephropathy: Historically, diabetic nephropathy would affect 30-50% of T1DM patients throughout the initial 20 years of the disease.[254-256] While there has been an evidence for a secular decline in the incidence of DM in T1DM,[255,257,258] this decline may have leveled off[259] and the cumulative risk remains at least 20%-30%. The incidence peaks 10-15 years after diagnosis and then declines, suggesting that only a subset of the patients is susceptible to diabetic nephropathy. Increased urinary albumin excretion or microalbuminuria (MA) has been shown to be an early manifestation of diabetic nephropathy, rather than merely a prognostic factor.[260-262] Since MA strongly predicts progression to later stages of diabetic nephropathy (overt proteinuria and end-stage renal disease),[263,264] current standards of care include screening for MA and treatment of those positive for MA with ACE inhibitors.[265-268] However, prediction of the progression to overt nephropathy based on the presence of MA and the other risk factors is imprecise. In addition, some patients in whom MA has not increased may nonetheless develop advanced renal lesions.[269] Previously, increased GFR,[270,271] increased ambulatory blood pressure,[272,273] and autonomic neuropathy[273,274] have shown some promise but have not gained wide acceptance.

Recent reports from patients with type 1 and type 2 diabetes suggest that podocyte loss is associated with progression of diabetic glomerulosclerosis.[275-279]

Risk Factors for Microvascular Complications in Type 1 Diabetes

Traditional epidemiological distinction between micro- and macrovascular complictations of diabetes is becoming more fluid with the recognition that similar processes (e.g., glycation, hypertension, inflammation, and endothelial dysfunction) affect both vascular beds. Accurate markers of early retinal microvascular alterations may offer a powerful predictive tool concerning the risk of cumulative vascular burden and future micro- and macrovascular clinical events elsewhere in the body.

Hyperglycemia: Chronic hyperglycemia is probably the most accepted risk factor for development of microvascular complications in type 1 diabetes mellitus.[242,280-283]

Blood pressure: Numerous previous studies have demonstrated that increased systolic or diastolic blood pressure is a powerful predicator of microvascular complications.[284-287]

Cigarette smoking is an established risk factor for diabetic nephropathy.[285,288,289] Interestingly, smoking does not appear to be an independent risk factor for diabetic retinopathy.[290]

Lipids and lipoproteins: Elevated TG and low HDL-cholesterol,[286,291,292] and increased total and LDL-cholesterol[286,293,294] have been reported to predict diabetic retinopathy and diabetic nephropathy.

Insulin resistance is an emergent risk factors for both macro- and microvascular complications in type 1 DM.[291,292] However, it is unclear if this phenomenon is mediated by associated risk factors[287] or a common genetic determinant.[295,296]

Visceral obesity, PAI-1 and markers of inflammation and endothelial dysfunction tend to cluster with microalbuminuria and insulin resistance in diabetic and nondiabetic persons and are clearly associated with macrovascular complications. The relationship of these markers to microvascular diabetic complications has been less studied. Visceral obesity,[291,292] vWF and fibrinogen[286,297] as well as PAI-1 levels[298] have been found to predict diabetic nephropathy and/or diabetic retinopathy. Most of the previous studies were cross-sectional and none has measured visceral obesity directly or analyzed all of these factors jointly. This prospective study will be in a unique position to shed new light on this cluster of novel microvascular risk factors.

Age at onset of diabetes, C-peptide: Greater severity of initial presentation, related largely to the HLA-DR,DQ genotype and age at diagnosis, predicts the faster loss of endogenous insulin production and poorer glycemic control. Lower endogenous insulin production (marked by undetectable C-peptide levels and associated with higher HbA1c) has been shown to predict progression of diabetic retinopathy[299] and diabetic nephropathy.[300] These observations are consistent with the association of microvascular complications with HLA-DR3/4 DR3/4 genotype[301,302] and faster loss of GAD autoantibodies.[303]

Genetics of Microvascular Complications

Despite the seemingly inevitable development of complications, at least half of the patients survive over 40 years and a quarter of these have no major complications.[304] Retinopathy develops in virtually all patients, given enough time, but the more severe proliferative retinopathy and visual impairment appear in only up to half of the patients.[304] Survival and avoidance of complications is related to better metabolic control, but patients developing microvascular complications, especially nephropathy, frequently have no identifiable classical risk factors, suggesting that the genetic component of renal disease is different from that of diabetes.

Diabetic nephropathy clearly clusters in certain families, apparently independent of glycemic control and both in type 1[305-307] and type 2 diabetes.[308-313] Familial clustering of diabetic nephropathy in T1DM families has been confirmed in a study incorporating the results of kidney biopsies in addition to more common measures of renal function such as urinary AER.[314] The clustering is more pronounced in African Americans than Caucasians.[309,311,315]

In segregation analysis of the adenine/creatinine ratio as a continuous trait in type 1 diabetes, adjusting for age, sex, and duration of diabetes, adenine/creatinine inheritance was most consistent with a Mendelian model with multifactorial inheritance,[316] i.e., determined by a mixture of genes, variable effects, and environment factors such as diabetes and hypertension. adenine/creatinine ratio heritability (*h2*) was estimated to be 0.27.

Inconsistent associations have been reported between diabetic nephropathy and HLA-DR4[317] and several genes involved in blood pressure regulation.[304,318-322] ApoE genotype may mediate well known association between dyslipidemia and development of diabetic nephropathy.[323-325] Matrix metalloproteinase-9 polymorphism may play a role[326] Additional genes (IL1RN, NHE5, NOS1, KLKB1, RAGE,) may be minor contributors to genetic risk. Extensive analysis of the chromosome 3 region suggested strong evidence of a nephropathy gene, but a detailed evaluation of the AGTR1 gene suggested that alleles of AGTR1 were not the source of the linkage.

The genetic components of retinopathy and other microvascular complications have been investigated to a lesser degree than nephropathy. Diabetic retinopathy clusters in families.[307]

Diabetic retinopathy has been associated with the presence of the HLA-DR3/4 genotype.[301,302] The associations between diabetic nephropathy, diabetic retinopathy, and markers of insulin resistance may be mediated by a polymorphism in the PC-1 gene coding region.[295,296]

Coronary Artery Disease

Coronary artery disease is the main cause of death in persons with type 1 diabetes and accounts for a large proportion of premature morbidity and mortality in the general population. Heart disease in type 1 diabetic patients occurs earlier in life, affects women as often as men, and associated mortality is dramatically higher than that in the general population.[327-330] Women with type 1 diabetes are 9 to 29 times more likely to die of coronary artery disease than nondiabetic women; the risk for men is increased 4 to 9-fold. Patients with proteinuria are at a 15-37 times increased risk of fatal coronary artery disease while the risk of those without proteinuria is 3-4-fold, compared to the general population.[327,331] However, it is far from clear whether the association of coronary artery disease with nephropathy is mediated by hypertension and dyslipidemia — features of renal failure — or rather by risk factors that predispose to both nephropathy and coronary artery disease. While conventional coronary artery disease risk factors (hypertension, smoking, low HDL cholesterol and high triglycerides) increase the risk, the role of hyperglycemia, autonomic neuropathy, endothelial dysfunction, insulin resistance and diabetes duration is less established.[332-335] In type 1 diabetic patients, atherosclerosis is more diffuse,[336,337] leading to higher case fatality,[338,339] higher cardiac failure[340] and restenosis rates,[341] and shorter survival,[341,342] compared to the general population. These poor outcomes emphasize the need for primary prevention of coronary artery disease in type 1 diabetic patients. Silent ischemia is common — 24% of asymptomatic patients older than 35 yrs had ischemia on exercise test, Holter monitoring or dynamic thallium scintigraphy and 10% had coronary stenosis greater than 50% by angiography.[343] Small clinical studies using B-mode imaging of carotid arteries have suggested that type 1 diabetic patients have significant atherosclerosis as early as at the age of 10-19 yrs and strongly associated with diabetes duration.[344,345] However, very little is know about the prevalence of subclinical coronary artery disease and risk factors for progression to clinical endpoints.

Summary

Type 1 diabetes can be diagnosed at any age, but clinical course, genetic, and environmental determinants appear to be heterogenous by age. The common pathway begins with preclinical β-cell autoimmunity with progressive defect of insulin secretion, followed by onset of hyperglycemia, transient usually partial remission, and finally complete insulinopenia associated with acute and chronic complications and premature death. Current research effort is focused on identification of the genetic and environmental determinants of this process and the ways they interact.

References

1. Rewers M. The changing face of the epidemiology of insulin-dependent diabetes mellitus (IDDM): Research designs and models of disease causation. Ann Med 1991; 23:419-426.
2. Libman I, Songer T, LaPorte R. How many people in the U.S. have IDDM? Diabetes Care 1993; 16:841-842.
3. Rewers M, LaPorte RE, King H et al. Trends in the prevalence and incidence of diabetes: Insulin-dependent diabetes mellitus in childhood. World Health Stat Q 1988; 41:179-189.
4. Patrick SL, Moy CS, LaPorte RE. The world of insulin-dependent diabetes mellitus: What international epidemiologic studies reveal about the etiology and natural history of IDDM. Diabetes Metab Rev 1989; 5:571-578.
5. Diabetes in America 1995; 2.
6. Songer TJ. Health services and costing in diabetes. Med J Aust 1990; 152:115-117.
7. US Public Health Service. Healthy people 2000: National health promotion and disease prevention objectives. MD, USA: US Department of Health and Human Services, 1991.
8. Skyler JS, Crofford OB, Dupre J et al. Prevention of Type 1 diabetes mellitus. Diabetes 1990; 39:1151-1152.

9. American Diabetes Association. Report of the Expert Committee on the Diagnosis and Classification of Diabetes Mellitus. Diabetes Care 1997; 20:1183-1197.
10. Alberti KG, Zimmet PZ. Definition, diagnosis and classification of diabetes mellitus and its complications. Part 1: Diagnosis and classification of diabetes mellitus provisional report of a WHO consultation. Diabet Med 1998; 15:539-553.
11. Atkinson MA, Eisenbarth GS. Type 1 Diabetes: New perspectives on disease pathogenesis and treatment. Lancet 2000.
12. Maugendre D, Chaillous L, Rohmer V et al. Multiple antibody status in type 1 diabetic patients and subjects at various risk with islet-cell antibodies. Diabetes Metab 1997; 23:320-326.
13. Imagawa A, Hanafusa T, Miyagawa J et al. A novel subtype of type 1 diabetes mellitus characterized by a rapid onset and an absence of diabetes-related antibodies. Osaka IDDM Study Group. New Engl J Med 2000; 342:301-307.
14. Bottazzo GF, Florin-Christensen A, Doniach D. Islet-cell antibodies in diabetes mellitus with autoimmune polyendocrine deficiencies. Lancet 1974; 2:1279-1283.
15. Palmer JP, Asplin CM, Clemons P et al. Insulin antibodies in insulin-dependent diabetics before insulin treatment. Science 1983; 222:1337-1339.
16. Vardi P, Dib SA, Tuttleman M et al. Competitive insulin autoantibody assay. Prospective evaluation of subjects at high risk for development of type I diabetes mellitus. Diabetes 1987; 36:1286-1291.
17. Williams AJK, Bingley PJ, Bonifacio E et al. A novel micro-assay for insulin autoantibodies. J Autoimmun 1997; 10:473-478.
18. Baekkeskov S, Nielsen JH, Marner B et al. Autoantibodies in newly diagnosed diabetic children immunoprecipitate specific human pancreatic islet cell proteins. Nature 1982; 298:167-169.
19. Baekkeskov S, Aanstoot H-J, Christgau S et al. Identification of the 64K autoantigen in insulin-dependent diabetes as the GABA-synthesizing enzyme glutamic acid decarboxylase [published erratum appears in Nature 1990 Oct 25;347(6295):782]. Nature 1990; 347:151-156.
20. Rabin DU, Pleasic SM, Shapiro JA et al. Islet cell antigen 512 is a diabetes-specific islet autoantigen related to protein tyrosine phosphatases. J Immunol 1994; 152:3183-3188.
21. Verge CF, Gianani R, Kawasaki E et al. Prediction of type I diabetes in first-degree relatives using a combination of insulin, GAD, and ICA512bdc/IA-2 autoantibodies. Diabetes 1996; 45:926-933.
22. Bingley PJ, Christie MR, Bonifacio E et al. Combined analysis of autoantibodies improves prediction of IDDM in islet cell antibody-positive relatives. Diabetes 1994; 43:1304-1310.
23. Bingley PJ, Bonifacio E, Williams AJ et al. Prediction of IDDM in the general population: Strategies based on combinations of autoantibody markers. Diabetes 1997; 46:1701-1710.
24. Bonifacio E, Bingley PJ, Shattock M et al. Quantification of islet-cell antibodies and prediction of insulin-dependent diabetes. Lancet 1990; 335:147-149.
25. Johnston C, Millward BA, Hoskins P et al. Islet-cell antibodies as predictors of the later development of type 1 (insulin-dependent) diabetes. A study in identical twins. Diabetologia 1989; 32:382-386.
26. McCulloch DK, Palmer JP. The appropriate use of B-cell function testing in the preclinical period of Type 1 diabetes. Diabetic Med 1991; 8:800-804.
27. Spinas GA, Matter L, Wilkin T et al. Islet-cell and insulin autoantibodies in first-degree relatives of type I diabetics: A 5-year follow-up study in a Swiss population. Adv Exp Med Biol 1988; 246:209-214.
28. Thivolet C, Beaufrere B, Betuel H et al. Islet cell and insulin autoantibodies in subjects at high risk for development of type 1 (insulin-dependent) diabetes mellitus: The Lyon family study. Diabetologia 1988; 31:741-746.
29. Maclaren N, Horne G, Spillar R et al. The feasibility of using ICA to predict IDDM in US school children. Diabetes 1990; 39 (suppl1):122A.
30. Landin-Olsson M, Palmer JP, Lernmark A et al. Predictive value of islet cell and insulin autoantibodies for Type 1 (insulin-dependent) diabetes mellitus in a population- based study of newly-diagnosed diabetic and matched control children. Diabetologia 1992; 35:1068-1073.
31. Karjalainen JK. Islet cell antibodies as predictive markers for IDDM in children with high background incidence of disease. Diabetes 1990; 39:1144-1150.
32. Yu J, Yu L, Bugawan TL et al. Transient anti-islet autoantibodies: Infrequent occurrence and lack of association with genetic risk factors. J Clin Endocrinol Metab 2000; 85:2421-2428.
33. Lindberg BA, Ericsson UB, Kockum I et al. Prevalence of beta-cell and thyroid autoantibody positivity in schoolchildren during three-year follow-up. Autoimmunity 1999; 31:175-185.
34. Yu J, Yu L, Bugawan TL et al. Transient anti-islet autoantibodies: Infrequent occurrence and lack of association with genetic risk factors. J Clin Endocrinol Metab 2000; 85:2421-2428.

35. Riley WJ, Maclaren NK, Krischer J et al. A prospective study of the development of diabetes in relatives of patients with insulin-dependent diabetes. N Engl J Med 1990; 323:1167-1172.
36. Groop LC, Bottazzo GF, Doniach D. Islet cell antibodies identify latent Type I Diabetes in patients aged 35-75 years at diagnosis. Diabetes 1986; 35:237-241.
37. Tuomi T, Groop LC, Zimmet PZ et al. Antibodies to glutamic acid decarboxylase reveal latent autoimmune diabetes mellitus in adults with a noninsulin- dependent onset of disease. Diabetes 1993; 42:359-362.
38. Zimmet PZ, Tuomi T, Mackay IR et al. Latent autoimmune diabetes mellitus in adults (LADA): The role of antibodies to glutamic acid decarboxylase in diagnosis and prediction of insulin dependency. Diabet Med 1994; 11:299-303.
39. Hamman RF, Cook M, Keefer S et al. Medical care patterns at the onset of insulin-dependent diabetes mellitus: Association with severity and subsequent complications. Diabetes Care 1985; 8 (Suppl1):94-100.
40. Karjalajnen S, Salema P, Ilonen J. A comparison of childhood and adult type 1 diabetes mellitus. N Engl J Med 1989; 320:881-886.
41. Levy-Marchal C, Papoz L, de Beaufort C et al. Clinical and laboratory features of type I diabetic children at time of diagnosis. Diabetic Med 1992; 9:279-284.
42. Pinkney JH, Bingley PJ, Sawtell PA et al. Presentation and progress of childhood diabetes mellitus: A prospective population-based study. Diabetologia 1994; 37:70-74.
43. Eberhardt MS, Wagener DK, Orchard TJ et al. HLA heterogeneity of insulin-dependent diabetes mellitus at diagnosis. The Pittsburgh IDDM Study. Diabetes 1985; 34:1247-1252.
44. Foulis AK, Liddle CN, Farquharson MA et al. The histopathology of the pancreas in type I diabetes (insulin dependent) mellitus: A 25-year review of deaths in patients under 20 years of age in the United Kingdom. Diabetologia 1986; 29:267-274.
45. Childhood Diabetes Research Committee, Polish Diabetes Research Group, The Netherlands Institute for Preventive Health Care, Diabetes Research Center of Pittsburgh. How frequently do children die at the onset of insulin-dependent diabetes? Analyses of registry data from Japan, Poland, the Netherlands, and Allegheny County, Pennsylvania. Diab Nutr Metab 1990; 3:57-62.
46. Kostraba JN, Gay EC, Rewers M et al. Increasing trend of outpatient management of children with newly diagnosed IDDM. Colorado IDDM registry, 1978-88. Diabetes Care 1992; 15:95-100.
47. Rewers M, LaPorte RE, King H et al. Trends in the prevalence and incidence of diabetes: Insulin-dependent diabetes mellitus in childhood. World Health Stat Q 1988; 41:179- 189.
48. Libman I, Songer T, LaPorte R. How many people in the U.S. have IDDM? Diabetes Care 1993; 16:841-842.
49. Diabetes Epidemiology Research International Group. Geographic patterns of childhood insulin-dependent diabetes mellitus. Diabetes 1988; 37:1113-1119.
50. Green A, Gale EAM, Patterson CC. Incidence of childhood-onset insulin-dependent diabetes mellitus: The EURODIAB ACE Study. Lancet 1992; 339:905-909.
51. Karvonen M, Viik-Kajander M, Moltchanova E et al. Incidence of childhood type 1 diabetes worldwide. Diabetes Mondiale (DiaMond) Project Group. Diabetes Care 2000; 23:1516-1526.
52. Rewers M, Kostraba JN. Epidemiology of Type I diabetes. 1994.
53. Adojaan B, Knip M, Vahasalo P et al. Relationship between the incidence of childhood IDDM and the frequency of ICA positivity in nondiabetic children in the general population. Diabetes Care 1996; 19:1452-1454.
54. Pilcher CC, Dickens K, Elliott RB. ICA only develop in early childhood. Diabetes Res Clin Pract 1991; 14(Suppl. 1):s82-s82.
55. Tajima N, LaPorte RE, Hibi I et al. A comparison of the epidemiology of youth-onset insulin-dependent diabetes mellitus between Japan and the United States (Allegheny County, Pennsylvania). Diabetes Care 1985; 8(Suppl1) :17-23.
56. Rewers M, Stone RA, LaPorte RE et al. Poisson regression modeling of temporal variation in incidence of childhood insulin-dependent diabetes mellitus in Allegheny County, Pennsylvania, and Wielkopolska, Poland, 1970-1985. Am J Epidemiol 1989; 129:569-581.
57. Pinhas-Hamiel O, Dolan LM, Daniels SR et al. Increased incidence of noninsulin-dependent diabetes mellitus among adolescents. J Pediatrics 1996; 128:(Pt1):608-615.
58. Rosenbloom AL, Joe JR, Young RS et al. Emerging epidemic of type 2 diabetes in youth. Diabetes Care 1999; 22:345-354.
59. Dabelea D, Hanson RL, Bennett PH et al. Increasing prevalence of Type II diabetes in American Indian children. Diabetologia 1998; 41:904-910.
60. Joner G, Sovik O. The incidence of type 1 (insulin-dependent) diabetes mellitus 15- 29 years in Norway 1978-1982. Diabetologia 1991; 34:271-274.

61. Muntoni S, Songini M. High incidence rate of IDDM in Sardinia. Sardinian Collaborative Group for Epidemiology of IDDM. Diabetes Care 1992; 15:1317-1322.
62. Bruno G, Merletti F, Vuolo A et al. Sex differences in incidence of IDDM in age group 15-29 yr. Higher risk in males in Province of Turin, Italy. Diabetes Care 1993; 16(1):133-136.
63. Christau B, Kromann H, Andersen OO et al. Incidence, seasonal and geographic patterns of juvenile-onset insulin-dependent diabetes mellitus is Denmark. Diabetologia 1977; 13(4):281-284.
64. Melton III LJ , Palumbo PJ, Chu CP. Incidence of diabetes mellitus by clinical type. Diabetes Care 1983; 6:75-86.
65. Turner R, Stratton I, Horton V et al. UKPDS 25: Autoantibodies to islet-cell cytoplasm and glutamic acid decarboxylase for prediction of insulin requirement in type 2 diabetes. UK Prospective Diabetes Study Group [published erratum appears in Lancet 1998 Jan 31;351(9099):376]. Lancet 1997; 350:1288-1293.
66. Horton V, Stratton I, Bottazzo GF et al. Genetic heterogeneity of autoimmune diabetes: Age of presentation in adults is influenced by HLA DRB1 and DQB1 genotypes (UKPDS 43). UK Prospective Diabetes Study (UKPDS) Group. Diabetologia 1999; 42:608-616.
67. Diabetes Epidemiology Research International Group. Geographic patterns of childhood insulin-dependent diabetes mellitus. Diabetes 1988; 37:1113-1119.
68. Kostraba JN, Gay EC, Cai Y et al. Incidence of insulin-dependent diabetes mellitus in Colorado. Epidemiology 1992; 3:232-238.
69. Siemiatycki J, Colle E, Campbell S et al. Case-control study of IDDM. Diabetes Care 1989; 12:209-216.
70. Zimmet PZ, Rowley MJ, Mackay IR et al. The ethnic distribution of antibodies to glutamic acid decarboxylase: Presence and levels of insulin-dependent diabetes mellitus in Europid and Asian subjects. J Diabetes Complications 1993; 7:1-7.
71. Gale EA, Gillespie KM. Diabetes and gender. Diabetologia 2001; 44:3-15.
72. Karvonen M, Pitkaniemi M, Pitkaniemi J et al. Sex difference in the incidence of insulin-dependent diabetes mellitus: An analysis of the recent epidemiological data. World Health Organization DIAMOND Project Group. Diabetes Metab Rev 1997; 13:275-291.
73. Serban V, Timar R, Dabelea D et al. The epidemiology of childhood-onset type 1 diabetes mellitus in Romania. ONROCAD Study Group. National Romanian Organisation for the Care of Diabetic Children and Adolescents. J Pediatr Endocrinol Metab 2001; 14:535-541.
74. Karvonen M, Pitkaniemi M, Pitkaniemi J et al. Sex difference in the incidence of insulin-dependent diabetes mellitus: An analysis of the recent epidemiological data. World Health Organization DIAMOND Project Group. Diabetes Metab Rev 1997; 13:275-291.
75. Green A, Hauge M, Holm NV et al. Epidemiological studies of diabetes mellitus in Denmark. II. A prevalence study based on insulin prescriptions. Diabetologia 1981; 20:468-470.
76. Bruno G, Merletti F, Vuolo A et al. Sex differences in incidence of IDDM in age group 15-29 yr. Higher risk in males in Province of Turin, Italy. Diabetes Care 1993; 16(1):133-136.
77. Ustvedt HJ, Olsen E. Incidence of diabetes mellitus in Oslo, Norway 1956-65. Br J Prev Soc Med 1977; 31:251-257.
78. Roglic G, Pavlic-Renar I, Sestan-Crnek S et al. Incidence of IDDM during 1988-1992 in Zagreb, Croatia. Diabetologia 1995; 38:550-554.
79. Vandewalle CL, Coeckelberghs MI, De Leeuw IH et al. Epidemiology, clinical aspects, and biology of IDDM patients under age 40 years. Comparison of data from Antwerp with complete ascertainment with data from Belgium with 40% ascertainment. The Belgian Diabetes Registry. Diabetes Care 1997; 20:1556-1561.
80. Risch N, Ghosh S, Todd JA. Statistical evaluation of multiple-locus linkage data in experimental species and its relevance to human studies: Application to nonobese diabetic (NOD) mouse and human insulin- dependent diabetes mellitus (IDDM). Am J Hum Genet 1993; 53:702-714.
81. Gale EA, Gillespie KM. Diabetes and gender. Diabetologia 2001; 44:3-15.
82. Gamble DR. The epidemiology of insulin-dependent diabetes, with particular reference to the relationship of virus infection to its etiology. Epidemiol Rev 1980; 2:49-70.
83. Kostraba JN, Gay EC, Cai Y et al. Incidence of insulin-dependent diabetes mellitus in Colorado. Epidemiology 1992; 3:232-238.
84. Dahlquist G, Gustavsson KH, Holmgren G et al. The incidence of diabetes mellitus in Swedish children 0-14 years of age. A prospective study 1977-1980. Acta Paediatr Scand 1982; 71:7-14.
85. Weinberg CR, Dornan TL, Hansen JA et al. HLA-related heterogeneity in seasonal patterns of diagnosis in type I (insulin-dependent) diabetes. Diabetologia 1984; 26(3):199-202.
86. Ludvigsson J, Afoke AO. Seasonality of type 1 (insulin-dependent) diabetes mellitus: Values of C-peptide, insulin antibodies and haemoglobin A_{1c} show evidence of a more rapid loss of insulin secretion in epidemic patients. Diabetologia 1989; 32:84-91.

87. Rewers M, LaPorte R, Walczak M et al. Apparent epidemic of insulin-dependent diabetes mellitus in midwestern Poland. Diabetes 1987; 36:106-113.
88. Diabetes Epidemiology Research International Group. Secular trends in incidence of childhood IDDM in 10 countries. Diabetes 1990; 39:858-864.
89. Nystrom L, Dahlquist G, Rewers M et al. The Swedish childhood diabetes study: An analysis of the temporal variation in diabetes incidence, 1978-1987. Int J Epidemiol 1990; 19:141-146.
90. Tuomilehto J, Rewers M, Reunanen A et al. Increasing trend in type I (insulin-dependent) diabetes mellitus in childhood in Finland. Analysis of age, calendar time, and birth cohort effects during 1965 to 1984. Diabetologia 1991; 34:282-287.
91. Dokheel TM. An epidemic of childhood diabetes in the United States? Diabetes Care 1993; 16:1606-1611.
92. . Variation and trends in incidence of childhood diabetes in Europe. EURODIAB ACE Study Group. Lancet 2000; 355:873- 876.
93. Karvonen M, Pitkaniemi J, Tuomilehto J. The onset age of type 1 diabetes in Finnish children has become younger. The Finnish Childhood Diabetes Registry Group. Diabetes Care 1999; 22:1066-1070.
94. Rosenbauer J, Herzig P, von Kries R et al. Temporal, seasonal, and geographical incidence patterns of type I diabetes mellitus in children under 5 years of age in Germany. Diabetologia 1999; 42:1055-1059.
95. Cinek O, Lanska V, Kolouskova S et al. Type 1 diabetes mellitus in Czech children diagnosed in 1990-1997: A significant increase in incidence and male predominance in the age group 0-4 years. Collaborators of the Czech Childhood Diabetes Registry. Diabetic Med 2000; 17:64-69.
96. Gardner SG, Bingley PJ, Sawtell PA et al. Rising incidence of insulin dependent diabetes in children aged under 5 years in the Oxford region: Time trend analysis. The Bart's-Oxford Study Group [see comments]. BMJ 1997; 315:713-717.
97. Schoenle EJ, Lang-Muritano M, Gschwend S et al. Epidemiology of type I diabetes mellitus in Switzerland: Steep rise in incidence in under 5 year old children in the past decade. Diabetologia 2001; 44:286-289.
98. Cordell HJ, Todd JA. Multifactorial inheritance in type 1 diabetes. [Review]. Trends Genet 1995; 11:499-504.
99. Rewers M, Stone RA, LaPorte RE et al. Poisson regression modeling of temporal variation in incidence of childhood insulin-dependent diabetes mellitus in Allegheny County, Pennsylvania, and Wielkopolska, Poland, 1970-1985. Am J Epidemiol 1989; 129:569-581.
100. Tuomilehto J, Rewers M, Reunanen A et al. Increasing trend in type 1 (insulin-dependent) diabetes mellitus in childhood in Finland. Analysis of age, calendar time and birth cohort effects during 1965 to 1984. Diabetologia 1991; 34:282-287.
101. Nystrom L, Dahlquist G, Rewers M et al. The Swedish childhood diabetes study. An analysis of the temporal variation in diabetes incidence 1978-1987. Int J Epidemiol 1990; 19:141-146.
102. Bruno G, Merletti F, Biggeri A et al. Increasing trend of type I diabetes in children and young adults in the province of Turin (Italy). Analysis of age, period and birth cohort effects from 1984 to 1996. Diabetologia 2001; 44:22-25.
103. Onkamo P, Vaananen S, Karvonen M et al. Worldwide increase in incidence of Type I diabetes—the analysis of the data on published incidence trends. Diabetologia 1999; 42:1395-1403.
104. Wagener DK, Sacks JM, LaPorte RE et al. The Pittsburgh study of insulin-dependent diabetes mellitus. Risk for diabetes among relatives of IDDM. Diabetes 1982; 31:136-144.
105. Allen C, Palta M, D'Alessio DJ. Risk of diabetes in siblings and other relatives of IDDM subjects. Diabetes 1991; 40:831-836.
106. Lorenzen T, Pociot F, Hougaard P et al. Long-term risk of IDDM in first-degree relatives of patients with IDDM. Diabetologia 1994; 37:321-327.
107. O'Leary LA, Dorman JS, LaPorte RE et al. Familial and sporadic insulin-dependent diabetes: Evidence for heterogeneous etiologies? Diab Res Clin Pract 1991; 14:183-190.
108. Erlich HA, Griffith RL, Bugawan TL et al. Implication of specific DQB1 alleles in genetic susceptibility and resistance by identification of IDDM siblings with novel HLA-DQB1 allele and unusual DR2 and DR1 haplotypes. Diabetes 1991; 40:478-481.
109. Bertrams J, Baur MP. Insulin-dependent diabetes mellitus.1984; 348-358.
110. Todd JA. The role of MHC class II genes in susceptibility to insulin- dependent diabetes mellitus. Curr Top Microbiol Immunol 1990; 164:17-40.
111. Rewers M, Bugawan TL, Norris JM et al. Newborn screening for HLA markers associated with IDDM: Diabetes Autoimmunity Study in the Young (DAISY). Diabetologia 1996; 39:807-812.

112. Pugliese A, Gianani R, Moromisato R et al. HLA-DQB1*0602 is associated with dominant protection from diabetes even among islet cell antibody-positive first-degree relatives of patients with IDDM. Diabetes 1995; 44:608-613.
113. Huang W, She J-X, Muir A et al. High risk HLA-DR/DQ genotypes for IDD confer susceptibility to autoantibodies but DQB1*0602 does not prevent them. J Autoimmun 1994; 7:889-897.
114. Greenbaum CJ, Cuthbertson D, Eisenbarth GS et al. Islet cell antibody positive relatives with HLA-DQA1*0102, DQB1*0602: Identification by the Diabetes Prevention Trial-1. J Clin Endocrinol Metab 2000; 85:1255-1260.
115. Chase HP, Voss MA, Butler-Simon N et al. Diagnosis of pretype I diabetes. J Pediatr 1987; 111:807-812.
116. Levy-Marchal C, Tichet J, Fajardy I et al. Follow-up of children from a background population with high ICA titres. Diabetologia 1992; 35(Suppl1):A32-A32.
117. Owerbach D, Gabbay KH. Localization of a Type I diabetes susceptibilty locus to the variable tandem repeat region flanking the insulin gene. Diabetes 1993; 42:1708-1714.
118. Davies JL, Kawaguchi Y, Bennett ST et al. A genome-wide search for human type 1 diabetes susceptibility genes. Nature 1994; 371:130-136.
119. Hashimoto L, Habita C, Beressi JP et al. Genetic mapping of a susceptibility locus for insulin-dependent diabetes mellitus on chromosome 11q. Nature 1994; 371:161-164.
120. Owerbach D, Gabbay KH. The HOXD8 locus (2q31) is linked to type I diabetes: Interaction with chromosome 6 and 11 disease susceptibility genes. Diabetes 1995; 44:132-136.
121. Field LL, Tobias R, Magnus T. A locus on chromosome 15q26(IDDM3) produces susceptibility to insulin-dependent diabetes mellitus. Nat Genet 1996; 8:189-194.
122. Copeman JB, Cucca F, Hearne CM et al. Linkage disequilibrium mapping of type 1 diabetes susceptibility gene *(IDDM7)* to chromosome 2q31-q33. Nat Genet 1995; 9:80-85.
123. Cox NJ, Wapelhorst B, Morrison VA et al. Seven regions of the genome show evidence of linkage to type 1 diabetes in a consensus analysis of 767 multiplex families. Am J Hum Genet 2001; 69:820-830.
124. Dahlman I, Eaves IA, Kosoy R et al. Parameters for reliable results in genetic association studies in common disease. Nat Genet 2002; 30:149-150.
125. Kaprio J, Tuomilehto J, Koskenvuo M et al. Concordance for type 1 (insulin-dependent) and type 2 (noninsulin-dependent) diabetes mellitus in a population-based cohort of twins in Finland. Diabetologia 1992; 35:1060-1067.
126. Pak CY, Hyone-Myong E, McArthur RG et al. Association of cytomegalovirus infection with autoimmune type 1 diabetes. Lancet 1988; 2:1-3.
127. Sairenji T, Daibata M, Sorli CH et al. Relating homology between the Epstein-Barr virus BOLF1 and HLA-DQw8 beta chain to recent onset type 1 -dependent) diabetes mellitus. Diabetologia 1991; 34:33-39.
128. Helmke K, Otten A, Willems WR et al. Islet cell antibodies and the development of diabetes mellitus in relation to mumps infection and mumps vaccination. Diabetologia 1986; 29:30-33.
129. Hyoty H, Hiltunen M, Reunanen A et al. Decline of mumps antibodies in type 1 (insulin-dependent) diabetic children and a plateau in the rising incidence of type 1 diabetes after introduction of the mumps-measles-rubella vaccine in Finland. Childhood Diabetes in Finland Study Group. Diabetologia 1993; 36:1303-1308.
130. Ginsberg-Fellner F, Witt ME, Yagihashi S et al. Congenital rubella syndrome as a model for type 1 (insulin- dependent) diabetes mellitus: Increased prevalence of islet cell surface antibodies. Diabetologia 1984; 27 Suppl:87-89.
131. Menser MA, Forrest JM, Bransby RD. Rubella infection and diabetes mellitus. Lancet 1978; 1:57-60.
132. Suenaga K, Yoon JW. Association of beta-cell-specific expression of endogenous retrovirus with development of insulitis and diabetes in NOD mouse. Diabetes 1988; 37:1722-1726.
133. Conrad B, Weidmann E, Trucco G et al. Evidence for superantigen involvement in insulin-dependent diabetes mellitus aetiology. Nature 1994; 371:351- 355.
134. Honeyman MC, Coulson BS, Stone NL et al. Association between rotavirus infection and pancreatic islet autoimmunity in children at risk of developing type 1 diabetes. Diabetes 2000; 49:1319-1324.
135. Oldstone MB, Nerenberg M, Southern P et al. Virus infection triggers insulin-dependent diabetes mellitus in a transgenic model: Role of anti-self (virus) immune response. Cell 1991; 65:319-331.
136. Ohashi PS, Oehen S, Buerki K et al. Ablation of "tolerance" and induction of diabetes by virus infection in viral antigen transgenic mice. Cell 1991; 65:305-317.
137. Bodansky HJ, Dean BM, Bottazzo GF et al. Islet-cell antibodies and insulin autoantibodies in association with common viral infections. Lancet Dec 1986; 13:1351-1353.

138. Champsaur HF, Bottazzo GF, Bertrams J et al. Virologic, immunologic, and genetic factors in insulin-dependent diabetes mellitus. J Pediatr 1982; 100:15-20.
139. Uriarte A, Cabrera E, Ventura R et al. Islet cell antibodies and ECHO-4 virus infection. Diabetologia 1987; 30:590A.
140. Blom L, Nystrom L, Dahlquist G. The Swedish childhood diabetes study. Vaccinations and infections as risk determinants for diabetes in childhood. Diabetologia 1991; 34:176-181.
141. Karounos DG, Wolinsky JS, Thomas JW. Monoclonal antibody to rubella virus capsid protein recognizes a β-cell antigen. J Immunol 1993; 150:3080-3085.
142. Rewers M, Atkinson M. The possible role of enteroviruses in diabetes mellitus. 1996; 353-385.
143. Graves PM, Norris JM, Pallansch MA et al. The role of enteroviral infections in the development of IDDM: Limitations of current approaches. Diabetes 1997; 46:161-168.
144. Jenson AB, Rosenberg HS, Notkins AL. Pancreatic islet-cell damage in children with fatal viral infections. Lancet 1980; 16:354-358.
145. Yoon JW, Austin M, Onodera T et al. Virus-induced diabetes mellitus: Isolation of a virus from the pancreas of a child with diabetic ketoacidosis. N Engl J Med 1979; 300:1173-1179.
146. Donsky J, Massad R. Community medicine in the training of family physicians. J Fam Pract 1979; 8:965-971.
147. Wagenknecht LE, Roseman JM, Herman WH. Increased incidence of insulin-dependent diabetes mellitus following an epidemic of coxsackievirus B5. Am J Epidemiol 1991; 133:1024-1031.
148. Kaufman DL, ClareSalzler M, Tian J et al. Spontaneous loss of T-cell tolerance to glutamic acid decarboxylase in murine insulin-dependent diabetes [see comments]. Nature 1993; 366:69-72.
149. Scherbaum WA, Hampl W, Muir P et al. No association between islet cell antibodies and coxsackie B, mumps, rubella and cytomegalovirus antibodies in nondiabetic individuals aged 7-19 years. Diabetologia 1991; 34:835-838.
150. Hyoty H, Hiltunen M, Knip M et al. A prospective study of the role of coxsackie B and other enterovirus infections in the pathogenesis of IDDM. Childhood Diabetes in Finland (DiMe) Study Group. Diabetes 1995; 44:652-657.
151. Lonnrot M, Korpela K, Knip M et al. Enterovirus infection as a risk factor for beta-cell autoimmunity in a prospectively observed birth cohort: The Finnish Diabetes Prediction and Prevention Study. Diabetes 2000; 49:1314-1318.
152. Lonnrot M, Salminen K, Knip M et al. Enterovirus RNA in serum is a risk factor for beta-cell autoimmunity and clinical type 1 diabetes: A prospective study. Childhood Diabetes in Finland (DiMe) Study Group. J Med Virol 2000; 61:214-220.
153. Dahlquist G, Frisk G, Ivarsson SA et al. Indications that maternal coxsackie B virus infection during pregnancy is a risk factor for childhood-onset IDDM. Diabetologia 1995; 38:1371-1373.
154. Dahlquist GG. Viruses and other perinatal exposures as initiating events for beta- cell destruction. Ann Med 1997; 29:413-417.
155. McKinney PA, Parslow R, Gurney KA et al. Perinatal and neonatal determinants of childhood type 1 diabetes. A case-control study in Yorkshire, UK. Diabetes Care 1999; 22:928-932.
156. Dahlquist G, Kallen B. Maternal-child blood group incompatibility and other perinatal events increase the risk for early-onset type 1 (insulin- dependent) diabetes mellitus. Diabetologia 1992; 35:671-675.
157. Blom L, Dahlquist G, Nystrom L et al. The Swedish childhood diabetes study—social and perinatal determinants for diabetes in childhood. Diabetologia 1989; 32:7-13.
158. Bache I, Bock T, Volund A et al. Previous maternal abortion, longer gestation, and younger maternal age decrease the risk of type 1 diabetes among male offspring. Diabetes Care 1999; 22:1063-1065.
159. Bock T, Pedersen CR, Volund A et al. Perinatal determinants among children who later develop IDDM. Diabetes Care 1994; 17:1154-1157.
160. Patterson CC, Carson DJ, Hadden DR et al. A case-control investigation of perinatal risk factors for childhood IDDM in northern Ireland and Scotland. Diabetes Care 1994; 17:376-381.
161. Rothwell PM, Staines A, Smail P et al. Seasonality of birth of patients with childhood diabetes in Britain. BMJ 1996; 312:1456-1457.
162. Oldstone MB. Viruses as therapeutic agents. I. Treatment of nonobese insulin- dependent diabetes mice with virus prevents insulin-dependent diabetes mellitus while maintaining general immune competence. J Exp Med 1990 171:2077-2089.
163. Wilberz S, Partke HJ, Dagnaes-Hansen F et al. Persistent MHV (mouse hepatitis virus) infection reduces the incidence of diabetes mellitus in nonobese diabetic mice. Diabetologia 1991; 34:2-5.
164. Like AA, Guberski DL, Butler L. Influence of environmental viral agents on frequency and tempo of diabetes mellitus in BB/Wor rats. Diabetes 1991; 40:259-262.
165. Kolb H, Elliott RB. Increasing incidence of IDDM a consequence of improved hygiene? Diabetologia 1994; 37:729.

166. McKinney PA, Okasha M, Parslow RC et al. Early social mixing and childhood Type 1 diabetes mellitus: A case- control study in Yorkshire, UK. Diabet Med 2000; 17:236-242.
167. Infections and vaccinations as risk factors for childhood type I (insulin-dependent) diabetes mellitus: A multicentre case-control investigation. EURODIAB Substudy 2 Study Group. Diabetologia 2000; 43:47-53.
168. Dahlquist G, Gothefors L. The cumulative incidence of childhood diabetes mellitus in Sweden unaffected by BCG-vaccination. Diabetologia 1995; 38:873-874.
169. Heijbel H, Chen RT, Dahlquist G. Cumulative incidence of childhood-onset IDDM is unaffected by pertussis immunization. Diabetes Care 1997; 20:173-175.
170. Lindberg B, Ahlfors K, Carlsson A et al. Previous exposure to measles, mumps, and rubella—but not vaccination during adolescence—correlates to the prevalence of pancreatic and thyroid autoantibodies. Pediatrics 1999; 104:e12.
171. Graves PM, Barriga KJ, Norris JM et al. Lack of association between early childhood immunizations and beta-cell autoimmunity. Diabetes Care 1999; 22:1694-1697.
172. Hummel M, Fuchtenbusch M, Schenker M et al. No major association of breast-feeding, vaccinations, and childhood viral diseases with early islet autoimmunity in the German BABYDIAB Study. Diabetes Care 2000; 23 :969-974.
173. Atkinson MA, Winter WE, Skordis N et al. Dietary protein restriction reduces the frequency and delays the onset of insulin dependent diabetes in BB rats. Autoimmunity 1988; 2:11-20.
174. Martin JM, Trink B, Daneman D et al. Milk proteins in the etiology of insulin-dependent diabetes mellitus (IDDM). Ann Med 1991; 23:447-452.
175. Atkinson MA, Bowman MA, Kuo-Jang K et al. Lack of immune responsiveness to bovine serum albumin in insulin- dependent diabetes. New Engl J Med 1993; 329:1853-1858.
176. Lampasona V, Ferrari M, Bosi E et al. Sera from patients with IDDM and healthy individuals have antibodies to ICA69 on western blots but do not immunoprecipitate liquid phase antigen. J Autoimmun 1994; 7:665-674.
177. Martin S, Lampasona V, Dosch M et al. Islet cell autoantigen 69 antibodies in IDDM. Diabetologia 1996; 39:747.
178. Krokowski M, Caillat-Zucman S, Timsit J et al. Anti-bovine serum albumin antibodies: Genetic heterogeneity and clinical relevance in adult-onset IDDM. Diabetes Care 1995; 18:170-173.
179. Norris JM, Beaty B, Klingensmith G et al. Lack of association between early exposure to cow's milk protein and beta-cell autoimmunity: Diabetes Autoimmunity Study in the Young (DAISY). JAMA 1996.
180. Virtanen SM, Hypponen E, Laara E et al. Cow's milk consumption, disease-associated autoantibodies and type 1 diabetes mellitus: A follow-up study in siblings of diabetic children. Childhood Diabetes in Finland Study Group. Diabet Med 1998; 15:730-738.
181. Couper JJ, Steele C, Beresford S et al. Lack of association between duration of breast-feeding or introduction of cow's milk and development of islet autoimmunity. Diabetes 1999; 48:2145-2149.
182. Borch-Johnsen K, Joner G, Mandrup-Poulsen T et al. Relation between breast-feeding and incidence rates of insulin-dependent diabetes mellitus. Lancet 1984; 2:1083-1086.
183. Mayer EJ, Hamman RF, Gay EC et al. Reduced risk of IDDM among breast-fed children. The Colorado IDDM Registry. Diabetes 1988; 37:1625-1632.
184. Kostraba JN, Dorman JS, LaPorte RE et al. Early infant diet and risk of IDDM in blacks and whites. A matched case-control study. Diabetes Care 1992; 15:626-631.
185. Nigro G, Campea L, De Novellis A et al. Breast-feeding and insulin-dependent diabetes mellitus. Lancet 1 1985; 467 (Letter).
186. Blom L, Dahlquist G, Nystrom L et al. The Swedish childhood diabetes study - social and perinatal determinants for diabetes in childhood. Diabetologia 1989; 32:7-13.
187. Kyvik KO, Green A, Svendsen A et al. Breast feeding and the development of type I diabetes mellitus. Diabetic Med 1992; 9:233-235.
188. Samuelsson U, Johansson C, Ludvigsson J. Breast-feeding seems to play a marginal role in the prevention of insulin-dependent diabetes mellitus. Diabetes Res Clin Pract 1993; 19:203-210.
189. Virtanen SM, Rasanen L, Aro A et al. The 'Childhood Diabetes in Finland' Study Group. Feeding in infancy and the risk of type 1 diabetes mellitus in Finnish children. Diabetic Med 1992; 9:815-819.
190. Gerstein HC. Cow's milk exposure and type I diabetes mellitus - a critical overview of the clinical literature. Diabetes Care 1994; 17:13-19.
191. Norris JM, Scott FW. A meta-analysis of infant diet and insulin-dependent diabetes mellitus: Do biases play a role? Epidemiology 1996; 7:87-92.
192. Savilahti E, Akerblom HK, Tainio VM et al. Children with newly diagnosed insulin dependent diabetes mellitus have incresed levels of cow's milk antibodies. Diabetes Res 1988; 7:137-140.

193. Karjalainen J, Martin JM, Knip M et al. A bovine albumin peptide as a possible trigger of insulin-dependent diabetes mellitus [see comments]. N Engl J Med 1992; 327:302-307.
194. Virtanen SM, Laara E, Hypponen E et al. Cow's milk consumption, HLA-DQB1 genotype, and type 1 diabetes: A nested case-control study of siblings of children with diabetes. Childhood diabetes in Finland study group [In Process Citation]. Diabetes 2000; 49:912-917.
195. Paronen J, Knip M, Savilahti E et al. Effect of cow's milk exposure and maternal type 1 diabetes on cellular and humoral immunization to dietary insulin in infants at genetic risk for type 1 diabetes. Finnish Trial to Reduce IDDM in the Genetically at Risk Study Group. Diabetes 2000; 49:1657-1665.
196. Elias D, Prigozin H, Polak N et al. Autoimmune diabetes induced by the beta-cell toxin STZ. Immunity to the 60-kDa heat shock protein and to insulin. Diabetes 1994; 43:992-998.
197. Rayfield EJ, Ishimura K. Environmetal factors and insulin dependent diabetes mellitus. Diabetes Metab Rev 1987; 3:925-957.
198. Helgason T, Jonasson MR. Evidence for a food additive as a cause of ketosis-prone diabetes. Lancet 1981; 2:716-720.
199. Dahlquist GG, Blom LG, Persson L-Å et al. Dietary factors and the risk of developing insulin dependent diabetes in childhood. Br Med J 1990; 300:1302-1306.
200. Kostraba JN, Gay EC, Rewers M et al. Nitrate levels in community drinking waters and risk of IDDM. Diabetes Care 1992; 15:1505-1508.
201. Toniolo A, Onodera T, Yoon JW et al. Induction of diabetes by cumulative environmental insults from viruses and chemicals. Nature 1980; 288:383-385.
202. Khoury MJ, Flanders WD, Greenland S et al. On the measurement of susceptibility in epidemiologic studies. Am J Epidemiol 1989; 129:183-190.
203. Fohlman J, Bohme J, Rask L et al. Matching of host genotype and serotypes of Coxsackie B virus in the development of juvenile diabetes. Scand J Immunol 1987; 26:105-110.
204. D'Alessio DJ. A case-control study of group B Coxsackievirus immunoglobulin M antibody prevalence and HLA-DR antigens in newly diagnosed cases of insulin-dependent diabetes mellitus. Am J Epidemiol 1992; 135:1331-1338.
205. Schernthaner G, Banatvala JE, Scherbaum W et al. Coxsackie-B-virus-specific IgM responses, complement-fixing islet-cell antibodies, HLA DR antigens, and C-peptide secretion in insulin-dependent diabetes mellitus. Lancet 1985; 2:630-632.
206. Kostraba JN, Cruickshanks KJ, Lawler-Heavner J et al. Early exposure to cow's milk and solid foods in infancy, genetic predisposition and risk of IDDM. Diabetes 1993; 42:288-295.
207. Norris JM, Beaty B, Klingensmith G et al. Lack of association between early exposure to cow's milk protein and β-cell autoimmunity: Diabetes Autoimmunity Study in the Young (DAISY). JAMA 1996; 276:609-614.
208. Knip M, Sakkinen A, Huttunen NP et al. Postinitial remission in diabetic children—an analysis of 178 cases. Acta Paediatr Scand 1982; 71:901-908.
209. Agner T, Damm P, Binder C. Remission in IDDM: Prospective study of basal C-peptide and insulin dose in 268 consecutive patients. Diabetes Care 1987; 10:164-169.
210. Sochett EB, Daneman D, Clarson C et al. Factors affecting and patterns of residual insulin secretion during the first year of type 1 (insulin-dependent) diabetes mellitus in children. Diabetologia 1987; 30 :453-459.
211. Ortqvist E, Falorni A, Scheynius A et al. Age governs gender-dependent islet cell autoreactivity and predicts the clinical course in childhood IDDM. Acta Paediatr 1997; 86:1166-1171.
212. Schiffrin A, Suissa S, Poussier P et al. Prospective study of predictors of beta-cell survival in type I diabetes. Diabetes 1988; 37:920-925.
213. Wallensteen M, Dahlquist G, Persson B et al. Factors influencing the magnitude, duration, and rate of fall of B-cell function in Type I (insulin-dependent) diabetic children followed for two years from their clinical diagnosis. Diabetologia 1988; 31:664-669.
214. Sabbah E, Savola K, Kulmala P et al. Diabetes-associated autoantibodies in relation to clinical characteristics and natural course in children with newly diagnosed type 1 diabetes. The Childhood Diabetes In Finland Study Group. J Clin Endocrinol Metab 1999; 84:1534-1539.
215. Petersen JS, Dyrberg T, Karlsen AE et al. Glutamic acid decarboxylase (GAD65) autoantibodies in prediction of B-cell function and remission in recent-onset IDDM after cyclosporin treatment. Diabetes 1994; 43:1291-1296.
216. Fukuda M, Tanaka A, Tahara Y et al. Correlation between minimal secretory capacity of pancreatic beta-cells and stability of diabetic control. Diabetes 1988; 37:81-88.
217. Kolb H, Dannehl K, Gruenklee D et al. Prospective analysis of islet cell antibodies in children with type I (insulin-dependent) diabetes. Diabetologia 1988; 31:189-194.

218. Knip M, Llonen J, Mustonen A et al. Evidence of an accelerated B-cell destruction in HLA-Dw3/Dw4 heterozygous children with type I (insulin-dependent) diabetes. Diabetologia 1986; 29:347-351.
219. Pipeleers D, Ling Z. Pancreatic beta cells in insulin-dependent diabetes. Diabetes Metab Rev 1992; 8:209-227.
220. Madsbad S, Faber OK, Binder C et al. Prevalence of residual beta-cell function in insulin-dependent diabetics in relation to age at onset and duration of diabetes. Diabetes 1978; 27(Suppl1):262-264.
221. Dorman JS, LaPorte RE, Kuller LH et al. The Pittsburgh insulin-dependent diabetes mellitus (IDDM) morbidity and mortality study. Mortality results. Diabetes 1984; 33:271-276.
222. Major cross-country differences in risk of dying for people with IDDM. Diabetes Epidemiology Research International Mortality Study Group. Diabetes Care 1991; 14:49-54.
223. Matsushima M, LaPorte RE, Maruyama M et al. For the DERI Mortality Study Group. Geographic variation in mortality among individuals with youth-onset diabetes mellitus across the world. Diabetologia 1997; 40:212-216.
224. International evaluation of cause-specific mortality and IDDM. Diabetes Epidemiology Research International Mortality Study Group. Diabetes Care 1991; 14:55-60.
225. Borch-Johnsen K, Nissen H, Henriksen E et al. The natural history of insulin-dependent diabetes mellitus in Denmark: 1. Long-term survival with and without late diabetic complications. Diabetic Med 1987; 4:201-210.
226. Chowdhury TA, Dyer PH, Mijovic CH et al. Human leucocyte antigen and insulin gene regions and nephropathy in Type I diabetes. Diabetologia 1999; 42:1017-1020.
227. Doria A, Warram JH, Krolewski AS. Genetic predisposition to diabetic nephropathy. Evidence for a role of the angiotensin I-converting enzyme gene. Diabetes 1994; 43:690-695.
228. Doria A, Onuma T, Gearin G et al. Angiotensinogen polymorphism M235T, hypertension, and nephropathy in insulin-dependent diabetes. Hypertension 1996; 27:1134-1139.
229. Fogarty DG, Harron JC, Hughes AE et al. A molecular variant of angiotensinogen is associated with diabetic nephropathy in IDDM. Diabetes 1996; 45:1204-1208.
230. Nagi DK, Mansfield MW, Stickland MH et al. Angiotensin converting enzyme (ACE) insertion/deletion (I/D) polymorphism, and diabetic retinopathy in subjects with IDDM and NIDDM. Diabet Med 1995; 12:997-1001.
231. Parving HH, Jacobsen P, Tarnow L et al. Effect of deletion polymorphism of angiotensin converting enzyme gene on progression of diabetic nephropathy during inhibition of angiotensin converting enzyme: Observational follow up study. BMJ 1996; 313:591-594.
232. Tarnow L, Cambien F, Rossing P et al. Angiotensinogen gene polymorphisms in IDDM patients with diabetic nephropathy. Diabetes 1996; 45:367-369.
233. Tarnow L, Cambien F, Rossing P et al. Lack of relationship between an insertion/deletion polymorphism in the angiotensin I-converting enzyme gene and diabetic nephropathy and proliferative retinopathy in IDDM patients. Diabetes 1995; 44:489-494.
234. Ruiz J, Blanche H, James RW et al. Gln-Arg192 polymorphism of paraoxonase and coronary heart disease in type 2 diabetes [see comments]. Lancet 1995; 346:869-872.
235. Garin MC, James RW, Dussoix P et al. Paraoxonase polymorphism Met-Leu54 is associated with modified serum concentrations of the enzyme. A possible link between the paraoxonase gene and increased risk of cardiovascular disease in diabetes. J Clin Invest 1997; 99:62-66.
236. Rewers M, Kamboh MI, Hoag S et al. Apolipoprotein A-IV polymorphism associated with myocardial infarction in obese NIDDM patients. The San Luis Valley Diabetes Study. Diabetes 1994; 43:1485-1489.
237. Retinopathy and nephropathy in patients with Type 1 Diabetes four years after a trial of intensive therapy. N Engl J Med 2000; 342(6):381-9.
238. Matsushima M, LaPorte RE, Maruyama M et al. For the DERI Mortality Study Group. Geographic variation in mortality among individuals with youth-onset diabetes mellitus across the world. Diabetologia 1997; 40:212-216.
239. Songer TJ, LaPorte R, Lave JR et al. Health insurance and the financial impact of IDDM in families with a child with IDDM. Diabetes Care 1997; 20:577-584.
240. Javitt JC, Chiang Y-P. Economic Impact of Diabetes 1995; 2:601-611.
241. Portuese E, Orchard TJ. Mortality in insulin-dependent diabetes. 1995; 2:221-232.
242. The Diabetes Control and Complications Trial Research Group. The effect of intensive treatment of diabetes on the development and progression of long-term complications in insulin-dependent diabetes mellitus. N Engl J Med 1993; 329:977-986.
243. The Diabetes Control and Complications Trial Research Group. The relationship of glycemic exposure (HbA1c) to the risk of development and progression of retinopathy in the diabetes control and complications trial. Diabetes 1995; 44:968-983.

244. Retinopathy and nephropathy in patients with type 1 diabetes four years after a trial of intensive therapy. The Diabetes Control and Complications Trial/Epidemiology of Diabetes Interventions and Complications Research Group. N Engl J Med 2000; 342:381-389.
245. DCCT. Epidemiology of severe hypoglycemia in the diabetes control and complications trial. The DCCT Research Group. Am J Med 1991; 90:450-459.
246. The Diabetes Control and Complications Trial Research Group. Adverse events and their association with treatment regimens in the diabetes control and complications trial. Diabetes Care 1995; 18:1415-1427.
247. Klein R, Klein BEK, Moss SE et al. The Wisconsin Epidemiologic Study of Diabetic Retinopathy. II. Prevalence and risk of diabetic retinopathy when age at diagnosis is less than 30 years. Arch Ophthalmol 1984; 102:520-526.
248. Mizutani M, Kern TS, Lorenzi M. Accelerated death of retinal microvascular cells in human and experimental diabetic retinopathy. J Clin Invest 1996; 97:2883-2890.
249. Konno S, Feke GT, Yoshida A et al. Retinal blood flow changes in type I diabetes. A long-term follow-up study. Invest Ophthalmol Vis Sci 1996; 37:1140-1148.
250. Bursell SE, Clermont AC, Kinsley BT et al. Retinal blood flow changes in patients with insulin-dependent diabetes mellitus and no diabetic retinopathy. Invest Ophthalmol Vis Sci 1996; 37:886-897.
251. Mori F, Konno S, Hikichi T et al. Pulsatile ocular blood flow study: Decreases in exudative age related macular degeneration. Br J Ophthalmol 2001; 85:531-533.
252. Feke GT, Buzney SM, Ogasawara H et al. Retinal circulatory abnormalities in type 1 diabetes. Invest Ophthalmol Vis Sci 1994; 35:2968-2975.
253. Clermont AC, Aiello LP, Mori F et al. Vascular endothelial growth factor and severity of nonproliferative diabetic retinopathy mediate retinal hemodynamics in vivo: A potential role for vascular endothelial growth factor in the progression of nonproliferative diabetic retinopathy. Am J Ophthalmol 1997; 124:433-446.
254. Andersen AR, Christiansen JS, Andersen JK et al. Diabetic nephropathy in Type 1 (insulin-dependent) diabetes: An epidemiological study. Diabetologia 1983; 25:496-501.
255. Krolewski AS, Warram JH, Christlieb AR et al. The changing natural history of nephropathy in type 1 diabetes. Am J Med 1985; 78:785-794.
256. Klein R, Klein BE, Moss SE. The incidence of gross proteinuria in people with insulin- dependent diabetes mellitus. Arch Intern Med 1991; 151:1344-1348.
257. Kofoed-Enevoldsen A, Borch-Johnsen K, Kreiner S et al. Declining incidence of persistent proteinuria in type I (insulin- dependent) diabetic patients in Denmark. Diabetes 1987; 36:205-209.
258. Bojestig M, Arnqvist HJ, Hermansson G et al. Declining incidence of nephropathy in insulin-dependent diabetes mellitus. N Engl J Med 1994; 330:15-18.
259. Tuomilehto J, Borch-Johnsen K, Molarius A et al. The unchanging incidence of hospitalization for diabetic nephropathy in a population-based cohort of IDDM patients in Finland. Diabetes Care 1997; 20:1081-1086.
260. Chavers BM, Bilous RW, Ellis EN et al. Glomerular lesions and urinary albumin excretion in type I diabetes without overt proteinuria. N Engl J Med 1989; 320:966-970.
261. Fioretto P, Steffes MW, Mauer M. Glomerular structure in nonproteinuric IDDM patients with various levels of albuminuria. Diabetes 1994; 43:1358-1364.
262. Adler SG, Kang SW, Feld S et al. Glomerular mRNAs in human type 1 diabetes: Biochemical evidence for microalbuminuria as a manifestation of diabetic nephropathy. Kidney Int 2001; 60:2330-2336.
263. Viberti GC, Hill RD, Jarrett RJ et al. Microalbuminuria as a predictor of clinical nephropathy in insulin- dependent diabetes mellitus. Lancet 1982; 1:1430-1432.
264. Parving HH, Oxenboll B, Svendsen PA et al. Early detection of patients at risk of developing diabetic nephropathy. A longitudinal study of albumin excretion. Acta Endocrinol 1982; 100:550-555.
265. Viberti GC, Mogensen CE, Groop LC et al. European Microalbuminuria Captopril Study Group. Effect of Captopril on progression to clinical proteinuria in patients with insulin-dependent diabetes mellitus and microalbuminuria. JAMA 1994; 271:275-279.
266. Laffel LM, McGill JB, Gans DJ. The beneficial effect of angiotensin-converting enzyme inhibition with captopril on diabetic nephropathy in normotensive IDDM patients with microalbuminuria. North American Microalbuminuria Study Group. Am J Med 1995; 99:497-504.
267. Randomised placebo-controlled trial of lisinopril in normotensive patients with insulin-dependent diabetes and normoalbuminuria or microalbuminuria. The EUCLID Study Group. Lancet 1997; 349:1787-1792.

268. Should all patients with type 1 diabetes mellitus and microalbuminuria receive angiotensin-converting enzyme inhibitors? A meta-analysis of individual patient data. Ann Intern Med 2001; 134:370-379.
269. Caramori ML, Fioretto P, Mauer M. The need for early predictors of diabetic nephropathy risk: Is albumin excretion rate sufficient? Diabetes 2000; 49:1399-1408.
270. Mogensen CE, Christensen CK. Predicting diabetic nephropathy in insulin-dependent patients. N Engl J Med 1984; 311:89-93.
271. Caramori ML, Gross JL, Pecis M et al. Glomerular filtration rate, urinary albumin excretion rate, and blood pressure changes in normoalbuminuric normotensive type 1 diabetic patients: An 8-year follow-up study. Diabetes Care 1999; 22:1512-1516.
272. Garg SK, Chase HP, Icaza G et al. 24-hour ambulatory blood pressure and renal disease in young subjects with type I diabetes. J Diabetes Complications 1997; 11:263-267.
273. Lafferty AR, Werther GA, Clarke CF. Ambulatory blood pressure, microalbuminuria, and autonomic neuropathy in adolescents with type 1 diabetes. Diabetes Care 2000; 23:533-538.
274. Sundkvist G, Lilja B. Autonomic neuropathy predicts deterioration in glomerular filtration rate in patients with IDDM. Diabetes Care 1993; 16:773-779.
275. Lemley KV, Blouch K, Abdullah I et al. Glomerular permselectivity at the onset of nephropathy in type 2 diabetes mellitus. J Am Soc Nephrol 2000; 11:2095-2105.
276. Lemley KV, Abdullah I, Myers BD et al. Evolution of incipient nephropathy in type 2 diabetes mellitus. Kidney Int 2000; 58:1228-1237.
277. Mifsud SA, Allen TJ, Bertram JF et al. Podocyte foot process broadening in experimental diabetic nephropathy: Amelioration with renin-angiotensin blockade. Diabetologia 2001; 44:878-882.
278. Steffes MW, Schmidt D, McCrery R et al. Glomerular cell number in normal subjects and in type 1 diabetic patients. Kidney Int 2001; 59:2104-2113.
279. Pagtalunan ME, Miller PL, Jumping-Eagle S et al. Podocyte loss and progressive glomerular injury in type II diabetes. J Clin Invest 1997; 99:342-348.
280. Chase HP, Jackson WE, Hoops SL et al. Glucose control and the renal and retinal complications of insulin-dependent diabetes. JAMA 1989; 261:1155-1160.
281. Klein R, Klein BE, Moss SE et al. Relationship of hyperglycemia to the long-term incidence and progression of diabetic retinopathy. Arch Intern Med 1994; 154:2169- 2178.
282. Krolewski AS, Laffel LMB, Krolewski M et al. Glycosylated hemoglobin and the risk of microalbuminuria in patients with insulin-dependent diabetes mellitus. N Engl J Med 1995; 332:1251-1255.
283. Orchard TJ, Forrest KY, Ellis D et al. Cumulative glycemic exposure and microvascular complications in insulin- dependent diabetes mellitus. The glycemic threshold revisited. Arch Intern Med 1997; 157:1851-1856.
284. Chase HP, Garg SK, Harris S et al. High-normal blood pressure and early diabetic nephropathy. Arch Intern Med 1990; 150:639-641.
285. Klein R, Klein BE, Moss SE et al. Ten-year incidence of gross proteinuria in people with diabetes. Diabetes 1995; 44:916-923.
286. Lloyd CE, Klein R, Maser RE et al. The progression of retinopathy over 2 years: The Pittsburgh Epidemiology of Diabetes Complications (EDC) Study. J Diabetes Complications 1995; 9:140-148.
287. Porta M, Sjoelie AK, Chaturvedi N et al. Risk factors for progression to proliferative diabetic retinopathy in the EURODIAB Prospective Complications Study. Diabetologia 2001; 44:2203-2209.
288. Chase HP, Garg SK, Marshall G et al. Cigarette smoking increases the risk of albuminuria among subjects with type I diabetes. JAMA 1991; 265:614-617.
289. Scott LJ, Warram JH, Hanna LS et al. A nonlinear effect of hyperglycemia and current cigarette smoking are major determinants of the onset of microalbuminuria in Type 1 Diabetes. Diabetes 2001; 50:2842-2849.
290. Moss SE, Klein R, Klein BE. Cigarette smoking and ten-year progression of diabetic retinopathy. Ophthalmology 1996; 103:1438-1442.
291. Chaturvedi N, Sjoelie AK, Porta M et al. Markers of insulin resistance are strong risk factors for retinopathy incidence in type 1 diabetes. Diabetes Care 2001; 24:284-289.
292. Chaturvedi N, Bandinelli S, Mangili R et al. Microalbuminuria in type 1 diabetes: Rates, risk factors and glycemic threshold. Kidney Int 2001; 60:219-227.
293. Watts GF, Powrie JK, O'Brien SF et al. Apolipoprotein B independently predicts progression of very-low-level albuminuria in insulin-dependent diabetes mellitus. Metabolism 1996; 45:1101-1107.
294. Lam KS, Cheng IK, Janus ED et al. Cholesterol-lowering therapy may retard the progression of diabetic nephropathy. Diabetologia 1995; 38:604-609.
295. De Cosmo S, Argiolas A, Miscio G et al. A PC-1 amino acid variant (K121Q) is associated with faster progression of renal disease in patients with type 1 diabetes and albuminuria. Diabetes 2000; 49:521-524.

296. Lustman PJ, Freedland KE, Griffith LS et al. Fluoxetine for depression in diabetes: A randomized double-blind placebo-controlled trial. Diabetes Care 2000; 23:618-623.
297. Jensen T. Increased plasma concentration of von Willebrand factor in insulin dependent diabetics with incipient nephropathy. BMJ 1989; 298:27-28.
298. Maser RE, Ellis D, Erbey JR et al. Do tissue plasminogen activator-plasminogen activator inhibitor-1 complexes relate to the complications of insulin-dependent diabetes mellitus? Pittsburgh Epidemiology of Diabetes Complications Study. J Diabetes Complications 1997; 11:243-249.
299. Klein R, Klein BE, Moss SE. The Wisconsin Epidemiologic Study of Diabetic Retinopathy. XVI. The relationship of C-peptide to the incidence and progression of diabetic retinopathy. Diabetes 1995; 44:796-801.
300. Lustman PJ, Skor DA, Carney RM et al. Stress and diabetic control. Lancet 1983; 1:588.
301. Rand LI, Krolewski AS, Aiello LM et al. Multiple factors in the prediction of risk of proliferative diabetic retinopathy. N Engl J Med 1985; 313:1433-1438.
302. Lustman PJ, Clouse RE. Relationship of psychiatric illness to impotence in men with diabetes. Diabetes Care 1990; 13:893-895.
303. Goldston DB, Kelley AE, Reboussin DM et al. Suicidal ideation and behavior and noncompliance with the medical regimen among diabetic adolescents. J Am Acad Child Adolesc Psychiatry 1997; 36:1528-1536.
304. Borch-Johnsen K, Nissen H, Henriksen E et al. The natural history of insulin-dependent diabetes mellitus in Denmark: 1. Long-term survival with and without late diabetic complications. Diabetic Med 1987; 4:201-210.
305. Seaquist ER, Goetz FC, Rich S et al. Familial clustering of diabetic kidney disease. Evidence for genetic susceptibility to diabetic nephropathy. N Engl J Med 1989; 320:1161-1165.
306. Borch-Johnsen K, Norgaard K, Hommel E et al. Is diabetic nephropathy an inherited complication? Kidney Int 1992; 41:719-722.
307. The Diabetes Control and Complications Trial Research Group. Clustering of long-term complications in families with diabetes in the diabetes control and complications trial. The Diabetes Control and Complications Trial Research Group. Diabetes 1997; 42:1829-1839.
308. Pettitt DJ, Saad MF, Bennett PH et al. Familial predisposition to renal disease in two generations of Pima Indians with type 2 (noninsulin-dependent) diabetes mellitus. Diabetologia 1990; 33:438-443.
309. Freedman BI, Tuttle AB, Spray BJ. Familial predisposition to nephropathy in African-Americans with non insulin-dependent diabetes mellitus. Am J Kidney Dis 1995; 25:710-713.
310. Bergman S, Key BO, Kirk KA et al. Kidney disease in the first-degree relatives of African-Americans with hypertensive end-stage renal disease. Am J Kidney Dis 1996; 27:341-346.
311. Spray BJ, Atassi NG, Tuttle AB et al. Familial risk, age at onset, and cause of end-stage renal disease in white Americans. J Am Soc Nephrol 1995; 5:1806-1810.
312. O'Dea DF, Murphy SW, Hefferton D et al. Higher risk for renal failure in first-degree relatives of white patients with end-stage renal disease: A population-based study. Am J Kidney Dis 1998; 32:794-801.
313. Canani LH, Gerchman F, Gross JL. Familial clustering of diabetic nephropathy in Brazilian type 2 diabetic patients. Diabetes 1999; 48:909-913.
314. Fioretto P, Steffes MW, Barbosa J et al. Is diabetic nephropathy inherited? Studies of glomerular structure in type 1 diabetic sibling pairs. Diabetes 1999; 48:865-869.
315. Brancati FL, Whittle JC, Whelton PK et al. Excess incidence of diabetic end-stage renal disease among blacks: A population-based study. Diabetes 1992; 41(suppl1):128A-128A.
316. Fogarty DG, Rich SS, Hanna L et al. Urinary albumin excretion in families with type 2 diabetes is heritable and genetically correlated to blood pressure. Kidney Int 2000; 57:250-257.
317. Chowdhury TA, Dyer PH, Mijovic CH et al. Human leucocyte antigen and insulin gene regions and nephropathy in Type I diabetes. Diabetologia 1999; 42:1017-1020.
318. Doria A, Warram JH, Krolewski AS. Genetic predisposition to diabetic nephropathy. Evidence for a role of the angiotensin I-converting enzyme gene. Diabetes 1994; 43:690-695.
319. Doria A, Onuma T, Gearin G et al. Angiotensinogen polymorphism M235T, hypertension, and nephropathy in insulin-dependent diabetes. Hypertension 1996; 27:1134-1139.
320. Fogarty DG, Harron JC, Hughes AE et al. A molecular variant of angiotensinogen is associated with diabetic nephropathy in IDDM. Diabetes 1996; 45:1204-1208.
321. Nagi DK, Mansfield MW, Stickland MH et al. Angiotensin converting enzyme (ACE) insertion/deletion (I/D) polymorphism, and diabetic retinopathy in subjects with IDDM and NIDDM. Diabet Med 1995; 12:997-1001.
322. Tarnow L, Cambien F, Rossing P et al. Angiotensinogen gene polymorphisms in IDDM patients with diabetic nephropathy. Diabetes 1996; 45:367-369.

323. Lustman PJ, Anderson RJ, Freedland KE et al. Depression and poor glycemic control: A meta-analytic review of the literature. Diabetes Care 2000; 23:934-942.
324. Hamilton CL, Brobeck JR. Control of food intake in normal and obese monkeys. Ann NY Acad Sci 1965; 131:583-592.
325. Lustman PJ, Griffith LS, Clouse RE et al. Effects of alprazolam on glucose regulation in diabetes. Results of double-blind, placebo-controlled trial. Diabetes Care 1995; 18:1133-1139.
326. Maeda S, Haneda M, Guo B et al. Dinucleotide repeat polymorphism of matrix metalloproteinase-9 gene is associated with diabetic nephropathy. Kidney Int 2001; 60:1428-1434.
327. Krolewski AS, Kosinski EJ, Warram JH et al. Magnitude and determinants of coronary artery disease in juvenile- onset, insulin-dependent diabetes mellitus. Am J Cardiol 1987; 59:750-755.
328. Moss SE, Klein R, Klein BE. Cause-specific mortality in a population-based study of diabetes. Am J Public Health 1991; 81:1158-1162.
329. Dorman JS, LaPorte RE, Kuller LH et al. The Pittsburgh insulin-dependent diabetes mellitus (IDDM) morbidity and mortality study. Mortality results. Diabetes 1984; 33:271-276.
330. Manson JE, Colditz GA, Stampfer MJ et al. A prospective study of maturity-onset diabetes mellitus and risk of coronary heart disease and stroke in women. Arch Intern Med 1991; 151:1141-1147.
331. Borch-Johnsen K, Kreiner S. Proteinuria: Aalue as predictor of cardiovascular mortality in insulin dependent diabetes mellitus. Br Med J [Clin Res] 1987; 294:1651-1654.
332. Martin FI, Hopper JL. The relationship of acute insulin sensitivity to the progression of vascular disease in long-term type 1 (insulin-dependent) diabetes mellitus. Diabetologia 1987; 30:149-153.
333. Orchard TJ, Dorman JS, Maser RE et al. Factors associated with avoidance of severe complications after 25 yr of IDDM. Pittsburgh Epidemiology of Diabetes Complications Study I. Diabetes Care 1990; 13:741-747.
334. Orchard TJ. From diagnosis and classification to complications and therapy. DCCT. Part II? Diabetes Control and Complications Trial. Diabetes Care 1994; 17:326-338.
335. Lloyd CE, Kuller LH, Ellis D et al. Coronary artery disease in IDDM. Gender differences in risk factors but not risk. Arterioscler Thromb Vasc Biol 1996; 16:720-726.
336. Crall Jr FV, Roberts WC. The extramural and intramural coronary arteries in juvenile diabetes mellitus: Analysis of nine necropsy patients aged 19 to 38 years with onset of diabetes before age 15 years. Am J Med 1978; 64:221-230.
337. Valsania P, Zarich SW, Kowalchuk GJ et al. Severity of coronary artery disease in young patients with insulin- dependent diabetes mellitus. Am Heart J 1991; 122:695-700.
338. Miettinen H, Lehto S, Salomaa V et al. Impact of diabetes on mortality after the first myocardial infarction. The FINMONICA Myocardial Infarction Register Study Group [see comments]. Diabetes Care 1998; 21:69-75.
339. Chun BY, Dobson AJ, Heller RF. The impact of diabetes on survival among patients with first myocardial infarction. Diabetes Care 1997; 20:704-708.
340. Savage MP, Krolewski AS, Kenien GG et al. Acute myocardial infarction in diabetes mellitus and significance of congestive heart failure as a prognostic factor. Am J Cardiol 1988; 62:665-669.
341. Rozenman Y, Sapoznikov D, Mosseri M et al. Long-term angiographic follow-up of coronary balloon angioplasty in patients with diabetes mellitus: A clue to the explanation of the results of the BARI study. Balloon Angioplasty Revascularization Investigation. J Am Coll Cardiol 1997; 30:1420-1425.
342. Weintraub WS, Stein B, Kosinski A et al. Outcome of coronary bypass surgery versus coronary angioplasty in diabetic patients with multivessel coronary artery disease. J Am Coll Cardiol 1998; 31:10-19.
343. Koistinen MJ. Prevalence of asymptomatic myocardial ischaemia in diabetic subjects. BMJ 1990; 301:92-95.
344. Yamasaki Y, Kawamori R, Matsushima H et al. Atherosclerosis in carotid artery of young IDDM patients monitored by ultrasound high-resolution B-mode imaging. Diabetes 1994; 43:634-639.
345. Kanters SD, Algra A, Banga JD. Carotid intima-media thickness in hyperlipidemic type I and type II diabetic patients. Diabetes Care 1997; 20:276-280.

CHAPTER 10

Humoral Autoimmunity

Liping Yu and George S. Eisenbarth

A little more than a decade ago, screening for autoantibodies associated with Type 1A (immune mediated) diabetes was limited to measuring cytoplasmic islet cell autoantibodies (ICAs) and insulin antibodies. Currently, sequenced autoantigens with recombinant autoantibody assays are available[1-7] and include insulin,[8] glutamic acid decarboxylase (GAD),[9] ICA512/IA-2,[10-12] I-A2b (phogrin)[13,14] and carboxypeptidase H.[15] In addition, there are many other autoantigens in a variety of stages of characterization, including molecules with characterized sequences but without fully developed assays and proteins with only known molecular weights of 155 kd[16,17] 52 kd,[18] 37-38 kd.[19,20] Other molecules have been described but their association with Type 1A diabetes has either not been evaluated in man, has not been substantiated, or follow-up studies have not been published. These molecules include anti-bovine serum albumin antibodies (anti-BSA),[21] antibodies reacting with ICA69,[22] anti-mouse heat shock protein antibodies,[23] anti-insulin receptor antibodies,[24] antibodies to heat shock proteins,[25,26] and antibodies to a series of autoantigens identified by screening islet libraries. Finally, a subset of autoantibodies termed anti-islet cell surface antibodies are currently rarely measured as most recent studies have failed to demonstrate disease specificity.[27]

Despite the importance of autoantibodies for disease prediction, it is likely that anti-islet autoantibodies do not by themselves cause the beta cell destruction that leads to Type 1A diabetes. The most cogent evidence of lack of direct damage by autoantibodies is the lack of diabetes in infants born to antibody positive mothers, or women who developed Type 1A diabetes during pregnancy. Autoantibodies (ICA, anti-insulin, and anti-GAD autoantibodies) are readily detected in the serum of infants of such antibody-positive mothers in the absence of diabetes in the child. It will take careful epidemiological studies to determine if specific autoantibodies (e.g., spontaneous and not insulin therapy induced insulin antibodies) predispose children to develop type 1 diabetes. Of note, a child with genetic B cell deficiency and no antibodies progressed to type 1 diabetes.[28]

Studies by groups associated with Blizzard, Nerup, and Irvine in the early 1970s demonstrated that Type 1A diabetes was specifically associated with idiopathic Addison's disease.[29] This was in contrast to tuberculous Addison's disease. Idiopathic Addison's disease with lymphocytic infiltrates and destruction of the adrenal cortex was considered an organ-specific autoimmune disorder. In addition, insulitis in the pancreatic islets of new-onset diabetic patients who died shortly after diagnosis was reported. These two findings and discovery of human leukocyte (HLA) alleles associated with Type 1 but not Type 2 diabetes led to the hypothesis that in a subset of patients with Type 1 diabetes, the disease resulted from autoimmune beta cell destruction. Patients with Addison's disease, important for recognizing the association of Type 1A diabetes with autoimmunity, provided the sera for the discovery of cytoplasmic islet cell autoantibodies.[30] Patients with polyendocrine autoimmunity had high titers of antibodies that reacted with frozen sections of human pancreas. As is currently recognized, such patients with Type 1A diabetes have ICAs that frequently persist for decades, in contrast to the majority

Immunology of Type 1 Diabetes, Second Edition, edited by George S. Eisenbarth.

of patients with Type 1A diabetes where the ICA autoantibodies usually disappear within 5 years. The discovery of cytoplasmic islet cell antibodies in this population facilitated their identification in patients with the more common form of Type 1A diabetes. Nevertheless, the utilization of immunohistochemical tests that are dependent upon human pancreas (namely the ICA test) has contributed to debate as to whether anti-islet autoantibodies are predictive of diabetes in a general population and whether such antibodies frequently appear and disappear in relatives of patients with Type 1A diabetes.[31,32] We believe these tests can for the most part be replaced with "biochemical" autoantibody assays. In addition to utility in predicting type 1A diabetes, islet autoantibodies aid in the diagnosis of type 1 diabetes. Approximately 7% of patients with initially noninsulin dependent adult onset diabetes have GAD autoantibodies.[33] These GAD positive individuals almost always become insulin dependent.[32]

Islet Cell Autoantibodies (ICAs)

Since the discovery of cytoplasmic ICA multiple workshops have been held and a Proficiency Testing Service established for laboratories measuring cytoplasmic islet cell autoantibodies.[34] The Immunology of Diabetes Society (IDS) workshops rapidly demonstrated that there existed large variation between laboratories in terms of sera called ICA positive. A standard for ICA was established utilizing dilutions of sera and the units assigned were termed JDF units (named for the Juvenile Diabetes Foundation, which sponsored the meetings). With such a standard, differences in assay sensitivity could be rationalized and it became apparent that greater than 20 JDF units of cytoplasmic islet cell autoantibodies were highly predictive of diabetes among first-degree relatives. Laboratories attempting to measure cytoplasmic islet cell autoantibodies are strongly encouraged to take part in a proficiency service. Many laboratories throughout the world can achieve 100% specificity and sensitivity in the small panel of sera sent "blinded" to laboratories at regular intervals, but even as recently as the last proficiency exchange many laboratories had unacceptable sensitivity and specificity. The World Health Organization and Centers for Disease Control are developing panels of sera to serve as standards for anti-islet autoantibody determination. In addition the Centers for Disease Control have evaluated filter paper collection of sera for determination of autoantibodies with the finding that autoantibodies can readily be measured following elution from such blood spots.

Limits of the ICA test should be appreciated. Current ICA testing measures greater binding of immunoglobulin to islet versus acinar pancreas. Thus any antigen that is in greater density within islets, or that for a variety of reasons survives the preparation of frozen sections of pancreas better within islets than within acinar pancreas, may be recognized by antibodies and contribute to ICA positivity. Despite this caveat, in appropriate populations high titers of ICA are a very important risk factor for the development of Type 1A diabetes.[32] Prognostically significant heterogeneity of ICA has been appreciated and there exist first-degree relatives of patients with Type 1A diabetes or patients with polyendocrine autoimmunity with high titers of ICA with a very low risk of progression to overt diabetes.[35,36]

To date, three autoantigens have been defined that contribute to the ICA positivity (Table 1). The best-characterized subset of ICA autoantibodies is unusual (high titers and unique determinants) anti-GAD autoantibodies.[35-37] Such antibodies appear to be present in a small minority of individuals actually developing Type 1A diabetes but are present in a significant proportion of ICA-positive relatives or polyendocrine patients followed for more than 5 years without developing diabetes. Such ICA antibodies have been termed restricted/selective in that they fail to react with mouse islets and give a beta cell-specific pattern of staining. These ICA antibodies can be absorbed with GAD and are similar to the ICA antibodies found in patients with stiff-man syndrome.

Other defined antigens accounting for a subset of ICA staining are antibodies reacting with ICA512/IA-2 (or I-A2β) and glycolipids. Along with GAD65 autoantibodies, these antibodies appear to be the bulk of ICA reactivity in individuals progressing to diabetes.[38] It is also likely that the above autoantigens do not account for all ICA reactivity and in particular human-specific

Table 1. Biochemically characterized autoantigens

	Sensitivity	Comment
Insulin	49-92%	Higher levels young children
GAD (Glutamic a. Decarboxylase)	84%	Higher sensitivity adult onset type 1A
ICA512/IA-2	74%	Tyrosine Phosphatase like molecule
IA-2β/Phogrin	61%	Tyrosine Phosphatase like molecule
Carboxypeptidase H	10%	Infrequent
GLIMA38	14%	Not yet sequenced
GM2-1	?	Ganglioside: Chromatography assay
ICA69	?	Western blot assay with poor specificity
ICA12	< 20%	Diabetes relatedness requires further study

antibodies exist that fail to react with mouse or rat pancreas (Table 1) and that are not absorbed by GAD, ICA512 or gangliosides.

Sequenced Autoantigens

Glutamic Acid Decarboxylase (GAD)

The discovery that anti-GAD autoantibodies accounted for an important component of anti-64 kd autoantibodies and a subset of ICA began with the discovery that patients with stiff-man syndrome expressed autoantibodies that gave rise to ICA staining and had high titers of anti-GAD autoantibodies.[37] Glutamic acid decarboxylase (GAD) is the enzyme responsible for GABA synthesis within the nervous system and islet cells.[39] GAD is expressed within all human islet cells (e.g., α, β, δ). For rat islets GAD is beta cell specific and levels of GAD are extremely low in mouse islets.[40,41] In that only beta cells are destroyed in man, and insulitis of man and animal models disappears from islets once all the beta cells are destroyed, it is possible that anti-GAD autoantibodies and the autoimmune response to GAD is of secondary rather than of primary pathogenic importance.[40,42] In addition to islets, the two forms of GAD, termed GAD65 and GAD67, are synthesized in testes, ovary, and neurons. The anti-GAD autoantibodies of patients with stiff-man syndrome (SMS) (a rare neurological disorder characterized by spasms of muscular rigidity) are unusual in that they react with fixed sections of the brain and react on Western blots with purified GAD.[37] Prior to studies of SMS, Baekkeskov and coworkers[43,44] had described anti-64 kd autoantibodies utilizing an immunoprecipitation assay of metabolically labeled islet cells. Although a high percentage of patients with new-onset diabetes had anti-GAD autoantibodies with this assay, these antibodies do not Western blot GAD or show immunohistochemical reactivity with neurons. Furthermore, anti-GAD stiff-man antibodies recognized different epitopes in contrast to the anti-GAD antibodies usually found in new-onset diabetic patients. In 1990, the groups of DeCamilli and Baekkeskov[9] jointly discovered that the bulk of anti-64 kd autoantibodies reacted at low titer (relative to stiff-man syndrome antibodies) with GAD and that a small subset of the anti-GAD autoantibodies of new-onset diabetics were similar to anti-GAD stiff-man syndrome antibodies. In 1991, Gianani and coworkers[35] (for relatives of patients with Type 1A diabetes) and Genovese and coworkers[36] (for polyendocrine patients) discovered that individuals with high titers of anti-GAD antibodies reacted on sections and Western blot similar to SMS patients anti-GAD antibodies. These antibodies were termed "restricted" ICA or "selective" ICA in that the antibodies failed to react with islets of mouse pancreas and on rat islets stained only beta cells. This restricted reactivity is apparently due to the almost eight-fold less expression of GAD by mouse islets.[40] When ICA reactivity is due solely to autoantibodies reacting with GAD, absorption of sera with affinity purified GAD, recombinantly produced GAD, or absorption with a brain extract can be used to identify such antibodies (Table 1).[36,45] In studies utilizing brain extract, the specificity for GAD absorption is usually evaluated by preclearing the extract with a monoclonal anti-GAD antibody. Utilization of beta cell-specific staining to aid identification of restricted ICA is a problem with human pancreas in that GAD is expressed by a subset of nonbeta cells of human pancreas. In contrast, within rat islets GAD expression is beta cell specific.[41]

Several different radioassays for anti-GAD autoantibodies have been developed and the first Immunology of Diabetes Workshop (IDW) for measuring such antibodies was held. Four formats for determining anti-GAD autoantibodies include: (1) determination of antibody precipitation of GAD enzymatic activity, (2) radioassays utilizing affinity purified porcine brain GAD that has been labeled with I,[125] (3) radioassays utilizing endogenously labeled GAD produced by in vitro transcription and translation of GAD cDNA,[46] and (4) ELISA assay formats. These assay formats are likely to give very different results depending upon the population studied. In particular, the immunoenzymatic assay appears to be less sensitive than either radioassay. Radioassays utilizing in vitro transcribed and translated GAD65 or ^{125}I-labelled recombinant human GAD65 can give equivalent results with high sensitivity and specificity. The Combinatorial Autoantibody International workshop of the Immunology of Diabetes Society indicated that most laboratories (but not all) concordantly measured anti-GAD65 autoantibodies.[47]

The quantitation of anti-GAD autoantibodies can also aid in the detection of extremely high titers of such antibodies associated with restricted ICA. The highest levels of anti-GAD autoantibodies that we have observed among ICA-positive relatives occur in relatives with restricted ICA independent of the GAD autoantibody assay utilized.

Among ICA-positive first-degree relatives, the levels of anti-GAD autoantibodies are remarkably constant with up to a decade of prospective evaluation. It is noteworthy that Harrison and coworkers[48] have reported that there is an inverse correlation between levels of GAD autoantibodies and a T cell proliferative response to GAD. Thus it is hypothesized that high levels of GAD autoantibodies may be associated with a diminished anti-islet T cell response (Th 2 rather than Th1) and therefore less or no beta cell destruction.

Two isomers of GAD, GAD65 and GAD67, share an identical exon-intron structure and are 76% identical in amino-acid sequences with only 22% identity in N-terminal region and over 90% identity in C-terminal regions.[49] Antibodies to GAD65 were detected in over 80% of the patients with type 1A diabetes[50] versus only 11-18% having antibodies to GAD67.[51,52] The autoantibodies binding to GAD67 in IDDM seem to represent antibodies to shared epitopes with GAD65[51] although there was a report that the patients with Graves' disease have been found carrying specific GAD67 antibodies in the absence of GAD65 antibodies.[53]

GAD67 and GAD65 have similar protein conformations that are important for antibody binding. GAD65/GAD67 chimeric proteins were widely utilized in most of laboratories to define the epitope of GAD65 while maintaining the overall GAD protein conformation. The precise mapping of GAD epitopes targeted by patients with type 1A diabetes is difficult because serum contains polyclonal antibodies with multiple specificities. Monoclonal antibodies derived from individuals with new-onset diabetes have been useful in mapping GAD epitopes in type 1A diabetes.[54-56] Two major epitope regions targeted by human diabetic sera have been identified[57] and confirmed by other laboratories. One epitope is in the middle part of GAD65 between amino acids 240 – 435 (termed IDDM-E1) and one is at the C-terminal region of GAD65 between amino acids 451 – 570 (termed IDDM-E2). Recently, Baekkeskov, et al[58] used homolog-scanning mutagenesis to identify GAD65-specific amino acid residues which form autoreactive antibody epitopes in this molecule. Detailed mapping of 13 conformational epitopes, recognized by human monoclonal antibodies derived from patients, together with two and three dimensional structure predictions led to a model of the GAD65 dimer. The results revealed a remarkable spectrum of human autoreactivity to GAD65, targeting almost the entire surface of the molecule. This suggests that native folded GAD65 is the immunogen for autoreactive B-cells in IDDM.

Comparisons of the autoimmune repertoires to GAD65 in type 1 diabetes, Stiff Man Syndrome (SMS), and Autoimmune Polyendocrine Syndrome (APS) might provide insight into why some individuals develop one disease and other individuals develop the other disease. Most of type 1 sera only target GAD65 protein epitopes dependent on protein conformation and do not bind GAD65 protein fragments or react with denatured GAD65 protein with immunoblotting.[9,59] In contrast, SMS sera have antibodies that bind GAD65 protein fragments consisting of N-terminal, middle, and C-terminal regions.[60-62] Characteristically, SMS sera target an epitope contained in the N-terminal first eight amino acids and neither IDDM and APS sera have such antibodies. Despite clear differences in the GAD65 antibody profile in SMS and IDDM, an important similarity exists. Both SMS and IDDM sera contain antibodies that bind a conformation-dependent region in the middle part of the protein and a second set of antibodies that binds a conformation-dependent epitope in the C-terminal regions of the protein.[62] Antibodies specific for these two regions of GAD65 are present in approximately equivalent titers in most sera although a single specificity may predominant. Like SMS, APSI sera have antibodies that inhibit the enzymatic activity of GAD65, while this is not a property of type 1 diabetes autoantibodies.[63] The similarity and difference in the GAD65 antibody repertoire in the Autoimmune Polyendocrine Type II Syndrome (APSII) and diabetes has been studied[64]. Neither diabetes nor APS2 sera bind the N-terminal third of GAD65, but instead target the C-terminal two-thirds of GAD65, IDDM-E1 and IDDM-E2. More detailed mapping has found differences for APSII sera reactivity in the IDDM-E2 region.[64]

ICA512 (IA-2) and IA-2β (Phogrin)

Following the identification that a major component of the 64 kd antigen is GAD, many groups reported an association between Type 1A diabetes and anti-GAD antibodies. Evidence for heterogeneity of anti-64 kd autoantibodies came from the work of Christie,[65] who found that antigenic tryptic fragments were generated after mild trypsinization of 64 kd molecule(s). In particular, precipitation of human islets with sera from nondiabetic polyglandular failure individuals, followed by trypsinization, yielded a 50 kd fragment whereas a 38-40 kd fragment was precipitated by new-onset and prediabetic sera. Of note, mild trypsinization of the GAD molecule yields a 50 kd fragment. Reports[65-65] indicated that autoantibodies to the 37-38 kd fragment had a higher positive predictive value for diabetes than anti-GAD antibodies. It appeared that there existed two different 64 kd autoantigens in human islets, one being GAD and another two molecules that after trypsinization yield a fragment of 37-40 kd. Christie reported that the 40 kd fragment is related to the protein tyrosine phosphatase IA-2 (ICA512)[67] and the 37Kd fragment to the molecule IA-2β.[14]

The autoantigens ICA512 (IA-2) and subsequently IA-2β (phogrin) were isolated independently by several investigators probably reflecting the importance of these related molecules. Rabin and coworkers[10] screened an islet expression library with sera from patients with type 1A diabetes. The molecule he isolated was termed ICA512. The same molecule has been termed IA-2. The original sequence in GenBank from Rabin and coworkers lacked several nucleotides out of approximately 2,000, but with a frame shift predicted a shortened protein, which was originally utilized in autoantibody assays. A related protein with homology in a putative tyrosine phosphatase region (though to date no enzymatic activity has been demonstrated for either molecule) was termed I-A2β by Notkins and colleagues,[13] and phogrin (phosphatase of granules of rat insulinoma) by Hutton and coworkers.[14] Both molecules are most homologous in their C-terminal intracytoplasmic domains to which essentially all of the autoantibodies are directed. Almost all autoantibodies which react with I-A2β also react with ICA512, while approximately 10% of patients developing type 1A diabetes have autoantibodies reacting with ICA512 but not with IA-2β. Thus we do not routinely assay for IA-2β autoantibodies.

There is additional heterogeneity of the autoantibodies which react with ICA512. A large number of epitopes (>10) within the intracytoplasmic C-terminus of the molecule are targets of autoantibodies.[68] In addition islet cells express differentially spliced messenger RNA for ICA512, with one form lacking exon 13 which includes the transmembrane region of the

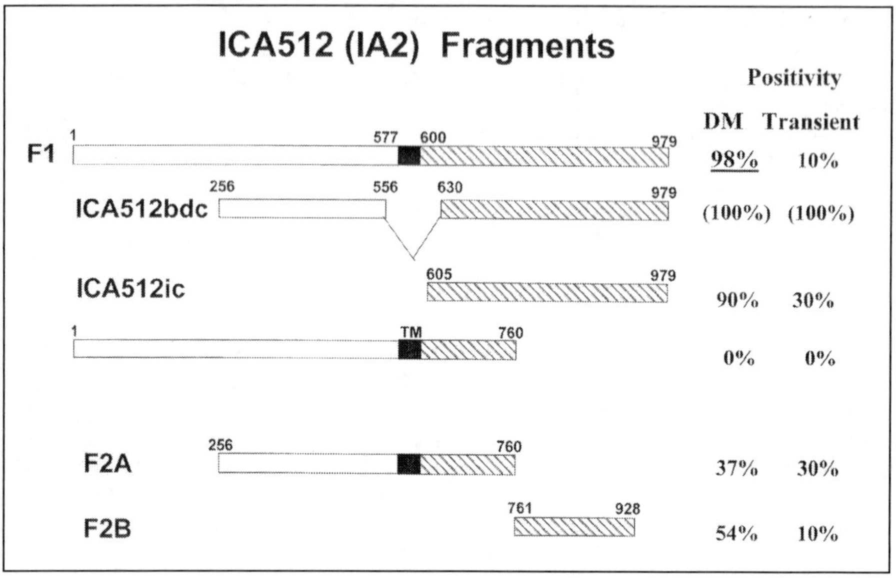

Figure 1. In vitro transcription and translation of fragments of ICA512 produce peptides recognized by ICA512 autoantibody positive sera. Modified from Miao et al. J Autoimmun 2002. F1= full length.

molecule. In our laboratory by screening an islet expression library we obtained an ICA512 clone (ICA512bdc) which was used to develop an autoantibody radioassay.[69,70] This clone lacks exon 13 [construct termed ICA512bdc-(bdc for Barbara Davis Center)]. Approximately 10% of patients developing type 1 diabetes have autoantibodies reacting with only one or the other of full-length ICA512 or ICA512bdc. Thus to fully define all individuals expressing these autoantibodies one needs to apply both assays. It appears that with exon 13 missing, unique epitopes recognized by autoantibodies are created with the joining of exon 12 to exon 14. Other patients have autoantibodies to the transmembrane/juxtamembrane region of the molecule which is lacking in ICA512bdc. With multiple autoantigens tested (e.g., insulin, GAD65 and ICA512) very few patients developing type 1A diabetes have autoantibodies to only a single construct of ICA512 and lack all other autoantibodies. Thus, testing for only ICA512bdc or ICA512 autoantibodies is usually sufficient.

Two epitopes of particular interest within the ICA512 molecule are a mini-epitope described by Dotta and coworkers at the C-terminus of the molecule[71] and a juxtamembrane epitope described by Bonifacio and coworkers.[72] The mini-epitope is of interest in that despite in vitro transcribing and translating only amino acids 761-928 of the ICA512 molecule, approximately 54% of patients who have ICA512 autoantibodies react with this epitope.[4] Bonifacio and coworkers have evidence that autoantibodies to the juxtamembrane region (amino acids) are amongst the first to appear in individuals developing type 1 diabetes. In our studies of young infants we have not been able to confirm this finding but the number of such children we have studied developing antibodies de novo is small.

Glycolipid Autoantigens

The initial studies suggesting that antibodies directed against islet glycolipids may be a component of anti-islet autoantibodies came from evaluation of monoclonal antibodies that reacted with neuronal gangliosides and that also react with islets. Monoclonal antibodies A2B5,[73] 3G5,[74] R2D6,[75] and tetanus toxin all react with complex gangliosides and all give an ICA pattern of staining that involves all cells within islets.[73,75] In contrast to anti-islet monoclonal antibodies that react with proteins (e.g., HISL19,[76] 4F2[77]), the above anti-ganglioside antibodies were not species specific.[76] In that gangliosides are composed of a ceramide group connected to a complex polysaccharide with at least one sialic acid group, gangliosides are amphiphilic. Thus, organic solvents such as methanol remove gangliosides from tissue sections, while gangliosides are stable to nonpolar solvents such as acetone. Stability to acetone but removal by methanol was a quality found for most ICA. Gangliosides can readily be extracted from tissues with organic solvents and then separated from nonpolar lipids by partitioning with chloroform. Extraction of human pancreas followed by the above Folch partition, followed by chromatography that separates gangliosides based on their hydrophobicity (thin-layer chromatography, TLC) or number of sialic acid residues (DEAE chromatography) allowed the isolation of ganglioside preparations that, when incubated with ICA-positive sera, removed all reactivity.[38] It is noteworthy that the absorption assay utilized mouse pancreas.[38] It is likely (in retrospect) that the choice of mouse pancreas allowed absorption to be demonstrated even if sera contained high-titer anti-GAD autoantibodies that are not absorbed by glycolipid preparations (and that do not give ICA-positive staining with mouse islets; e.g., restricted ICA).

Which glycolipids are autoantigens and which are the target autoantigens for ICA? Three candidate molecules are GT3,[78] sulphatides,[79] and what has been termed a GM2-1[80-82] ganglioside (on thin-layer chromatography it has a mobility between GM2 and GM1 standards). Answers to the above questions will almost by necessity lead to the development of a biochemical assay for anti-glycolipid autoantibodies to aide in the prediction of Type 1A diabetes. Buschard and coworkers have reported in abstract form that sulphatides may be a relevant autoantigen as they detect anti-sulphatide antibodies (utilizing immunochemical detection of the staining of sulphatides following thin-layer chromatography) in the majority of patients with recent-onset diabetes.[79] Marcus and coworkers, utilizing a liposome-based assay, detected antibodies to GT3.[78] Finally, Dotta and coworkers have analyzed islets of man[83] and NOD mice for gangliosides. They have isolated and sequenced in molecule termed GM2-1,[81] isolated from human pancreas that contains a terminal sialic acid residue. In addition, they have developed a TLC-based assay for autoantibodies reacting with GM2-1 and find the majority of prediabetics express antibodies reacting with this molecule.[84] Isolation and/or synthesis of GM2-1 should allow the development of a standardized assay for autoantibodies and the direct testing by absorption of the frequency of antibodies to this molecule.

Insulin

Insulin is a 51-amino acid disulphide-linked heterodimer that is specifically produced by beta cells of islets. Proinsulin which is processed within the secretory granule is the precursor to insulin. To date insulin and proinsulin are the only known beta cell-specific autoantigens within human islets (all other putative autoantigens are produced by human nonbeta islet cells). Of note, both in mouse and man, proinsulin messenger RNA and proinsulin are present in the thymus.[85] Transgenic mice with transgenes with the rat insulin promoter coupled to several proteins indicates that thymic transcription of the insulin gene is possible.[86] Insulin is produced as a preprohormone that is processed to pro-insulin and then, with removal of connecting peptide (C peptide), to mature insulin. Processing occurs within beta cell secretory granules, where insulin is packaged in a crystal form.[87] Equimolar concentrations of insulin and C peptide are secreted by beta cells.

In the 1950s, Berson and Yallow developed the first radioimmunoassay utilizing sera with insulin antibodies from patients treated with bovine insulin.[88] Bovine insulin differs from human insulin at 3 of 51 amino acids and porcine insulin differs from human insulin by one amino acid (at the terminus of the B chain). For most patients treated with subcutaneous insulin, the presence of anti-insulin antibodies does not interfere with insulin therapy. A subset of insulin-treated patients with extremely high levels of insulin antibodies are insulin resistant with mean insulin binding capacities greater than 216 nM (30,000 microunits of insulin/ml serum).[89] For such patients, the species of insulin used for therapy is usually changed (e.g., from bovine to human insulin) or to humalogue insulin and occasionally to sulfated insulin.

Anti-insulin antibodies during pregnancy may facilitate transplacental passage of animal insulins. A very rare, but interesting syndrome, is the Insulin Autoimmune Syndrome, also termed Hirata syndrome.[90,91] Patients with this syndrome (almost all with DRB1*0406) after exposure to sulfhydryl containing medications (e.g., methimizole) develop extremely high levels of insulin autoantibodies. These patients usually present with hypoglycemia which resolves with discontinuation of the medication. In addition to this MHC restricted syndrome, some patients have monoclonal insulin autoantibodies produced by B lymphocyte tumors.[92]

The production of antibodies to animal insulin was not unexpected. It was, however, discovered that individuals treated with human insulin also produce anti-human insulin antibodies (though at a lower concentration).[89] The development of anti-insulin antibodies following subcutaneous therapy with human recombinant insulin has been ascribed to the possible presence of "denatured" insulin molecules. It is, however, very likely that self proteins when administered subcutaneously, especially in a depot form (e.g., complexed with the positively charged protein NPH (neutral protein Hagedorn), or zinc, as in the two most common pharmacologic long-acting insulin preparations), will induce antibodies. Antibodies also develop following therapy within a number of other recombinant proteins including human growth hormone and factor VIII. To test the generality of production of insulin antibodies following subcutaneous insulin administration, we transplanted a pituitary cell line producing rat insulin (which has the same sequence as mouse insulin) into histoincompatible mice. With rejection of the tissue, anti-insulin "autoantibodies" were induced.[93]

In 1983 Palmer and coworkers[94] discovered the presence of anti-insulin antibodies in patients with new-onset Type 1A diabetes prior to the administration of exogenous insulin. Subsequently a large number of studies have demonstrated that anti-insulin antibodies are present for years before the development of Type 1A diabetes.[93] Initially, two different assay formats were utilized to detect such antibodies (an ELISA format with insulin immobilized on plates and fluid phase radioassays). The two different formats gave very different results relative to the positive predictive value for development of Type 1A diabetes. A series of international workshops and sera exchanges led to the observation that the two formats measure different antibodies, and that only the antibodies detected with the radioassay formats were associated with risk for Type 1A diabetes.[8] Both assay formats detected antibodies following insulin therapy and rarely were positive in normal controls.

As more has been learned concerning insulin autoantibodies, the inability of standard ELISA assays to detect disease-relevant antibodies is understandable. In particular, the anti-insulin autoantibodies associated with diabetes risk are all of extremely high affinity and, most important, of very low capacity (10^{-12} M), making their detection with plate-binding assays problematic.[8] The epitope of the insulin molecule recognized with fluid phase assays appears to be homogeneous.[95] In particular, the antibodies from all antibody-positive relatives we have studied react with a conformational epitope (do not react with either the A or B chain) and react equally well with insulin and proinsulin. To date, all insulin autoantibodies of at-risk relatives react with des B23 to B30 insulin lacking the terminal eight amino acids of the B chain. Des B23 to B30 insulin fails to bind to the insulin receptor. The A13 leucine is in the base of a pocket of the insulin molecule, and if this leucine is replaced by a tryptophan that extends out of the pocket, insulin autoantibody reactivity is abrogated. The reason for this marked specificity of insulin autoantibodies reacting with the face opposite the insulin-binding domain is unknown. It is, however, very likely that the antibody response to insulin is "antigen driven," giving rise to high-affinity autoantibodies. Insulin may react with B cells while bound to insulin receptors, either of neighboring immunocytes or perhaps following its transport on insulin receptors through the endothelium. The ability to produce insulin analogues that are recognized by insulin autoantibodies but not by the insulin receptor suggests that one can selectively target B cells producing such antibodies. Whether elimination of such B cells which are probably extremely efficient at presenting insulin to T cells, would influence beta cell destruction is unknown. Autoantibodies of prediabetics which react with proinsulin are all absorbed with insulin, and assays for insulin autoantibodies are reported to have higher disease specificity compared to proinsulin autoantibodies.

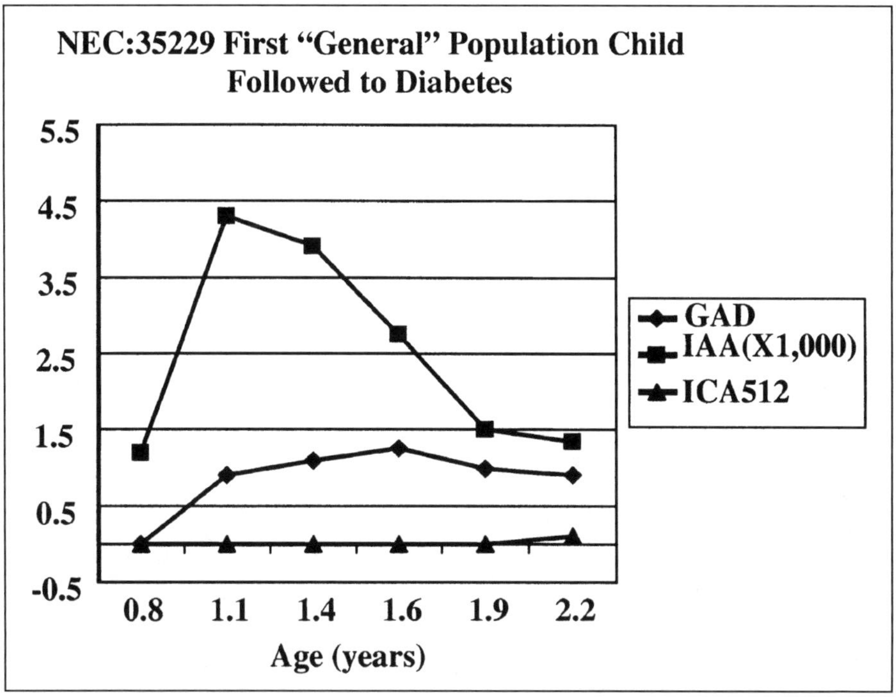

Figure 2. Development of GAD and insulin autoantibodies in child from the general population followed to the development of diabetes. This child did not express ICA512 autoantibodies.

Levels of insulin autoantibodies, similar to GAD autoantibodies, appear to be regulated at different levels over long periods of time in prediabetic first-degree relatives (Fig 2).

The levels of antibodies correlate inversely with the age at which type 1 diabetes develops. Thus levels greater than 2000 nU/ml are almost exclusively found in patients who progress to Type 1A diabetes prior to age 5, and less than half of individuals developing Type 1A diabetes after age 15 have levels of anti-insulin autoantibodies distinguished from controls. The levels of such antibodies are to some extent genetically influenced, are associated with DR4,[96] and are also associated with other haplotypes with DQA1 alleles of lineage 2, DQA1*0102, 0201,0301, 0401.[96] Presence of anti-insulin autoantibodies and their levels appear to be associated with such lineage 2 alleles in a dominant manner. In particular, individuals developing type 1A diabetes who are homozygous for DQA1*0501/DQB1*0201 (DR3 homozygotes) either are negative for anti-insulin autoantibodies or have very low levels. There are suggestive data that the levels of anti-insulin autoantibodies among ICA-positive first-degree relative correlate with the rate at which individuals progress to overt diabetes.[97] Relatives, however, who only express anti-insulin autoantibodies infrequently progress to overt diabetes. A high proportion of such anti-insulin autoantibody-positive, ICA-negative relatives under the age of 10 convert to ICA positivity.

Williams and coworkers modified the insulin autoantibody assay by utilizing protein A to precipitate autoantibodies[98] rather than polyethylene glycol used in standard insulin autoantibody assays for the past two decades.[94] The assay (termed micro-insulin autoantibody assay) was modified such that rather than 600ul of sera/sample, the assay utilized approximately 25ul of sera per sample for duplicate determination with and without competition with nonlabeled insulin. Not only did the assay utilize less sera, it also eliminated two artifacts apparently related to the use of polyethylene glycol (polyethylene glycol precipitates many proteins based on a protein's solubility). Both cord blood and hemolyzed sera have factors, which bind labeled

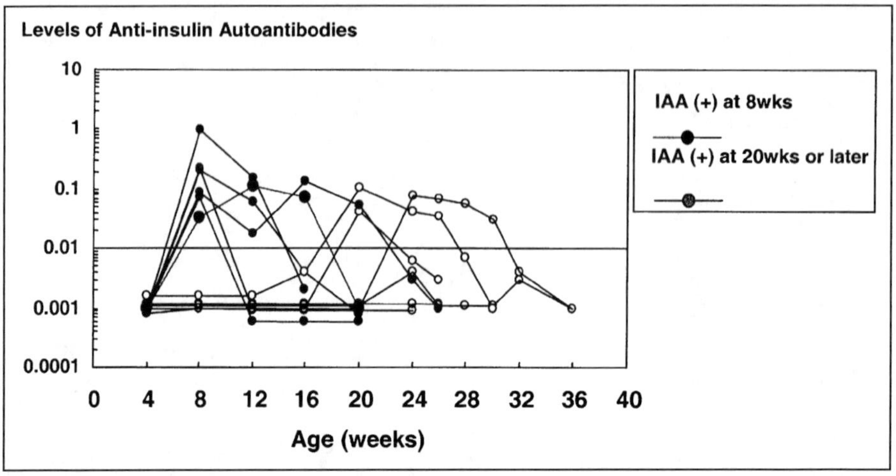

Figure 3. Development of insulin autoantibodies in individual NOD mice. Mice which expressed insulin autoantibodies at 8 weeks of age developed diabetes prior to 16 weeks, while mice expressing insulin autoantibodies after 20 weeks developed diabetes later. Yu et al. Proc Natl Acad Sci 2000; 97:1701. Copyright © 2000 National Academy of Sciences, U.S.A.

insulin and are precipitated by polyethylene glycol while both types of sera are negative with assays utilizing protein A precipitation.[99,100] We have further modified the Williams assay by utilizing a 96-well plate format with membrane filtration to separate bound from free autoantibodies, and direct counting on a beta counter of the precipitated ^{125}I-insulin.[101] The counting in a 96-well plate, and handling in 96-well plates allows semi-automated determination of insulin autoantibodies in a format similar to that for the GAD65 and ICA512 autoantibody assays.

Insulin autoantibodies are usually the first autoantibody to appear in young children developing type 1 diabetes.[102] This is particularly true for infants less than 1 year of age who begin to express autoantibodies. Of five such children with persistent insulin autoantibodies we have observed to date, four have progressed to overt diabetes prior to age 3. In addition the micro-IAA assay readily detects insulin autoantibodies in NOD mice and the early expression of insulin autoantibodies of NOD mice correlates with the early development of type 1 diabetes. Approximately 90% of NOD mice expressing insulin autoantibodies at 8 weeks of age develop diabetes by 16 weeks of age (Fig. 3). Insulin autoantibodies can be rapidly induced in normal Balb/c mice with the administration of the autoantigenic peptide, residues B:9-23 of insulin. These antibodies do not react with the immunizing peptide but with intact insulin.[103] This surprising finding clearly indicates that normal mice have autoreactive T and B cells able to respond to the peptide (T cell) and to intact insulin (B cell). Further studies using the B:9-23 peptide and poly-IC in Balb/c mice have generated insulitis.[103a]

A recent small workshop has evaluated insulin, GAD65, and ICA512 (IA-2) autoantibodies in NOD mice. The workshop report concludes that insulin autoantibodies measured by a sensitive radioassay are strongly associated with autoimmunity while neither GAD65 nor ICA512 (IA-2) specific autoantibodies were demonstrated.[103b]

Thomas and coworkers have produced a series of monoclonal insulin autoantibodies and recently transgenic mice producing an insulin autoantibody.[104] These mice are tolerant to insulin, suggesting that even low levels of insulin can induce B cell tolerance. With recent evidence that proinsulin is present in thymus and lymph node, the influence of such extra-islet insulin requires further study.[105,106]

ICA69

Pietropaolo and coworkers identified a novel islet protein termed ICA69[107] through the screening of a lambda gt11 human islet expression library with ICA-positive sera.[22] This protein, which on SDS gel migrates at 69 kd, has been sequenced and found to have in molecular weight of 54,600 (the aberrant gel migration is probably explained by the presence of several highly charged region in the molecule, as ICA 69 is not glycosylated). ICA69 is present, at least at the levels of messenger RNA, in brain, lung, kidney, and heart, with high levels in islet and other neuroendocrine tissues. In the rat, ICA69 is beta cell specific.[108] The subcellular localization of ICA69 has been elucidated.[108a]

ICA69 is identical in sequence to the cow's milk-related protein p69 described by Dosch in coworkers.[109] It has been reported that diabetic individuals have antibodies to bovine serum albumin and there are a number of epidemiological studies indicating that the neonatal ingestion of cow's milk increases the development of Type 1A diabetes. ICA69 and bovine serum albumin have two short regions (five amino acids) of identity and in the region of the ABBOS albumin peptide (potential T cell epitope), four of nine amino acids are identical. Such identity may be enough to stimulate cross-reactive T cell clones.

In the initial assay format, antibodies to ICA69 were measured by Western blot utilizing recombinant ICA69. The Western blot assay format relative to radioassays for antibodies reacting with insulin, GAD65, and ICA512 is inadequate, with more than 5% of normal controls reacting on Western blots. Anti-ICA69 autoantibodies are diabetes-related autoantibodies, but are present also in rheumatoid arthritis patients.[110] Further studies are needed to elucidate the role of this molecule in the pathogenesis of Type 1A diabetes. A report by Atkinson and coworkers questions the association with Type 1A diabetes of anti-albumin antibodies and of T cell responses to albumin.[111] Dosch and coworkers however in studies of mice can accelerate or inhibit development of diabetes with a peptide of the mouse molecule.[112]

Carboxypeptidase H

Carboxypeptidase H is another autoantigen discovered by the screening of islet expression libraries with prediabetic sera. Within islets, carboxypeptidase H cleaves carboxyterminal amino acids during the processing of proinsulin to insulin. This molecule is not islet specific, being present also in bovine adrenal, pituitary, and kidney.[87] In beta cells, carboxypeptidase H is localized in the insulin secretory granule, where it exists both in a membrane bound (52 kd) and soluble form (50 kd). Carboxypeptidase H is probably the most abundant islet protein after proinsulin-insulin. A radioassay or an ELISA for anticarboxypeptidase antibodies has not been developed. In the original assay format utilizing reactivity with recombinant Escherichia coli plaques, one of four prediabetics had anti-carboxypeptidase H antibodies.

ICA12

In the screening of islet expression libraries with patient autoantibodies Rabin and coworkers identified a molecule they termed ICA12[10] in addition to ICA512, discussed above. The complete sequence of ICA12 is now known and radioassays have been developed with in vitro transcription and translation of ICA12. Only a small percentage of patients with new onset diabetes express ICA12 autoantibodies. It is not clear whether the percentage of patients positive for ICA12 autoantibodies, compared to controls with other autoimmune disorders, is high enough for ICA12 autoantibodies to be truly type 1 diabetes related.[112a]

Partially Characterized Autoantigens

52 kd

A 52 kd antigen distinct from carboxypeptidase has been identified by Western blotting of human islet extracts with diabetic sera.[113] Interestingly, this antigen that appears to be islet cell-specific shares an antigenic determinant with the protein PC2, a component of the rubella virus capsid. The rubella virus is as yet to the only virus that has been clearly associated with the development of diabetes and only congenital rubella infection is associated with diabetes. The 52 kd antigen appears to be expressed within beta cell granules[113] and with future characterization may contribute to detection of anti-islet autoimmunity.

37-38 kd

A number of groups have reported antibodies to 38 kd molecules in patients with Type 1A diabetes and their first-degree relatives. In the original description of antibodies precipitating the 64 kd molecule form islets, it was shown that some sera also precipitated a 38 kd molecule. Honeymann and coworkers[114] identified a 38 kd autoantigen as the nuclear transcription protein jun-B from an islet expression library. The relevance of this antigen to autoantibodies of prediabetics is unclear. An additional 38 kd molecule has been described by Pak and coworkers.[115] Following their observation that some new-onset diabetic patients have cytomegalovirus infection, they immunized BALB/c mice with cytomegalovirus and obtained islet-reactive monoclonal antibodies that by Western blot recognize a 38 kd human islet protein.

Roep and coworkers[116] have identified a 38 kd islet granule autoantigen to which T cells react. The relationship between this antigen recognized by T cells and autoantibodies to 38 kd molecules is unknown.

155 kd

In 1990, McEvoy and coworkers used a novel method to detect diabetes-associated autoantigens. The developed a panel of mouse monoclonal antibodies reacting with the rat insulinoma cell line RIN5F and displaced the binding of the monoclonals to the tumor cells with diabetic sera. The binding of one of these monoclonals, termed 1A2, was selectively displaced by sera from diabetic children. The 1A2 monoclonal binds specifically to the insulinoma cell line RIN5F but not the other rat or human tissues. In a subsequent communication,[117] the authors reported that monoclonal 1A2 recognizes by Western blot a 150 kd membrane-pound protein DAP1 (diabetes associated protein 1). Several sera from IDDM patients also identified a 150 kd band by Western blotting of rat brain homogenates. Using an assay based on displacement of the radiolabeled monoclonal binding to rat insulinoma cells, 159 of 169 sera of children with IDDM versus 10 of 351 age-matched control sera were positive for autoantibodies to DAP1. Autoantibodies to this antigen are, however, also present in a large percentage of nondiabetic first-degree relatives (> 60%).

GLIMA38

Baekkeskov and coworkers have immunoprecipitated a glycosylated islet cell membrane antigen which they have termed GLIMA 38. The molecule of 38 Kd was immunoprecipitated with sera from 19% of patients with new onset diabetes.[118,118a]

Other Autoantigens

Reports from Elias and coworkers[119] suggested a role for T cell clones reacting to the heat shock protein (HSP)65 in the pathogenesis of diabetes in the NOD mouse. In particular, these authors identified a peptide within the sequence of human HSP that was recognized by T cell clones that could transfer a transient form of diabetes and hyperglycemia.[120] The administration of irradiation-attenuated HSP-specific T cells, or of the HSP peptide, prevented diabetes in the NOD mouse. Despite initial reports, HSP65 is apparently not an autoantigen of humoral autoimmunity in man.[121] Studies of heat shock protein in animal models have progressed[122] to phase I/phase II clinical trials.

In 1990, Johnson and coworkers[123,124] suggested that an islet-specific glucose transporter is an autoantigen in Type 1A diabetes. The evidence supporting this conclusion derived from the ability of IgG fractions from diabetic patients to inhibit glucose uptake from rat islet cells. This inhibitory activity could be removed by incubation of the IgG fraction with islet cell and hepatocyte membranes (which express two types of glucose transporter, GLUT-1 and GLUT-2), but not with erythrocytes or brush border membranes (which express a only GLUT-2). Sera from 26 of 27 patients with IDDM inhibited the rat islet glucose uptake versus 0 of 5 sera from patients with NIDDM. Interestingly, the uptake of L-leucine was not affected by the IgG fractions. Though no direct biochemical data are provided for reaction of antibodies with GLUT-2, a publication indicates that introduction of the GLUT-2 gene into cell lines results in detection of surface autoantibodies,[124] implying that GLUT-2 is recognized by autoantibodies of diabetic patients.

T Cell Reactivity

Some of the target molecules of autoantibodies described in this chapter are also able to drive proliferation of T cells from animal models and patients with Type 1A diabetes or from ICA-positive individuals.

The study to date of T cell responses of man to islet autoantigens has been particularly difficult with relatively low levels of stimulation, when stimulation is observed. This may relate to a very low frequency of autoreactive T cells outside of the islets. In an international workshop with multiple laboratories studying a series of "blinded" islet antigens and peptides none of the laboratories could distinguish a small panel of patients with diabetes from control individuals[125]. We believe it is likely that the assay formats will have to be greatly improved to be able to reliably detect T cell responses.

Atkinson and coworkers[126,127] studied the reactivity of peripheral blood mononuclear cells to recombinant glutamic acid decarboxylase in newly diagnosed type 1 patients, ICA-positive and ICA-negative relatives of individuals with diabetes, and in healthy controls. A higher proportion of subjects gave a positive proliferative response to GAD among the newly diagnosed patients than among the controls. Interestingly, nondiabetic ICA-positive relatives were also more likely to express T cell reactivity to GAD (63% versus 11%). Harrison and coworkers[128] reported proliferation of peripheral blood T cells in presence of the central region of GAD67 (aa 208-404) in 38% of newly diagnosed IDDM patients and in 41% of ICA-positive relatives (> 20 JDF units). Only 4% of HLA-matched controls responded to GAD67. In a subsequent report, Harrison and coworkers[48] described an inverse relationship between humoral and cellular immunity to glutamic acid decarboxylase, suggesting that ICA-positive relatives who have increased T cell reactivity to GAD (and lower titer of anti-GAD antibody) represent a group with higher diabetes risk.

Assays for reactivity of peripheral blood lymphocytes with GAD give relatively low stimulation indices.[126] In the NOD mouse, utilizing spleen cells, Kaufman and coworkers[129] and Tisch and coworkers[130] reported that T cell proliferative responses to GAD precede responses to other antigens. The stimulation indices again are low but of interest and T cell clones to GAD have been derived.[131] Nepom and coworkers have isolated a peptide of GAD65 and are developing T cell tetramer (DRB1*0401) assays for reactivity with this epitope.[132,132a]

Dosch and coworkers[133] have reported that T cells from new-onset diabetic children proliferate in the presence of the "ABBOS peptide, "a bovine albumin-specific region that has homology with the previously described islet molecule p69, and it is also a target of autoantibodies in IDDM. However, a study by Atkinson and coworkers[111] failed to confirm these results. Atkinson and coworkers speculate that anti-BSA autoantibodies are a marker of altered immunity rather than specific marker for Type 1A diabetes.

A number of groups have found T cell responses to ICA512 (IA-2) and IA-2β[134,135] and T cells recognizing IA-2β are able to transfer diabetes in a transplant model.[135]

Roep[136] reported that cell lines from newly diagnosed patients selectively respond to insulin secretory granule proteins. Such a response was identified in 14 diabetes-derived cell lines and in only two cell lines of controls. Interestingly, two nondiabetic ICA-positive individuals produced T cell lines responsive to insulin granule proteins. A 38 kd mitochondrial protein was eluted from gels after electrophoresis of insulin secretory granule membranes. Stimulation of insulin granule-responsive T cell lines with this proteins generated a response in 80% of the lines tested.[137] Reactivity to this protein (Imogen 38) was restricted to type 1 patients and to ICA-positive individuals.[116]

T cell proliferative responses to insulin can be detected utilizing peripheral blood of prediabetics or splenocytes of NOD mice, but again stimulation indices are low.[138] Wegmann and coworkers have found that a large percentage of T cells infiltrating islets of NOD mice strongly proliferate to insulin and a B-chain peptide of insulin(B:9-23).[139-141] In contrast, a minimal response of splenocytes to insulin is found. A subset of these T cell clones are able to transfer diabetes into immunodeficient NOD mice and thus are sufficient for disease pathogenesis. A response to insulin is seen as early as 4 weeks of age. This study in NOD mice, where islets are available for analysis, highlights a potential difficulty of studies in man utilizing peripheral lymphocytes. The T cell clones of Wegmann and coworkers recognizing insulin B-chain peptide B:9-23 recognize one of two distinct peptides, B:9-16 or B:9-23. Despite recognizing two different peptides both sets of T cell clones utilize a dominant Vα chain, Vα 13.3.[142] Alvia and coworkers have analyzed stimulation of human lymphocytes with the B:9-23 insulin peptide (identical sequence in mouse and man insulin) and find reactivity amongst patients but not controls.[143]

As discussed, Nepom and coworkers are developing class II tetramer reagents and in particular detecting autoreactive cells with a DRB1*0401 or 0404 tetramer loaded with GAD peptide 555 to 567[132] (and oral communication). There is certainly considerable further development needed but it is hoped that ELISPOT or tetramer assays will eventually allow quantitation of autoreactive T cells.[132a]

Conclusions

During the past several years, investigators have defined a growing family of islet autoantigens recognized by autoantibodies of prediabetics. The ability to produce many of the antigens with recombinant technology (insulin, GAD, ICA512) will facilitate the prediction of Type 1A diabetes. In addition, availability of such autoantigens will be important to elucidate the pathogenesis of Type 1A diabetes. It is likely that such a large number of autoantibodies are secondary to beta cell destruction and the presentation of normal antigens in an inflammatory environment. Nevertheless, studies in the NOD mouse suggest that influencing the immune response to several autoantigens (e.g., oral insulin, intravenous GAD, subcutaneous insulin) will prevent diabetes. At present, the primacy in disease pathogenesis of any of the defined autoantigens is unknown.

Despite uncertainty about pathogenic importance, the ability to biochemically detect autoantibodies to a series of autoantigens greatly facilitate the prediction of Type 1A diabetes[47] (Fig. 4). Utilizing only three radioassays for autoantibodies (insulin, GAD, and ICA512) 95% of prediabetic or recent-onset diabetic children express one or more antibodies (versus only 3 out of 100 controls). Eighty percent express two or more autoantibodies, versus "no" controls. Twenty-five percent of diabetics or prediabetics express all three of the above autoantibodies. The expression of autoantibodies assorts independently in prediabetic individuals and thus, by utilizing the three assays, a high positive predictive value for disease is possible while retaining disease sensitivity (Combinatorial Autoantibody Prediction[144]). Such "biochemical" autoantibody screening will likely have its largest impact for the prediction of Type 1A diabetes in the general population where the positive predictive value of ICA positivity is less than 10%, while presence of two or more of three autoantibodies should give positive predictive values greater than 70%. Screening of the general population for diabetes risk is likely to be a major importance if trials for the prevention of Type 1A diabetes among first-degree relatives are successful. Approximately 90% of individuals who develop Type 1A diabetes have no first-degree relatives with the disease and thus such screening may be essential for the large-scale prevention of Type 1A diabetes.

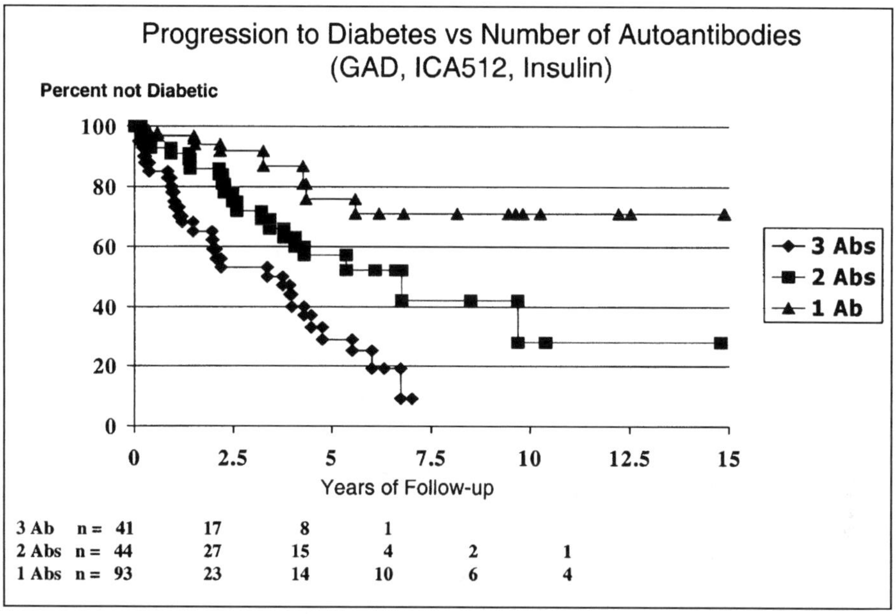

Figure 4. Combinatorial Autoantibody prediction. Verge et al. Diabetes 1996: 45:926-933. Reprinted with permission from The American Diabetes Association. Copyright © 1996 American Diabetes Association.

References

1. Leslie RD, Atkinson MA, Notkins AL. Autoantigens IA-2 and GAD in Type I (insulin-dependent) diabetes. Diabetologia 1999; 42:3-14.
2. Bach JF. Etiology and Pathogenesis of Human Insulin-Dependent Diabetes Mellitus 1999; 293-307.
3. Abiru N, Eisenbarth GS. Autoantibodies and autoantigens in type 1 diabetes: Role in pathogenesis, prediction and prevention. Can J Diabetes Care 1999; 23:59-65.
4. Bonifacio E, Scirpoli M, Kredel K et al. Early autoantibody responses in prediabetes are IgG1 dominated and suggest antigen-specific regulation. J Immunol 1999; 163:525-532.
5. Colman PG, McNair P, Margetts H et al. The Melbourne PreDiabetes Study: Prediction of type 1 diabetes mellitus using antibody and metabolic testing. Med J Aust 1998; 169:81-84.
6. Kulmala P, Savola K, Petersen JS et al. Prediction of insulin-dependent diabetes mellitus in siblings of children with diabetes: A population-based study. J Clin Invest 1998; 101:327-336.
7. Bingley PJ, Bonifacio E, Williams AJK et al. Prediction of IDDM in the general population: Strategies based on combinations of autoantibody markers. Diabetes 1997; 46:1701-1710.
8. Greenbaum C, Palmer JP, Kuglin B et al. Insulin autoantibodies measured by radioimmunoassay methodology are more related to insulin-dependent diabetes mellitus than those measured by enzyme-linked immunosorbent assay: Results of the Fourth International Workshop on the Standardization of Insulin Autoantibody Measurement. J Clin Endocrinol Metab 1992; 74:1040-1044.
9. Baekkeskov S, Aanstoot H-J, Christgau S et al. Identification of the 64K autoantigen in insulin-dependent diabetes as the GABA-synthesizing enzyme glutamic acid decarboxylase [published erratum appears in Nature 1990 Oct 25;347(6295):782]. Nature 1990; 347:151-156.
10. Rabin DU, Pleasic SM, Palmer-Crocker R et al. Cloning and expression of IDDM-specific human autoantigens. Diabetes 1992; 41:183-186.
11. Rabin DU, Pleasic SM, Shapiro JA et al. Islet cell antigen 512 is a diabetes-specific islet autoantigen related to protein tyrosine phosphatases. J Immunol 1994; 152:3183-3188.
12. Lan MS, LU J, Goto Y et al. Molecular cloning and identification of a receptor-type protein tyrosine phosphatase, IA-2, from human insulinoma. DNA Cell Biol 1994; 13:505-514.
13. Wasmeier C, Hutton JC. Molecular cloning of phogrin, a protein-tyrosine phosphatase homologue localized to insulin secretory granule membranes. J Biol Chem 1996; 271:18161-18170.

14. LU J, Li Q, Xie H et al. Identification of a second transmembrane protein tyrosine phosphatase, IA-2β, as an autoantigen in insulin-dependent diabetes mellitus: Precursor of the 37-kDa tryptic fragment. Proc Natl Acad Sci USA 1996; 93:2307-2311.
15. Castano L, Russo E, Zhou L et al. Identification and cloning of a granule autoantigen (carboxypeptidase-H) associated with type I diabetes. J Clin Endocrinol Metab 1991; 73:1197-1201.
16. Thomas NM, Ginsberg-Fellner F, Mcevoy RC. Strong association between diabetes and displacement of mouse anti-rat insulinoma cell monoclonal antibody by human serum in vitro. Diabetes 1990; 39:1203-1211.
17. Pan Y-X, Thomas NM, Mcevoy RC. Purification and partial characterization of 160 kD and 155 kD B cell membrane antigens detected by MAb 1A2. Autoimmunity 1993; 15:Suppl:77A.
18. Karounos DG, Wolinsky JS, Thomas JW. Monoclonal antibody to rubella virus capsid protein recognizes a β-cell antigen. J Immunol 1993; 150:3080-3085.
19. Christie MR, Hollands JA, Brown TJ et al. Detection of pancreatic islet 64,000 M_r autoantigens in insulin-dependent diabetes distinct from glutamate decarboxylase. J Clin Invest 1993; 22:240-248.
20. Roep BO, Kallan AA, Hazenbos WLW et al. T-cell reactivity to 38kD insulin-secretory-granule protein in patients with recent-onset type 1 diabetes. Lancet 1991; 337:1439-1441.
21. Pietropaolo M, Gianani R, Yu L et al. Independent expression of biochemically determined autoantibodies (can we do away with ICA?). Autoimmunity 1993.
22. Pietropaolo M, Castano L, Babu S et al. Islet cell autoantigen 69 kDa (ICA69): Molecular cloning and characterization of a novel diabetes associated autoantigen. J Clin Invest 1993; 92:359-371.
23. Birk OS, Elias D, Weiss AS et al. NOD mouse diabetes: The ubiquitous mouse hsp60 is a beta-cell target antigen of autoimmune T cells. J Autoimmun 1996; 9:159-166.
24. Elias D, Cohen I, Shechter Y et al. Antibodies to insulin receptor followed by anti-idiotype. Diabetes 1987; 36:348-354.
25. Georgopoulos C, McFarland H. Heat shock proteins in multiple sclerosis and other autoimmune diseases. Immunol Today 1993; 14:373-375.
26. Atkinson MA, Maclaren NK, Scharp DW. No evidence for HSP65 in insulin dependent diabetes. Lancet 1990; 337:263-264.
27. Lernmark Å, Freedman ZR, Hofmann C et al. Islet-cell-surface antibodies in juvenile diabetes mellitus. N Engl J Med 1978; 299:375-380.
28. Martin S, Wolf-Eichbaum D, Duinkerken G et al. Development of type 1 diabetes despite severe hereditary B-lymphocyte deficiency. N Engl J Med 2001; 345:1036-1040.
29. Nerup J. Addison's disease – clinical studies: A report of 108 cases. Acta Endocrinol 1974; 76:127-141.
30. Bottazzo GF, Florin-Christensen A, Doniach D. Islet-cell antibodies in diabetes mellitus with autoimmune polyendocrine deficiencies. Lancet 1974; 2:1279-1283.
31. Eizirik DL, Sandler S, Palmer JP. Repair of pancreatic β-cells: A relevant phenomenon in early IDDM? Diabetes 1993; 42:1383-1391.
32. Thai A-C, Eisenbarth GS. Natural history of IDDM. Diabetes Rev 1993; 1:1-14.
33. Turner R, Stratton I, Horton V et al. UKPDS 25: Autoantibodies to islet-cell cytoplasm and glutamic acid decarboxylase for prediction of insulin requirement in type 2 diabetes. UK Prospective Diabetes Study Group. Lancet 1997; 350:1288-1293.
34. Greenbaum CJ, Palmer JP, Nagataki S et al. Improved specificity of ICA assays in the Fourth International Immunology of Diabetes Exhange Workshop. Diabetes 1992; 41:1570-1574.
35. Gianani R, Pugliese A, Bonner-Weir S et al. Prognostically significant heterogeneity of cytoplasmic islet cell antibodies in relatives of patients with type I diabetes. Diabetes 1992; 41:347-353.
36. Genovese S, Bonifacio E, McNally JM et al. Distinct cytoplasmic islet cell antibodies with different risks for type I (insulin-dependent) diabetes mellitus. Diabetologia 1992; 35:385-388.
37. Solimena M, Folli F, Denis-Donini S et al. Autoantibodies to glutamic acid decarboxylase in a patient with stiff-man syndrome, epilepsy and type I diabetes mellitus. N Engl J Med 1988; 318:1012-1020.
38. Colman PG, Nayak RC, Campbell IL et al. Binding of cytoplasmic islet cell antibodies is blocked by human pancreatic glycolipid extracts. Diabetes 1988; 37:645-652.
39. Erdo SL, Wolff JR. Gamma-aminobutyric acid outside the mammalian brain. J Neurochem 1990; 54:363-372.
40. Petersen JS, Russel S, Marshall MO et al. Differential expression of glutamic acid decarboxylase in rat and human islets. Diabetes 1993; 42:484-495.
41. Kim J, Richter W, Aanstoot HJ et al. Differential expression of GAD65 and GAD67 in human, rat and mouse pancreatic islets. Diabetes 1993; 42:1799-1808.
42. Eisenbarth GS. Mouse or man: Is GAD the cause of type I diabetes? Diabetes Care 1994; 17:1-3.

43. Christie M, Landin-Olsson M, Sundkvist G et al. Antibodies to a Mr-64,000 islet cell protein in Swedish children with newly diagnosed type I (insulin-dependent) diabetes. Diabetologia 1988; 31:597-602.
44. Baekkeskov S, Landin M, Kristensen JK et al. Antibodies to a 64,000 Mr human islet cell antigen precede the clinical onset of insulin-dependent diabetes. J Clin Invest 1987; 79:926-934.
45. Gianani R, Jackson R, Eisenbarth GS. Evidence that the autoantigen of restricted ICA is GAD. Diabetes Res Clin Prac 1991; 14(suppl1):S13.
46. Velloso LA, Kampe O, Hallberg A et al. Demonstration of GAD-65 as the main immunogenic isoform of glutamate decarboxylase in type 1 diabetes and determination of autoantibodies using a radioligand produced by eukaryotic expression. J Clin Invest 1993; 91:2084-2090.
47. Verge CF, Stenger D, Bonifacio E et al. Combined use of autoantibodies (IA-2ab, Gadab, IAA, ICA) in type 1 diabetes: Combinatorial islet autoantibody workshop. Diabetes 1998; 47:1857-1866.
48. Harrison LC, Honeyman MC, DeAizpurua HJ et al. Inverse relation between humoral and cellular immunity to glutamic acid decarboxylase in subjects at risk of insulin-dependent diabetes. Lancet 1993; 341:1365-1369.
49. Bu DF, Tobin AJ. The exon-intron organization of the genes (GAD1 and GAD2) encoding two human glutamate decarboxylases (GAD67 and GAD65) suggests that they derive from a common ancestral GAD. Genomics 1994; 21:222-228.
50. Verge CF, Gianani R, Kawasaki E et al. Prediction of type I diabetes in first-degree relatives using a combination of insulin, GAD and ICA512bdc/IA-2 autoantibodies. Diabetes 1996; 45:926-933.
51. Hagopian WA, Michelsen B, Karlsen AE et al. Autoantibodies in IDDM primarily recognize the 65,000-M(r) rather than the 67,000-M(r) isoform of glutamic acid decarboxylase. Diabetes 1993; 42:631-636.
52. Falorni A, Ackefors M, Carlberg C et al. Diagnostic sensitivity of immunodominant epitopes of glutamic acid decarboxylase (GAD65) autoantibodies in childhood IDDM. Diabetologia 1996; 39:1091-1098.
53. Hallengren B, Falorni A, Landin-Olsson M et al. Islet cell and glutamic acid decarboxylase antibodies in hyperthyroid patients: At diagnosis and following treatment. J Intern Med 1996; 239:63-68.
54. Richter W, Endl J, Eiermann TH et al. Human monoclonal islet cell antibodies from a patient with insulin-dependent diabetes mellitus reveal glutamate decarboxylase as the target antigen. Proc Natl Acad Sci USA 1992; 89:8467-8471.
55. Richter W, Seissler J, Northemann W et al. Cytoplasmic islet cell antibodies recognize distinct islet antigens in IDDM but not in stiff man syndrome. Diabetes 1993; 42:1642-1648.
56. Syren K, Lindsay L, Stoehrer B et al. Immune reactivity of diabetes-associated human monoclonal autoantibodies defines multiple epitopes and detects two domain boundaries in glutamate decarboxylase. J Immunol 1996; 157:5208-5214.
57. Daw K, Powers AC. Two distinct glutamic acid decarboxylase autoantibody specificities in IDDM target different epitopes. Diabetes 1995; 44:216-220.
58. Schwartz HL, Chandonia JM, Kash SF et al. High-resolution autoreactive epitope mapping and structural modeling of the 65 kDa form of human glutamic acid decarboxylase. J Mol Biol 1999; 287:983-999.
59. Ujihara N, Daw K, Gianani R et al. Identification of glutamic acid decarboxylase autoantibody heterogeneity and epitope regions in type I diabetes. Diabetes 1994; 43:968-975.
60. Butler MH, Solimena M, Dirkx R et al. Identification of a dominant epitope of glutamic acid decarboxylase (GAD-65) recognized by autoantibodies in stiff-man syndrome. J Exp Med 1993; 178:2097.
61. Kim J, Namchuk M, Bugawan T et al. Higher autoantibody levels and recognition of a linear NH2-terminal epitope in the autoantigen GAD65, distinguish stiff-man syndrome from insulin-dependent diabetes mellitus. J Exp Med 1994; 180:595-606.
62. Daw K, Ujihara N, Atkinson M et al. Glutamic acid decarboxylase autoantibodies in Stiff-man syndrome and insulin-dependent diabetes mellitus exhibit similarities and differences in epitope recognition. J Immunol 1996; 156:818-825.
63. Bjork E, Velloso LA, Kampe O et al. GAD autoantibodies in IDDM, stiff-man syndrome, and autoimmune polyendocrine syndrome type I recognize different epitopes. Diabetes 1994; 43:161-165.
64. Powers AC, Bavik K, Tremble J et al. Comparative analysis of epitope recognition of glutamic acid decarboxylase (GAD) by autoantibodies from different autoimmune disorders. Clin Exp Immunol 1999; 118:349-356.
65. Christie MR, Tun RYM, Lo SSS et al. Antibodies to GAD and tryptic fragments of islet 64K antigen as distinct markers for development of IDDM: Studies with identical twins. Diabetes 1992; 41:782-787.

66. Christie MR, Genovese S, Cassidy D et al. Antibodies to islet 37k antigen, but not to glutamate decarboxylase, discriminate rapid progression to IDDM in endocrine autoimmunity. Diabetes 1994; 43:1254-1259.
67. Payton MA, Hawkes CJ, Christie MR. Relationship of the 37,000- and 40,000-M_r tryptic fragments of islet antigens in insulin-dependent diabetes to the protein tyrosine phosphatase-like molecule IA-2 (ICA512). J Clin Invest 1995; 96:1506-1511.
68. Kawasaki E, Yu L, Rewers MJ et al. Definition of multiple ICA512/phogrin autoantibody epitopes and detection of intramolecular epitope spreading in relatives of patients with type 1 diabetes. Diabetes 1998; 47:733-742.
69. Gianani R, Rabin DU, Verge CF et al. ICA512 autoantibody radioassay. Diabetes 1995; 44:1340-1344.
70. Kawasaki E, Yu L, Gianani R et al. Evaluation of islet cell antigen (ICA) 512/IA-2 autoantibody radioassays using overlapping ICA512/IA-2 constructs. J Clin Endocrinol Metab 1997; 82:375-380.
71. Farilla L, Tiberti C, Luzzago A et al. ICA512bdc cDNA expression library displayed on the capsid protein D of bacteriophage lambda: A novel strategy to identify major antigenic epitopes in autoimmune diseases.1999.
72. Ziegler A-G, Hummel M, Schenker M et al. Autoantibody appearance and risk for development of childhood diabetes in offspring of parents with type 1 diabetes. The 2-year analysis of the German BABYDIAB study. Diabetes 1999; 48:460-468.
73. Kundu SK, Pleatman MA, Redwine WA et al. Binding of monoclonal antibody A2B5 to gangliosides. Biochem Biophys Res Commun 1983; 116:836-842.
74. Nayak RC, Attawia NA, Cahill CJ et al. Expression of a monoclonal antibody (3G5) defined ganglioside antigen in renal cortex. Kidney Int 1992; 41:1638-1645.
75. Alejandro R, Shienvold FL, Hajek SAV et al. A ganglioside antigen on the rat pancreatic B cell surface identified by monoclonal antibody R2D6. J Clin Invest 1984; 74:25-38.
76. Srikanta S, Krisch K, Eisenbarth GS. Islet cell proteins defined by monoclonal islet cell antibody HISL-19. Diabetes 1986; 35:300-305.
77. Srikanta S, Telen M, Posillico JT et al. Monoclonal antibodies to a human islet cell surface glycoprotein: 4F2 and LC7-2. Endocrinology 1987; 120:2240-2244.
78. Gillard BK, Thomas JW, Nell LJ et al. Antibodies against ganglioside GT3 in the sera of patients with type I diabetes mellitus. J Immunol Methods 1989; 142:3826-3832.
79. Buschard K, Josefsen K, Horn T et al. Sulphatide and anti-sulphatide antibodies in type I diabetes mellitus. Diabetes 1993; 42(Suppl 1):5A.
80. Dotta F, Previti M, Lenti L et al. GM2-1 pancreatic islet ganglioside: Identification and characterization of a novel islet-specific molecule. Diabetologia 1995; 38:1117-1121.
81. Dotta F, Gianani R, Dionisi S et al. Expression of anti-GM2-1 islet ganglioside antibodies in ICA+ relatives of type I diabetics. Autoimmunity 1993; 15(suppl):70-70.
82. Folkman S, Lazarus RS. 1988; 1.
83. Dotta F, Colman PG, Lombardi D et al. Ganglioside expression in human pancreatic islets. Diabetes 1989; 38:1478-1483.
84. Dotta F, Previti M, Tiberti C et al. Identification of the GM2-1 islet ganglioside: Similarities with a major neuronal autoantigen. Diabetes 1992; 41:365A-
85. Pugliese A, Zeller M, Fernandez A et al. The insulin gene is transcribed in the human thymus and transcription levels correlate with allelic variation at the INS VNTR-IDDM2 susceptibility locus for type I diabetes. Nat Genet 1997; 15:293-297.
86. Hanahan D. Peripheral-antigen-expressing cells in thymic medulla: Factors in self- tolerance and autoimmunity. Curr Opin Immunol 1998; 10:656-662.
87. Davidson HW, Hutton JC. The insulin-secretory-granule carboxypeptidase H. Biochem J 1993; 245:575-582.
88. Berson SA, Yallow RS. Quantitative aspects of the reaction between insulin and insulin-binding antibody. J Clin Invest 1959; 38:1996-2016.
89. Davidson JK, DeBra DW. Immunologic insulin resistance. Diabetes 1978; 27:307-318.
90. Hirata Y, Tominaga M, Ito J et al. Spontaneous hypoglycaemia with insulin autoimmunity in Graves' disease. Ann Intern Med 1974; 81:214-218.
91. Uchigata Y, Kuwata S, Tsushima T et al. Patients with Graves' disease who developed insulin autoimmune syndrome (Hirata disease) possess HLA-Bw62/Cw4/DR4 carrying DRB1*0406. J Clin Endocrinol Metab 1993; 77:249-254.
92. Uchigata Y, Hirata Y. Insulin Autoimmune Syndrome (IAS, Hirata Disease) 1999; 133-148.
93. Ziegler AG, Vardi P, Gross DJ et al. Production of insulin antibodies by mice rejecting insulin transfected cells. J Autoimmun 1989; 2:219-227.

94. Palmer JP, Asplin CM, Clemons P et al. Insulin antibodies in insulin-dependent diabetics before insulin treatment. Science 1983; 222:1337-1339.
95. Castano L, Ziegler AG, Ziegler R et al. Characterization of insulin autoantibodies in relatives of patients with type 1 diabetes. Diabetes 1993; 42:1202-1209.
96. Pugliese A, Bugawan T, Moromisato R et al. Two subsets of HLA-DQA1 alleles mark phenotypic variation in levels of insulin autoantibodies in first degree relatives at risk for insulin-dependent diabetes. J Clin Invest 1994; 93:2447-2452.
97. Vardi P, Ziegler AG, Matthews JH et al. Concentration of insulin autoantibodies at onset of type I diabetes. Inverse log-linear correlation with age. Diabetes Care 1988; 11:736-739.
98. Williams AJK, Bingley PJ, Bonifacio E et al. A novel micro-assay for insulin autoantibodies. J Autoimmun 1997; 10:473-478.
99. Bilbao JR, Calvo B, Urrutia I et al. Anti-insulin activity in normal newborn cord-blood serum
100. Absence of IgG-mediated insulin binding. Diabetes 1997; 46:713-716.
101. Naserke HE, Bonifacio E, Ziegler A-G. Immunoglobulin G insulin autoantibodies in BABYDIAB offspring appear postnatally: Sensitive early detection using a protein A/G-based radiobinding assay. J Clin Endocrinol Metab 1999; 84:1239-1243.
102. Rosenbauer J, Herzig P, von Kries R et al. Temporal, seasonal, and geographical incidence patterns of type I diabetes mellitus in children under 5 years of age in Germany. Diabetologia 1999; 42:1055-1059.
103. Yu L, Robles DT, Abiru N et al. Early expression of anti-insulin autoantibodies of man and the NOD mouse: Evidence for early determination of subsequent diabetes. Proc Natl Acad Sci USA 2000; 97:1701-1706.
103a. Moriyama H, Wen L, Abiru N et al. Induction and acceleration of insulitis/diabetes in mice with a viral mimic (polyinosinic-polycytidylic acid) and an insulin self-peptide. Proc Natl Acad Sci USA 2002; 99(8):5539-5544.
103b. Bonifacio E, Atkinson M, Eisenbarth GS et al. International workshop on autoimmunity in animal models of autoimmune diabetes identifies insulin, but not GAD or IA-2 as specific antigens of humoral autoimmunity in the nonobese diabetic mouse. Diab 2001; 50:2451-2458.
104. Abiru N, Maniatis AK, Yu L et al. Peptide and MHC Specific Breaking of Humoral Tolerance to Native Insulin with the B:9-23 Peptide in Diabetes Prone and Normal Mice. Diabetes 2001; 50:1274-1281.
105. Rojas M, Hulbert C, Thomas JW. Anergy and not clonal ignorance determines the fate of B cells that recognize a physiological autoantigen. J Immunol 2001; 166:3194-3200.
106. Bellward GD, Chang T, Rodrigues B et al. Hepatic cytochrome P-450j induction in the spontaneously diabetic BB rat. Mol Pharmacol 1988; 33:140-143.
107. Pugliese A, Brown D, Garza D et al. Self-Antigen Presenting Cells Expressing Islet Cell Molecules in Human Thymus and Peripheral Lymphoid Organs: Phenotypic Characterization and Implications for Immunological Tolerance and Type 1 Diabetes. J Clin Invest 2001; 107:555-564.
107. Li L, Hagopian WA, Brashear HR et al. Identification of autoantibody epitopes of glutamic acid decarboxylase in stiff-man syndrome patients. J Immunol 1994; 152:930-934.
108. Stassi G, Schloot N, Pietropaolo M. Islet cell autoantigen 69 kDa (ICA69) is preferentially expressed in the human islets of Langerhans than exocrine pancreas. Diabetologia 1997; 40:120-121.
108a. Spitzenberger F, Pietropaolo S, Verkade P et al. Islet cell autoantigen of 69 kDa is an arfaptin-related protein associated with the golgi complex of insulinoma INS-1 cells. J Biol Chem 2003; 278(28):26166-26173.
109. Karges W, Gaedigk R, Hui MF et al. Molecular cloning of murine ICA69: Diabetes-prone mice recognize the human autoimmune-epitope, Tep69, conserved in splice variants from both species. Biochim Biophys Acta 1997; 1360:97-101.
110. Martin S, Kardorf J, Schulte B et al. Autoantibodies to the islet antigen ICA69 occur in IDDM and in rheumatoid arthritis. Diabetologia 1995; 38:351-355.
111. Atkinson MA, Bowman MA, Kao K et al. Lack of immune responsiveness to bovine serum albumin in insulin- dependent diabetes. N Engl J Med 1993; 329:1853-1858.
112. Winer S, Gunaratnam L, Astsatourov I et al. Peptide dose, MHC affinity, and target self-antigen expression are critical for effective immunotherapy of nonobese diabetic mouse prediabetes. J Immunol 2000; 165:4086-4094.
112a. Fida S, Myers MA, Whittingham S et al. Autoantibodies to the transcriptional factor SOX13 in primary biliary cirrhosis compared with other diseases. J Autoimmun 2002; 19(4):251-257.
113. Karounos DG, Simmerman L, Hickman SL et al. Identification of the p52-Rubella related autoantigen as an insulin secretory granule protein. Diabetes 1993; 42(Suppl 1):221A.
114. Honeyman MC, Cram DS, Harrison LC. Transcription factor jun-B is target of autoreactive T-cells in IDDM. Diabetes 1993; 42:626-630.

115. Pak CY, Cha CY, Rajotte RV et al. Human pancreatic islet cell specific 38 kilodalton autoantigen identified by cytomegalovirus-induced monoclonal islet cell autoantibody. Diabetologia 1990; 33:569-572.
116. Arden SD, Roep BO, Neophytou PI et al. Imogen 38: A novel 38-kD islet mitochondrial autoantigen recognized by T cells from a newly diagnosed type I diabetic patient. J Clin Invest 1996; 97:551-561.
117. Mcevoy RC, Thomas NM, Greig F et al. Anti-islet autoantibodies detected by monoclonalantibody 1A2: Further studies suggesting a role in the pathogenesis of IDDM. Diabetologia 1996; 39:1365-1371.
118. Aanstoot H-J, Kang S-M, Kim J et al. Identification and characterization of glima 38, a glycosylated islet cell membrane antigen, which together with GAD65 and IA2 marks the early phases of autoimmune response in type I diabetes. J Clin Invest 1996; 97:2772-2783.
118a. Roll U, Turck CW, Gitelman SE et al. Peptide mapping and characterisation of glycation patterns of the glima 38 antigen recognised by autoantibodies in Type I diabetic patients. Diabetol 2000; 43(5):598-608.
119. Birk OS, Elias D, Weiss AS et al. NOD mouse diabetes: The ubiquitous mouse hsp60 is a beta-cell target antigen of autoimmune T cells. J Autoimmun 1996; 9:159-166.
120. Elias D, Reshef T, Birk OS et al. Vaccination against autoimmune mouse diabetes with a T-cell epitope of the human 65-kDa heat shock protein. Proc Natl Acad Sci USA 1991; 88:3088-3091.
121. Jones DB, Hunter NR, Duff GW. Heat-shock protein 65 as a B cell antigen of insulin-dependent diabetes. Lancet 1990; 336:583-585.
122. Elias D, Meilin A, Ablamunits V et al. Hsp60 peptide therapy of NOD mouse diabetes induces a Th2 cytokine burst and downregulates autoimmunity to various β-cell antigens. Diabetes 1997; 46:758-764.
123. Johnson JH, Crider BP, McCorkle K et al. Inhibition of glucose transport into rat islet cells by immunoglobulins from patients with new-onset insulin-dependent diabetes mellitus. N Engl J Med 1990; 322:653-659.
124. Inman LR, McAllister CT, Chen L et al. Autoantibodies to the GLUT-2 glucose transporter of β cells in insulin-dependent diabetes mellitus of recent onset. Proc Natl Acad Sci USA 1993; 90:1281-1284.
125. Roep BO, Atkinson MA, Van Endert PM et al. Autoreactive T cell responses in Insulin-dependent (type 1) diabetes mellitus. Report of the first international workshop for standardization of T cell assays. J Autoimmun 1999; 13:267-282.
126. Atkinson MA, Kaufman DL, Campbell L et al. Response of peripheral-blood mononuclear cells to glutamate decarboxylase in insulin-dependent diabetes. Lancet 1992; 339:458-459.
127. Atkinson MA, Bowman MA, Campbell L et al. Cellular immunity to a determinant common to glutamate decarboxylase and coxsackie virus in insulin-dependent diabetes. J Clin Invest 1994; 94:2125-2129.
128. Harrison LC, Chu SX, DeAizpurua HJ et al. Islet-reactive T cells are a marker of preclinical insulin-dependent diabetes. J Clin Invest 1992; 89:1161-1165.
129. Kaufman DL, ClareSalzler M, Tian J et al. Spontaneous loss of T-cell tolerance to glutamic acid decarboxylase in murine insulin dependent diabetes. Nature 1993; 366:69-72.
130. Tisch R, Yang X-D, Singer SM et al. Immune response to glutamic acid decarboxylase correlates with insulitis in nonobese diabetic mice. Nature 1993; 366:72-75.
131. Tisch R, Wang B, Serreze DV. Induction of glutamic acid decarboxylase 65-specific Th2 cells and suppression of autoimmune diabetes at late stages of disease is epitope dependent. J Immunol 1999; 163:1178-1187.
132. Nepom GT, Lippolis JD, White FM et al. Identification and modulation of a naturally processed T cell epitope from the diabetes-associated autoantigen human glutamic acid decarboxylase 65 (hGAD65). Proc Natl Acad Sci USA 2001; 98:1763-1768.
132a. Lieberman SM, Evans AM, Han B et al. Identification of the {beta} cell antigen targeted by a prevalent population of pathogenic CD8+ T cells in autoimmune diabetes. Proc Natl Acad Sci USA 2003; 100(14):8384-8388.
133. Karjalainen J, Cheung R, Vandermeulen JM et al. T cell proliferation response against bovine albumin (BSA) follows the natural course and is restricted to the linear ABBOS sequence in insulin-dependent diabetes mellitus. Autoimmunity 1993; 15 Supplement:62A.
134. Hummel M, Durinovic-Bello I, and Ziegler A-G. Relation between cellular and humoral immunity to islet cell antigens in type I diabetes. J Autoimmun 1996; 9:427-430.
135. Kelemen K, Crawford ML, Gill RG et al. Cellular immune response to phogrin in the NOD mouse: Cloned T-cells cause destruction of islet transplants. Diabetes 1999; 48:1529-1534.
136. Roep BO, Arden SD, deVries RP et al. T-cell clones from a type-1 diabetes patient respond to insulin secretory granule proteins. Nature 1990; 345:632-634.

137. Van Vliet E, Roep BO, Meulenbroek L et al. Human T cell clones with specificity for insulinoma cell antigens. Eur J Immunol 1989; 19:213-216.
138. Keller RJ. Cellular immunity to human insulin in individuals at high risk for the development of type 1 diabetes mellitus. J Autoimmun 1990; 3:321-327.
139. Abiru N, Wegmann D, Kawasaki E et al. Dual overlapping peptides recognized by insulin peptide B:9-23 reactive T cell receptor AV13S3 T cell clones of the NOD mouse. J Autoimmun 2000; 14:231-237.
140. Daniel D, Gill RG, Schloot N et al. Epitope specificity, cytokine production profile and diabetogenic activity of insulin-specific T cell clones isolated from NOD mice. Eur J Immunol 1995; 25:1056-1062.
141. Daniel D, Wegmann DR. Protection of nonobese diabetic mice from diabetes by intranasal or subcutaneous administration of insulin peptide B-(9-23). Proc Natl Acad Sci USA 1996; 93:956-960.
142. Simone E, Daniel D, Schloot N et al. T cell receptor restriction of diabetogenic autoimmune NOD T cells. Proc Natl Acad Sci USA 1997; 94:2518-2521.
143. Alleva DG, Crowe PD, Jin L et al. A disease-associated cellular immune response in type 1 diabetics to an immunodominant epitope of insulin. J Clin Invest 2001; 107:173-180.
144. Bonifacio E, Genovese S, Braghi S et al. Islet autoantibody markers in IDDM: Risk assessment strategies yielding high sensitivity. Diabetologia 1995; 38:816-822.

Chapter 11

Prediction of Type 1 Diabetes:
The Natural History of the Prediabetic Period

George S. Eisenbarth

Since the first edition[1] of this book improvements in autoantibody determination[2,3] and large prospective studies following children from birth and adults with LADA (latent autoimmune diabetes of adults) have extended and improved our understanding of the variable "natural history" of type 1A diabetes. The importance of understanding the natural history of immune mediated prediabetes lies in the development of prevention strategies.[4] Several initial randomized clinical intervention trials have concluded and the next generation of such trials will rely upon improved and simplified identification of individuals who are at high risk of progression to diabetes. This is essential to ensure that trials will have sufficient statistical power to detect a given effect of the intervention (if it exists) within the time available for the study. Such understanding is also needed to avoid exposing those who will not develop diabetes to the risk of adverse effects of the intervention. In addition, it is likely that many interventions will be more effective if given early, while many beta cells remain intact.

First-degree relatives of individuals with type 1 diabetes have an approximate 5% risk of also developing the disease. Longitudinal studies of autoantibody-positive relatives have provided a wealth of information on the natural history of autoimmunity during prehyperglycemic phase of the disease (prediabetes). They have established the predictive value of age,[5] islet cell antibodies (ICAs),[6] multiple "biochemical" autoantibodies[7-10] first phase insulin release (FPIR),[11] and human leukocyte (HLA) haplotypes.[12] Increasingly, combinations of markers are being used to better define the risk of diabetes.[7,13-16] Prediction is not absolute, but can be expressed as a percentage risk of developing diabetes within a given time period. This chapter will review predictive factors currently in use and discuss some of the unanswered questions on the natural history of prediabetes.

Predictive Factors in Relatives

The first large scale studies of the prediction of type 1A diabetes relied upon the detection of cytoplasmic islet cell autoantibodies (ICA). These studies framed much of our knowledge concerning progression to diabetes but suffered from the difficulty of the cytoplasmic ICA assay, including difficulty in quantitation and standardization.[8] As will subsequently be emphasized determination of three major islet autoantigens, insulin, GAD65, ICA512 (IA-2) have become central components of studies of the natural history of diabetes. Cytoplasmic ICA autoantibodies represents antibodies reacting with GAD65, ICA512, and other unknown antigens, but not insulin autoantibodies. High titer cytoplasmic ICA is most often associated with the presence of multiple anti-islet autoantibodies (of GAD65, ICA512, or insulin) and thus is associated with a high risk of progression to diabetes.[17]

Riley and coworkers reported on the University of Florida, Gainesville, family study in 1990.[5] At that time 3413 first-degree relatives had been screened for ICA, of whom 3.3% were positive. Positive sera were defined as those with 10 or more Juvenile Diabetes Foundation

Immunology of Type 1 Diabetes, Second Edition, edited by George S. Eisenbarth.

(JDF) units. ICAs are more frequent among the siblings of diabetic probands than among parents. They were also more frequent among relatives from multiplex families and among relatives less than 20 years old. After a maximum follow-up of 10 years (median 3.5 years), 40 relatives developed type 1 diabetes. The risk of diabetes was significantly higher in relatives with ICAs of 20 or more JDF units at the time of initial screening, those aged under 10 years (at initial screening), and those from multiplex families. Each of these risk factors was independent of the others. The presence of IAA was not an independent risk factor, after allowing for ICA. However, a later report with further follow-up came to the opposite conclusion, with evidence that IAA add independently to prediction.[18] The presence of ICA in titers of less then 20 JDF units did not confer significantly increased or risk of diabetes, but increasing titers above this level were associated with progressively increasing risk. Nevertheless, one third of the relatives progressing to diabetes were ICA negative on the first test and, of these, 62% remained ICA negative at the onset of diabetes. With analysis of "biochemical" autoantibodies it has become evident that the presence of ICA in the absence of GAD65 or ICA512 autoantibodies is associated with a low risk of progression to diabetes. Verge et al[7] proposed the "general" rule that presence of autoantibodies reacting with >=2 of the biochemical autoantibodies is associated with greatly increased risk of type 1 diabetes. The same was found with further follow up of the Gainesville natural history studies, with the concordant observation that in the absence of biochemical autoantibodies ICA alone is associated with little risk.[19]

The Bart's-Windsor family study, conducted in England from 1978 onwards, also found that higher ICA titer was associated with shorter diabetes-free survival.[6] Seven hundred and nineteen first-degree relatives were followed for up to 10.5 years. ICAs were tested every 4 to 6 months, using a sensitive assay with a detection limit of four JDF units. Detectable ICAs were found in the sera of 3.3% of the relatives at initial testing, compared with 2.2% of 540 healthy child and adult controls. However, only one control had ICA of 20 or more JDF units, compared with 10 relatives. ICAs were detected in an additional 14 relatives on follow-up samples and follow-up time survival analysis was calculated from the time of ICA detection. With increasing ICA cutoff the positive predictive value for future diabetes rose, and the sensitivity fell. The risk of diabetes within 10 years associated with ICAs of 80 or more JDF units, was a 100% (95% confidence interval: 52%-100%). In comparison, ICAs of 20 or more JDF units were associated with a 73% risk (95% confidence interval: 45%-100%), and ICAs of four or more JDF units with a 40% risk (95% confidence interval: 23%-57%). Utilization of biochemical autoantibody assays in these cohorts also demonstrated the importance of multiple autoantibodies to identify high risk individuals.

Data from the Joslin Diabetes Center family study indicate that impaired FPIR is an additional risk factor.[11] Thirty-five first-degree relatives with high-titer ICAs ($\geq$ 40 JDF units) underwent serial intravenous glucose tolerance testing (IVGTT). The age of the subjects ranged from 2.6 to 66 years, the mean follow-up was 3.6 years from the first test, and 18 progressed to overt diabetes. The FPIR was calculated as the sum of the 1- and 3-minute insulin levels after a standard bolus of intravenous glucose (0.5 g per kg body weight, infused as a 20%-25% solution over 2 to 4 minutes). Percentiles for the FPIR were determined in 225 healthy, nonobese, control subjects. Even in control subjects the FPIR showed wide within-subject variation (discussed in detail below). Nevertheless, relatives with an initial FPIR below the first percentile (48 mU/l) had significantly reduced diabetes-free survival. Importantly, the presence of FPIR below the first percentile did not signify that diabetes was already present. For most of the relatives, oral glucose tolerance tests (OGTT) were also performed during follow-up to detect asymptomatic diabetes. For the survival analysis, the onset of diabetes was defined by a diabetic OGTT or the occurrence of symptomatic hyperglycemia, whichever came first. The average time from the discovery FPIR below the first percentile to the onset of diabetes was 1.8 years.

Updated data from the Joslin family study, with longer follow-up and larger numbers of relatives, have confirmed these findings (Fig. 1). Among 79 relatives with high titer ICAs ($\geq$ 40 JDF units), those with FPIR below the first percentile on the first test had 3-year diabetes-free

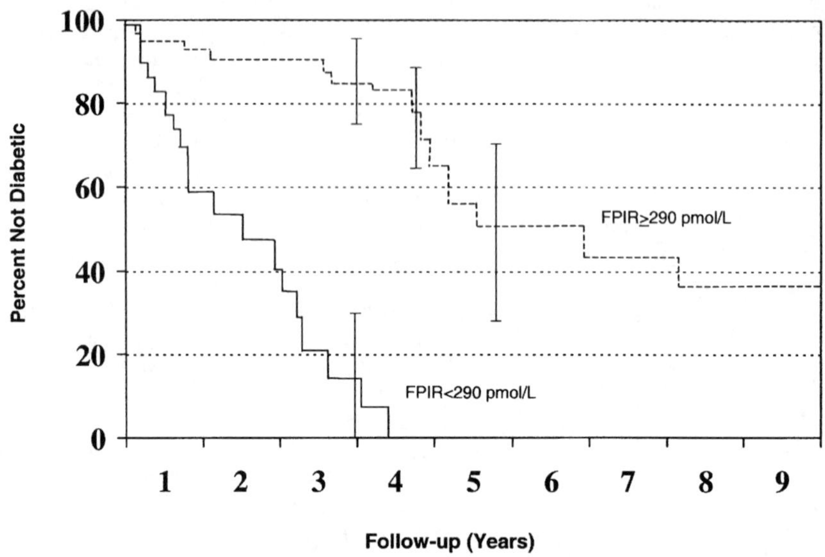

FPIR ≥290 pmol/L	n=55	46	38	27	17	8	8	6	4	3
FPIR <290 pmol/L	n=24	13	7	2	0					

Figure 1. Survival analysis of the progression to diabetes in ICA-positive first-degree relatives followed in the Joslin family study. The relatives were subdivided by the FPIR on the earliest IVGTT performed, above or below the first percentile (48 mU/L or 290 pmol/L). Error bars show the 95% confidence intervals for survival estimates. Survival was significantly different between the two groups ($p<0.0001$, log-rank test). Reproduced with permission from: Eisenbarth GS et al. Diabetes 1993; 42:941-947. Copyright © 1993 by the American Diabetes Association.

survival of 13% (95% confidence interval: 0%-30%) compared with a 78% (95% confidence interval: 63%-93%) for the group with higher FPIR.[20] Studies at the Barbara Davis Center have confirmed the predictive value of FPIR measurements as has studies from the Melbourne family study, studies from Finland, and the DPT-1 (Diabetes Prevention Trial) North American study[21] and analysis of the combined ICARUS database.[22]

Data from the Joslin study also suggest that IAA add to the prediction of type 1 diabetes, but with a weaker effect than ICA.[23] Forty-two ICA-positive (≥ 20 JDF units) relatives and 1670 ICA-negative relatives (representing a subset of all relatives found to be ICA negative) were tested for IAA. Among the ICA-negative relatives, 2.7% were IAA positive, whereas among ICA-positive relatives, 45% were IAA positive. IAA alone had less predictive value than ICA, but the combination of IAA and ICA was useful. The risk of diabetes within 5 years was 17% for the IAA-positive/ICA-negative group, increasing to 42% for the ICA-positive/IAA-native group and to 77% for double antibody positive relatives.

Reports that identifying ICA subtypes improves the predictive value of ICA can now be put in the general context of the rule that multiple biochemical autoantibodies are associated with high risk. The restricted ICA subtype, defined according to the pattern of sustaining on pancreatic sections, confers a significantly lower risk of progression to diabetes than a nonrestricted subtype.[24] Preabsorption of sera with glutamate decarboxylase (GAD) blocks the ICA staining of restricted ICA-positive sera[25] and reduces the ICA staining of most nonrestricted ICA-positive sera.[26] This suggests that restricted in ICA are due to antibodies directed against a single antigen (GAD).

The ICA assay is difficult to standardize, is labor intensive, and requires human pancreas.[7] At the Barbara Davis Center, we utilize a combination of three assays employing recombinant antigens (IAA, anti-GAD, and anti-ICA512[27]) and no longer determine cytoplasmic ICA. For children the ICA assay provides only marginal additional information, compared with the combination of defined-antigen assays. Relatives expressing two or more of IAA, anti-GAD, and anti-ICA512 have overall risk of diabetes within 5 years of more than 68% by life table analysis. Once these three antibodies are considered, the presence or absence of ICA does not influence this risk. However, the addition of intravenous glucose tolerance testing does improve prediction of the time to overt diabetes. Among relatives with two or more antibodies, those with an FPIR less than the first percentile have a 50% risk of diabetes within 1 year; those with higher FPIR have a 50% risk within 3 years.[7]

Genetic factors should also be considered in assessing diabetes risk. Deschamps and coworkers examine the predictive value of HLA typing in a study of 536 siblings of diabetic probands in France.[12] The risk of type 1 diabetes after 8 years, estimated by life table analysis, was 10% for siblings who were HLA identical with the proband, 3%-4% for siblings with either DR3 or DR4, and 16% for those with DR3/DR4. This compares with 56% for those with ICA greater than 4 JDF units and 70% for those with the combination of ICA and the highest risk HLA type, DR3/DR4. In other studies, molecular typing has revealed that the HLA haplotype DQA1*0102 DQB1*0602 confers strong protection from type 1 diabetes, in a dominant fashion (Chapter 7). In our experience, autoantibody-positive relatives with this haplotype have a very low risk of progression to diabetes[28] and usually express only a single autoantibody, namely anti-GAD, although a few also express IAA. Such protection is however not absolute, and approximately 1% of children developing type 1A diabetes (versus 20% of the general U.S. population) and 3% of adults with type 1 diabetes have DQB1*0602 (DQB1*0602 is usually part of the haplotype DRB1*1501, DQA1*0102, DQB1*0602).

Prediction in the General Population

It is likely that genetic typing will have an even greater impact on assessing diabetes risk in the general population. Most studies of prediabetic subjects have involved the screening of first-degree relatives of diabetic probands, rather than the general population. However, less than 10% of new cases of type 1 diabetes have an affected relative, so the general population will need to be screened eventually if an effective intervention is to have a major impact. Screening the general population is likely to be more difficult than screening relatives. Bayes' theorem states that a screening test will have a lower positive predictive value in the general population than in a selected group with a higher prevalence of disease, such as first-degree relatives. One approach toward solving this problem is to screen the general population with markers of genetic susceptibility first, followed by autoantibody testing of susceptible individuals. For example, among the general Denver population 2.4% of individuals express both DR3 and DR4 (with associated DQ2 and DQ8). Of this subgroup, it is predicted that approximately 6% will develop type 1 diabetes, similar to the risk among first-degree relatives.

Unexpectedly, studies performed in Florida suggest that ICA have a predictive value in the general population similar to that in relatives.[29] In contrast, studies in England suggest that ICA will have a lower positive predictive value in the general population. The prevalence of ICAs of 20 or more JDF units was only two to three times higher in siblings than in the general population, compared with a 13 times greater risk of diabetes in the siblings.[30] Many population studies have been initiated (Table 1). The difference between the two studies probably relates to differences in the ICA assays, with primarily individuals with higher levels of ICA (that associated with multiple biochemical autoantibodies) followed in the Gainesville general population studies.

Several studies have now been initiated where children are followed from birth for the development of anti-islet autoantibodies. The three studies with the longest followup are the BabyDiab study from Germany, the DAISY study from Denver Colorado, and major studies in Finland.[15,31-34] The BabyDiab study evaluates offspring of patients with diabetes while the

Table 1. The prevalence of islet cell antibodies (ICA) among schoolchildren in the general population of different countries compared with the incidence of type 1 diabetes

Country	Age Range	Number Screened	Prevalence of ICA	Cutoff (ICA units)	Incidence[a] of Type 1 Diabetes	References
Japan	7-18	473	0.4%	–	2.0	43,44
Germany	6-21	4,287	1.05%	≥5	7.0	45,46
			0.4%	≥20		
France	6-17	8,363	1.79%	≥4.5	7.1	47,48
			0.2%	≥24		
Netherlands	5-19	4,806	0.42%	≥0.6	11.0	49,50
Spain	14-17	2,291	0.35%	≥5	11.3	51,52
New Zealand	5-7	20,000	1.2%	≥10	12.8	53,54
USA						
Florida	5-18	9,696	0.59%	≥10	13.7[b]	29,55,56
Washington State	12-18	3,992	1.7%	≥1		
Australia	4-18	1,900	0.68%	≥20	14.5	57,58
England	9-13	2,925	2.8%	≥4	15.6	59,60
			0.8%	≥20		
Sweden	0-14	420	3%	≥5	23.6	61,62
Finland	3-18	1,212	4.1%	≥3	35.3	63,64

[a]Per 100,00 per year in the 0-14 age group; [b]Incidence in Pennsylvania

DAISY study has prospectively followed both first degree relatives (offspring and siblings) as well as children stratified by HLA type from the general population. All three studies are providing generally concordant results. Anti-islet autoimmunity frequently develops in the first year of life. Often insulin autoantibodies appear first, but GAD65 and less often ICA512 autoantibodies can also be the first to develop. The presence of multiple anti-islet autoantibodies in young children portends a high risk of progression to diabetes for both relatives and the general population.[7,9] It is likely that there is no age at which a genetically susceptible individual has no chance of converting to anti-islet autoantibody positivity, but progression to diabetes is usually less rapid in older individuals.

Prediction/Diagnosis in Adults

Both patients with gestational diabetes and adults with a diagnosis of type 2 diabetes have a significant risk of having type 1A diabetes.[35] Between 5 and 10% of patients with a diagnosis of gestational diabetes express anti-islet autoantibodies and the great majority progress to type 1A diabetes. In a similar manner, the diagnosis of type 1 diabetes is associated with expression of islet autoantibodies in adults thought initially to have type 2 diabetes.[36,37] Assays for GAD65 autoantibodies are the most useful in both patients with gestational or adult onset type 1 diabetes.[38,39]

At What Stage in Life Is Beta Cell Autoimmunity Initiated? What Is the Initial Target Antigen, and in What Sequence Do Different Autoantibodies Appear?

Studies in first-degree relatives have detected evidence of beta cell autoimmunity many years before clinical presentation with overt hyperglycemia.[5,6,40] However, the existence of young infants with type 1 diabetes indicates that beta cell destruction may also occur rapidly in some individuals. A plausible hypothesis is that beta cell autoimmunity begins early in life, with variation in the age of onset of overt diabetes explained by differences in the rate of beta cell destruction from one individual to another. In support of this hypothesis, a major peak in the incidence of type 1 diabetes occurs during early adolescence in nearly all populations studied.[41]

After the age of about 15 years, measuring the incidence of type 1 diabetes is complicated by misclassification of some patients as type II diabetics. Nevertheless, it has been estimated that at least 37% of type 1 diabetes occurs after the age of 19 years and 15% after the age of 30 years.[42]

New cases of type 1 diabetes presenting in adult life tend to have a longer duration of symptoms before diagnosis and higher C peptide levels remaining at diagnosis compared with those presenting in childhood,[65] suggesting a slower rate of beta cell destruction. "Slow onset" type 1 diabetes, presenting in late adult life, may be diagnosed initially as type II diabetes. Tuomi and coworkers studied 102 adults who were treated for type II diabetes, of whom one third had low stimulated C peptide levels and two thirds had normal levels.[66] Of the group with low C peptide, 76% were positive for anti-GAD (similar to the frequency among children with newly diagnosed type 1 diabetes). In comparison, only 12% of this group with normal C peptide were positive for anti-GAD.

Most of the at-risk relatives followed to date were initially identified by testing for autoantibodies. It is unknown how long autoantibodies were present in these individuals before they were discovered. However, studies by Pilcher and Elliott in New Zealand suggest that ICAs develop early in childhood and are usually preceded by IAA.[67] In the studies 666 first-degree relatives, aged less than 20 years and initially ICA negative (<10 JDF units), were followed longitudinally. Those age under 5 years were retested annually, with older subjects retested every 3 to 5 years. Sixteen (2.4%) became ICA positive during follow-up, with seroconversion occurring at an average age of 3.2 years, ranging from 1.6 to 7.1 years. All except the two oldest seroconverters were IAA positive at the time of seroconversion and the appearance of IAA was documented to precede ICA in 6 (46%) of 13 seroconverters who could be retrospectively checked for IAA.

A cohort of infants of mothers with type 1 or gestational diabetes has been studied by Ziegler and coworkers in Germany.[68] Data from the study also indicate that autoantibodies may appear very early in life, with IAAs appearing first. A high frequency of both IAAs and ICAs was found in cord blood samples. This was attributable transplacental passage of maternal IgG and to falsely positive IAA results (that may be obtained in up to one half of cord blood samples from normal pregnancies).[69] However, by 9 months of age all became ICA negative and only 3 of 90 infants remained IAA positive. By 2 years of age, one of these infants became IAA negative, two remained IAA positive, and another child became persistently IAA positive for the first time. It is not possible to absolutely distinguish in the first year of life transplacental anti-islet autoantibodies from autoantibodies produced by the infant. Measurement of cord blood or mothers serum at birth allow distinction, with extremely high levels of the autoantibodies usually associated with autoantibodies in the infant that persist for up to 12 months. Transplacental anti-insulin autoantibodies are usually of the IgG4 subclass in contrast the dominance of IgG1 spontaneous autoantibodies.[70] A rising level of autoantibodies is almost always associated with spontaneous autoantibodies.

Both insulin and GAD have been suggested as candidates for the initiating antigen but another (unidentified) autoantigen could be responsible.[71,72] Neither IAA nor anti-GAD is universally present among patients with newly diagnosed type 1 diabetes[73-80] GAD is an antigen recognized by the ICA assay,[26] so the findings in very young children discussed above favor insulin as an initiating autoantigen rather than GAD. However, only small numbers have been studied so far and it should be recognized that the assay for ICA by indirect immunofluorescence is likely to have lower sensitivity than the radioassay for IAA. Biochemical characterization of the antigen responsible for ICA[26,81] will allow the development of more precise assays for these sera.

Several other observations provide circumstantial evidence for insulin as an initiating autoantigen. Insulin is the only antigen unique to the beta cell and it is present on the cell surface.[82,83] In contrast, GAD is also present in alpha cells,[84] which are not destroyed,[85] and insulitis disappears when all insulin-containing cells have been destroyed.[86] IAAs occur with greater frequency and at higher level in younger newly diagnosed patients[87-89] and the level of IAA appears to correlate with the rate of beta cell destruction in prediabetic individuals.[90] When 29 patients were studied, their IAAs recognized the same epitope.[91] The assay for insulin autoantibodies has been recently improved with the development of a microassay by Williams and coworkers.[92,93] This assay utilizes protein A or protein A/protein G for autoantibody precipitation and thus avoids the false autoantibody positives for cord or hemolyzed blood associated with the polyethylene glycol based assays. We have modified the Williams assay such that it is performed in 96-well microtiter plates utilizing membrane filtration and direct β-counting in the microtiter plate.[94] With this assay not only do prediabetic individuals express readily detectable anti-insulin autoantibodies but NOD mice express high levels of the autoantibody.[94] Of note administration of a dominant peptide of insulin (insulin B chain peptide B:9-23) induces autoantibodies to insulin, even in Balb/c normal mice that react with intact insulin and are not absorbed by the immunizing peptide.[95] This autoantibody induction is MHC restricted, suggesting that with the MHC of either Balb/c or NOD mice, T cell activation by an insulin peptide activates already sensitized B lymphocytes recognizing insulin.

Are Environmental Factors Important Either in the Initiation of Beta Cell Autoimmunity or in Modulating the Process Once it Has Begun?

The diabetes susceptibility genes identified so far are present in a high proportion of the general population, indicating that they are not sufficient to cause the disease. Furthermore, the concordance rate observed in HLA-identical siblings is lower than that in identical twins,[12,96-98] suggesting that other important loci remain to be discovered. As type 1 diabetes is associated with other autoimmune diseases,[99,100] it is likely that at least some of the important genes will not be specific to type 1 diabetes but also involved in the susceptibility to these other autoimmune diseases.

Despite the importance of genetic factors, the low concordance rate (<40%) seen in identical twins[96-98] implies a critical role for nongenetic random rearrangement of antigen receptor genes during the differentiation of T lymphocytes,[101] somatic mutation, and environmental factors. Epidemiological studies with validated case ascertainment support a role for environmental factors. A rising incidence of type 1 diabetes over time, too rapid to be explained by an increase in the frequency of diabetogenic genes in the population, has been reported in multiple countries.[102-111] In addition, increased rates have been observed in migrant populations compared with their countries of origin.[112,113]

An environmental factor could be involved either by initiating autoimmunity or by altering its progression, once established. There is a seasonal variation in the onset of type 1 diabetes,[57,102,103,114,115] with more cases in the winter months, associated with viral infections. It is likely that such viral infections alter the process late in its course by increasing insulin requirements, thereby advancing the time of diabetes onset. In humans, the best evidence for the involvement of the specific viral agent in the initiation of autoimmunity comes from the observation that patients with congenital rubella have an increased prevalence of type 1 diabetes,[116,117] as well as other autoimmune diseases.[118] Coxsackievirus B infection has also been suggested as an environmental trigger,[119] although a large Swedish study found no significant difference in the levels of IgM antibodies against Coxsackievirus B types 1-5 between newly diagnosed diabetic children and population controls.[120] There is likely to be a long lag time between a triggering infection and the onset of diabetes but IgG levels were not measured in this study. It has been suggested that an infectious agent might activate autoimmunity by molecular mimicry, whereby the immune response to a viral antigen, such as the P2-C protein of Coxsackievirus

B4[121] or the rubella capsid,[122] might cross-react with a beta cell antigen. Some evidence against a significant role for an infectious agent comes from Japan. Despite wide variation in the incidence rates for infectious diseases in different parts of the country, the incidence of type 1 diabetes is uniformly low.[123]

At present studies of potential environmental factors triggering the development of anti-islet autoantibodies in children followed from birth have not led to the identification of major precipitating factors, and contradictory results are a problem.[124] Studies from Finland have implicated both enterovirus infection and ingestion of bovine milk formula.[34,125] In contrast studies from DAISY and BabyDiab have failed to associate these factors with the initiation of autoimmunity[126-128] A report of rotavirus infection[129] associated with development of anti-islet autoantibodies from the Australian Baby study has not been replicated with analysis of children from the DAISY study. It is likely that environmental factors contributing to the initiation of anti-islet autoimmunity are relatively ubiquitous and multiple factors may contribute. This is readily demonstrated in animal models where injection of poly-IC into multiple rat strains with the HLA haplotype RT1-U develop either insulitis and/or diabetes. Poly-IC (inosinic cytodylic acid) is a mimic of viral RNA and a potent inducer of interferon-α through Toll receptors of the innate immune system.[130]

In addition, in animal models of type 1 diabetes, exposure to infectious agents may suppress rather than trigger autoimmunity. NOD mice raised in a sterile environment had an increased incidence of diabetes. Similarly diabetes-prone BB rat housed in "viral antibody free" conditions have a higher frequency of diabetes than those housed in less stringent "specific pathogen free" conditions.[131] Specific infection with mouse hepatitis virus is associated with reduced incidence of diabetes in the NOD mouse,[132] and infection with lymphocytic choriomeningitis virus is associated with reduced incidence in both the NOD mouse and BB rat.[133,134] However, in at least one instance infection is implicated as a trigger. Kilham's rat virus, a parvovirus, has been shown to trigger beta cell autoimmunity and diabetes in the diabetes-resistant BB rat.[135] It appears that the Kilham virus acts similar to poly-IC rather than by infecting islet cells. Other reports of viruses triggering diabetes in animals may involve direct viral damage of beta cells without an autoimmune mechanism.[136,137]

In humans, associations with infant feeding have been reported. Several population-based case-control studies found a protective effect associated with breast-feeding[138-140] and an increased risk associated with the early introduction of supplemental feeding.[138,139,141] A Danish study, which eliminated several possible sources of bias by using the records of postnatal health visitors, found no evidence to support an associated with the total duration of breast feeding,[142] but did not examine the timing of introduction of supplemental feeding. Both BabyDiab and DAISY found no evidence for infant bovine milk ingestion[126] but a pilot study of the elimination of bovine milk from infant formula was associated with a reduction of development of cytoplasmic ICA, with little reduction of GAD65 autoantibodies.

The reported associations with infant feeding are weak (odds ratios ~1.5),[143] but this is consistent with the effect of a common exposure acting only on genetically susceptible individuals. Several mechanisms could explain the associations with infant feeding. Breast-feeding could protect the infant from an infection. The increased caloric intake and weight gain associated with artificial feeding might cause increased insulin secretion and presentation of beta cell autoantigens.[144] Karjalainen and coworkers have suggested that intact cow's milk peptides might cross the immature gut, initiating an immune response to cross-reacts with a beta cell surface antigen.[145] Higher levels of antibodies against several components of cow's milk have been reported in children with newly diagnosed type 1 diabetes than in controls.[146-150] T cell proliferation in response to the ABBOS peptide was higher in newly diagnosed patients than in controls.[151] However, these results could not be reproduced by Atkinson and coworkers,[152] who found no difference between newly diagnosed patients and controls for either anti-BSA antibodies or T cell proliferation to ABBOS. In animal studies, the addition of cow's milk to the diet of the BB rat at the time of weaning increases the incidence of diabetes,[153,154] but there are conflicting reports on the effect of cow's milk in the diet of the NOD.[155,156] Of interest

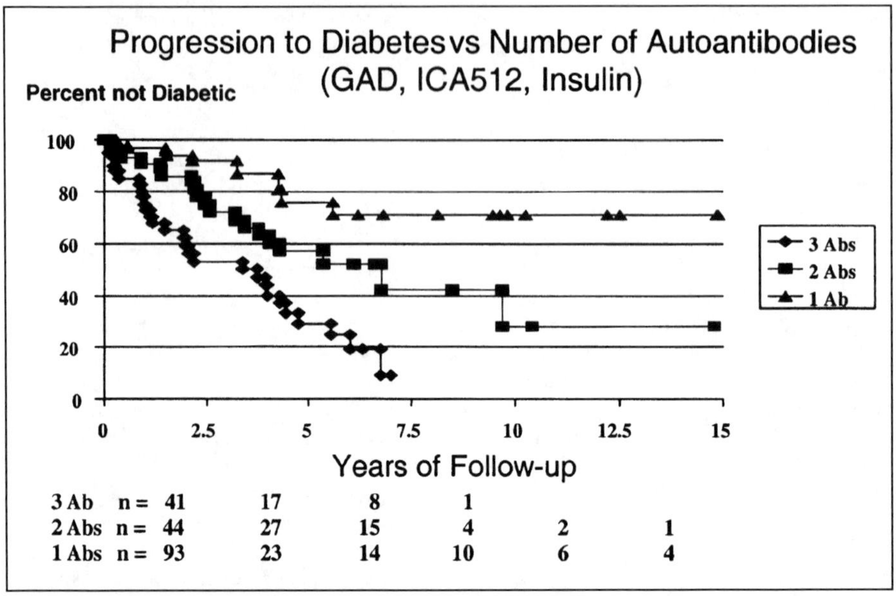

Figure 2. Combinatorial autoantibody prediction. (Modified from Verge et al. Diabetes 1998; 47:1857-1866. Copyright © 1998 The American Diabetes Association)

there is insulin in both human milk and bovine milk and a reported induction of anti-bovine insulin antibodies with bovine milk ingestion.[157]

Once Autoimmunity Is Initiated Are Beta Cells Destroyed in a Progressive Linear Fashion or Are There Spontaneous Remissions?

The long-term follow-up of high-titer ICA-positive relatives in the Joslin family study with serial IVGTTs suggested a model in which beta cell destruction is a progressive, linear process,[158,159] (Fig. 2). The rate of beta cell destruction may vary widely from one individual to another. It may be so slow in some individuals that overt diabetes does not develop during the person's lifetime. Such individuals might appear to have decreased but stable beta cell function when assessed with serial IVGTTs over many years. A correlation between serial FPIR levels and the number of years to diabetes in ICA-positive relatives,[160] and the addition of IAA as a risk marker,[23] led to the development of a "dual parameter" linear regression model.[90] In this model the time remaining before the onset of diabetes is predicted by a combination of the FPIR (reflecting the remaining beta cell mass) and the IAA level (reflecting the rate of loss of beta cells).

Other studies, including relatives with lower titer ICA, have suggested that progression to diabetes may be less predictable.[161-163] They have suggested an alternative model, in which the rate of beta cell destruction may fluctuate, with spontaneous remission in some individuals. This model is similar to that suggested for autoimmune thyroiditis.[164] Reports of transient ICA in relatives,[161,165-167] in discordant identical twins,[168] in the general population,[47,63,169,170] and in patients initially treated for type II diabetes[163] lend support to this hypothesis. Most, but not all[161,167,169,170] of these transient ICAs have been reported in subjects with initially low titers, raising the possibility that the phenomenon may frequently be explained by variability in the ICA assay.[171] Some of these studies did not report the levels of ICA in terms of endpoint titer or JDF units.[163,165,166] However, several confirmed the change in ICA status by retesting this sera from all time points for a given subjects in one assay.[63,161,163,166] Even with the utilization of "biochemical" autoantibody assays there is some conflicting data concerning the persistence of anti-islet autoantibodies, but a number of prospective studies suggest that individuals with readily detectable levels of islet autoantibodies, high risk HLA genotypes of multiple autoantibodies, have extremely persistent autoantibodies.[172-174] It is likely that even with "biochemical" autoantibody assays, a subset of autoantibody positivity is not associated with islet autoimmunity. For instance we have recently found that for the rare individual with transient ICA512 autoantibodies, the autoantibodies do not recognize multiple epitopes of the molecules in sharp contrast to the persistent autoantibodies associated with diabetes risk.

Low-titer ICAs carry a reduced risk of type 1 diabetes, compared with high titer,[5,6] and it may be that fluctuating levels do not signify a disease process, especially unaccompanied by other autoantibodies. Nonpersistent ICAs may represent a nonspecific immunological response associated with infection, as Helmke and coworkers reported the transient appearance of ICAs in normal schoolchildren during mumps infection.[169] The development of more reproducible assays for antibodies against biochemically defined autoantigens may improve the assessment of diabetes risk. Among those relatives followed in the Joslin family study who express more than one autoantibody, levels of IAA,[23] anti-GAD, and anti-ICA512 (unpublished data) have been remarkably stable over time.[172]

Carel and coworkers reported that ICA-negative siblings of diabetic children have significantly lower mean FPIR than controls.[175] They suggested that autoimmune beta cell destruction may have been initiated in some of these siblings, but spontaneously remitted. However, other autoantibodies may have been present in some of the ICA-negative siblings in this study. At most, only 94% of children with newly diagnosed type 1 diabetes are ICA positive, using a very sensitive assay[176] but the addition of other autoantibodies may reduce the autoantibody-negative proportion. In addition, although the diet was standardized for the 3 days prior to the IVGTT, there may be differences over the longer term in the diets of children with and without a diabetic sibling that could have affected the IVGTT results.

Decreased insulin secretion measured with an IVGTT could be due to reduced beta cell mass or (reversible) impairment of beta cell function, or both. Inflamed islets from diabetes-prone BB rats have an enhanced insulin secretory response to glucose,[177] suggesting that reduced beta cell mass is the dominant mechanism.

Experiments with streptozocin-treated baboons indicate that beta cell function (assessed by IVGTT) correlates with beta cell mass.[178] In humans, however, the IVGTT has several limitations. The results of beta cell function tests may vary with insulin sensitivity according to age,[175,179] physical fitness,[180] body mass index,[179,181] pubertal status,[181,182] and physical or psychological stress. Different protocols for the IVGTT give different results[183] but a standard protocol has been developed to allow meta-analyses.[184]

The interpretation of serial FPIR data is complicated by within-subject variability in the measurement. Figures 3A and 3B illustrate the course of the FPIR over time for two ICA-positive relatives. One has a clearly linear loss of secretion (Fig. 3A). In contrast, the other relative (Fig. 3B) has what appears to be a stepwise loss of secretion, with remissions and periods of recovery. However, this second pattern could equally well result from wide within-subject measurement variation, superimposed on an underlying linear loss of secretion. In the large DPT-1 study loss of first phase insulin secretion was strongly associated with diabetes risk factors[21] as it was for the combined ICARUS dataset.[22]

Several centers have performed duplicate IVGTTs, spaced 1 or more weeks apart, on healthy volunteers. The median within-subject coefficients of variation (CV) reported for FPIR varied from 4% to 36%[179,185,186] but the CV was as high as 56% for some subjects. There are conflicting reports about whether calculating the area under the curve for insulin release improves reproducibility, compared with a sum of the 1- and 3-minute insulin levels.[179,185,186] McCulloch and coworkers studied 18 normal postpubertal subjects on two occasions separated by an average of 1 year.[161] They used an IVGTT protocol with frequent sampling and a bolus of tolbutamide at 20 minutes, allowing the calculation of the insulin sensitivity index. The within-subject CV for the incremental area under the insulin curve (above fasting levels) from 0 to 10 minutes

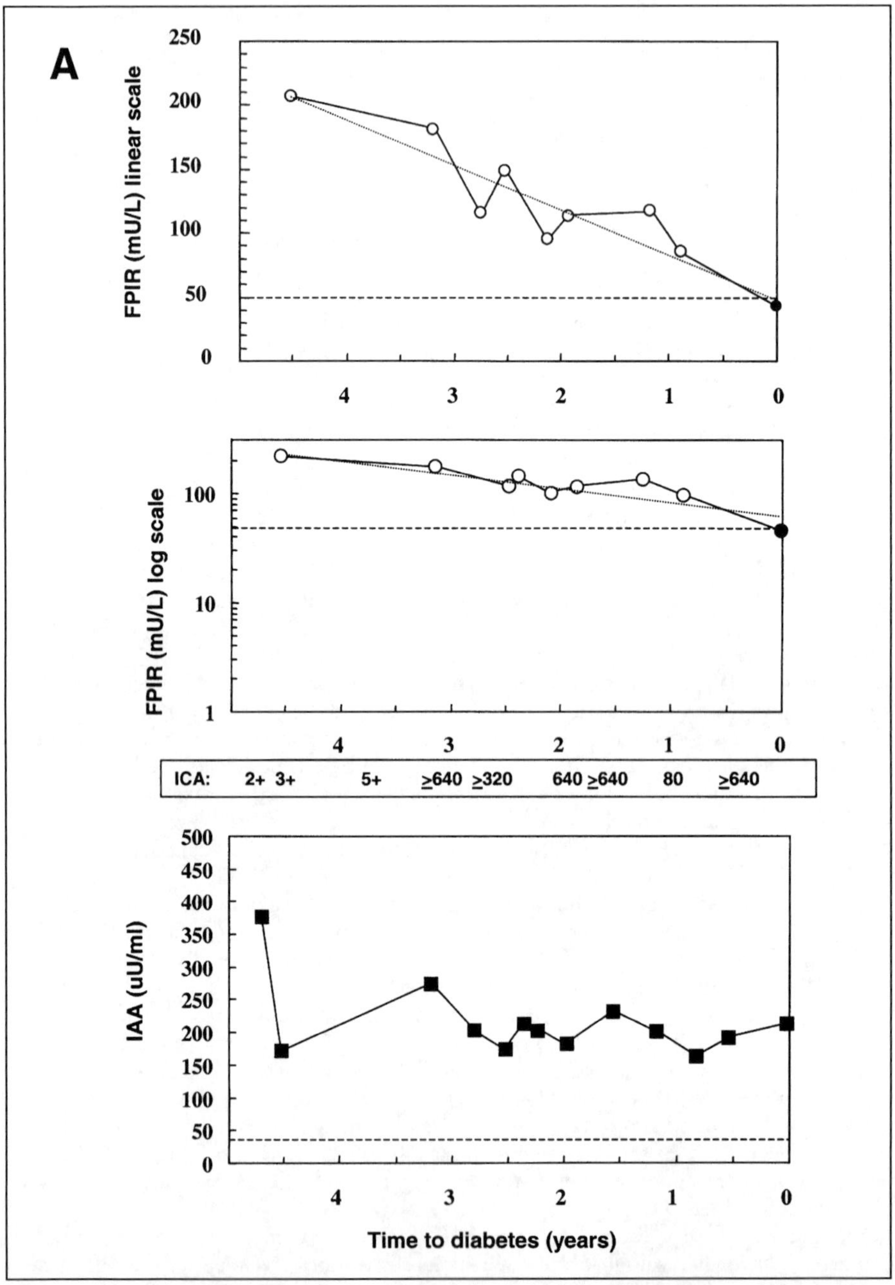

Figure 3A. Changes in the FPIR, ICA titer, and IAA level during longitudinal follow-up of two ICA-positive first-degree relatives prior to the onset of overt diabetes. The top two panels show the FPIR plotted on linear and logarithmic scales. The dashed line indicates the 1st percentile for FPIR among controls (48 mU/L) and the linear regression through the data is shown with a dotted line. (A) plots the time-course of the subject with the most linear loss of FPIR followed in the Joslin family study (onset of diabetes at age 14.4 years).

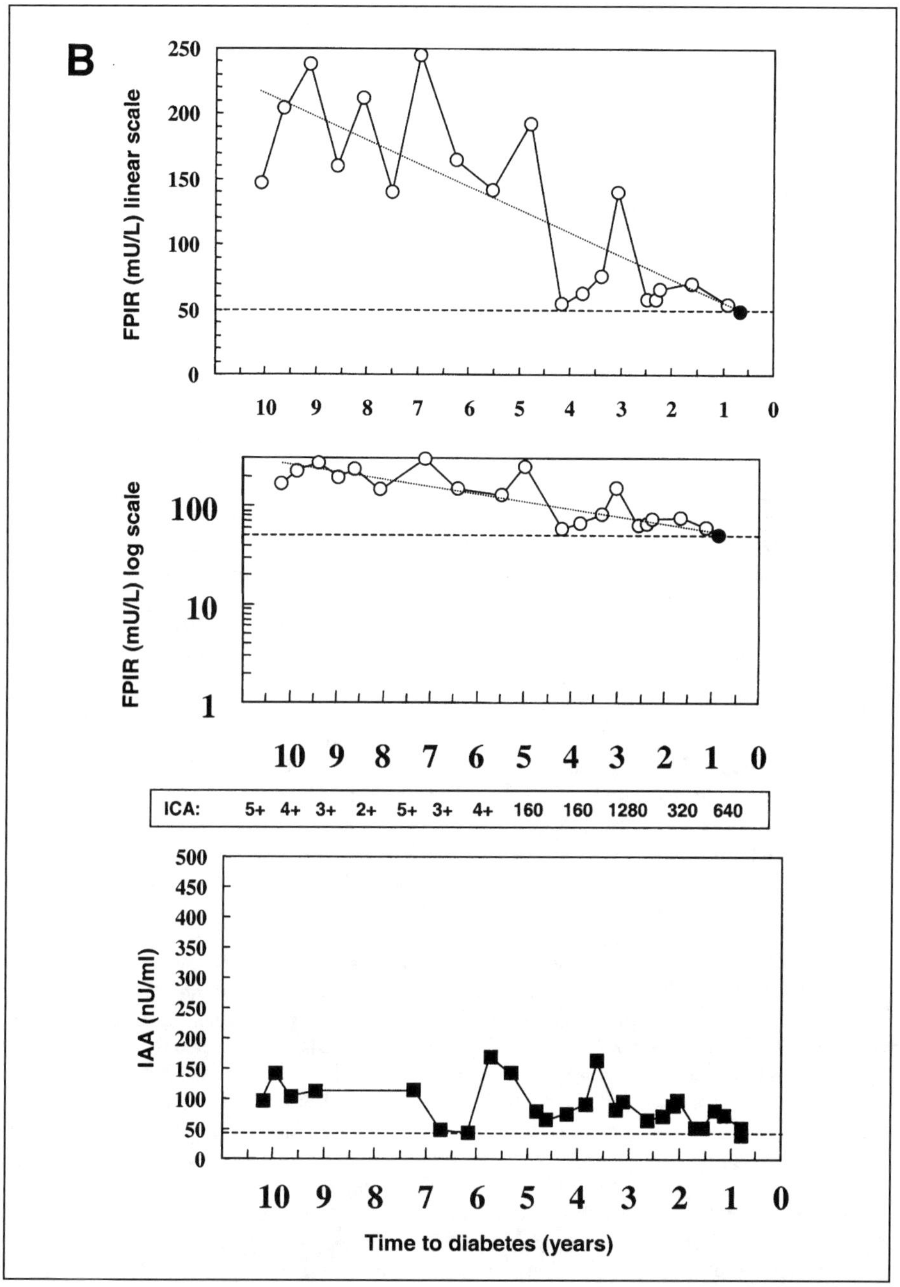

Figure 3B. Plots the time course of the least linear subject (onset of diabetes at age 17.8 years). Positive ICA levels determined before the introduction of JDF units are shown on a scale of 1+ to 6+ depending on the intensity of immunofluorescence. In the lowest panel, the dashed line indicates the upper limit of normal for IAA (mean +3 SD among controls).

was 11.3%. Adjusting for the insulin sensitivity index resulted in a similar CV (10.5%). However, it is likely that adjusting for insulin sensitivity may improve reproducibility in pubertal subjects and those followed over a longer period. The subjects also underwent arginine stimulation tests (both at physiological plasma glucose concentration and hyperglycemic clamp). The reproducibility of the response to arginine was similar to that of the response to glucose, indicating little advantage gained by the use of arginine.

Rayman and coworkers achieved highly reproducible results using retrograde venous cannulation and arterialization of the hand from which the samples are drawn,[185] but recent studies found no improvement with these techniques.[187,188] Other maneuvers that have been evaluated to improve reproducibility include placement of the intravenous line 1 hour before the test to minimize the effects of stress hormones. Allen and coworkers tested the hypothesis that catecholamine release, associated with anxiety about the test, may inhibit insulin release but found no correlation between FPIR and plasma catecholamine levels or standardized anxiety scores.[179] However, a significant "first test effect" was observed. Among 11 normal subjects who underwent two tests, 9 had a higher FPIR on the second test.[179] One subject who fainted during the first test had a very low FPIR (2 mU/L) that became normal on a subsequent occasion.

The FPIR is a measure of beta cell function, but a measure of beta cell mass might be more useful. In the future, new ways of assessing the beta cells may be developed, including high-performance glycated hemoglobin assays and imaging with nuclear magnetic resonance (MRI) or isotopic scans. Leech and coworkers measured the HbA1c by high-performance liquid chromatography and healthy teenagers who were also tested for ICA.[189] The mean HbA1c was slightly, but significantly, higher in the ICA-positive subjects, suggesting that elevation of the HbA1c within the normal range may be an early marker of loss of beta cell function. Signore and coworkers reported preliminary data suggesting that scans following the injection of ^{177}I-labeled interleukin-2 (IL-2) may be useful in the detection of beta cell autoimmunity.[190] A significant accumulation of tracer was detected in the pancreatic region of four newly diagnosed and two prediabetic subjects, compared with four controls. MRI has been used to detect pancreatic graft rejection.[191-193] With refinements in MRI techniques it may be possible in the future to detect insulitis or to approximate beta cell mass.

There is little information on changes in islet histology before the onset of overt diabetes in humans. However, Sutherland and coworkers studied patients with longstanding type 1 diabetes who received pancreatic transplants from their nondiabetic identical twins.[86] Of these, three received either delayed immunosuppression or no immunosuppression after the transplants and all three became insulin dependent again within 5 to 12 weeks. Serial biopsies of the graft revealed T cell infiltration with selective beta cell destruction, unlike graft rejection. Rather, these histological findings were consistent with reactivation of beta cell autoimmunity. Of note, the insulitis disappeared after the insulin-containing cells were destroyed.

Foulis and coworkers studied the pancreatic histology of 119 patients with type 1 diabetes who had died of ketoacidosis, of whom 60 had died within 1 year of diagnosis.[85] More recent studies from Japan indicate that the prescence of anti-islet autoantibodies correlate with insulitis on biopsy. They found that insulitis is a patchy process, in which islets with or without insulitis tend to be grouped separately in different lobules. At diagnosis, the beta cells in some islets were destroyed while other islets were left untouched, but most islets were insulin deficient after a 1-year duration of diabetes. Hanafusa and coworkers performed pancreatic biopsy on seven Japanese adults, 2-4 months after the diagnosis of type 1 diabetes. The islets were atrophic, with reduced or absent insulin staining, but surprisingly there was hardly any lymphocytic infiltration.[194]

The diagnosis of diabetes is to some extent an arbitrary point in the natural history of the autoimmune disease. The timing of overt diabetes may be influenced by factors that increase insulin requirements, such as infections and puberty. A significant proportion of the beta cell mass may remain at diagnosis. Experiments with streptozotocin-treated baboons suggest that the FPIR may be undetectable even when about 20%-50% of beta cell mass remains,[178] although pancreatic insulin content was close to zero at this point. Autoimmune beta cell destruction continues after the diagnosis of diabetes. A temporary remission from insulin dependency may occur in up to 27% of patients, soon after diagnosis,[195] and is attributed to the beta cell rest caused by insulin treatment. Younger age at onset, male sex, high-titer ICA, severe ketoacidosis at diagnosis, and a short duration of symptoms prior to diagnosis are associated with a more rapid loss of C peptide secretion.[196] In one study a more rapid loss was seen in patients heterozygous for DR3 and DR4,[197] although another study came to the opposite conclusion,[195] and another found no significant DR associations.[196]

Conclusions

We propose a model in which immunological abnormalities appear early in life in genetically susceptible individuals. This is marked by the appearance of low level, or low affinity, single autoantibodies. Autoimmunity may be activated by somatic mutation or environmental triggers. Alternatively it may be aborted by environmental factors (e.g., various infections). In most individuals, these abnormalities are transient and are not associated with beta cell destruction. In some, beta cell destruction is initiated, marked by the appearance of multiple immunological abnormalities and later by measurable loss of FPIR. Genetic and environmental factors may modulate the process, affecting the rate of beta cell destruction. The loss of beta cells may be so slow in some individuals that overt diabetes does not occur during the person's lifetime. Once tolerance is broken to more than one of autoantigen, the levels of autoantibodies (IAA, anti-ICA512, and anti-GAD) appear to be stable over time. The presence of more than one of these autoantibodies, combined with FPIR measurements and typing for HLA-DQ alleles, allows the identification of a subset of relatives with sufficiently high-risk for type 1 diabetes to begin preventive trials.

The hypothesis that type 1 diabetes developed in a chronic and predictable manner forms the basis for prevention trials. It is equally important to identify those autoantibody-positive relatives who are unlikely to progress to diabetes. Knowledge about the natural history of the prediabetic period may also facilitate the diagnosis of type 1diabetes. Such information becomes clinically relevant in genetic counseling, the diagnosis of diabetes in adults (type 1 versus type II), and in children with an unusual clinical course. It is also relevant in the evaluation of individuals with transient hyperglycemia and of potential renal donors to relatives with type 1 diabetes.

Evaluation of each of the above the circumstances will usually include genetic, immunological, and metabolic information. The presence of the HLA allele DQA1*0102/DQB1*0602 is so rare among children with type 1 diabetes[198,199] that an alternative disorder should be considered as the cause of diabetes if it's haplotype is present. Other diseases that can present as insulin-requiring diabetes in childhood include Wolfram's syndrome (DIDMOAD-diabetes insipidus, diabetes mellitus, optic atrophy, and nerve deafness),[200] mitochondrial mutations (often associated with nerve deafness),[201] and MODY[202] (maturity onset diabetes of youth, that may be caused by mutations in the glucokinase and HLA genes).[203] The DQB1*0602 protective haplotype is likely to be present in these disorders at the normal population frequency of approximately 20%.

In adults, the presence of specific autoantibodies in a diabetic individual is probably the best available evidence of type 1 diabetes. At an epidemiological level, the diagnosis of type 1 diabetes by presence of ketoacidosis, insulin requirement, and young age of onset is useful, but for an individual may be of little value. In the past, determination of ICA was the only immunological test available. In most series high titers of ICA are present in less than 70% of new-onset patients with classic type 1 diabetes.[7] The use of three recombinant antigen based assays (insulin, anti-GAD, and anti-ICA512) appears to detect more than 90% of individuals with type 1 diabetes. Two or more of these antibodies are present in approximately 80% of patients. As such assays become available, the detection of slow-onset type 1 diabetes among adults (LADA: Latent Autoimmune Diabetes of Adults) should be enhanced. Such patients will likely compose more than 10% of adults developing diabetes as has now been confirmed in multiple studies[36,37,204] Whether this will lead to the immediate institution of insulin therapy rather than oral hypoglycemic agents requires clinical trials of the potential risks and benefits of such therapy.

Of 63 children with transient hyperglycemia we have evaluated, most (89%) remain non-diabetic.[205] All of the few with ICAs or IAAs became permanently diabetic within 18 months. This likely that the measurement of antibodies against multiple biochemically defined antigens will also improve diagnosis of this group. The IVGTT is the diagnostic test choice in this group. Among children with resolved transient hyperglycemia, none of those with normal FPIR (greater than the 5th percentile) has progressed to diabetes. Of those with low FPIR, a subset normalizes upon repeat testing, but those remaining abnormal have all become overtly diabetic within 1 year.

An individual contemplating the donation of a kidney to a diabetic relative should have an OGTT to rule out the presence of overt diabetes, an IVGTT by ICARUS criteria, and determination of autoantibodies with assays that have high specificity and sensitivity (e.g., the combination of anti-GAD, anti-insulin, and anti-ICA512). Most of these assays are available now in commercial laboratories.

Finally, the most important reason for detecting individuals at risk for type 1 diabetes is the potential for preventive therapy. Large trials of preventive therapies are now underway in the United States and Europe. It is likely that with more detailed analysis and improvements in the assays available, the prediction of type 1 diabetes will improve further. It will be very important to begin such prediction in the general population, as the majority of new cases have no affected relative. The large DPT-1 trial (namely parenteral therapy) is now complete and this study with more than 90,000 first degree relatives screened for islet autoantibodies will provide a wealth of information. The parenteral or oral insulin therapy did not slow progression to diabetes though it did induce insulin antibodies even though human insulin was administered. Initial analysis of more than 70,000 relatives tested for GAD65 and ICA512 autoantibodies confirms the predominant antibodies for prediction.[17] One half of cytoplasmic ICA positive (approximately 4% ICA positive) relatives did not express any of the biochemical autoantibodies, and just as important the cytoplasmic ICA assay failed to detect GAD65 or ICA512 autoantibodies in approxiamtely 2% of the relatives.[17] The presence of biochemical anti-islet autoantibodies at the initial screening was associated with high risk for eventual eligibility for the DPT trial (e.g., low first phase insulin secretion). Further analyses will be forthcoming, but it is likely that relatives expressing multiple biochemical autoantibodies will be high risk individuals without the need for cytoplasmic ICA testing or even intravenous glucose tolerance testing. The presence of a single biochemical autoantibody and low first phase insulin secretion will likely identify an additional high risk group. Following analysis of the large DPT-1 dataset more efficient paradigms for defining diabetes risk should be available for the next generation of trials for the prevention of diabetes.

A very large NIH effort is underway to prevent type 1A diabetes, including a North American TrialNet, the Immune Tolerance Network, Autoimmunity Centers of Excellence and Autoimmunity Prevention Centers. TrialNet will evaluate multiple therapies in new onset patients (aimed at preserving C-peptide secretion) and selective studies of diabetes prevention. This organization is encouraging proposals from the diabetes and immunology community, whether or not individuals are members of the core set of Centers. In a similar manner the Immune Tolerance Network is seeking protocols, and has a web site, (www.immunetolerance.org) to receive concept proposals. The Immune Tolerance Network will support core laboratories and trials and its funding is available to international studies. A major deficiency in our current efforts to prevent type 1A diabetes is a lack of proven quantitative assays of the antigen specific T cells that destroy islet B-cells. Further basic work in parallel with clinical trials is essential to take advantage of the international effort to prevent type 1 diabetes.

References

1. Verge CF, Eisenbarth GS. Prediction of type I diabetes: The natural history of the prediabetic period. 1996; 1st:230-258.
2. Kawasaki E, Eisenbarth G. High-Throughput radioassays for autoantibodies to recombinant autoantigens. Frontiers in Bioscience 2000; 5:181-190.
3. Leslie RD, Atkinson MA, Notkins AL. Autoantigens IA-2 and GAD in Type I (insulin-dependent) diabetes. Diabetologia 1999; 42:3-14.
4. Flanders G, Graves P, Rewers M. Prevention of type 1 diabetes from laboratory to public health. [Review] [91 refs]. Autoimmunity 1999; 29:235-246.
5. Riley WJ, Maclaren NK, Krischer J et al. A prospective study of the development of diabetes in relatives of patients with insulin-dependent diabetes. N Engl J Med 1990; 323:1167-1172.
6. Bonifacio E, Bingley PJ, Shattock M et al. Quantification of islet-cell antibodies and prediction of insulin-dependent diabetes. Lancet 1990; 335:147-149.
7. Verge CF, Gianani R, Kawasaki E et al. Prediction of type I diabetes in first-degree relatives using a combination of insulin, GAD and ICA512bdc/IA-2 autoantibodies. Diabetes 1996; 45:926-933.
8. Verge CF, Stenger D, Bonifacio E et al. Combined use of autoantibodies (IA-2ab, Gadab, IAA, ICA) in type 1 diabetes: Combinatorial islet autoantibody workshop. Diabetes 1998; 47:1857-1866.
9. Strebelow M, Schlosser M, Ziegler B et al. Karlsburg Type I diabetes risk study of a general population: Frequencies and interactions of the four major Type I diabetes- associated autoantibodies studied in 9419 schoolchildren. Diabetologia 1999; 42:661-670.
10. Christie MR, Roll U, Payton MA et al. Validity of screening for individuals at risk for type I diabetes by combined analysis of antibodies to recombinant proteins. Diabetes Care 1997; 20:965-970.
11. Vardi P, Crisa L, Jackson RA et al. Predictive value of intravenous glucose tolerance test insulin secretion less than or greater than the first percentile in islet cell antibody positive relatives of type I (insulin-dependent) diabetic patients. Diabetologia 1991; 34:93-102.
12. Deschamps I, Khalil I. The role of DQ alpha—beta heterodimers in genetic susceptibility to insulin-dependent diabetes. Diabetes Metab Rev 1993; 9:71-92.
13. Bingley PJ, Christie MR, Bonifacio E et al. Combined analysis of autoantibodies improves prediction of IDDM in islet cell antibody-positive relatives. Diabetes 1994; 43:1304-1310.
14. Naserke HE, Ziegler AG, Lampasona V et al. Early development and spreading of autoantibodies to epitopes of IA-2 and their association with progression to type 1 diabetes. J Immunol 1998; 161:6963-6969.
15. Kimpimaki T, Kulmala P, Savola K et al. Disease-Associated autoantibodies as surrogate markers of type 1 diabetes in young children at increased genetic risk. Childhood Diabetes in Finland Study Group [In Process Citation]. J Clin Endocrinol Metab 2000 Mar; 85(3):1126-32 2000; 85:1126-1132.
16. Pietropaolo M, Peakman M, Pietropaolo SL et al. Combined analysis of GAD65 and ICA512(IA-2) autoantibodies in organ and nonorgan-specific autoimmune diseases confers high specificity for insulin-dependent diabetes mellitus. J Autoimmun 1998; 11:1-10.
17. Yu L, Cuthbertson DD, Maclaren N et al. Expression of GAD65 and Islet Cell Antibody (ICA512) autoantibodies amongst cytoplasmic ICA+ relatives is associated with eligibility for Diabetes Prevention Trial Type-1. Diabetes 2001; 50:1735-1740.
18. Krischer JP, Schatz D, Riley WJ et al. Insulin and islet cell autoantibodies as time-dependent covariates in the development of insulin-dependent diabetes: A prospective study in relatives. J Clin Endocrinol Metab 1993; 77:743-749.
19. Maclaren N, Lan M, Coutant R et al. Only multiple autoantibodies to islet cells (ICA), insulin, GAD65, IA-2 and IA-2beta predict immune-mediated (Type 1) diabetes in relatives. J Autoimmun 1999; 12:279-287.
20. Eisenbarth GS, Verge CF, Allen H et al. The design of trials for the prevention of IDDM. Diabetes 1993; 42:941-947.
21. Chase HP, Dolan LM, Krischer JP et al. First phase insulin release during the intravenous glucose tolerance test as a risk factor for type 1 diabetes. J Pediatr 2001; 138:244-249.
22. Bingley PJftIG. Interactions of age, islet cell antibodies, insulin autoantibodies, and first-phase insulin response in predicting risk of progression to IDDM in ICA+ relatives: The ICARUS data set. Diabetes 1996; 45:1720-1728.
23. Ziegler AG, Ziegler R, Vardi P et al. Life table analysis of progression to diabetes of anti-insulin autoantibody-positive relatives of individuals with type I diabetes. Diabetes 1989; 38:1320-1325.
24. Gianani R, Pugliese A, Bonner-Weir S et al. Prognostically significant heterogeneity of cytoplasmic islet cell antibodies in relatives of patients with type I diabetes. Diabetes 1992; 41:347-353.

25. Marshall MO, Hoyer PE, Petersen JS et al. Contribution of glutamate decarboxylase antibodies to the reactivity of islet cell cytoplasmic antibodies. J Autoimmun 1994; 7:497-508.
26. Gianani R, Jackson R, Eisenbarth GS. Evidence that the autoantigen of restricted ICA is GAD. Diabetes Res Clin Prac 1991; 14(suppl. 1):S13.
27. Gianani R, Rabin DU, Verge CF et al. ICA512 autoantibody radioassay. Diabetes 1995; 44:1340-1344.
28. Pugliese A, Gianani R, Moromisato R et al. HLA-DQB1*0602 is associated with dominant protection from diabetes even among islet cell antibody-positive first-degree relatives of patients with IDDM. Diabetes 1995; 44:608-613.
29. Schatz D, Krischer J, Horne G et al. Islet cell antibodies predict insulin-dependent diabetes in United States school age children as powerfully as in unaffected relatives. J Clin Invest 1994; 93:2403-2407.
30. Bingley PJ, Bonifacio E, Shattock M et al. Can islet cell antibodies predict IDDM in the general population? Diabetes Care 1993; 16:45-50.
31. Kupila A, Muona P, Simell T et al. Feasibility of genetic and immunological prediction of type I diabetes in a population-based birth cohort. Diabetologia 2001; 44:290-297.
32. Rewers M, Norris JM, Eisenbarth GS et al. Beta-cell autoantibodies in infants and toddlers without IDDM relatives: Diabetes autoimmunity study in the young (Daisy). J Autoimmun 1996; 9:405-410.
33. Schenker M, Hummel M, Ferber K et al. Early expression and high prevalence of islet autoantibodies for DR3/4 heterozygous and DR4/4 homozygous offspring of parents with Type I diabetes: The German BABYDIAB study. Diabetologia 1999; 42:671-677.
34. Virtanen SM, Laara E, Hypponen E et al. Cow's milk consumption, HLA-DQB1 genotype, and type 1 diabetes: A nested case-control study of siblings of children with diabetes. Childhood diabetes in Finland study group. Diabetes 2000; 49:912-917.
35. Naserke HE, Schenker M, Ziegler AG. Frequency of islet autoantibodies in relatives of patients with Type II diabetes or gestational diabetes. Diabetologia 1999; 42:1377-
36. Turner R, Stratton I, Horton V et al. UKPDS 25: Autoantibodies to islet-cell cytoplasm and glutamic acid decarboxylase for prediction of insulin requirement in type 2 diabetes. UK Prospective Diabetes Study Group. Lancet 1997; 350:1288-1293.
37. Horton V, Stratton I, Bottazzo GF et al. Genetic heterogeneity of autoimmune diabetes: Age of presentation in adults is influenced by HLA DRB1 and DQB1 genotypes (UKPDS 43). UK Prospective Diabetes Study (UKPDS) Group. Diabetologia 1999; 42:608-616.
38. Zimmet PZ, Tuomi T, Mackay IR et al. Latent autoimmune diabetes mellitus in adults (LADA): The role of antibodies to glutamic acid decarboxylase in diagnosis and prediction of insulin dependency. Diabet Med 1994; 11:299-303.
39. Carlsson A, Sundkvist G, Groop L et al. Insulin and glucagon secretion in patients with slowly progressing autoimmune diabetes (LADA). J Clin Endocrinol Metab 2000; 85(1):76-80.
40. Srikanta S, Ganda OP, Rabizadeh A et al. First-degree relatives of patients with type I diabetes mellitus. Islet cell antibodies and abnormal insulin secretion. N Engl J Med 1985; 313:461-464.
41. Rewers M, LaPorte RE, King H et al. Trends in the prevalence and incidence of diabetes: Insulin-dependent diabetes mellitus in childhood. World Health Stat Q 1988; 41:179-189.
42. Laakso M, Pyorala K. Age at onset and type of diabetes. Diabetes Care 1985; 8:114-117.
43. Japan IDDM Epidemiological Study Group. Lack of regional variation in IDDM risk in Japan. Diabetes Care 1993; 16:796-800.
44. Notsu K, Oka N, Note S et al. Islet cell antibodies in the Japanese population and subjects with type I (insulin-dependent) diabetes. Diabetologia 1985; 28:660-662.
45. Boehm BO, Manfras B, Seibler J et al. Epidemiology and immunogenetic background of islet cell antibody positive nondiabetic schoolchildren. Diabetes 1991; 40:1435-1439.
46. Diabetes Epidemiology Research International (DERI) Group. Geographic patterns of childhood insulin-dependent diabetes mellitus. Diabetes 1988; 37:1113-1119.
47. Levy-Marchal C, Tichet J, Fajardy I et al. Islet cell antibodies in normal French schoolchildren. Diabetologia 1992; 35:577-582.
48. Levy-Marchal C, Papoz L, de Beaufort C et al. Incidence of juvenile type I (insulin-dependent) diabetes mellitus in France. Diabetologia 1990; 33:465-469.
49. Bruining GJ, Molenaar JL, Grobbee DE et al. Ten-year follow-up study of islet-cell antibodies and childhood diabetes mellitus. Lancet 1989; 1:1100-1103.
50. Green A, Gale EAM, Patterson CC. Incidence of childhood-onset insulin-dependent diabetes mellitus: The EURODIAB ACE Study. Lancet 1992; 339:905-909.
51. Bergua M, Sole J, Marion G et al. Prevalence of islet cell antibodies, insulin antibodies and hyperglycemia in 2291 schoolchildren. Diabetologia 1987; 30:724-726.

52. Serrano Rios M, Moy CS, Martin-Serrano R et al. Incidence of type I (insulin-dependent) diabetes mellitus in subjects 0-14 years of age in the Comunidad of Madrid, Spain. Diabetologia 1990; 33:422-424.
53. Elliott RB, Pilcher CC, McGregor MA. Protection from diabetes in normal school children using nicotinamide. Autoimmunity 1993; 15 (Suppl):57
54. Scott RS, Brown LJ, Darlow BA et al. Temporal variation in incidence of IDDM in Canterbury, New Zealand. Diabetes Care 1992; 15:895-899.
55. Rowe RE, Leech NJ, Nepom GT et al. High genetic risk for IDDM in the Pacific Northwest. First report from the Washington State Diabetes Prediction Study. Diabetes 1994; 43:87-94.
56. Dokheel TM, The Pittsburgh Epidemiology Research Group. An epidemic of childhood diabetes in the United States? Diabetes Care 1993; 16:1606-1611.
57. Verge CF, Silink M, Howard NJ. The incidence of childhood IDDM in New South Wales, Australia. Diabetes Care 1994; 17:693-696.
58. Colman PG, McNair PD, Caudwell JP et al. Screening of an apparent 'hot spot' of IDDM for markers of preclinical IDDM. Autoimmunity 1993; 15(Suppl):31
59. Bingley PJ, Bonifacio E, Shattock M et al. Can islet cell antibodies predict IDDM in the general population? Diabetes Care 1993; 16(1):45-50.
60. Bingley PJ, Gale EA. Incidence of insulin dependent diabetes in England: A study in the Oxford region, 1985-6. BMJ 1989; 298:558-560.
61. Dahlquist G, Blom L, Holmgren G et al. The epidemiology of diabetes in Swedish children 0-14 years - a six year prospective study. Diabetologia 1985; 28:802-808.
62. Landin-Olsson M, Palmer JP, Lernmark A et al. Predictive value of islet cell and insulin autoantibodies for Type 1 (insulin-dependent) diabetes mellitus in a population- based study of newly-diagnosed diabetic and matched control children. Diabetologia 1992; 35:1068-1073.
63. Karjalainen JK. Islet cell antibodies as predictive markers for IDDM in children with high background incidence of disease. Diabetes 1990; 39:1144-1150.
64. Tuomilehto J, Lounamaa R, Tuomilehto-Wolf E et al. Epidemiology of childhood diabetes mellitus in Finland - background of a nationwide study of type I (insulin-dependent) diabetes mellitus. Diabetologia 1992; 35:70-76.
65. Karjalainen J, Salmela P, Ilonen J et al. A comparison of childhood and adult type 1 diabetes mellitus. N Engl J Med 1989; 320:881-886.
66. Tuomi T, Groop LC, Zimmet PZ et al. Antibodies to glutamic acid decarboxylase reveal latent autoimmune diabetes mellitus in adults with a noninsulin- dependent onset of disease. Diabetes 1993; 42:359-362.
67. Pilcher CC, Elliott RB. Ontogeny of islet cell and insulin autoantibodies in first degree relatives of IDDM. Autoimmunity 1993; 15:78
68. Ziegler AG, Hillebrand B, Rabl W et al. On the appearance of islet associated autoimmunity in offspring of diabetic mothers: A prospective study from birth. Diabetologia 1993; 36:402-408.
69. Vardi P, Soloveicik L, Barzilai D et al. Insulin binding activity in cord blood of normal pregnancies. Autoimmunity 1993; 50.
70. Ziegler A-G, Hummel M, Schenker M et al. Autoantibody appearance and risk for development of childhood diabetes in offspring of parents with type 1 diabetes. The 2-year analysis of the German BABYDIAB study. Diabetes 1999; 48:460-468.
71. Baekkeskov S, Kanaani J, Jaume JC et al. Does GAD have a unique role in triggering IDDM? J Autoimmun 2000; 15:279-286.
72. Wegmann DR, Eisenbarth GS. It's Insulin. J Autoimmun 2000; 15:286-291.
73. Vardi P, Dib SA, Tuttleman M et al. Competitive insulin autoantibody assay. Prospective evaluation of subjects at high risk for development of type I diabetes mellitus. Diabetes 1987; 36:1286-1291.
74. Vardi P, Ziegler AG, Matthews JH et al. Concentration of insulin autoantibodies at onset of type I diabetes. Inverse log-linear correlation with age. Diabetes Care 1988; 11:736-739.
75. Martino GV, Tappaz ML, Braghi S et al. Autoantibodies to glutamic acid decarboxylase (GAD) detected by an immuno-trapping enzyme activity assay: Relation to insulin- dependent diabetes mellitus and islet cell antibodies. J Autoimmun 1991; 4:915-923.
76. DeAizpurua HJ, Harrison LC, Cram DS. An ELISA for antibodies to recombinant glutamic acid decarboxylase in IDDM. Diabetes 1992; 41:1182-1187.
77. DeAizpurua HJ, Harrison LC. Glutamic acid decarboxylase in insulin-dependent diabetes mellitus [published erratum appears in Diabetes Metab Rev 1992 Oct; 8(3):following table of contents]. Diabetes Metab Rev 1992; 8:133-147.

78. Thivolet CH, Tappaz M, Durand A et al. Glutamic acid decarboxylase (GAD) autoantibodies are additional predictive markers of Type 1 (insulin-dependent) diabetes mellitus in high risk individuals. Diabetologia 1992; 35:570-576.
79. Sohnlein P, Muller M, Syren K et al. Epitope spreading and a varying but not disease-specific GAD65 antibody response in Type I diabetes. The Childhood Diabetes in Finland Study Group. Diabetologia 2000; 43:210-217.
80. Hawa M, Fava D, Medici F et al. Antibodies to IA-2 and GAD65 in type 1 and type 2 diabetes. Diabetes Care 2000; 23:228.
81. Dotta F, Previti M, Tiberti C et al. Identification of the GM2-1 islet ganglioside: Similarities with a major neuronal autoantigen. Diabetes 1992; 41:365A.
82. Larsson L-I, Nielsen JH, Hutton JC et al. Pancreatic hormones are expressed on the surface of human and rat islet cells through exocytotic sites. Eur J Cell Biol 1989; 48:45-51.
83. Aguilar-Diosdado M, Parkinson D, Corbett JA et al. Potential autoantigens in IDDM. Expression of carboxypeptidase-H and insulin but not glutamate decarboxylase on the beta-cell surface. Diabetes 1994; 43(3):418-425.
84. Petersen JS, Russel S, Marshall MO et al. Differential expression of glutamic acid decarboxylase in rat and human islets. Diabetes 1993; 42:484-495.
85. Foulis AK, Liddle CN, Farquharson MA et al. The histopathology of the pancreas in type 1 (insulin-dependent) diabetes mellitus: A 25-year review of deaths in patients under 20 years of age in the United Kingdom. Diabetologia 1986; 29:267-274.
86. Sutherland DE, Sibley R, Xu XA et al. Twin-to-twin pancreas transplantation: Reversal and reenactment of the pathogenesis of type I diabetes. Trans Assoc Am Physicians 1984; 97:80-87.
87. Arslanian SL, Becker DJ, Rabin B et al. Correlates of insulin antibodies in newly diagnosed children with insulin-dependent diabetes before insulin therapy. Diabetes 1985; 34:926-930.
88. Karjalainen J, Knip M, Mustonen A et al. Relation between insulin antibody and complement-fixing islet cell antibody at clinical diagnosis of IDDM. Diabetes 1986; 35:620-622.
89. Mcevoy RC, Witt ME, Ginsberg-Fellner F et al. Anti-insulin antibodies in children with type I diabetes mellitus. Genetic regulation of production and presence at diagnosis before insulin replacement. Diabetes 1986; 35:634-641.
90. Jackson RA, Vardi P, Herskowitz RD et al. Dual parameter linear model for prediction of onset of type I diabetes in islet cell antibody positive relatives. Clin Res 1988; 36:585A
91. Castano L, Ziegler AG, Ziegler R et al. Characterization of insulin autoantibodies in relatives of patients with type 1 diabetes. Diabetes 1993; 42:1202-1209.
92. Williams AJK, Bingley PJ, Bonifacio E et al. A novel micro-assay for insulin autoantibodies. J Autoimmun 1997; 10:473-478.
93. Naserke HE, Dozio N, Ziegler A-G et al. Comparison of a novel micro-assay for insulin autoantibodies with the conventional radiobinding assay. Diabetologia 1998; 41:681-683.
94. Yu L, Robles DT, Abiru N et al. Early expression of antiinsulin autoantibodies of humans and the NOD mouse: Evidence for early determination of subsequent diabetes. Proc Natl Acad Sci USA 2000; 97:1701-1706.
95. Abiru N, Maniatis AK, Yu L et al. Peptide and MHC specific breaking of humoral tolerance to native insulin with the B:9-23 peptide in diabetes prone and normal mice. Diabetes 2001; 50:1274-1281.
96. Olmos P, A'Hern R, Heaton DA et al. The significance of the concordance rate for type 1 (insulin- dependent) diabetes in identical twins. Diabetologia 1988; 31:747-750.
97. Soltesz G, Madacsy L, Bekefi D et al. The hungarian childhood diabetes epidemiology group: Rising incidence of type 1 diabetes in hungarian children. Diabetic Med 1994; 7:111-114.
98. Verge CF, Gianani R, Yu L et al. Late progression to diabetes and evidence for chronic beta cell autoimmunity in identical twins of patients with type I diabetes. J Investig Med 1995; 43:227A-
99. Savilahti E, Simell O, Koskimies S et al. Celiac disease in insulin-dependent diabetes mellitus. J Pediatr 1986; 108:690-693.
100. Maclaren NK, Riley WJ. Thyroid, gastric, and adrenal autoimmunities associated with insulin-dependent diabetes mellitus. Diabetes Care 1985; 8(Suppl1):34-38.
101. Kronenberg M, Siu G, Hood LE et al. The molecular genetics of the T-cell antigen receptor and T-cell antigen recognition. Ann Rev Immunol 1986; 4:529-591.
102. Joner G and Sovik O. Increasing incidence of diabetes mellitus in norwegian children 0-14 years of age, 1973-1982. Diabetologia 1989; 32:79-83.
103. Green A, Andersen PK, Svendsen AJ et al. Increasing incidence of early onset type I (insulin-dependent) diabetes mellitus: A study of Danish male birth cohorts. Diabetologia 1992; 35:178-182.

104. Soltesz G, Madacsy L, Bekefi D et al. Rising incidence of type 1 diabetes in Hungarian children (1978- 1987). Hungarian Childhood Diabetes Epidemiology Group. Diabetic Med 1990; 7:111-114.
105. Tuomilehto J, Rewers M, Reunanen A et al. Increasing trend in type I (insulin-dependent) diabetes mellitus in childhood in Finland. Analysis of age, calendar time, and birth cohort effects during 1965 to 1984. Diabetologia 1991; 34:282-287.
106. Drykoningen CEM, Mulder ALM, Vaandrager GJ et al. The incidence of male childhood type 1 (insulin-dependent) diabetes mellitus is rising rapidly in the Netherlands. Diabetologia 1992; 35:139-142.
107. Nystrom L, Dahlquist G, Rewers M et al. The Swedish childhood diabetes study: An analysis of the temporal variation in diabetes incidence, 1978-1987. Int J Epidemiol 1990; 19:141-146.
108. Rewers M, LaPorte RE, Walczak M et al. Apparent epidemic of insulin dependent diabetes mellitus in Midwestern Poland. Diabetes 1987; 36:106-113.
109. Hayashi T, Faustman D. Essential role of human leukocyte antigen-encoded proteasome subunits in NF-kappaB activation and prevention of tumor necrosis factor-alpha-induced apoptosis. J Biol Chem 2000; 275:5238-5247.
110. Onkamo P, Vaananen S, Karvonen M et al. Worldwide increase in incidence of Type I diabetes—the analysis of the data on published incidence trends. Diabetologia 1999; 42:1395-1403.
111. Variation and trends in incidence of childhood diabetes in Europe. EURODIAB ACE Study Group. Lancet 2000; 355:873-876.
112. Elliott RB, Pilcher C, Edgar BW. Geographic IDDM in Polynesia and Macronesia: The epidemiology of insulin dependent diabetes in Polynesian children born and reared in Polynesia, compared with Polynesian children resident in Auckland, New Zealand. Diabetes in the Young Bulletin (Proc of the ISGD) 1989; 20:16.
113. Bodansky HJ, Staines A, Stephenson C et al. Evidence for an environmental effect in the aetiology of insulin dependent diabetes in a transmigratory population. BMJ 1992; 304:1020-1022.
114. Dahlquist G, Blom L, Holmgren G et al. The epidemiology of diabetes in Swedish children 0-14 years - a six year prospective study. Diabetologia 1985; 28:802-808.
115. Kostraba JN, Gay EC, Cai Y et al. Incidence of insulin-dependent diabetes mellitus in Colorado. Epidemiology 1992; 3:232-238.
116. Menser MA, Forrest JM, Bransby RD. Rubella infection and diabetes mellitus. Lancet 1978; 1:57-60.
117. Rubinstein P, Walker ME, Fedun B et al. The HLA system in congenital rubella patients with and without diabetes. Diabetes 1982; 31:1088-1091.
118. Clarke WL, Shaver KA, Bright GM et al. Autoimmunity in congenital rubella syndrome. J Pediatr 1984; 104:370-373.
119. Yoon J-W, Morishima T, McClintock PR et al. Virus-induced diabetes mellitus: Mengovirus infects pancreatic beta cells in strains of mice resistant to the diabetogenic effect of encephalomyocarditis virus. J Virol 1984; 50:684-690.
120. Tuvemo T, Dahlquist G, Frisk G et al. The Swedish childhood diabetes study III: IgM against coxsackie B viruses in newly diagnosed type 1 (insulin-dependent) diabetic children - no evidence of increased antibody frequency. Diabetologia 1989; 32:745-747.
121. Kaufman DL, Erlander MG, ClareSalzler M et al. Autoimmunity to two forms of glutamate decarboxylase in insulin-dependent diabetes mellitus. J Clin Invest 1992; 89:283-292.
122. Karounos DG, Simmerman L, Hickman SL et al. Identification of the p52-Rubella related autoantigen as an insulin secretory granule protein. Diabetes 1993; 42 (Suppl 1):221A
123. Anonymous. Lack of regional variation in IDDM risk in Japan. Japan IDDM Epidemiology Study Group. Diabetes Care 1993; 16:796-800.
124. Dahlquist GG. Primary and secondary prevention strategies of pretype 1 diabetes. Potentials and pitfalls. [Review]. Diabetes Care 1999; 22 Suppl 2:B4-B6.
125. Lonnrot M, Korpela K, Knip M et al. Enterovirus infection as a risk factor for beta-cell autoimmunity in a prospectively observed birth cohort: The Finnish Diabetes Prediction and Prevention Study. Diabetes 2000; 49:1314-1318.
126. Norris JM, Beaty B, Klingensmith G et al. Lack of association between early exposure to cow's milk protein and β-cell autoimmunity: Diabetes Autoimmunity Study in the Young (DAISY). JAMA 1996; 276:609-614.
127. Graves PM, Rewers M. The role of enteroviral infections in the development of IDDM: Limitations of current approaches. Diabetes 1997; 46:161-168.
128. Graves PM, Barriga KJ, Norris JM et al. Lack of association between early childhood immunizations and beta-cell autoimmunity. Diabetes Care 1999; 22:1694-1697.
129. Honeyman MC, Coulson BS, Stone NL et al. Association between rotavirus infection and pancreatic islet autoimmunity in children at risk of developing type 1 diabetes. Diabetes 2000; 49:1319-1324.

130. Medzhitov R, Preston-Hurlburt P, Janeway CA. A human homologue of the Drosophila Toll protein signals activation of adaptive immunity. Nature 1997; 388:394-397.
131. Like AA, Guberski DL, Butler L. Influence of environmental viral agents on frequency and tempo of diabetes mellitus in BB/Wor rats. Diabetes 1991; 40:259-262.
132. Wilberz S, Partke HJ, Dagnaes-Hansen F et al. Persistent MHV (mouse hepatitis virus) infection reduces the incidence of diabetes mellitus in nonobese diabetic mice. Diabetologia 1991; 34:2-5.
133. Oldstone MB. Prevention of type I diabetes in nonobese diabetic mice by virus infection. Science 1988; 239:500-502.
134. Dyrberg T, Schwimmbeck PL, Oldstone MB. Inhibition of diabetes in BB rats by virus infection. J Clin Invest 1988; 81:928-931.
135. Guberski DL, Thomas VA, Shek WR et al. Induction of Type 1 diabetes by Kilham's rat virus in diabetes-resistant BB/Wor rats. Science 1991; 254:1010-1013.
136. Yoon JW. Role of viruses in the pathogenesis of IDDM. Ann Med 1991; 23:437-445.
137. Yoon J-W, McClintock PR, Bachurski CJ et al. Virus-induced diabetes mellitus. No evidence for immune mechanisms in the destruction of beta-cells by the D variant of encephalyomyocarditis virus. Diabetes 1985; 34:922-925.
138. Virtanen S, Rasanen L, Aro A et al. Infant feeding in finnish children less than 7 yr of age with newly diagnosed IDDM. Childhood Diabetes in Finland Study Group. Diabetes Care 1991; 14:415-417.
139. Virtanen SM, Rasanen L, Aro A et al. Feeding in infancy and the risk of type 1 diabetes mellitus in Finnish children. The 'Childhood Diabetes in Finland' Study Group. Diabetic Med 1992; 9:815-819.
140. Mayer EJ, Hamman RF, Gay EC et al. Reduced risk of IDDM among breast-fed children. The Colorado IDDM Registry. Diabetes 1988; 37:1625-1632.
141. Verge CF, Howard NJ, Irwig L et al. Environmental factors in childhood IDDM: A population-based, case-control study. Diabetes Care 1994; 17:1381-1389.
142. Kyvik KO, Green A, Svendsen A et al. Breast feeding and the development of type I diabetes mellitus. Diabetic Med 1992; 9:233-235.
143. Gerstein HC. Cow's milk exposure and type I diabetes mellitus – a critical overview of the clinical literature. Diabetes Care 1994; 17:13-19.
144. Johansson C, Samuelsson U, Ludvigsson J. A high weight gain early in life is associated with an increased risk of type I (insulin-dependent) diabetes mellitus. Diabetologia 1994; 37:91-94.
145. Karjalainen J, Martin JM, Knip M et al. A bovine albumin peptide as a possible trigger of insulin-dependent diabetes mellitus [see comments]. N Engl J Med 1992; 327:302-307.
146. Savilahti E, Akerblom HK, Tainio V-M et al. Children with newly diagnosed insulin dependent diabetes mellitus have increased levels of cow's milk antibodies. Diabetes Res 1988; 7:137-140.
147. Yokota A, Yamaguchi Y, Ueda Y et al. Comparison of islet cell antibodies, islet cell surface antibodies and anti-bovine serum albumin antibodies in type 1 diabetes. Diabetes Res Clin Prac 1990; 9:211-217.
148. Pocecco M, Nicoloso F, Tonini G et al. Increased levels of cow's milk antibodies in children with newly diagnosed insulin-dependent diabetes mellitus (IDDM). Horm Metab Res 1991; 35:67-67.
149. Dahlquist G, Savilahti E, Landin-Olsson M. An increased level of antibodies to beta-lactoglobulin is a risk determinant of early-onset Type I (insulin-dependent) diabetes mellitus independent of islet cell antibodies and early introduction of cow's milk. Diabetologia 1992; 35:980-984.
150. Karjalainen J, Saukkonen T, Savilahti E et al. Disease-associated anti-bovine serum albumin antibodies in type I (insulin-dependent) diabetes mellitus are detected by particle concentration fluoroimmunoassay, and not by enzyme linked immunoassay. Diabetologia 1992; 35:985-990.
151. Karjalainen J, Cheung R, Vandermeulen JM et al. T cell proliferation response against bovine albumin (BSA) follows the natural course and is restricted to the linear ABBOS sequence in insulin-dependent diabetes mellitus. Autoimmunity 1993; 15(Suppl):62A
152. Atkinson MA, Bowman MA, Kuo-Jang K et al. Lack of immune responsiveness to bovine serum albumin in insulin- dependent diabetes. N Engl J Med 1993; 329:1853-1858.
153. Elliott RB, Martin JM. Dietary protein: A trigger of insulin-dependent diabetes in the BB rat? Diabetologia 1984; 26:297-299.
154. Daneman D, Fishman L, Clarson C et al. Dietary triggers of insulin-dependent diabetes in the BB rat. Diabetes Res 1987; 5:93-97.
155. Coleman DL, Kuzava JE, Leiter EH. Effect of diet on incidence of diabetes in nonobese diabetic mice. Diabetes 1990; 39:432-436.
156. Elliott RB, Reddy SN, Bibby NJ et al. Dietary prevention of diabetes in the nonobese diabetic mouse. Diabetologia 1988; 31:62-64.

157. Paronen J, Knip M, Savilahti E et al. Effect of cow's milk exposure and maternal type 1 diabetes on cellular and humoral immunization to dietary insulin in infants at genetic risk for type 1 diabetes. Finnish Trial to Reduce IDDM in the Genetically at Risk Study Group. Diabetes 2000; 49:1657-1665.
158. Srikanta S, Ganda OP, Gleason RE et al. Pretype I diabetes. Linear loss of beta cell response to intravenous glucose. Diabetes 1984; 33:717-720.
159. Eisenbarth GS. Type I diabetes mellitus. A chronic autoimmune disease. N Engl J Med 1986; 314:1360-1368.
160. Bleich D, Jackson RA, Soeldner JS et al. Analysis of metabolic progression to type I diabetes in ICA+ relatives of patients with type I diabetes. Diabetes Care 1990; 13:111-118.
161. McCulloch DK, Klaff LJ, Kahn SE et al. Nonprogression of subclinical B-cell dysfunction among first degree relatives of IDDM patients. 5 year follow-up of the Seattle Family Study. Diabetes 1990; 39:549-556.
162. Beer SF, Heaton DA, Alberti KG et al. Impaired glucose tolerance precedes but does not predict insulin-depedendent diabetes mellitus: A study of identical twins. Diabetologia 1990; 33:497-502.
163. Kobayashi T, Itoh T, Kosaka K et al. Time course of islet cell antibodies and beta-cell function in noninsulin-dependent stage of type I diabetes. Diabetes 1987; 36:510-517.
164. Tunbridge WMG, Brewis M, French JM et al. Natural history of autoimmune thyroiditis. BMJ 1981; 282:258-262.
165. Spencer KM, Tarn A, Dean BM et al. Fluctuating islet-cell autoimmunity in unaffected relatives of patients with insulin-dependent diabetes. Lancet 1984; 1:764-766.
166. Chase HP, Voss MA, Butler-Simon N et al. Diagnosis of pretype I diabetes. J Pediatr 1987; 111:807-812.
167. Thivolet C, Beaufrere B, Geburher L et al. Autoantibodies and genetic factors associated with the development of type I (insulin-dependent) diabetes in first degree relatives of diabetic patients. Diabetologia 1991; 34:186-191.
168. Millward BA, Alviggi L, Hoskins PJ et al. Immune changes associated with insulin dependent diabetes may remit without causing the disease: A study in identical twins. Br Med J (Clin Res Ed) 1986; 292:793-796.
169. Helmke K, Otten A, Willems WR et al. Islet cell antibodies and the development of diabetes mellitus in relation to mumps infection and mumps vaccination. Diabetologia 1986; 29:30-33.
170. Landin-Olsson M, Karlsson A, Dahlquist G et al. Islet cell and other organ-specific autoantibodies in all children developing Type 1 (insulin-dependent) diabetes mellitus in Sweden during one year and in matched control children. Diabetologia 1989; 32:387-395.
171. Landin-Olsson M. Precision of the islet-cell antibody assay depends on the pancreas. J Clin Lab Anal 1990; 4:289-294.
172. Kulmala P, Rahko J, Savola K et al. Stability of autoantibodies and their relation to genetic and metabolic markers of Type I diabetes in initially unaffected schoolchildren. Diabetologia 2000; 43:457-464.
173. Yu J, Yu L, Bugawan TL et al. Transient anti-islet autoantibodies: Infrequent occurrence and lack of association with genetic risk factors. J Clin Endocrinol Metab 2000; 85:2421-2428.
174. Colman PG, Steele C, Couper JJ et al. Islet autoimmunity in infants with a Type I diabetic relative is common but is frequently restricted to one autoantibody. Diabetologia 2000; 43:203-209.
175. Carel JC, Boitard C, Bougneres PF. Decreased insulin response to glucose in islet cell antibody-negative siblings of type 1 diabetic children. J Clin Invest 1993; 92:509-513.
176. Pilcher CC, Elliott RB. A sensitive and reproducible method for the assay of human islet cell antibodies. J Immunol Methods 1990; 129:111-117.
177. Santiago JV. Insulin therapy in the last decade. A pediatric perspective. Diabetes Care 1993; 16 (Suppl 3):143-154.
178. McCulloch DK, Koerker DJ, Kahn SE et al. Correlations of in vivo beta-cell function tests with beta-cell mass and pancreatic insulin content in streptozocin-treated baboons. Diabetes 1991; 40:673-679.
179. Allen HF, Jeffers BW, Klingensmith GJ et al. First-phase insulin release in normal children. J Pediatr 1993; 123:733-738.
180. Arslanian S, Austin A. Determinants of first and second phase insulin secretion in healthy adolescents. Pediatr Res 1993; 3:S74.
181. Cutfield WS, Bergman RN, Menon RK et al. The modified minimal model: Application to measurement of insulin sensitivity in children. J Clin Endocrinol Metab 1990; 70:1644-1650.
182. Smith CP, Williams AJK, Thomas JM et al. The pattern of basal and stimulated insulin responses to intravenous glucose in first degree relatives of type I (insulin-dependent) diabetic children and unrelated adults aged 5 to 50 years. Diabetologia 1988; 31:430-434.

183. Colman PG, Stewart V, Kean J et al. Comparison of two commonly used standard IVGTTs. Diabetes Care 1992; 15:1053-1055.
184. Bingley PJ, Colman P, Eisenbarth GS et al. Standardization of IVGTT to predict IDDM. Diabetes Care 1992; 15:1313-1316.
185. Rayman G, Clark P, Schneider AE et al. The first phase insulin response to intravenous glucose is highly reproducible. Diabetologia 1990; 33:631-634.
186. Smith CP, Tarn AC, Thomas JM et al. Between and within subject variation of the first phase insulin response to intravenous glucose. Diabetologia 1988; 31:123-125.
187. McNair PD, Colman PG, Alford FP et al. Reproducibility of the first phase insulin response (FPIR) in the intravenous glucose tolerance test (IVGTT) is not improved by retrograde cannulation and arterialisation or by the use of a lower glucose dose. Autoimmunity 1993.
188. Rowe RE, Leech NJ, McCullock DK. Does drawing blood from a retrogradely cannulated hand vein improve reproducibility in the intravenous glucose tolerance test (IVGTT)? Autoimmunity 1993; 15(Suppl):79.
189. Leech NJ, Rowe RE, Bucksa J et al. HbA1c may be an early indicator of islet cell dysfunction. Autoimmunity 1993; 15(Suppl):75.
190. Signore A, Chianelli M, Ferretti E et al. New tools for imaging insulitis in pre diabetes. Autoimmunity 1993; 15(Suppl):64-64.
191. Yuh WTC, Wiese JA, Abu-Yousef MM et al. Pancreatic transplant imaging. Radiology 1988; 167:679-683.
192. Kelcz F, Sollinger HW, Pirsch JD. MRI of the pancreas transplant: Lack of correlation between imaging and clinical status. Mag Res Med 1991; 21:30-38.
193. Fernandez MP, Bernardino ME, Neylan JF et al. Diagnosis of pancreatic transplant dysfunction: Value of gadopentetate dimeglumine-enhanced MR imaging. AJR 1991; 156:1171-1176.
194. Hanafusa T, Miyazaki A, Miyagawa J et al. Examination of islets in the pancreas biopsy specimens from newly diagnosed type I (insulin-dependent) diabetic patients. Diabetologia 1990; 33:105-111.
195. Martin S, Pawlowski B, Greulich B et al. Natural course of remission in IDDM during 1st year after diagnosis. Diabetes Care 1992; 15:66-74.
196. Schiffrin A, Suissa S, Weitzner G et al. Factors predicting course of B-cell function in IDDM. Diabetes Care 1992; 15:997-1001.
197. Knip M, Llonen J, Mustonen A et al. Evidence of an accelerated B-cell destruction in HLA-Dw3/Dw4 heterozygous children with type I (insulin-dependent) diabetes. Diabetologia 1986; 29:347-351.
198. Baisch JM, Weeks T, Giles R et al. Analysis of HLA-DQ genotypes and susceptibility in insulin-dependent diabetes mellitus. N Engl J Med 1990; 322:1836-1841.
199. Erlich HA, Griffith RL, Bugawan TL et al. Implication of specific DQB1 alleles in genetic susceptibility and resistance by identification of IDDM siblings with novel HLA-DQB1 allele and unusual DR2 and DR1 haplotypes. Diabetes 1991; 40:478-481.
200. Karasik A, O'Hara C, Srikanta S et al. Genetically programmed selective islet beta-cell loss in diabetic subjects with Wolfram's syndrome. Diabetes Care 1989; 12:135-138.
201. Kadowaki T, Kadowaki H, Mori Y et al. A subtype of diabetes mellitus associated with a mutation of mitochondrial DNA. N Engl J Med 1994; 330:962-968.
202. Bowden DW, Gravius TC, Akots G et al. Identification of genetic markers flanking the locus for maturity-onset diabetes of the young on human chromosome 20. Diabetes 1992; 41:88-92.
203. Froguel P, Zouali H, Vionnet N et al. Familial hyperglycemia due to mutations in glucokinase. Definition of a subtype of diabetes mellitus [see comments]. N Engl J Med 1993; 328:697-702.
204. Carlsson A, Sundkvist G, Groop L et al. Insulin and glucagon secretion in patients with slowly progressing autoimmune diabetes (LADA). J Clin Endocrinol Metab 2000; 85:76-80.
205. Herskowitz-Dumont R, Wolfsdorf JI, Jackson RA et al. Distinction between transient hyperglycemia and early insulin-dependent diabetes mellitus in childhood: A prospective study of incidence and prognostic factors. J Pediatr 1993; 123:347-354.

CHAPTER 12

Clinical Trials for the Prevention of Type 1 Diabetes

H. Peter Chase, Anthony R. Hayward and George S. Eisenbarth

The real issues for strategies to prevent type 1 diabetes are the anticipated duration of the prophylaxis, the effectiveness of prophylaxis, the toxicity of treatment, and the complexity of the treatment regimen. If, as seems likely in man, the propensity to develop type 1A diabetes (immune mediated insulin-dependent diabetes mellitus (IDDM)) is lifelong, then young people at risk for the disease will have to be treated for many decades. If disease susceptibility, on the other hand, depends on specific environmental factors, then prophylaxis might be tailored to match these circumstances. Longstanding evidence that the complications of diabetes are delayed by good control has recently been confirmed by the Diabetes Control and Complications Trial. This means that the benefits, inconveniences, and toxicity of diabetes prophylaxis have to be weighed against the potential for good control. These are likely to be related issues, because adherence to a prophylactic regimen may well require the same degree of commitment (or compliance) as does maintaining a good glycosylated hemoglobin level using existing treatments for a person with diabetes. If this is the case, then the majority of patients may fail to benefit from a prophylaxis approach, just as they fail to achieve good control with the available treatments. Faced with this dilemma, a rational approach would be to determine how effectively diabetes could be prevented. There are no secure data in humans yet, while there is a superfluity of data from animal models.[1] The problem with these results, which are reviewed in the next section, is the ease with which it is possible to prevent diabetes in nonobese diabetic (NOD) and BioBreeding (BB) rats. The likelihood that interventions that work in rodents may not work in man must be borne in mind.

Animal Models for Type 1 Diabetes Prevention

In both BB rats and NOD mice, pancreatic beta cells are destroyed by immunological mechanisms. The strength of these models for human diabetes are discussed in Chapter 3 and extend to the characterization of several genetic predisposing factors and to the identification of pathogenic T cell clones. The weaknesses of the animal models stem mainly from our ignorance of the mechanisms by which autoimmunity is triggered or sustained in the experimental animals or, for that matter, in man. Both animal strains have been bred to maximize their incidence of diabetes and, in addition, most laboratories house their animals under conditions that yield the highest number of diabetics. Nevertheless, some 20% of female NOD mice escape even under "optimal" diabetogenic conditions. Whether the environmental or stochastic factors that operate to protect these mice have parallels in humans is unknown. This limits somewhat the conclusions that can be drawn from the diverse interventions that prevent IDDM in experimental models.[2] With this caveat, the general approaches that have proved effective in animals are summarized in Table 1.

Immunology of Type 1 Diabetes, Second Edition, edited by George S. Eisenbarth.

Table 1. Examples of interventions preventing type 1 diabetes in animal models

Strategy	Specificity	Animal	Conclusion
Immune Suppression	None	NOD and BB	Nonspecific suppression of Cell-mediated immunity prevents type 1 diabetes but long-term immunosuppression is unacceptable.
Anti-CD4	CD4 T Cells	NOD	Anti-CD4 monoclonals prevent diabetes.
Anti-CD3	CD3 T Cells	NOD	Anti-CD3 monclonals reverse diabetes at onset with long-duration of effect.
Transplantation	None	NOD and BB	Grafts of marrow, fetal dendritic cells, liver and thymus protect.
Immune Stimulation	None	NOD	Immune activation by agents such as BCG protect.
Immunologic Vaccination	Insulin/GAD/ HSP60	NOD	Multiple mechanisms, T potential activation regulatory cells.
Oral Tolerance	Insulin	NOD	Immune deviation delays diabetes
Diets	Unknown	NOD	Radical scavenging, nutritional specific (e.g., lack peptides) protect
Galactosyl-ceramide	NK T cells	NOD	Activation CD1 restricted cells
Cytokines	IL-10, TNF, IL-4	NOD	Complex effect dependent often upon timing of Rx
Gene Therapy	Antigens/ Cytokines	NOD	Multiple potential targets

NOD= nonobese diabetic mouse; BB= BioBreeding rat

Clinical Trials for the Prevention of Type 1 Diabetes

The recognition of insulin-dependent (type 1A) diabetes mellitus as an immunological disease led to the first controlled trials of interventions using nonspecific immunosuppressive treatments in subjects already diagnosed with diabetes. Cyclosporin A (CyA) and prednisone-azathioprine were tried initially because they had been effective in animals. These early trials were important because they showed that insulin dependency could be delayed for a period while immunosuppression was maintained. They also showed that the natural history of diabetes progression was reinstituted following withdrawal of immunosuppression.

Trials of prophylaxis in individuals with recent-onset diabetes are based on the view that beta cell destruction is a gradual process and that an intercurrent stress precipitates symptoms (and loss of glucose homeostasis) before all insulin-making potential is lost. Unfortunately, 80 to 90 percent of islets have already been destroyed by the time diabetes is diagnosed.[3] Thus, the best possible outcome of trials in people who already have diabetes is prolonging the time when small amounts of insulin continue to be made, rather than prevention of diabetes. Benefit to the subject under these circumstances might come not from discontinuing insulin shots but from simplifying daily blood glucose control, lowering glycosylated hemoglobin levels, and reducing the risk of hypoglycemia and diabetes complications. Benefit of this type is probable, but is unlikely to appeal to recent-onset diabetics unless the preventive regimen is simple and safe.

A different approach for prophylaxis trials is to concentrate on individuals identified as prediabetic by the criteria described below. The best possible outcome in prediabetic individuals (whose glucose homeostasis is not so impaired so as to make them hyperglycemic) would be the maintenance of insulin production and a life free of insulin replacement. Ascertainment of individuals with a high likelihood of developing diabetes is currently by screening first-degree relatives of affected individuals. The yield, about 3% of those screened, is very small in relation to the number of individuals developing type 1 diabetes per year. Prevention trials in recent-onset cases are therefore likely to continue to be important until better screening methods, applicable to entire populations are developed.

Trials in Newly Diagnosed Subjects

Cyclosporin A (CyA)

One method of prolonging insulin production has been successful to date, the administration of CyA.[4,5] The effects of this agent were greatest when started within 6 weeks of initiation of insulin therapy, and the treatment had to be continued for at least 6 months and, preferably, for 1 year to show benefit. Unfortunately, the effects of CyA on T cells are lost when the CyA is discontinued. As CyA is nephrotoxic, it has been concluded[5] "that additional studies with CyA alone, in subjects with newly diagnosed IDDM, are no longer warranted." Despite this rather negative conclusion, the impact of the trials with CyA should not be underestimated. The initial promise of immunosuppression as a means to prevent diabetes stimulated interest in the early diagnosis of diabetes and led, ultimately, to screening approaches by which prediabetes could be recognized. The fact that some participants in the trials could maintain glycosylated hemoglobin (HbA1) levels in the normal range for months while on CyA showed for the first time that the destruction of beta cells could be slowed even if not halted.[6] A major limitation of CyA therapy in new-onset diabetic patients is the relapse to hyperglycemia within 3 years even if CyA was continued. Relapses occurred despite the maintenance of C-peptide secretion. This suggested the relapses resulted from insufficient beta cell mass at the initiation of therapy rather than ineffective immunosuppression.[7] The trials also stimulated awareness of the cost/benefit ratio of interventions to prevent diabetes. The costs in the context of CyA were high, as the drug is expensive, and even with careful and expensive monitoring of blood levels, renal damage still occurred. Thus the costs were felt to outweigh the benefits.

Other Immunosuppressive Agents (Rapamycin, Mycophenolate Mofetil, Anti-IL-2 Antibodies, Methotrexate)

It is noteworthy that Shapiro and coworkers using an immunosuppressive regimen consisting of anti-IL-2 monoclonal antibodies, rapamycin (sirolimus) and low dose FK506 (tacrolimus, a calcineurin inhibitor similar to cyclosporine A) achieved insulin free status in patients receiving islet transplants for type 1 diabetes.[8,9] Of approximately 12 patients transplanted five now again require low dose insulin therapy, with the longest insulin free patient now followed for almost two years.[9a,9b] It is more difficult to prevent transplant rejection and apparent autoimmune destruction for islet transplants compared to pancreas transplants and these results suggest that the above immunosuppressive cocktail may be sufficient to abrogate anti-islet autoimmunity. Longer follow up will be necessary, in particular to determine whether all patients will eventually destroy their islet transplant, but the initial data suggests that in a subset this may not occur. There are obviously risks associated with the therapies, even beyond the usual risks associated with immunosuppression, including hyperlipidemia for rapamycin.

Mycophenolate mofetil (Cellcept),[10] an inhibitor of the synthesis of the purine guanosine monophosphotic (GMP) from inosine, prevents diabetes in BB rats.[11] It appears to be more potent in comparison to azathioprine, and when used with calcineurin inhibitors, for pancreas transplantation has had a major benefit allowing reduced steroid dosage. Trials of mycophenolate mofetil in patients with new onset diabetes are commencing (Peter Gottlieb, oral communication). In a very small trial methotrexate did not preserve C-peptide secretion in new onset patients.[12]

Anti-CD3 Therapy

In the NOD mouse monoclonal antibodies directed at the T cell receptor CD3 molecule are able to reverse hyperglycemia in acutely diabetic mice but the same antibodies do not prevent diabetes when administered earlier in the course of disease.[13,14] Human trials are underway in patients with recent onset diabetes using engineered anti-CD3 monoclonals (engineered to "abrogate" Fc binding, and "humanized").[14a] These antibodies do not induce severe cytokine release syndromes, in contrast to standard anti-CD3 monoclonal antibodies. Preliminary studies indicate some preservation of C-peptide secretion and limited toxicity (rash) with a subset of patients developing anti-idiotypic antibodies.[14a] It is likely that this form of therapy will require repeated administration of the monoclonal and trials are planned for longer term therapy of new onset patients.

High Dose Intravenous Insulin

Infusion of large quantities of exogenous insulin to control blood glucose levels tightly was reported to preserve C-peptide secretion post development of diabetes.[15] The stress of keeping the patient constantly in bed for 14 days with a large needle in place would probably have prevented widespread acceptance of this treatment, and there has been no replication of these studies in the subsequent decade.

Immunostimulation

A third approach to prolonging insulin production in the newly diagnosed diabetic is through immunostimulation.[16] This strategy is based on the lower incidence of diabetes in NOD mice stimulated immunologically with agents as diverse as allogeneic cells or bacillus Calmette-Guérin (BCG).[17] The mechanism for the protection is not clear, though stimulation and expansion of populations of T cells making IL4 or IL10 has been suggested.[17] In the absence of evidence linking BCG antigens with those expressed on the islet cell surface, the mechanism of action may well be immunologically nonspecific. It was a possibility that immune stimulation after the onset of diabetes might reduce insulin requirements or increase C-peptide synthesis when measured 1 year after vaccination with the attenuated mycobacterium, BCG. Vaccination with BCG is already known to be safe and immunogenic, and a preliminary human study gave some evidence for benefit. We have completed a double-blind study employing two medical teams, one to manage the diabetes and one to give and read the BCG and placebo vaccines. Entry was restricted to subjects within 8 weeks of clinical onset, or within 16 weeks if the islet cell antibody (ICA) test is positive, who are between ages 5 and 18 years and who have negative human immunodeficiency virus (HIV) and purified protein derivative (tuberculin) (PPD) tests. BCG vaccination did not preserve C-peptide secretion, as found in other studies.[18] In a similar manner, Q-fever vaccine apparently had no effect in new onset patients (Harrison, oral communication).

Treatments to halt islet cell destruction and prevent diabetes would ideally be started before the beta cell mass is significantly compromised. Nevertheless, detecting partial benefit in recent-onset diabetics would be useful if it is a precursor to an intervention that could be used with greater efficiency in subjects with pretype 1 diabetes. Ultimately, the focus of the prevention of diabetes must be in patients with pretype 1 diabetes (hereafter referred to as prediabetes), and the remainder of this chapter reviews progress that is being made toward this goal.

Identification of Subjects with PreType 1 Diabetes

Prediabetes describes the period when islet cell destruction is in progress but glucose homeostasis has not been lost. The best definition of prediabetes is an individual actually followed to the development of diabetes. Subjects are currently identified as being at high risk by the sequence of events summarized below:

1. They are ascertained by their relationship to a diabetic individual (usually a first-degree relative).
2. Their serum has tested positive for immunological markers of islet specific immunity (the presence of antibodies associated with islet beta cells).
3. Their first phase insulin production (see Metabolic Testing, below) is low enough to give at least a 50% or greater risk for diabetes within the next 5 years.

These selection criteria in effect reproduce two of the factors thought to be important in the etiology of a type 1 diabetes: genetic predisposition and the development of autoimmunity. The outcome of trials currently being started in high-risk individuals (to be referred to as prediabetics) are likely to influence treatment strategies for a decade, so it is worth considering the strengths and weaknesses of the selection criteria in detail.

Genetics

MHC Genes

When one pair of identical twins develops type 1 diabetes, the second twin has a 30-50% risk of also developing diabetes (see Chapter 7). The findings that fewer than 100% of second twins become diabetic suggests that additional factors, probably environmental, contribute to the pathogenesis of type 1 diabetes. Similarly, a higher risk for diabetes was shown for siblings (approximately a 5-10% risk) or for offspring when the father had diabetes (approximately 12% risk) or when the mother had diabetes (approximately a 6% risk).[19] The strongest genetic associations are with the human leukocyte antigen (HLA) system. Approximately 95% of people with type 1 diabetes have either DR3 or DR4 (compared to 50% of the general population), and 30 to 53% have the DR3 and DR4 combination (compared to 2.4% of the general Denver population).[20] An 8-year prospective study[21] found a 70% risk for developing diabetes in ICA-positive siblings with DR3, 4, compared with a 37% risk in ICA-positive siblings who were negative for DR3, 4.

Much of the contribution from HLA DR4 stems from linkage with a HLA DQ allele, DQ8 (DQB1*0302) with an influence of DR4 subtypes (e.g., DRB1*0403 decreases risk even on DQB1*0302 containing haplotypes). The allele DQB1*0302 lacks aspartic acid at position 57 of the DQB chain (Asp-negativity) and increases the risk for type 1 diabetes.[22,23] In Finland, Asp-negativity was found in 74.5% of patients with type 1 diabetes compared to 18% of controls.[24] It is important to realize that absence of asp57 is only an approximate rule and that alleles such as DQB1*0401 that provide high risk have asp at position 57, so typing for full DQ allele sequences is important.[25] Several HLA alleles are associated with a very low incidence of diabetes (see Chapter 7). Thus, genetic screening is likely to play an important part in identifying subjects who could be enrolled in prevention trials. The present generation of trials will enroll high-risk individuals identified among first-degree relatives of index cases. Because 70-90% of patients have no first-degree relative with diabetes,[26] this creates a potential for significant ascertainment bias. For example, first-degree relative of diabetics presenting in infancy are likely to be underrepresented merely because family sizes are small and parents may be discouraged from having more children where the first has a time-consuming and emotionally demanding condition. First-degree relatives of parents who develop diabetes after 40 years of age may be over represented merely because family size will not have been limited in response to parental disease. How important these sources of bias are will depend on the diversity of genes contributing to disease susceptibility.

Studies are underway following HLA typed infants from birth. In Denver Colorado (DAISY study) 2.4% of newborns have the genotype DR3/4, DQ2/8(DQ8=DQB1*0302). Approximately 50% of children presenting with diabetes before age five have this genotype. It is estimated that children from the general population (no relative with type 1 diabetes) with

DR3/4, DQ2/8 have a risk of diabetes of approximately 1/16, similar to the overall risk of first degree relatives. In contrast a sibling of a patient with type 1 diabetes with DQ2/DQ8 has a risk of activating autoimmunity (e.g., Islet cell autoantibodies) of approximately 50%. Though high risk can now be identified even in the general population with antibody testing, HLA evaluation has a relatively low sensitivity compared to conventional newborn screening programs. Studies however can be designed to evaluate novel therapies, even in newborns (e.g., Finnish studies currently underway: DIPP) and if a safe and effective therapy is documented it would likely find wide use.

NonMHC Genes

Animal studies indicate multiple genes contribute to susceptibility to diabetes. An advantage to identifying prediabetics by screening first-degree relatives is that selection for nonHLA genes is likely to occur in parallel. The development of multiple markers for polymorphisms on each human chromosome is making the identification of nonMHC genes predisposing to diabetes possible, so that screening of the general population will be possible in the long-term.

As more genetic markers associated with susceptibility and protection are identified it will be important to apply them retrospectively to the prevention protocols where high-risk people are initially identified with immunological tests. For example, if immunological and metabolic tests indicate a high risk for diabetes and the subject is entered into a prevention protocol, it will be important to make sure that relevant genes are similarly divided in numbers of subjects with high risk and lack of protective HLA alleles. Cells might also be saved for other markers that may be shown to good predictors in future years, as this area of research is advancing rapidly.

Immunological

Because the incidence of diabetes in identical twins of type 1 diabetics is less than 50%, additional selection criteria are needed to identify subjects for most prevention studies. Some reports have suggested that the degree of ICA positivity is the best predictor for advancement to insulin dependence.[27-29] It is likely that the ICA test will be replaced by "biochemical autoantibody" determination.[30]

Others consider ICA positivity only to be an indicator that damage has occurred to the islet, and insist that a metabolic test (see below) be done to determine the amount of damage. We previously noted that nine of 10 first-degree relatives advancing to diabetes had ICA titers greater than or equal to 80 JDF units, making this a high-risk finding.[31] However, six other first-degree relatives of diabetics had similar titers of ICA had not advanced to diabetes over a 6 year period. Thus, the high titer results might have a relatively poor positive predictive value (10 of 16 subjects). Also, subjects with lower titers of ICA obviously also develop diabetes, although the percentages are not as high. Bruining and coworkers[32] noted that only four of eight children found to have positive ICAs in a general population (not first-degree relatives) developed diabetes in a 10 year period. Another 3 of 4806 children initially negative for ICA also developed diabetes. Thus, whether screening first-degree relatives or the general population, it would seem that a second test, more specific for quantity of damage, would be important.

ICA assays are difficult to quantitate and this leads to interlaboratory differences. Factors that are likely to limit the reproducibility of the assay between laboratory include (1) quality of the human pancreas chosen for the assay and (2) interobserver differences for a subjective endpoint. Nevertheless, the Third International Workshop on ICA concluded, "The indirect immunofluorescent of (IFL) test for ICA is still the most useful screen for identifying people at risk for type 1 diabetes, and, until such time as the relevant antigens are identified in purified in large quantities, it remains the method of choice for its detection".[33] More recently, other antibodies have been identified in the blood of newly diagnosed subjects (and pretype 1 subjects). These include autoantibodies to insulin (IAA),[34-36] glutamic acid decarboxylase (GAD),[37-39] and ICA512.[40,41] With the three islet autoantigens that have been isolated there

***Table 2. FPIR* and time to diabetes*[31]**

FPIR* (μU/ml)	Time to Diabetes (Months) Mean time	95th percentile
<67	19	57
<46	14	21
<30	4.6	18

*FPIR = First phase insulin response (sum of 1-minute +3-minute insulin levels) on intravenous glucose tolerance testing

are standard radioassays; namely assays for autoantibodies reacting with insulin, with glutamic acid decarboxylase, and with the molecule ICA512 (IA-2). These radioassays are typically performed in 96-well filtration plates and are also commercially available. A high-throughput laboratory can probably perform each assay for a cost less than $4 per sample and thus large scale screening just using the biochemical assays is feasible.

The presence of >=2 of the three autoantibodies (insulin, GAD65, ICA512) indicates a high risk of progression to type 1 diabetes (see Chapter 10). The risk appears to exceed 50% in five years even without the additional information provided by intravenous glucose tolerance testing. Nevertheless, for any combination of autoantibodies, loss of first phase insulin secretion, decreases the time to diabetes. Of note, an individual with a single anti-islet autoantibody (overall risk approximately 20%) who has first phase insulin secretion below the first percentile has a diabetes risk exceeding 50% in five years.

Metabolic Testing

The interval between finding an autoantibody in a first-degree relative of a diabetic and the development of diabetes can exceed 10 years. Prospective studies where the success of an intervention will not been known for more than 10 years are clearly not practical, so a further selection for subjects with extensive beta cell destruction is usual. Direct measurement of the functional mass of islet cells is not possible and the best approximation comes from the intravenous glucose tolerance test (IVGTT). Srikanta and coworkers[42] were the first to show that the sum of the 1- and 3-minute insulin levels had some predictive value. The Barbara Davis Center experience confirmed the predictive value of a very low first-phase insulin response (FPIR) on IVGTT (<25 μU/ml) in children.[43] A later report(17) noted predictability for different levels of FPIR in first-degree relatives under age 18 years who had two or more positive ICA tests (Table 2). Vardi and coworkers[44] found 15 of 17 (88%) ICA-positive first-degree relatives to develop insulin dependence if their FPIR fell to <48 μU/ml (the first percentile at the Joslin Diabetes Center). In contrast, only 3 of 18 (17%) with values above this level developed insulin dependence.

Little is known about the replicate errors, either within subject or between subjects, of the IVGTT. The test needs to performed with the utmost accuracy in timing and consistency.[45,46] It should be performed after an overnight fast of uniform duration. Some people vary in their response and the downward progression of the FPIR is not always linear.[43] Finally, the results are age-dependent, particularly in prepubertal versus pubertal children.[47] Despite the problems listed above, the FPIR test is of value in predicting time to insulin dependence in first-degree relatives of diabetics who are ICA positive. It will therefore be important to relate the outcomes of diabetes prevention trials with the FPIR level required for the study entry. Comparison between trials with different FPIR entry criteria may prove to be impossible, particularly if the efficacy of different prevention strategies is affected by the amount of damage already sustained by beta cells.

Trials Using Subjects with PreType 1 Diabetes

An NIH consensus report on prevention of type 1 diabetes concluded: " The general consensus from the meeting was that indeed there is methodology that can identify, with near certainty among first-degree relatives of type 1 diabetic patients, those who will develop diabetes and that immune intervention therapy before the onset of symptoms might prevent the disease from occurring."[48]

It is now generally accepted that prediabetes can be accurately diagnosed in select high-risk groups of subjects. The justification for identifying this high-risk group comes from their potential recruitment into protocols to test strategies to prevent diabetes. These high-risk subjects are, however, so rare that no center is likely to identify more than 20 subjects suitable to enter a prevention trial in any one year. Our own power estimates indicated that high-risk prediabetics (with two ICA tests positive at 40 JDF units and two FPIRS <60 μU/ml) would have to be randomized into control and test groups of 60 subjects each in order to have an 80% chance for detecting a statistically significant difference after 3 years. Such sample sizes are a major complication in planning interventions for subjects who are likely to develop diabetes within 3 to 5 years. Identifying prediabetics is very expensive (we estimate $1,200 per subject, allowing for autoantibody screening and two IVGTT tests), so a cost-effective approach requires extensive and international collaboration.

The largest prediabetic prevention studies have evaluated insulin (parenteral and oral administration) and nicotinamide.

Nicotinamide

The mechanism of nicotinamide protection is currently unknown.[49] Nicotinamide was initially shown to delay diabetes in NOD mice.[50] Nicotinamide inhibits poly (ADP-ribose) synthetase[51] and at high concentrations, can act as a free radical scavenger.[52] Pretreatment with nicotinamide prevents the decrease in proinsulin synthesis in islets from rats treated with alloxan or streptozocin.[51] It also inhibits the activated macrophage killing of beta cells in vitro,[53] and the expression of class 2 major histocompatibility complex (MHC) on beta cells caused by the cytokines, tumor necrosis factor (TNF), and interferon gamma.[54] These are all processes thought to be a part of the immune-mediated destruction of beta cells found in the human disease. A workshop report summarizes the biological action and therapeutic potential of nicotinamide.[55] A number of studies have evaluated nicotinamide in new onset patients with reports that some preservation of C-peptide may occur, but only in older individuals[56] as in well controlled trials in children where there was no effect.[57]

A pilot study of nicotinamide was undertaken jointly between Auckland, NZ, and Denver, U.S.A.[58] Among 14 treated subjects, only one developed diabetes after a mean of 17 months of treatment, whereas eight untreated subjects were all insulin-dependent after a mean of 17 months. All subjects were less than 16 years old when diabetes was detected by ICA screening and all had ICA titers equal to or greater than 80 JDF units. All subjects had a FPIR of less than 67 μU/ml (the 5th percentile of normal). On further (unpublished) follow-up of approximately 5 years, 3 of 4 subjects in Denver (all of who had a FPIR <30 μU/ml) and 5 of 10 Auckland subjects (most with the higher FPIR levels) had become insulin dependent. This pilot trial was weakened by the fact that all the control subjects were followed in Denver.

Nicotinamide was evaluated in a German prevention trial headed by Dr. Hubert Kolb (Diabetes-Forchunginstitut an der Universitat Dusseldorf), and evaluated in the European Nicotinamide Diabetes Intervention Trial (ENDIT) study under the direction of Dr. Edwin Gale (Department of Endocrinology and Diabetes at the University of Bristol). The German trial is named "DENIS" for Deutsche Nicotinamide Intervention Study. The DENIS trial was ended with no difference in progression to diabetes detected comparing the placebo to nicotinamide treated group.[59] The trial was relatively small and had criteria for early discontinuation, which were met.

The ENDIT trial was a double-blind, randomized, and placebo-controlled.[60] Enrollment was open to first-degree relatives of patients with type 1 diabetes (sibling, child, parent) who had at least two ICA-positive tests, one of which is equal to or greater than 20 JDF units, drawn 3-12 months apart. An IVGTT was performed by the ICARUS protocol[46] and was used for the subsequent meta-analysis but not for randomization. An oral glucose tolerance test (OGTT) was performed as a two-point test and (0 min and 2 h capillary blood glucose) to rule out diabetes mellitus by World Health Organization (WHO) criteria.[61] Follow-up procedures included yearly FPIR with ICA and IAA tests. The dose of nicotinamide was 1.2 g/M^2 per day.

Unfortunately, the ENDIT trial did not show an effect from nicotinamide in preventing the onset of diabetes (Edwin Gale, oral communication). Thus, most investigators now accept that there is no benefit to giving nicotinamide to individuals at high risk for type 1 diabetes.

Insulin

The mechanism by which insulin might delay or prevent diabetes is complex. It was initially hypothesized that the beta cells were being allowed to "rest" and might therefore be less prone to immunological attack. This may be the case for the BB rat model where only biologically active insulin prevents disease and insulin must be administered in amounts that produce hypoglycemia.[62,63] Additional mechanisms, such as induction of immunoregulation/ "tolerance", contribute to protection in the NOD mouse model.[64-70] The initial study of insulin in subjects with prediabetes was reported from the Joslin Clinic.[71] Five subjects, ages 7 to 14 years, were treated with 5 days of IV insulin aimed at keeping blood glucose levels between 3.3 and 4.2 mmol/L (60-76 mg/dl). Outpatient daily insulin was divided into two subcutaneous injections consisting of human regular and ultralente insulin in the morning and lente insulin at bedtime. The total dose was approximately 0.2 units/kg body weight per day. The children's lifestyles were not restricted during the outpatient treatment and severe hypoglycemia was not to a problem. After the mean of 3.04 years, only one of the five treated subjects was insulin-dependent, compared to all seven subjects who had refused the treatment.

The very small number of subjects treated is a weakness of this pilot trial, as was the inclusion of two subjects with FPIR levels of 76 and 83 μU/ml in the treatment arm. The use of subjects who refused treatment as a control group is not ideal and none of the controls had FPIR above 49 μU/ml. Additional points of concern relate to the difficulty of treatment and the reluctance of ICA-positive parents to enter into the study. Measurement of blood glucose on subjects every 15 to 20 minutes for 5 days while IV insulin is given is a level of intensiveness outside routine care. Nevertheless, a group of investigators in United States opted to evaluate IV and subcutaneous insulin, as well as oral insulin (for induction of oral tolerance) with a multicenter trial under the auspices of the National Institutes of Health (NIH). The criteria for randomization into the parenteral DPT-1 trial are summarized in Table 3.

For this DPT-1 trial (Diabetes Prevention Trial Type 1) more than 100,000 first degree relatives of patients with type 1 diabetes were screened for the expression of cytoplasmic islet cell antibodies (ICA). There were two separate trials in DPT-1; parenteral insulin administration and a trial of oral insulin. Approximately 4% of the screened were ICA positive. One half of the ICA positive however did not express either GAD65 or ICA512 (IA-2) autoantibodies, and these individuals relatively infrequently were found to be eligible for the intervention phase of DPT-1[30] (eligibility for parenteral trial dependent upon loss of first phase insulin secretion on IVGTT (FPIR= first phase insulin secretion)) or presence of insulin autoantibodies in addition to ICA for the oral trial. The parenteral insulin prophylaxis study randomized 339 subjects (Skyler, oral communication) to either close observation (no therapy) or intravenous insulin for 4 days once a year and 0.125 U/kg body weight of subcutaneous ultralente insulin twice daily. Analysis of progression to diabetes (as defined by oral glucose tolerance testing) indicated that the insulin therapy did not delay progression to diabetes.[70a] The DPT-1 trial was designed in the early 90's and it has subsequently become apparent that in animal models such as the NOD a major effect of insulin appears to relate to induction of immunoregulatory T cells, and

Table 3. U.S. Diabetes Prevention Trial (DPT 1) eligibility criteria for high-risk (parenteral insulin) protocol

First-degree relatives: 4 to 45 years (inclusive)

Second-degree relatives: 4 to 20 years (inclusive)

ICA ≥ 10 JDF units (2 of 3 times in series)

Not DQA1 0102, DQB1 0602

IVGTT criteria:

A. Siblings and offspring 4 to 45 years (or 2nd-degree relatives 4-20 years):
Age ≤ 7: <60 μU/ml (sum 1+3 min [2 of 3])
Age ≥ 8: <100 μU/ml (2 of 3)
B. Parents (≤ 45 y): <60 μU/ml (2 of 3)

OGTT not diabetic (ADA Expert Committee Criteria)

ICA= Islet cell antibodies; IVGTT= Intravenous glucose tolerance test; OGTT= Oral glucose tolerance test; NDDG= National Diabetes Data Group

the dose chosen in man (of necessity for safety in terms of risk of hypoglycemia) is ineffective in the NOD model (Karounos, oral communication). Despite the lack of therapeutic benefit, the completed parenteral DPT-1 trial is providing a wealth of information concerning prediction of type 1A diabetes and trial design. It is a testament to the seriousness of the disorder that so many individuals were screened and willing to participate for years in such an intensive trial regimen. It appears that future trials will be greatly streamlined by screening with "biochemical autoantibody" assays. Almost 1/2 of the relatives with GAD or ICA512 autoantibodies do not have cytoplasmic ICA, and the ICA positive relatives lacking GAD and ICA512 autoantibodies appear to have a low risk of eligibility or progression to diabetes.[30] Thus, we would suggest that future prevention trials will not even utilize the cytoplasmic ICA assay that has proven difficult to standardize. The presence of multiple biochemical anti-islet autoantibodies (of GAD65, ICA512 or insulin) is likely (but further analysis is necessary) to provide enough prognostic information that even the intravenous glucose tolerance test may be optional for assigning high risk.

The second trial within DPT-1, namely the oral insulin trial has been now completed. Oral insulin delays the development of diabetes in NOD mice[73] but several studies in patients with new onset type 1 diabetes have not demonstrated preservation of C-peptide.[74,75] More than 200 individuals were entered into this randomized, double blind trial. Entry criteria were similar to those for the parenteral DPT-1 trial with the exception that first phase insulin secretion had to exceed those assigned for high risk in the parenteral trial and the oral glucose tolerance could not be impaired, and insulin autoantibodies had to be present. This identified a subgroup with a predicted risk of diabetes between 25 and 50% within five years. Unfortunately, there was no effect of oral insulin in delaying or preventing type 1 diabetes (Jay Skyler, oral communication).

The DPT-1 trial groups are expanding their efforts with the formation of TrialNet. TrialNet will create an international network for conducting trials for the prevention of type 1 diabetes. It is likely that many of the initial trials carried out by TrialNet will evaluate therapies in new onset patients. There will almost certainly be an open call for suggested therapies and protocols and with such a collaborative network it is hoped that the discovery of effective and safe therapies will be hastened. There are additional major efforts to develop therapies for diabetes prevention and in particular, the Immune Tolerance Network (www.immmunetolerance.org) is accepting applications to support therapies aimed at tolerance induction and assays of tolerance.[72] The Immune Tolerance Network, sponsored by the National Institutes of Health funds studies both within and outside of the United States. Nasal insulin is being evaluated in a novel and ambitious study from Finland (DIPP – Diabetes Prediction and Prevention[76]) where infants are followed from birth and when ICA is found, randomized to therapy or placebo. Another Finnish study will evaluate exclusion of bovine milk in early childhood.[77]

Additional Clinical Trials of Antigen Based Therapies

At least three islet antigens have entered clinical trials for preservation of insulin secretion at diabetes onset: an altered peptide ligand of insulin peptide B:9-23 (Neurocrine),[78,79] the GAD65 molecule (Diamyd),[80,81] and a heat shock protein peptide (Peptor).[82,83]

A phase I/II trial is underway evaluating an altered peptide ligand of the immunodominant insulin B:9-23 peptide that Daniel and coworkers found to be a major target of intra-islet T cells and a peptide that could, by multiple routes of administration prevent diabetes.[84-86] The peptide is identical in man and mice and recent studies using ELISPOT assays indicate that patients with or developing type 1 diabetes also recognize this molecule.[78] There are several important caveats relative to peptide immunotherapy. Even with a self-peptide one can induce anaphylaxis, and we have induced anaphylaxis in NOD mice with the native B:9-23 peptide.[78a] A self peptide may activate disease and we have induced insulin autoantibodies in NOD mice with the insulin B:9-23 peptide.[79]

The whole GAD65 molecule is entering clinical trails in Scandinavia apparently in patients with LADA (latent autoimmune diabetes of adults). This usually slower form of diabetes may be a difficult therapeutic target.

The p277 peptide of HSP60 is being studied in multiple small clinical trials including one from Israel with intriguing preliminary data relative to C-peptide preservation.[79a]

Conclusions

The fact that diabetes can be successfully managed, and the development of complications delayed by good control of blood glucose has placed significant constraints on treatment intended to prevent diabetes. Global immunosuppression protocols, of the types used for transplant recipients, clearly carry an unacceptable morbidity. Immunologists caring for patients with autoimmune diseases have a need to establish organ-specific immunological tolerance that is just as pressing as that facing transplant surgeons. The approaches that are currently being explored in trials seem very speculative when viewed in relation to the complexity of immunoregulation or the development of autoimmunity.[87] The animal models of type 1 diabetes should be a fruitful ground for treatments designed to prevent diabetes in humans. In reality it has been difficult to realize this expectation. One obstacle has been the ease with which diabetes is prevented in NOD mice.[1] For example, interventions that have succeeded in mice include special diets, injection with male sex hormones, and even the failure to exclude common mouse viruses from colonies. It is very difficult to judge the cost-effectiveness of preventive strategies in mouse colonies.

With the limitations of animal models in mind, it seems likely that much of the development of prevention approaches will have to be done in humans. Constraints on research funding on both sides of the Atlantic have increased the pressure for collaboration beyond local and national boundaries. All in all, the initiation of TrialNet and large multicenter trials to determine whether type 1 diabetes can be prevented or ameliorated at onset promises to make this decade an exciting one.

References

1. Atkinson MA, Leiter EH. The NOD mouse model of type 1 diabetes: As good as it gets? Nat Med 1999; 5:601-604.
2. Haskins K, Atkinson M. Prediction of diabetes and response to therapy in individual animals: Are there lessons for Man? Report on the JDF/Snow Mountain Meeting, 5-7 October 2000, Winter Park, CO, USA. J Autoimmun 2001; 16:15-19.
3. Gepts W, De Mey J. Islet cell survival determined by morphology. An immunocytochemical study of the islets of Langerhans in juvenile diabetes mellitus. Diabetes 1978; 27 Suppl:251-261.
4. Stiller CR, Dupre J, Gent M et al. Effects of cyclosporine immunosuppression in insulin-dependent diabetes mellitus of recent onset. Science 1984; 223:1362-1367.
5. Chase HP, Butler-Simon N, Garg SK et al. Cyclosporine A for the treatment of new-onset insulin-dependent diabetes mellitus. Pediatrics 1990; 85:241-245.
6. Carel J-C, Boitard C, Eisenbarth G et al. Cyclosporine delays but does not prevent clinical onset in glucose intolerant pretype I diabetic children. J Autoimmun 1996; 9:739-745.
7. Tuvemo T, Frisk G, Friman G et al. IgM against Coxsackie B viruses in children developing Type I diabetes mellitus – a seven year retrospective study. Diabetes Res 1988; 9:125-129.
8. Shapiro AM, Lakey JR, Ryan EA et al. Islet transplantation in seven patients with Type 1 Diabetes Mellitus using a Glucocorticoid-free immunosuppressive regimen. N Engl J Med 2000; 343:230-238.
9. Ryan EA, Lakey JRT, Rajotte RV et al. Clinical outcomes and insulin secretion after islet transplantation with the Edmonton protocol. Diabetes 2001; 50:710-719.

9a. Ryan EA, Lakey JR, Paty BW et al. Successful islet transplantation: continued insulin reserve provides long-term glycemic control. Diab 2002; 51(7):2148-2157.

9b. Ryan EA, Lakey JRT, Rajotte RV et al. Clinical outcomes and insulin secretion after islet transplantation with the Edmonton Protocol. Diab 2001; 50:710-719.

10. Vu MD, Qi S, Xu D et al. Synergistic effects of mycophenolate mofetil and sirolimus in prevention of acute heart, pancreas, and kidney allograft rejection and in reversal of ongoing heart allograft rejection in the rat. Transplantation 1998; 66:1575-1580.
11. Hao L, Chan S-M, Lafferty KJ. Mycophenolate mofetil can prevent the development of diabetes in BB rats. Ann NY Acad Sci 1994; 696:328-332.
12. Buckingham BA, Sandborg CI. A randomized trial of methotrexate in newly diagnosed patients with type 1 diabetes mellitus. Clin Immunol 2000; 96:86-90.
13. Chatenoud L, Thervet E, Primo J et al. Anti-CD3 antibody induces long-term remission of overt autoimmunity in nonobese diabetic mice. Proc Natl Acad Sci USA 1994; 91:123-127.
14. Chatenoud L, Primo J, Bach JF. CD3 antibody-induced dominant self tolerance in overtly diabetic NOD mice. J Immunol 1997; 158:2947-2954.

14a. Herold KC, Hagopian W, Auger JA et al. Anti-CD3 monoclonal antibody in new-onset type 1 diabetes mellitus. N Engl J Med 2002; 346(22):1692-1698.

15. Shah SC, Malone JI, Simpson NE. A randomized trial of intensive insulin therapy in newly diagnosed insulin-dependent diabetes mellitus. N Engl J Med 1989; 320:550-554.
16. Qin HY, Singh B. BCG vaccination prevents insulin-dependent diabetes mellitus (IDDM) in NOD mice after disease acceleration with cyclophosphamide. J Autoimmun 1997; 10:271-278.
17. Harada M, Kishimoto Y, Makino S. Prevention of overt diabetes and insulitis in NOD mice by a single BCG vaccination. Diabetes 1990; 8:85-90.
18. Allen HF, Klingensmith GJ, Jensen P et al. Effect of BCG vaccination on new-onset insulin-dependent diabetes mellitus: A randomized clinical study. Diabetes Care 1998; 22:1703-1707.
19. Warram JH, Krolewski AS, Gottlieb MS et al. Differences in risk of insulin-dependent diabetes in offspring of diabetic mothers and diabetic fathers. N Engl J Med 1984; 311:149-152.
20. Ludvigsson J, Lindblom B. Human lymphocyte antigen DR types in relation to early clinical manifestations in diabetic children. Pediatr Res 1984; 18(12):1239-1241.
21. Deschamps I, Boitard C, Hors J et al. Life table analysis of the risk of type I (insulin-dependent) diabetes mellitus in siblings according to islet cell antibodies and HLA markers. An 8 year prospective study. Diabetologia 1992; 35:951-957.
22. Morel PA, Dorman JS, Todd JA et al. Aspartic acid at position 57 of the HLA-DQ beta chain protects against type I diabetes: A family study. Proc Natl Acad Sci USA 1988; 85:8111-8115.
23. Baisch JM, Weeks T, Giles R et al. Analysis of HLA-DQ genotypes and susceptibility in insulin-dependent diabetes mellitus. N Engl J Med 1990; 322:1836-1841.
24. Reijonen H, Ilonen J, Knip M et al. HLA-DQB1 alleles and the absence of Asp 57 as susceptibility factors of IDDM in Finland. Diabetes 1991; 40:1640-1644.
25. Lee KH, Wucherpfennig KW, Wiley DC. Structure of a human insulin peptide/HLA-DQ8 complex and susceptibility to type 1 diabetes. Nature Immunology 2001; 2:501-507.
26. Bingley PJ, Bonifacio E, Gale EAM. Can we really predict IDDM? Diabetes 1993; 42:213-220.

27. Bonifacio E, Bingley PJ, Shattock M et al. Quantification of islet-cell antibodies and prediction of insulin-dependent diabetes. Lancet 1990; 335:147-149.
28. Riley WJ, Maclaren NK, Krischer J et al. A prospective study of the development of diabetes in relatives of patients with insulin-dependent diabetes. N Engl J Med 1990; 323:1167-1172.
29. Karjalainen JK. Islet cell antibodies as predictive markers for IDDM in children with high background incidence of disease. Diabetes 1990; 39:1144-1150.
30. Yu L, Cuthbertson DD, Maclaren N et al. Expression of GAD65 and Islet Cell Antibody (ICA512) autoantibodies amongst cytoplasmic ICA+ relatives is associated with eligibility for Diabetes Prevention Trial Type-1. Diabetes 2001; 50:1735-1740.
31. Chase HP, Garg SK, Butler-Simon N et al. Prediction of the course of pretype I diabetes. J Pediatr 1991; 118:838-841.
32. Bruining GJ, Molenaar JL, Grobbee DE et al. Ten-year follow-up study of islet-cell antibodies and childhood diabetes mellitus. Lancet 1989; 1:1100-1103.
33. Bonifacio E, Boitard C, Gleichman H et al. Assessment of precision, concordance, specificity, and sensitivity of islet cell antibody measurements in 41 assays. Diabetologia 1990; 33:731-736.
34. Palmer JP, Asplin CM, Clemons P et al. Insulin antibodies in insulin-dependent diabetics before insulin treatment. Science 1983; 222:1337-1339.
35. Mcevoy RC, Witt ME, Ginsberg-Fellner F et al. Anti-insulin antibodies in children with type I diabetes mellitus. Genetic regulation of production and presence at diagnosis before insulin replacement. Diabetes 1986; 35:634-641.
36. Ziegler AG, Ziegler R, Vardi P et al. Life-table analysis of progression to diabetes of anti-insulin autoantibody-positive relatives of individuals with type I diabetes. Diabetes 1989; 38:1320-1325.
37. Baekkeskov S, Aanstoot H-J, Christgau S et al. Identification of the 64K autoantigen in insulin-dependent diabetes as the GABA-synthesizing enzyme glutamic acid decarboxylase [published erratum appears in Nature 1990 Oct 25; 347(6295):782]. Nature 1990; 347:151-156.
38. Atkinson MA, Maclaren NK, Scharp DW et al. 64,000 Mr autoantibodies as predictors of insulin-dependent diabetes. Lancet 1990; 335:1357-1360.
39. Barmeier H, McCulloch DK, Neifing JL et al. Risk for developing type I (insulin-dependent) diabetes mellitus and presence of islet 64K antibodies. Diabetologia 1991; 34:727-733.
40. Hawa M, Fava D, Medici F et al. Antibodies to IA-2 and GAD65 in type 1 and type 2 diabetes. Diabetes Care 2000; 23:228.
41. Yu L, Cuthbertson DD, Eisenbarth GS et al. DPT-1: Prevalence of GAD and ICA512 (IA-2) Autoantibodies by Relationship to Proband. Ann NY Acad Sci 2001.
42. Srikanta S, Ganda OP, Gleason RE et al. Pretype I diabetes. Linear loss of beta cell response to intravenous glucose. Diabetes 1984; 33:717-720.
43. Chase HP, Voss MA, Butler-Simon N et al. Diagnosis of pretype I diabetes. J Pediatr 1987; 111:807-812.
44. Vardi P, Crisa L, Jackson RA et al. Predictive value of intravenous gluocse tolerance test insulin secretion less than or greater than the first percentile in islet cell antibody positive relatives of Type 1 (insulin-dependent) diabetic patients. Diabetologia 1991; 34:93-102.
45. Ad Hoc Expert Committee. Prevention of type 1 diabetes mellitus. Diabetes Care 1990; 13:1026-1027.
46. Bingley PJ, Colman P, Eisenbarth GS et al. Standardization of IVGTT to predict IDDM. Diabetes Care 1992; 15:1313-1316.
47. Allen HF, Jeffers BW, Klingensmith GJ et al. First-phase insulin release in normal children. J Pediatr 1993; 123:733-738.
48. Ad Hoc Expert Committee. Position statement: Prevention of type 1 diabetes mellitus. Diabetes 1990; 39:1151-1152.
49. Kolb H, Burkart V. Nicotinamide in type 1 diabetes. Mechanism of action revisited. Diabetes Care 1999; 22 Suppl 2:B16-B20.
50. Yamada K, Nonaka K, Miyazaki A et al. Preventive and therapeutic aspects of large dose nicotinamide injections on diabetes associated with insulitis: An observation in ion injection of alloxan and streptozotocin on islet proinsulin synthesis. Diabetes 1982; 31:749-753.
51. Uchigata Y, Yamamoto H, Nagai H et al. Effect of poly(ADP-ribose) synthetase inhibitor administration to rats before and after injection of alloxan and streptozotocin on islet proinsulin synthesis. Diabetes 1983; 32:316-318.
52. LeDoux SP, Hall CR, Forbes PM et al. Mechanisms of nicotinamide and thymidine protection from alloxan and streptozotocin toxicity. Diabetes 1988; 37:1015-1019.
53. Kolb H, Burkart V, Appels B et al. Essential contribution of macrophages to islet cell destruction in vivo and in vitro. J Autoimmun 1990; 3:117-120.

54. Yamada K, Miyajima E, Nonaka K. Inhibition of cytokine-induced MHC class II but not class I molecule expression on mouse islet cells by niacinamide and 3- aminobenzamide. Diabetes 1990; 39:1125-1130.
55. Pociot F, Reimers JI, Andersen HU. Nicotinamide – biological actions and therapeutic potential in diabetes prevention. Diabetologia 1993; 36:574-576.
56. Visalli N, Cavallo MG, Signore A et al. A multi-centre randomized trial of two different doses of nicotinamide in patients with recent-onset type 1 diabetes (the IMDIAB VI). Diabetes Metab Res Rev 1999; 15:181-185.
57. Chase HP, Butler-Simon N, Garg S et al. A trial of nicotinamide in newly diagnosed patients with type 1 (insulin-dependent) diabetes mellitus. Diabetologia 1990; 33:444-446.
58. Elliott RB, Chase HP. Prevention or delay of type I (insulin-dependent) diabetes mellitus in children using nicotinamide. Diabetologia 1991; 34:362-365.
59. Lampeter EF, Klinghammer A, Scherbaum WA et al. The Deutsche Nicotinamide Intervention Study: An attempt to prevent type 1 diabetes. DENIS Group. Diabetes 1998; 47:980-984.
60. Gale EA. Theory and practice of nicotinamide trials in pretype 1 diabetes. J Pediatr Endocrinol Metab 1996; 9:375-379.
61. World Health Organization. Report of a WHO Study Group. Diabetes Mellitus. Technical Report Series 1985; 727:1-99.
62. Gottlieb PA, Handler ES, Appel MC et al. Insulin treatment prevents diabetes mellitus but not thyroiditis in RT 6-depleted diabetes resistant BB/Wor rats. Diabetologia 1991; 34:296-300.
63. Gottlieb PA, Rossini AA. The BB Rat Models of IDDM 1994; 163-174.
64. Salomon B, Lenschow DJ, Rhee L et al. B7/CD28 costimulation is essential for the homeostasis of the CD4+CD25+ immunoregulatory T cells that control autoimmune diabetes. Immunity 2000; 12:431-440.
65. Roncarolo MG and Levings MK. The role of different subsets of T regulatory cells in controlling autoimmunity. Curr Opin Immunol 2000; 12:676-683.
66. Weiner HL. Oral tolerance, an active immunologic process mediated by multiple mechanisms. J Clin Invest 2000; 106:935-937.
67. Bergerot I, Arreaza GA, Cameron MJ et al. Insulin B-chain reactive CD4+ regulatory T-cells induced by oral insulin treatment protect from type 1 diabetes by blocking the cytokine secretion and pancreatic infiltration of diabetogenic effector T-cells. Diabetes 1999; 48:1720-1729.
68. Ploix C, Bergerot I, Durand A et al. Oral administration of cholera toxin B-insulin conjugates protects NOD mice from autoimmune diabetes by inducing CD4+ regulatory T-cells. Diabetes 1999; 48:2150-2156.
69. French MB, Allison J, Cram DS et al. Transgenic expression of mouse proinsulin II prevents diabetes in nonobese diabetic mice. Diabetes 1996; 46:34-39.
70. Harrison LC, Dempsey-Collier M, Kramer DR et al. Aerosol insulin induces regulatory CD8 gamma delta T cells that prevent murine insulin-dependent diabetes. J Exp Med 1996; 184:2167-2174.

70a. Diabetes Prevention Trial – Type 1 Diabetes Study Group. Effects of insulin in relatives of patients with type 1 diabetes mellitus. N Engl J Med 2002; 346(22):1685-1691.

71. Keller RJ, Eisenbarth GS, Jackson RA. Insulin prophylaxis in individuals at high risk of type I diabetes. Lancet 1993; 341:927-928.
72. Bluestone JA, Matthews JB. The immune tolerance network—an NIH/JDF-supported initiative to bring tolerance research into the clinic: A major new resource for clinical immunologists. Clin Immunol 2000; 96:171-173.
73. Zhang ZJ, Davidson L, Eisenbarth G et al. Suppression of diabetes in nonobese diabetic mice by oral administration of porcine insulin. Proc Natl Acad Sci USA 1991; 88:10252-10256.
74. Pozzilli P, Pitocco D, Visalli N et al. No effect of oral insulin on residual beta-cell function in recent-onset Type I diabetes (the IMDIAB VII). IMDIAB Group [In Process Citation]. Diabetologia 2000; 43:1000-1004.
75. Chaillous L, Lefevre H, Thivolet C et al. Oral insulin administration and residual beta-cell function in recent-onset type 1 diabetes: A multicentre randomised controlled trial. Lancet 2000; 356:545-549.
76. Kupila A, Muona P, Simell T et al. Feasibility of genetic and immunological prediction of type I diabetes in a population-based birth cohort. Diabetologia 2001; 44:290-297.
77. Kimpimaki T, Kulmala P, Savola K et al. Disease-associated autoantibodies as surrogate markers of type 1 diabetes in young children at increased genetic risk. Childhood Diabetes in Finland Study Group. J Clin Endocrinol Metab 2000; 85(3):1126-1132.
78. Alleva DG, Crowe PD, Jin L et al. A disease-associated cellular immune response in type 1 diabetics to an immunodominant epitope of insulin. J Clin Invest 2001; 107:173-180.

78a. Liu E, Moriyama H, Abiru N et al. Anti-peptide autoantibodies and fatal anaphylaxis in NOD mice in response to insulin self-peptides B:9-23 and B:13-23. J Clin Invest 2002; 110(7):1021-1027.
79. Abiru N, Maniatis AK, Yu L et al. Peptide and MHC specific breaking of humoral tolerance to native insulin with the B:9-23 peptide in diabetes prone and normal mice. Diabetes 2001; 50:1274-1281.
79a. Raz I, Elias D, Avron A et al. Beta-cell function in new-onset type 1 diabetes and immunomodulation with a heat-shock protein peptide (DiaPep277): a randomised, double-blind, phase II trial. Lancet 2001; 358(9295):1749-1753.
80. Tisch R, Wang B, Serreze DV. Induction of glutamic acid decarboxylase 65-specific Th2 cells and suppression of autoimmune diabetes at late stages of disease is epitope dependent. J Immunol 1999; 163:1178-1187.
81. Tisch R, Liblau RS, Yang X-D et al. Induction of GAD65-specific regulatory T-cells inhibits ongoing autoimmune diabetes in nonobese diabetic mice. Diabetes 1998; 47:894-899.
82. Elias D, Cohen IR. Treatment of autoimmune diabetes and insulitis in NOD mice with heat shock protein 60 peptide p277. Diabetes 1995; 44:1132-1138.
83. Ablamunits V, Elias D, Reshef T et al. Islet T cells secreting IFN-gamma in NOD mouse diabetes: Arrest by p277 peptide treatment. J Autoimmun 1998; 11:73-81.
84. Wegmann DR, Norbury-Glaser M, Daniel D. Insulin-specific T cells are a predominant component of islet infiltrates in prediabetic NOD mice. Eur J Immunol 1994; 24:1853-1857.
85. Daniel D, Wegmann DR. Protection of nonobese diabetic mice from diabetes by intranasal or subcutaneous administration of insulin peptide B-(9-23). Proc Natl Acad Sci USA 1996; 93:956-960.
86. Crawford M, Daniel D, Wegmann D et al. Autoimmune islet damage mediated by insulin-specific T cells. Transplant Proc 1997; 29:758-759.
87. Wucherpfennig KW, Eisenbarth GS. Type 1 Diabetes. Nat Immunol 2001; 2:1-3.

Theoretical Essay A

Type 1 Diabetes:
Molecular, Cellular and Clinical Immunology

George S. Eisenbarth

Theoretical essays provide an opportunity for "going" out on a limb, but hopefully not so far on the "limb" as to hang oneself. With recent information I will go further than I ventured in the first edition of this essay.

There is now compelling evidence in rats with the RT1-U MHC haplotype, and mice with the H-2d MHC haplotype and individuals with DR3/4, DQ2/8 haplotypes that developing diabetes is "EASY".[1-3] Given the right ("wrong") major histocompatibility genotype, induction of insulin autoantibodies, insulitis and even diabetes is a relatively common phenomenon. At the time of the first edition it was known that for BB rats and the related strain of DR (Diabetes Resistant BB rats that lack the lymphopenia gene) diabetes was dependent upon RT-U. Of interest was the infection of the DR strain with Kilham rat virus induced diabetes. Stimulation of immunity with poly-IC (poly-inosinic, cytidylic acid: a mimic of viral RNA that activates Toll receptors and induces interferon-alpha) was also sufficient to induce diabetes. It now appears that Kilham rat viral infection is similar to poly-IC induction and the virus does not infect islets. Many strains of "normal" rats that share RT1-U develop either insulitis or diabetes following poly-IC administration.[1]

In a similar manner, mice homozygous for H-2d (the MHC of the Balb/c mouse) with islet B7-1 expression (RIP-B7-1: Rat Insulin Promoter Transgene inducing beta cell expression of the T stimulatory molecule B7-1), following the administration of poly-IC develop diabetes.[4] In these mice there is a predominant CD8 T cell islet infiltrate. Even in the absence of B7 in islets, administration of the insulin peptide B:9-23 to normal Balb/c mice induces insulin autoantibodies,[3] and if combined with poly-IC, insulitis. The insulin autoantibodies are not absorbed by the peptide but react with intact insulin. Thus normal Balb/c mice have autoreactive B-lymphocytes reacting with intact insulin, and more important, administration of a specific peptide of insulin (other insulin peptides do not induce insulin autoantibodies) induces autoantibodies and insulitis. The "simple" combination of human DQ8 and the islet B7-1 transgene also leads to diabetes in mice.[5]

These observations lead to a series of hypotheses whose central component is autoimmunity directed against insulin with the concept that anti-islet autoimmunity (presumably anti-insulin) is "hard-wired" into the immune system. The "hard-wiring" is predominantly dependent upon MHC alleles (This is reminiscent of Irun Cohen's immunologic "homunculus"). One could substitute other islet antigens and other peptides for insulin for most of the discussion that follows, but as has been reviewed we favor insulin as a primary autoantigen.[6,6a] A large number of observations concerning the pathogenesis of type 1 diabetes that should guide pathogenic hypotheses are now available.[7,8] Beginning at the end of the process that leads to type 1A diabetes, it is clear that patients with type 1 diabetes and animals such as the

Immunology of Type 1 Diabetes, Second Edition, edited by George S. Eisenbarth.

nonobese diabetic (NOD) mouse produce autoantibodies and T cells reactive with a relatively large series of autoantigens including insulin, glutamic acid decarboxylase (GAD) and ICA512 (IA-2), etc. What is more striking is that T cell clones derived from NOD mice recognize multiple proteins and that different CD4 clones recognizing different antigens are able to transfer diabetes. This statement can be derived by combining studies of Haskins and coworkers on the initial BDC diabetogenic clones (e.g., BDC 2.5), Wegmann's studies of insulin and in particular insulin peptide B:9-23 reactive clones, as well as the clones of Wong, Serreze, and Santamaria.[8a] Both CD4 and CD8 clones reacting with different peptides and molecules can produce beta cell destruction and diabetes. For the NOD mouse the B cell repertoire at least as represented by high affinity autoantibodies is more limited and at the recent Snow Mountain Ranch workshop, only specific insulin autoantibodies were demonstrated.[9] Given so many different specificities for T cells in the NOD and for autoantibodies in man, the simplest hypothesis is that many clones arise subsequent to beta cell destruction and are "antigen" driven. It is also likely that at least a subset of clones are not secondary to islet destruction and may be the rate-limiting factor for development of type 1 diabetes. The clones isolated to date follow the classic rules of T cell recognition (e.g., trimolecular complex of major histocompatibility molecule, peptide and T cell receptor for recognition). There may however be some unique aspects to recognition of "pathogenic" self peptides. For example we have found a marked restriction of T cell receptor alpha chain usage for clones recognizing insulin peptide B:9-23 with no junctional conservation.[10] Overall, diabetogenic clones recognize peptides presented by classic MHC molecules, and can be deleted within the thymus by their cognate peptide.

Type 1A diabetes develops only in the setting of genetic susceptibility, but the genetic determinants vary. The major determinants are alleles of the major histocompatibility complex and in particular class II HLA alleles. It is however very likely that class I alleles will (as for a usual immune response) present peptides to cytotoxic T lymphocytes that will participate in islet destruction. There is some evidence that HLA-A2 and A24 enhances diabetes risk. In addition to HLA alleles increasing risk, a series of HLA alleles provide dominant protection. This is well illustrated by the NOD mouse, which have a "natural" mutation such that I-E is not expressed. Transgenic introduction of I-E prevents diabetes. For man DQA1*0102,DQB1*0602; DQA1*0201, DQB1*0303, and DRB1*1401 all appear to provide dominant protection.[11] Approximately 20% of patients from the general population have DQB1*0602, while less than 1% of children developing type 1 diabetes have this allele. DQB1*0602 does not prevent all autoimmunity and for instance is part of the high-risk haplotype for multiple sclerosis. Thus it is unlikely that DQB1*0602 has a global effect on autoimmunity, but rather its influence is likely to be specific for anti-islet T cell clones. Thus MHC alleles both protect and "cause" type 1 diabetes. One can speculate that these alleles, by the islet peptides that they bind and the type of immune response they engender (e.g., Th1 versus Th2, T regulatory cells, etc.) either increase or decrease the risk of type 1 diabetes.

In man, in addition to MHC alleles, polymorphisms of regulatory sequences and in particular a 3' variable nucleotide tandem repeat, of the insulin gene influence disease risk. Insulin is the only known beta cell specific islet autoantigen (e.g., GAD is expressed in non-beta cells of human islets). The insulin VNTR polymorphism associated with protection from diabetes is also associated with enhanced insulin messenger RNA production within the thymus. A significant number of cells both within the thymus and other lymphoid organs produce proinsulin.[12] Given this observation and the functional data from Hanahan's transgenics concerning Peripheral Antigen Expressing Cells,[13] it is hypothesized that insulin within the thymus alters diabetes risk.

In animal models and man there are multiple genes outside of the MHC and insulin locus that influence disease. Several "autoimmune genes" have been identified, including mutations in AIRE (Autoimmune Regulator for the APS-I syndrome; see Chapter 8) and the Scurfy gene of the XPID syndrome. Both of these syndromes are very rare with multiple autoimmune manifestations. A major effort is underway to define other loci and genes influencing diabetes risk, with two complementary hypotheses:

1. Type 1 diabetes is polygenic with multiple small influences of polymorphisms of many genes determining disease in each individual or
2. Type 1 diabetes is oligogenic but heterogeneous with a few major genes determining risk in individual families, but different genes for different families.

Both hypotheses may be true for different families, but until actual genes are better defined we will lack essential information. It is however very clear that the MHC locus on chromosome 6 provides the bulk of genetic susceptibility. It is a striking observation that approximately 50% of siblings of patients with type 1 diabetes with DR3/4, DQ2/8 genotype (and also HLA identical to their sibling with diabetes, thus sharing all the same class I and other MHC locus polymorphisms) develop islet autoantibodies prior to age 3, with the great majority progressing to diabetes by age 5.[3] Such a potent influence of this MHC genotype leaves little room for influence of non-MHC alleles in such children with the highest risk HLA genotype.

The above observations in man and mouse suggest that given specific HLA haplotypes the immune system is poised to destroy islet beta cells. Such destructive potential is probably a function of the islets possessing one or more peptides that T cells are either likely to target or unable to appropriately regulate an immune response directed at those peptides. For example, it appears that recognition of the B:9-23 peptide of insulin is dependent primarily upon the presence of a specific T cell alpha chain (AV13.3).[10] If recognition is not dependent upon the complex of TCR A and B chains and specific junctional sequences, one would expect a huge number of T cells able to recognize this peptide. In terms of protection from diabetes one can imagine MHC alleles that delete or "deviate" AV13.3 T cell receptors and thereby provide dominant suppression of disease. These self-reactive T lymphocytes are present in normal mice and presumably normal man, and in mice/rats can be activated with non-specific stimuli such as poly-IC. If this viral RNA mimic is able to induce disease in rodents, it is likely multiple relatively innocuous (and multiple different) viral infections may in a similar manner activate disease in man. This may explain the difficulty in isolating specific environmental factors responsible for triggering anti-islet autoimmunity.

Hypotheses are only useful if they can be tested. To begin to test the hypothesis that a single peptide of insulin may be central to disease, we are developing NOD mice lacking the two insulin genes,[6a] but with a transgene with an altered insulin sequence, namely an alanine at position 16 of the insulin B chain. We hypothesize that such an alteration will abrogate the development of diabetes if the specific sequence of the B:9-23 insulin peptide is essential. Of note this specific sequence has been modified to produce an altered peptide ligand of the B:9-23 molecule that is currently in clinical trials.

The Accompanying Figure Illustrates a View of Pathogenesis

MHC alleles of normal individuals allow the selection (positive and lack of negative) of T cells with receptors that recognize islet autoantigens. Given a number of potentially non-specific stimuli (e.g., poly-IC), such T cells are expanded and activated. At the same time regulatory T cells recognizing the same peptide (e.g., insulin peptide B:9-23) are selected, expanded and activated. Finally the balance of pathogenic and regulatory T cells, both potentially recognizing the same peptide, determines whether islet destruction begins. Once islet destruction begins many islet determinants are presented and pathogenic T lymphocytes recognizing a multitude of antigens produce enough beta cell destruction, over time (regulation is still active over time) to lead to diabetes. This hypothesis suggests that the prevention of type 1 diabetes is likely to be best accomplished in an antigen specific manner, potentially with vaccination with the very peptides central to pathogenesis. This is a hypothesis we are pursuing.

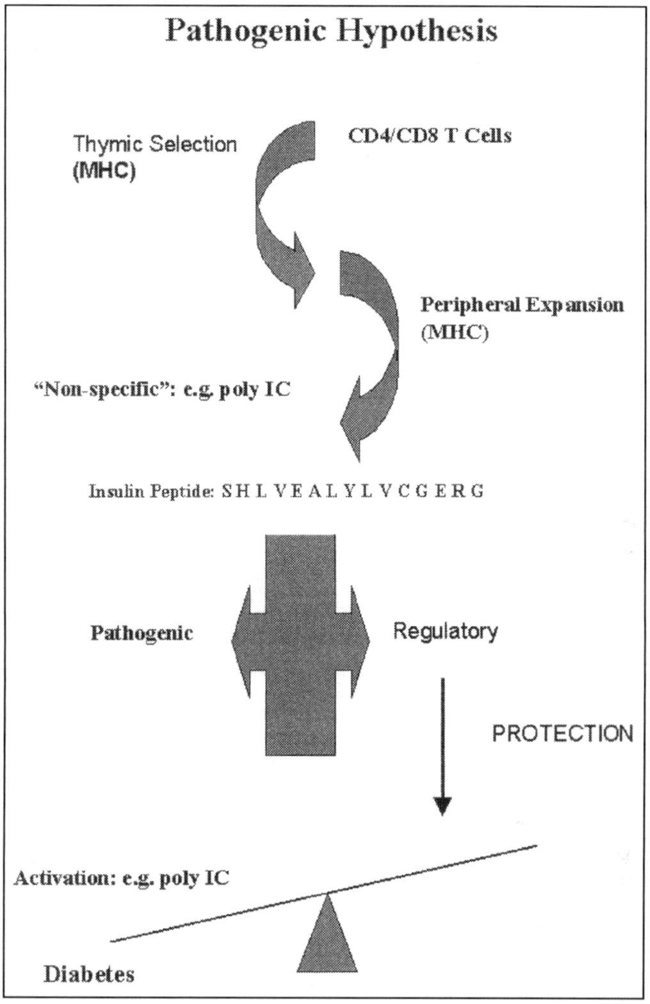

References

1. Ellerman KE and Like AA. Susceptibility to diabetes is widely distributed in normal class IIu haplotype rats. Diabetologia 2000; 43:890-898.
2. Abiru N, Maniatis AK, Yu L et al. Peptide and MHC Specific Breaking of Humoral Tolerance to Native Insulin with the B:9-23 Peptide in Diabetes Prone and Normal Mice. Diabetes 2001; 50:1274-1281.
3. Yu L, Robles DT, Abiru N et al. Early expression of antiinsulin autoantibodies of humans and the NOD mouse: evidence for early determination of subsequent diabetes. Proc Natl Acad Sci USA 2000; 97:1701-1706.
4. Moriyama H, Wen L, Abiru N et al. Induction and acceleration of insulitis/diabetes in mice with a viral mimic (Poly-IC) and an insulin self-peptide. Proc Natl Acad Sci USA 2002;
5. Wen L, Wong FS, Tang J et al. In vivo evidence for the contribution of human histocompatibility leukocyte antigen (HLA)-DQ molecules to the development of diabetes. J Exp Med 2000; 191:97-104.
6. Wegmann DR and Eisenbarth GS. It's Insulin. J Autoimmun 2000; 15:286-291.

6a. Moriyama H, Abiru N, Paronen J et al. Evidence for a primary islet autoantigen (preproinsulin I) for insulitis and diabetes in the NOD mouse. Proc Natl Acad Sci USA 2003; 100(18):10376-81.
7. Atkinson MA and Eisenbarth GS. Type 1 Diabetes: New Perspectives on Disease Pathogenesis and Treatment. Lancet 2001; 358(9277):221-229.
8. Wucherpfennig KW and Eisenbarth GS. Type 1 Diabetes. Nat Immunol 2001; 2:1-3.
8a. Lieberman SM, Evans AM, Han B et al. Identification of the β cell antigen targeted by a prevalent population of pathogenic CD8+ T cells in autoimmune diabetes. Proc Natl Acad Sci USA 2003; 100:8384-8388.
9. Bonifacio E, Atkinson MA, Eisenbarth GS et al. International workshop on lessons from animal models of human type 1 diabetes: analyzing target autoantigens of humoral immunity in nonobese diabetic mice. Am NY Acad Sci 2002; 958:1-2.
10. Simone E, Daniel D, Schloot N et al. T cell receptor restriction of diabetogenic autoimmune NOD T cells. Proc Natl Acad Sci USA 1997; 94:2518-2521.
11. Redondo MJ, Kawasaki E, Mulgrew CL et al. DR and DQ associated protection from type 1 diabetes: comparison of DRB1*1401 and DQA1*0102-DQB1*0602. J Clin Endocrinol Metab 2000; 85:3793-3797.
12. Pugliese A, Brown D, Garza D et al. Self-Antigen Presenting Cells Expressing Islet Cell Molecules in Human Thymus and Peripheral Lymphoid Organs: Phenotypic Characterization and Implications for Immunological Tolerance and Type 1 Diabetes. J Clin Invest 2001; 107:555-564.
13. Hanahan D. Peripheral-antigen-expressing cells in thymic medulla: Factors in self- tolerance and autoimmunity. Curr Opin Immunol 1998; 10:656-662.

Theoretical Essay B

Can Getting Excited Over Nothing Provoke Autoimmunity?

Donald Bellgrau

In this article I will present a hypothesis regarding the role of a diabetogenic gene in pro moting diabetes. Our concepts are formulated around experimental observations from our group regarding a novel T cell activation state found among a majority of peripheral T cells in the diabetes-prone Biobreeding rat. We hypothesize that the novel activation is promoted via an indirect effect wherein a diabetogenic gene creates a unique stimulus in the periphery, space.

The Biobreeding (BB) rat has provided an excellent animal model to study the causes and cures of type I diabetes (reviewed in this publication). Diabetogenicity is controlled by at least four disease susceptibility genes in the rat,[1] the most prominent in this species and in all diabetes-prone animals maps to the Class II region of the Major Histocompatibility Complex (MHC). The BB rat was derived from an outbred colony of Wistar rats.[2] Selective breeding lead to the development of two sublines, one diabetes-prone (DP) and one diabetes-resistant (DR). The BB-DP and BB-DR sublines appear to differ in only one disease susceptibility locus,[3] mapping to a syntenic region of rat chromosome four.[4-6] This (Ian gene) locus is commonly referred to as lyp for the phenotype that it produces, peripheral T cell lymphopenia.[7,7a] Congenic BB-DR rats bred for homozygosity at lyp demonstrate essentially 100% penetrance of the disease while non-lymphopenic BB-DR animals do not develop diabetes spontaneously;[4] with the proviso that the animals are maintained under specific pathogen free conditions. For example, infection with Kilham's rat virus can cause BB-DR animals to become diabetic.[8]

Lymphopenia is dramatic in the diabetes-prone BB rat (BB-DP). The peripheral T cell pool is diminished by as much as 80%. While both CD4+ and CD8+ T cells are affected the reduction is most dramatic among CD8+ T cells.[9] Lymphopenic animals bear a peripheral T cell pool that is virtually devoid of CD8+ T cells and the few remaining T cells of this phenotype express reduced levels of this co-receptor.[10] CD4 co-receptor and T cell receptor expression levels appear normal, however. The T cells from the lymphopenic animal also express low or undetectable levels (henceforth referred to as RT6-) of the RT6 rat differentiation antigen.[11] Reconstitution of BB-DP rats with RT6+ T cells from non-lymphopenic BB-DR donors protects the lymphopenic BB-DP recipient from disease.[12]

These observations indicated that the diabetogenicity of the lyp gene could be related to its effect on targeting RT6+ regulatory T cells. In the rat, Thy1 is expressed on immature T cells and is coordinately downregulated with the expression, post thymically of RT6. The frequency of Thy1+ T cells is markedly increased in BB-DP rat. However, the BB-DP RT6 gene is intact as the BB allelic product is expressed in a co-dominant fashion on T cells from BB-DP/Wistar F1 animals.[13] Therefore the Thy1+ RT6- phenotype that dominates the periphery of the BB-DP rat has been explained as due to a lyp gene mediated block in T cell development. RT6+ regulatory T cells never mature and the periphery of the lymphopenic animals is populated by

Immunology of Type 1 Diabetes, Second Edition, edited by George S. Eisenbarth.

RT6- immature T cells not regulated by RT6+ mature T cells. This immature, regulatory T cell deficient environment is favorable for the development of diabetes.

Therefore, contained within the RT6- T cells are at least some with autoreactive potential. Why do RT6- T cells with autoreactive potential exist? It is known that the deletion of high affinity anti-self reactive T cells occurs in the thymus during the negative selection process but that peripheral tolerance mechanisms also contribute to maintaining a T cell repertoire that is devoid of overt autoreactivity. Therefore lyp could be exerting its effect by interfering with either or both thymic and peripheral tolerance mechanisms. Peripheral tolerance mechanisms appear abnormal in the BB-DP. We defined one abnormality as an inability of T cells from the BB-DP to be tolerized following superantigen exposure in vivo.[14] The thymic infrastructure of the BB-DP rat also has been reported to be abnormal[15] and our laboratory reported a thymic dysfunction that mapped to thymic APC.[16]

Thymic and peripheral tolerance induction each involve the paradigm commonly described as the two signal model for T cell activation. A basic tenet of the model is that T cells are signaled not only by an interaction through the antigen specific T cell receptor (TCR) but also through various costimulatory interactions, provided most commonly by costimulatory molecules expressed on antigen presenting cells (APC).[17] It is widely held that signal one (TCR) without signal two (costimulation) can tolerize peripheral T cells either by anergizing them or causing them to be deleted. Thus, in the periphery signal one plus two is generally considered as an activating impetus while signal one alone is a tolerizing signal.[18] In the thymus, signal one and two appear to function differently compared to the periphery. Evidence indicates that signal one plus two when provided by thymic APC to immature T cells (CD4+CD8+ cortical thymocytes[19] and a subset of "single positive" more mature, medullary thymocytes[20]) initiates deletion or negative selection. The two signal model for negative selection in the thymus helps explain earlier observations proposing that negative selection appeared to be associated with the medulla rather than the cortical region of the thymus.[21] While it has been concluded that negative selection can occur in both cortical and medullary regions the perceived dominance of negative selection in the latter may simply reflect an increased frequency of thymic APC in this region.

If thymic negative selection and therefore tolerance is a two signal process in the thymus, while tolerance induction in the periphery is a one signal process, during T cell maturation it becomes possible that the same T cell can respond differently to identical signals. We hypothesized that defects in the way in which T cells respond to these two signals may permit them to escape tolerizing signals, thus promoting the survival of autoreactive T cells. To examine this we developed a two signal activation model in vitro to test T cells at various stages in development. A goal was to determine if T cells from lymphopenic animals responded similarly or not compared to T cells from non-lymphopenic animals. Our approach was to purify T cells such that endogenous sources of costimulation were absent. This could be accomplished by vigorous depletion of antigen presenting cells, resulting in a concomitant enrichment for T cells. Using the periphery as a source of lymphocytes we observed that this purified T cell population from normal rats underwent apoptosis if the TCR was engaged without provision of an exogenous source of costimulation. This was the predicted result based on published data using mice as the T cell source. Costimulation rescued the rat T cells from apoptosis and lead to progression into cell cycle followed by clonal expansion, also as predicted from studies performed with T cells from mice. However, when T cells from lymphopenic BB-DP or lymphopenic Fischer rats were purified and provided with costimulation the T cells were not rescued from apoptosis and did not clonally expand. Instead the stimulus lead to their death by apoptosis.[22] The provision of signal one alone was sufficient to push the T cells into cell cycle.[23] Such progression into cell cycle did not occur if the T cells were from non-lymphopenic donors. We ascribed the term promiscuous activation to this costimulation-independent progression into cell cycle of T cells from lymphopenic donors.

Costimulation-independent cell cycle progression could be attributed to antigen experienced T cells i.e., T cells that had undergone activation previously. However, T cells from lymphopenic animals did not express a cell surface phenotype consistent with activation. Activation markers such as OX40 and IL-2 receptor were absent from the cell surface of the vast majority of T cells from lymphopenic donors, unless TCR stimulation was provided. The QCA antigen, a resting T cell marker, was also expressed.[24,25] Therefore the costimulation-independent progression into cell cycle i.e., promiscuous activation, did not correlate with a cell surface phenotype commonly attributed to an activated T cell. An examination of cell cycle machinery was more informative. Here the phenotype was more consistent with the cell cycle progression data. The presumably "resting" T cells isolated from the animal and tested without further in vitro manipulation expressed low levels of the cyclin dependent kinase inhibitor p27kip.[23,26] Kip has been shown to act as the primary regulatory molecule controlling progression through the G1 restriction checkpoint in numerous cells including T cells.[27] Reduction in kip protein expression was accompanied by increased basal and serum stimulated gene transcription of the gene products cyclin D1, cyclin A and DFHR which are involved in the progression from G1 to S phase.[28]

Kip binding to cyclin prevents T cells from progressing from G1 into S phase and the ubiquitin dependent degradation of kip is held to be accomplished via the receipt of costimulation. As expected, T cells from normal, non-lymphopenic donors expressed high kip levels and were prevented from entering cell cycle without the provision of costimulation. Only in the presence of two signals was kip degraded and cell cycle progression initiated. The T cells from lymphopenic animals also expressed high levels of PCNA a proliferating cell nuclear antigen which functions as a requisite processivity factor for DNA polymerase.[29,30] Kip degradation is accompanied by upregulation of several cell cycle proteins that act as effectors downstream of kip in the coordination of cell cycle progression. One such molecule is Rb.[31] Rb has been shown to exist in an unphosphorylated, rapidly migrating 110kd species in resting T cells.[32] Upon activation two higher molecular weight species were observed that were found to contain multiple phosphorylation sites post-translationally modified by cyclin/cdk complexes that are active in G1 and G1/S phase transition.

We observed increased levels PCNA and hyperphosphorylated Rb, which we interpreted as additional indicators of activation events and cell cycle progression. Therefore the promiscuous activation phenotype, defined by entry into cell cycle without need of costimulation is associated with internal indicators of T cell activation defined by the molecular phenotypes of low Kip, high PCNA and activated Rb. We have coined the phrase "novel activation state" to identify T cells with characteristics of both resting (on the cell surface) but activated (in cell cycle machinery) phenotypes.

An examination of T cells at various stages of maturation indicated that the novel proliferative state originated at the recent thymic emigrant (RTE) stage. Medullary thymocytes but not RTE from lymphopenic donors could respond to costimulation and enter cell cycle comparably to their counterparts in non-lymphopenic animals. Medullary thymocytes from lymphopenic and non-lymphopenic donors also expressed comparable kip levels. The precipitous reduction in kip levels among peripheral T cells from lymphopenic versus non-lymphopenic donors only occurred when RTE were examined. This provided evidence that the promiscuous activation phenotype and novel activation state both had their origins as the medullary thymocytes were leaving the thymus and began entering the peripheral T cell pool.

We propose that promiscuous activation and the novel activation state are related. Cell cycle progression in T cells from lymphopenic donors occurs without the necessity of costimulation because the (normal) costimulation-dependent degradation of the cyclin dependent kinase inhibitor p27kip has already occurred. However, lyp does not appear to be targeting kip directly because kip levels are normal in both B lymphocytes and non lymphoid cells from lymphopenic donors. Also the murine kip gene does not map to the syntenic lyp locus.

If lyp is not directly involved in promiscuous T cell activation and the generation of the novel proliferative state then what is the nature of the stimulus that causes this activation to occur and what role does lyp play in this process that leads to the development of autoimmunity? We believe that the answers to these questions rest with another phenotype of T cells from lymphopenic animals. The majority appear to have an abnormally short life span, measured in days not weeks.[33,34] Therefore a widely held view of how the lymphopenic environment is created is that some medullary thymocytes[35] and most recent thymic emigrants die[36] before they have the opportunity to become long lived peripheral T cells. How this lyp gene mediated short life span relates to promiscuous activation and the novel activation state forms the basis around which our proposal is centered.

We reason that the direct effect of lyp involves the shortened life span of recent thymic emigrants and hypothesize that the shortened life span is caused by the lyp gene mediated death of recent thymic emigrants as they enter cell cycle and attempt to transit the G1 to S boundary. The impetus for entering cell cycle comes from two sources. First, we hypothesize that requisite peripheral survival signals (referred to recently as peripheral positive selection) provide an activation impetus. Recently, peripheral T cell homeostasis has received attention. Long term peripheral T cell survival in the absence of conventional antigenic stimulation is dependent on continuous contact with the thymic selecting MHC and peptides.[37-42] Several reports have shown that the peptide ligands that mediate thymic positive selection control both peripheral T cell survival and homeostatic proliferation in the peripheral compartments of T cell lymphopenic hosts, in the absence of conventional antigenic stimulation.[41,43,44] The proliferation that accompanied homeostatic expansion had two prerequisites, the first being expression of appropriate cognate MHC/peptide ligands and the second being T cell "space" in the periphery.[45] Lack of contact induces a short life span in both CD4+ and CD8+ T cell subsets.

However, these signals for survival need not lead to clonal expansion. We propose that the clonal expansion signals are provided by the indirect effect of the lyp gene. Lyp causes T cells to die in cell cycle arrest and thus creates the lymphopenic periphery. The lymphopenic periphery provides the stimulus for clonal expansion; space. It is the presence of space created by lymphopenia that signals the recent thymic emigrants to clonally expand in an attempt to fill up the void in the periphery.

The model for the diabetogenic lyp gene involvement is summarized as follows. The RTE are activated by peripheral positive selection. They attempt to divide as a result of the stimulus of space created by lymphopenia. The lyp gene aborts this process and the RTE die. The cycle continues. Peripheral positive selection provides a T cell survival signal. Lymphopenia provides a stimulus for clonal expansion and lyp perpetuates lymphopenia by interfering with T cell clonal expansion. In essence it is the excitement created by space that may provide an activation stimulus to generate autoreactivity and autoimmune disease.

References

1. Martin AM, Maxson MN, Leif J et al. Diabetes-prone and diabetes-resistant BB rats share a common major diabetes susceptibility locus, iddm4: additional evidence for a "universal autoimmunity locus" on rat chromosome 4. Diabetes 1999; 48:2138-2144.
2. Nakhooda AF, Like AA, Chappel CI et al. The spontaneously diabetic Wistar rat. Metabolic and morphologic studies. Diabetes 1977; 26:100-112.
3. Markholst H, Eastman S, Wilson D et al. Diabetes segregates as a single locus in crosses between inbred BB rats prone or resistant to diabetes. J Exp Med 1991; 174:297-300.
4. Jacob HJ, Pettersson A, Wilson D et al. Genetic dissection of autoimmune type I diabetes in the BB rat. Nat Genet 1992; 2:56-60.
5. Bieg S. Differential expression of p95vav in primary lymphoid tissue of BB rats congenic for the lymphopenia gene. Autoimmunity 1999; 30:37-42.
6. Hornum L, Markholst H. A sequence-ready PAC contig of a 550-kb region on rat chromosome 4 including the diabetes susceptibility gene Lyp. Genomics 2000; 69:305-313.

7. Jackson R, Rassi N, Crump A et al. The BB diabetic rat. Profound T-cell lymphocytopenia. Diabetes 1981; 30:887-889.
7a. Alper G, Jarjour IT, Reyes JD et al. Outcome of children with cerebral edema caused by fulminant hepatic failure. Pediatr Neurol 1998; 18:299-304.
8. Ellerman KE, Richards CA, Guberski DL et al. Kilham rat virus triggers T-cell-dependent autoimmune diabetes in multiple strains of rat. Diabetes 1996; 45:557-562.
9. Woda BA, Like AA, Padden C et al. Deficiency of phenotypic cytotoxic-suppressor T lymphocytes in the BB/W rat. J Immunol 1986; 136:856-859.
10. Bellgrau D and Lagarde A-C. Cytotoxic T-cell precursors with low-level CD8 in the diabetes-prone biobreeding rat: implications for the generation of an autoimmune T-cell repertoire. Proc Natl Acad Sci USA 1990; 87:313-317.
11. Greiner DL, Handler ES, Nakano K et al. Absence of the RT-6 T cell subset in diabetes-prone BB/W rats. J Immunol 1986; 136:148-151.
12. Rossini AA, Mordes JP, Pelletier AM et al. Transfusions of whole blood prevent spontaneous diabetes mellitus in the BB/W rat. Science 1983; 219:975-977.
13. Crisa L, Sarkar P, Waite DJ et al. An RT6 gene is transcribed and translated in lymphopenic diabetes prone BB rats. Diabetes 1993; 42:688-695.
14. Sellins KS, Gold DP, and Bellgrau D. Resistance to tolerance induction in the diabetes-prone biobreeding rat as one manifestation of abnormal responses to superantigens. Diabetologia 1996; 39:28-36.
15. Rozing J, Coolen C, Tielen FJ et al. Defects in the thymic epithelial stroma of diabetes prone BB rats. Thymus 1989; 14:125-135.
16. Georgiou HM, Lagarde AC, Bellgrau D. T cell dysfunction in the diabetes-prone BB rat: A role for thymic migrants that are not T cell precursors. J Exp Med 1988; 167:132-148.
17. Lafferty KJ and Cunningham AJ. A new analysis of allogeneic interactions. Aust J Exp Biol Med Sci 1975; 53:27-42.
18. Schwartz RH. Costimulation of T lymphocytes: the role of CD28, CTLA-4, and B7/BB1 in interleukin-2 production and immunotherapy. Cell 1992; 71:1065-1068.
19. Page DM, Kane LP, Allison JP et al. Two signals are required for negative selection of CD4+CD8+ thymocytes. J Immunol 1993; 151:1868-1880.
20. Kishimoto H and Sprent J. Negative selection in the thymus includes semimature T cells. J Exp Med 1997; 185:263-271.
21. von Boehmer H and Schubiger K. Thymocytes appear to ignore class I major histocompatibility complex antigens expressed on thymus epithelial cells. Eur J Immunol 1984; 14:1048-1052.
22. Moore JK, Gold DP, Dreskin SC et al. A diabetogenic gene prevents T cells from receiving costimulatory signals. Cell Immunol 1999; 194:90-97.
23. Moore JK and Bellgrau D. Promiscuous activation and cell cycle entry in T cells from autoimmune animals. Transplant Proc 1999; 31:1606-1610.
24. Groen H, Van Der Berk J, Nieuwenhuis P et al. Peripheral T cells in diabetes prone (DP) BB rats are CD45R-negative. Thymus 1989; 14:145-150.
25. Aspinall R and Kampinga J. A novel lymphocyte surface antigen recognized by the monoclonal antibody HIS45. Thymus 1989; 13:245-252.
26. Moore JK, Scheinman RI, and Bellgrau D. The identification of a novel T cell activation state controlled by a diabetogenic gene. J Immunol 2001; 166:241-248.
27. Sherr CJ and Roberts JM. Inhibitors of mammalian G1 cyclin-dependent kinases. Genes Dev 1995; 9:1149-1163.
28. Rivard N, L'Allemain G, Bartek J et al. Abrogation of p27Kip1 by cDNA antisense suppresses quiescence (G0 state) in fibroblasts. J Biol Chem 1996; 271:18337-18341.
29. Bravo R, Frank R, Blundell PA et al. Cyclin/PCNA is the auxiliary protein of DNA polymerase-delta. Nature 1987; 326:515-517.
30. Prelich G and Stillman B. Coordinated leading and lagging strand synthesis during SV40 DNA replication in vitro requires PCNA. Cell 1988; 53:117-126.
31. Weinberg RA. The retinoblastoma protein and cell cycle control. Cell 1995; 81:323-330.
32. Furukawa Y, DeCaprio JA, Freedman A et al. Expression and state of phosphorylation of the retinoblastoma susceptibility gene product in cycling and noncycling human hematopoietic cells. Proc Natl Acad Sci USA 1990; 87:2770-2774.
33. Zadeh HH, Greiner DL, Wu DY et al. Abnormalities in the export and fate of recent thymic emigrants in diabetes-prone BB/W rats. Autoimmunity 1996; 24:35-46.
34. Hernandez-Hoyos G, Joseph S, Miller NG et al. The lymphopenia mutation of the BB rat causes inappropriate apoptosis of mature thymocytes. Eur J Immunol 1999; 29:1832-1841.

35. Plamondon C, Kottis V, Brideau C et al. Abnormal thymocyte maturation in spontaneously diabetic BB rats involves the deletion of CD4-8+ cells. J Immunol 1990; 144:923-928.
36. Groen H, Klatter FA, Brons NH et al. High-frequency, but reduced absolute numbers of recent thymic migrants among peripheral blood T lymphocytes in diabetes-prone BB rats. Cell Immunol 1995; 163:113-119.
37. Kisielow P, Miazek A. Positive selection of T cells: rescue from programmed cell death and differentiation require continual engagement of the T cell receptor. J Exp Med 1995; 181:1975-1984.
38. Takeda S, Rodewald HR, Arakawa H et al. MHC class II molecules are not required for survival of newly generated CD4+ T cells, but affect their long-term life span. Immunity 1996; 5:217-228.
39. Kirberg J, Berns A, and von Boehmer H. Peripheral T cell survival requires continual ligation of the T cell receptor to major histocompatibility complex-encoded molecules. J Exp Med 1997; 186:1269-1275.
40. Brocker T. Survival of mature CD4 T lymphocytes is dependent on major histocompatibility complex class II-expressing dendritic cells. J Exp Med 1997; 186:1223-1232.
41. Rooke R, Waltzinger C, Benoist C et al. Targeted complementation of MHC class II deficiency by intrathymic delivery of recombinant adenoviruses. Immunity 1997; 7:123-134.
42. Viret C, Wong FS, and Janeway CA, Jr. Designing and maintaining the mature TCR repertoire: the continuum of self-peptide:self-MHC complex recognition. Immunity 1999; 10:559-568.
43. Goldrath AW and Bevan MJ. Low-affinity ligands for the TCR drive proliferation of mature CD8+ T cells in lymphopenic hosts. Immunity 1999; 11:183-190.
44. Ernst B, Lee DS, Chang JM et al. The peptide ligands mediating positive selection in the thymus control T cell survival and homeostatic proliferation in the periphery. Immunity 1999; 11:173-181.
45. Kieper WC, Jameson SC. Homeostatic expansion and phenotypic conversion of naive T cells in response to self peptide/MHC ligands. Proc Natl Acad Sci USA 1999; 96:13306-13311.

Theoretical Essay C

Pathogenesis of Type 1 Diabetes: A Viewpoint

Matthias G. von Herrath

Understanding the etiology and pathogenesis of human type 1 diabetes is a difficult task. The main issue is that we are probably dealing with a highly localized, slowly progressive inflammatory reaction that is mostly confined to the islets and pancreatic draining lymph node. Since the human pancreas is located retro-peritoneally, it is inaccessible and samples or live cells cannot be recovered easily and certainly not on a routine basis. Non-invasive imaging does not have the necessary resolution at this point to allow precise analysis of the cellular immunological process. Thus, our immunopathological knowledge of the human disease is limited and largely based on the assumption that animal models allow us to draw parallels. Therefore, it is important to recapitulate here, what we really know with certainty.

1. Autoantibodies: Autoantibodies to islet antigens can be detected accurately in serum samples from pre-diabetic individuals.[1-3] They predict the risk to develop disease well and antibodies to multiple β-cell antigens (insulin, GAD, others such as IA-2) are associated with a very high likelihood of clinical T1D. However, it is still controversial, whether such autoantibodies play a pathogenetic role.[4] Plasmapheresis in humans only temporarily clears islet-antibodies from the circulation and no effect on T1D development was consequentially observed.[5] Recent studies in the NOD mouse model suggest an instrumental role for maternal antibodies in initiating diabetes development in offspring.[6] In humans, the role of maternally transmitted islet antibodies is less clear, because no differentiation between maternal antibodies generated as a consequence of insulin therapy versus islet-autoimmunity was possible in previous studies.[2] However, no direct association has been established at this point and offspring from diabetic fathers carry a higher diabetes risk than those from diabetic mothers.[7] One could assume that such maternally transmitted antibodies would possibly encounter an immunologically active field, since lymphoid structures are routinely seen in fetal pancreata during development (P. Leenen et al, unpublished). Future investigations should focus on correlating antibody-isotypes and avidity with disease risk. This will bring increasing clarity to their possible pathogenetic role in particular in young children.[8]
2. T lymphocytes: Predominantly based on animal experimentation in the NOD mouse or transgenic models, T1D is thought to be a T cell mediated autoimmune disorder.[9] However, one has to acknowledge that infiltrates found in human islets are less T-cell rich than those present in NOD and, for example, RIP-LCMV mice.[10] Clearly, T cells (in contrast to autoantibodies) as well as clones with specificity for islet antigens can transfer disease in animal models consistently.[11] Comparable studies are of course impossible in humans and we must admit that no definite proof in respect to the importance of T lymphocytes has been obtained. However, indirect proof can be attributed to temporary success of systemic immunosuppressive regimens such as antibodies to CD4, CD3 and cyclosporine A.[12] Interestingly, certain human islet infiltrates exhibited a predominance of distinct Vβ T-cell

Immunology of Type 1 Diabetes, Second Edition, edited by George S. Eisenbarth.

receptor families, raising the possibility for an oligoclonal repertoire with defined antigen or super-antigen specificity.[13] Generally, autoreactive T cells are difficult to reliably detect, in particular in human peripheral blood. Even in NOD mouse spleens, there is a significant variability between colonies when comparing spontaneously arising islet specific T cells.[14] At this point one can therefore assume that numbers of autoreactive lymphocytes in T1D are low and that they are mostly localized in the pancreas and draining lymph node.[15] Improved tracking reagents such as tetramers and ELISPOT assays might elucidate their pathogenetic roles, because they include a more precise assessment of effector functions.[16,16a] These could then be correlated with progression and severity of the diabetogenic process on an individual basis. Current evidence from animal models indicates that autoreactive lymphocytes can have destructive or regulatory effector functions depending on which cytokines they produce.[15,17] One issue that is currently just emerging is to determine, which factors or cells provide the continuous drive for autoreactive T cells during a chronic autoimmune process that can take place over many years. Studies using adoptive transfers or one-time immunization with dendritic cells in animal models have shown that autoreactive T cells can easily run out of steam and that local inflammation might be required to initiate and maintain autoaggression.[18] It is therefore important to consider the role for antigen presenting cells in T1D.

3. Antigen presenting cells: Due to the relative paucity of lymphocytes, human islet infiltrates are exhibiting a higher degree of monocytes and possibly other APCs compared to mouse insulitis.[19] Their role is unclear, but it is an intriguing finding that activated APCs are detectable early in human pancreata. In animal models, there is no doubt that they are of critical importance for driving the local process and providing continued opportunity for autoreactive T cells to become activated. In support, one can cite numerous reports. For example, loss of APC or costimulatory function reduces diabetes incidence in NOD and other diabetes models, unless generation of regulatory cells are affected.[20] Conversely, adding B7 to the pancreatic equation' aggravates disease and, if expressed on β-cells, can precipitate spontaneous autoimmunity.[21] Once, destruction of some islet cells has occurred one has to assume that local APCs will increasingly present β-cell antigens. This leads to antigenic and epitope spreading, if non-tolerant T cells to the respective antigen are available in the periphery. Furthermore, one could hypothesize that activated APCs might be required to provide a 'fertile field' for the arrival of autoreactive lymphocytes, without which these would not be capable of causing chronic inflammation and disease. This consideration is relevant to the next topic discussed here. On the other hand, however, one must acknowledge that NOD mice as well as diabetic patients have distinct defects in APC activation and stimulatory capacity.[19] How can this be reconciled with the notion that they are required to drive autoaggression? This issue is unclear at this point, but it will be important to distinguish between local inflammation and systemic dysregulation to better understand experimental results.
4. Genetic and environmental triggers: The genetics of T1D are quite well understood. It is clear at this point that multiple genes can have predisposing as well as protective function resulting in a complex overall interaction. This is true for human[22,23] as well as NOD diabetes.[24,25] The MHC class II complex for sure plays a central role and it is intriguing that the NOD MHC class II that is required for spontaneous disease bears structural similarities to the human susceptibility allele HLA-DQ8[26,27] Importantly, overall contribution of genetic factors will not explain the etiology of disease alone, since there is a significant discordance in monozygotic twins.[28] Other environmental factors are therefore important, and many potential candidates are under evaluation, among them dietary cow's milk in young infants and viral infections. Viruses have been shown to influence the diabetic process in a positive or negative way in animal models.[29,30] In humans, no association has been proven, although Coxsackie and rota-viruses have been reported to precede autoantibody

sero-conversion in young infants.[31] Positive association could occur through mimicry, enhanced local inflammation and bystander activation. Negative association could involve apoptosis of aggressive lymphocytes and immune modulation. Very precise epidemiological studies will be required to establish a firm cause, it is well possible, however, that viruses contribute to T1D pathogenesis in a multifactorial way similar to genes.

5. Dysregulation: Does a possible systemic defect in immune regulation underlie or enhance the development of T1D? There is relatively good evidence from human studies that NK T cell activity, APC function as well as generation of CD25+ regulatory lymphocytes are all reduced in human diabetic patients.[32] This finding coupled with strong evidence from animal models that enhancing NK T cell activity,[33] systemic infections leading to immune activation or transfer of CD25+ cells[32] can prevent T1D, makes generalized immune dysregulation a likely underlying cause for T1D. Genetic factors, exposure to infections and other environmental influences might all be able to enhance or alleviate this systemic dysregulation. It is intriguing that most defects that have been found constitute a lack of certain immune functions including generation of molecules such as interferons by NK T cells that are known to be detrimental to islets. However, lack of NK T cell activity also results in less IL-4 production and lower degrees of APC activation might result in lower levels of CD25+ or TH2-like regulatory cells than needed to maintain a healthy immune equilibrium. Therapeutically, this situation can possibly be exploited, but one will have to proceed with caution, because any type of systemic correction will affect other immune functions such as host defense.

A hypothesis for T1D pathogenesis: A primary requirement for the development of islet autoimmunity is a non-tolerant peripheral repertoire containing lymphocytes that bear T cell receptors that can react with islet antigens.[34] This is not an uncommon occurrence, because thymic negative selection is not complete[35] and naïve autoreactive lymphocytes are present in peripheral lymphoid organs. This situation does normally not pose any problems and will not lead to autoimmunity, unless these cells become activated and gain access to the target organ.[34] I would like to propose that local events are more likely than systemic events to trigger T1D, because systemic activation of autoreactive lymphocytes requires large numbers of activated cells that are unlikely to be generated to a sufficient extent via molecular mimicry or bystander effects.[36] As a possible local precipitator, viral infections, APC dysfunction, or even maternal antibodies could all play a role. It is well possible that T1D is etiologically a quite heterogeneous disease and that all of the above will occur in conjunction with a varying genetic predisposition. Once a sufficient momentum of APCs that drive autoreactive lymphocytes has been reached in islets and draining lymph node, development of T1D is well on its way. As islet destruction progresses, more antigens will become targeted reflected in increased levels and varieties of islet autoantibodies. It is important that not all autoreactive lymphocytes will have detrimental effector functions and certain cytokines such as IL-4 or IL-10 can have positive effects,[15] if expressed constitutively at the right level and time prior to massive islet loss. This can be therapeutically exploited and maybe the ideal intervention would do a combination of both, augment regulatory responses and curb aggressive anti-islet activities. This could be achieved by combining systemic modulators such as anti-CD3[37] or Vitamin D3 analogs[37a] with islet-antigen specific DNA vaccines that induce regulatory lymphocytes.[38] Regardless of the etiology, the crucial effector pathways in islet destruction are well known and exhibit a significant amount of plasticity and redundancy. Cytotoxic T cells, type 2 interferon,[39] TNFα, IL-1βad nitric oxide generation are all detrimental to β-cells.[40] Since these can be produced by various effector lymphocytes including APCs and B cells, the above-mentioned redundancy would be expected. These considerations make effective late interventions harder to achieve. However, one must not forget, that successful islet destruction will rely on a critical mass of inflammation. Thus, acutely 'resetting' the system can brake the vicious cycle (for example by anti-CD3 or similar agents). Long term tolerance, however, should be maintained in an antigen specific way avoiding prolonged systemic immunosuppression.[38,41]

References

1. Pietropaolo M, Eisenbarth GS. Autoantibodies in human diabetes. Curr Dir Autoimmun 2001; 4:252-282.
2. Naserke HE, Bonifacio E, Ziegler AG. Prevalence, characteristics and diabetes risk associated with transient maternally acquired islet antibodies and persistent islet antibodies in offspring of parents with type 1 diabetes. J Clin Endocrinol Metab 2001; 86:4826-4833.
3. Bach J-F. Predictive medicine in autoimmune diseases: From the identification of genetic predisposition and environmental influence to precocious immunotherapy. Clin Immunol Immunopath 1994; 72:156-161.
4. Wrede G, Kiesel U, Freytag G et al. Chronic insulitis after partial islet damage by passive insulin antibody transfer. Horm Metab Res 1984; 16(1):97-101.
5. Seidel DK, Geiss HC, Donner MG et al. Course of islet autoantibody titers during Ig-immunoadsorption in a patient with newly diagnosed type 1 diabetes. J Autoimmun 1998; 11:273-277.
6. Greeley SAW, Katsumata M, Yu L et al. Elimination of maternally-transmitted autoantibodies prevents diabetes in nonobese diabetic mice. Nat Med 2002; In press.
7. Warram JH, Krolewski AS, Gottlieb MS et al. Differences in risk of insulin-dependent diabetes in offspring of diabetic mothers and diabetic fathers. N Engl J Med 1984; 311:149-152.
8. Rulli M, Simell O. Avidity of islet cell autoantibodies in non-diabetic children and children with insulin-dependent diabetes. Autoimmunity 1999; 31:187-193.
9. Bendelac A, Boitard C, Bedossa P et al. Adoptive T cell transfer of autoimmune nonobese diabetic mouse diabetes does not require recruitment of host B lymphocytes. J Immunol 1988; 141:2625-2628.
10. Jansen A, Homo-Delarche F, Hooijkaas H et al. Immunohistochemical characterization of monocytes-macrophages and dendritic cells involved in the initiation of the insulitis and beta-cell destruction in NOD mice. Diabetes 1994; 43:667-675.
11. Haskins K, McDuffie M. Acceleration of diabetes in young NOD mice with a CD4+ islet-specific T cell clone. Science 1990; 249:1433-1436.
12. Bach J-F. Insulin dependent diabetes mellitus as an autoimmune disease. Endocrine Reviews 1994; 15:516-542.
13. Conrad B, Weidmann E, Trucco G et al. Evidence for superantigen involvement in insulin-dependent diabetes mellitus aetiology. Nature 1994; 371:351-355.
14. Kaufmann SH, von Herrath MG et al. 2nd IDS NOD T-cell Workshop. Diabetes 2001; In press.
15. Homann D, Holz A, Bot A et al. Autoreactive CD4+ lymphocytes protect from autoimmune diabetes via bystander suppression using the IL-4/STAT6 pathway. Immunity 1999; 11:463-472.
16. Reijonen H, Daniels TL, Lernmark A et al. GAD65-specific autoantibodies enhance the presentation of an immunodominant T-cell epitope from GAD65. Diabetes 2000; 49:1621-1626.

16a. Anthony DD, Lehmann PV. T-cell epitope mapping using the ELISPOT approach. Methods 2003; 260-269.

17. Charlton B, Lafferty KJ. The Th1/Th2 balance in autoimmunity. Curr Opin Immunol 1995; 7:793-798.
18. Ludewig B, Odermatt B, Landmann S et al. Dendritic cells induce autoimmune diabetes and maintain disease via De Novo formation of local lymphoid tissue. J Exp Med 1998; 188:1493-1501.
19. Jansen A, van Hagen M, Drexhage HA. Defective maturation and function of antigen-presenting cells in type 1 diabetes. Lancet 1995; 345:491-492.
20. Lenschow DJ, Ho SC, Sattar H et al. Differential effects of anti-B7-1 and anti-B7-2 monoclonal antibody treatment on the development of diabetes in the nonobese diabetic mouse. J Exp Med 1995; 181:1145-1155.
21. de Seze J, Dubucquoi S, Lefranc D et al. IgG reactivity against citrullinated myelin basic protein in multiple sclerosis. J Neuroimmunol 2001; 117:149-155.
22. Todd JA, Wicker LS. Genetic protection from the inflammatory disease type 1 diabetes in humans and animal models. Immunity 2001; 15:387-395.
23. Merriman TR, Todd JA. Genetics of autoimmune disease. Curr Opin Immunol 1995; 7:786-792.
24. Todd JA, Aitman TJ, Cornall RJ et al. Genetic analysis of autoimmune type 1 diabetes mellitus in mice. Nature 1991; 351:542-547.
25. Wicker LS, Todd JA, Peterson LB. Genetic control of autoimmune diabetes in the NOD mouse. Annu Rev Immunol 1995; 13:179-200.
26. McDevitt H, Singer S, Tisch R. The role of MHC class II genes in susceptibility and resistance to type I diabetes mellitus in the NOD mouse. Horm Metab Res 1996; 28:287-288.
27. Sonderstrup G, McDevitt HO. DR, DQ, and you: MHC alleles and autoimmunity. J Clin Invest 2001; 107:795-796.

28. Redondo MJ, Rewers M, Yu L et al. Genetic determination of islet cell autoimmunity in monozygotic twin, dizygotic twin, and non-twin siblings of patients with type 1 diabetes: prospective twin study. BMJ 1999; 318:698-702.
29. Serreze DV, Ottendorfer EW, Ellis TM et al. Acceleration of type 1 diabetes by a coxsackievirus infection requires a preexisting critical mass of autoreactive T-cells in pancreatic islets. Diabetes 2000; 49:708-711.
30. Schwimmbeck PL, Dyrberg T, Oldstone MBA. Abrogation of diabetes in BB rats by acute virus infection. Association of viral-lymphocyte interactions. J Immunol 1988; 140:3394-3400.
31. Honeyman MC, Coulson BS, Stone NL et al. Association between rotavirus infection and pancreatic islet autoimmunity in children at-risk for type 1 diabetes. Diabetes 2000; 49:1319-1324.
32. Kukreja A, Cost G, Marker J et al. Multiple immuno-regulatory defects in type-1 diabetes. J Clin Invest 2002; 109:131-140.
33. Sharif S, Arreaza GA, Zucker P et al. Activation of natural killer T cells by alpha-galactosylceramide treatment prevents the onset and recurrence of autoimmune Type 1 diabetes. Nat Med 2001; 7:1057-1062.
34. von Herrath MG, Dockter J, Oldstone MBA. How virus induces a rapid or slow onset insulin-dependent diabetes mellitus in a transgenic model. Immunity 1994; 1:231-242.
35. Pugliese A, Brown D, Garza D et al. Self-antigen-presenting cells expressing diabetes-associated autoantigens exist in both thymus and peripheral lymphoid organs. J Clin Invest 2001; 107:555-564.
36. von Herrath MG, Coon B, Lewicki H et al. In vivo treatment with a MHC class I-restricted blocking peptide can prevent virus-induced autoimmune diabetes. J Immunol 1998; 161:5087-5096.
37. Chatenoud L, Thervet E, Primo J et al. Anti-CD3 antibody induces long-term remission of overt autoimmunity in nonobese diabetic mice. Proc Natl Acad Sci USA 1994; 91:123-127.
37a. Adorini L. 1,25-Dihydroxyvitamin D3 analogs as potential therapies in transplantation. Curr Opin Invest Drugs 2002; 3:1458-1463.
38. Coon B, An L-L, Whitton JL et al. DNA immunization to prevent autoimmune diabetes. J Clin Invest 1999; 104:189-194.
39. Thomas HE, Parker JL, Schreiber RD et al. IFN-gamma action on pancreatic beta cells causes class I MHC upregulation but not diabetes. J Clin Invest 1998; 102:1249-1257.
40. Seewaldt S, Thomas HE, Ejrnaes M et al. Virus-induced autoimmune diabetes: Most beta-cells die through inflammatory cytokines and not perforin from autoreactive (anti-viral) CTL. Diabetes 2000; 49:1801-1809.
41. Tisch R, Wang B, Serreze DV. Induction of glutamic acid decarboxylase 65-specific Th2 cells and suppression of autoimmune diabetes at late stages of disease is epitipe dependent. J Immunol 1999; 163:1178-1187.

Theoretical Essay D

Was There Type 1 Diabetes in the Olduvai Gorge?

David V. Serreze and Derry C. Roopenian

The T cell mediated autoimmune destruction of insulin producing pancreatic beta cells that leads to type 1 diabetes (T1D) in both humans and NOD mice is under complex polygenic control.[1,2] Insulin treatment, without which T1D is a lethal disease, was not developed until the 1920s. Furthermore, T1D often strikes individuals at a young pre-reproductive age. Hence, if genes predisposing to T1D were inherently deleterious, it might be expected that they would long ago have been eliminated by natural selection from the human gene pool. This has clearly not occurred since in many parts of the developed world the frequency of T1D is increasing.[3] The question then becomes why are genes contributing to T1D still maintained in the human gene pool? Perhaps the answer is that at some time these genes actually conferred a selective advantage to humans under environmental stresses differing from those of the present, and the induction of such alternative gene functions suppressed T1D development. One such beneficial function of "T1D genes" may be an ability to generate highly effective immunological responses against pathogens that are no longer commonly encountered by humans living in aseptically privileged societies. If this were true, then it is possible genes currently thought to only exert sinister diabetogenic effects were present in the earliest hominids residing in the Olduvai Gorge, but instead of eliciting T1D, they exerted other functions that allowed such individuals to ward off a wide range of infectious agents, and thus survive to become our ancestors.

Studies in the NOD mouse have provided insight to how particular genes might reciprocally contribute to T1D susceptibility or an ability to mount potent immunological responses to infectious agents. It appears that susceptibility genes both within and outside the major histocompatibility complex (MHC) interactively regulate the extent to which antigen presenting cells (APC) which are comprised of B-lymphocytes, macrophages, and dendritic cells can activate T cell responses.[2] T cells display different functional outcomes when achieving various activation thresholds. These functional outcomes include positive selection, induction of effector activity, or apoptotic deletion either in the thymus (negative selection) or in the periphery.[4-6] Intrathymic or peripheral deletion of T cells expressing a T cell receptor (TCR) that recognizes a peptide derived from a normal endogenous protein bound to a "self" MHC molecule on APC represents an important mechanism for preventing autoimmunity. The level of APC mediated T cell activation required to trigger such a deletional response is higher than that needed to induce positive selection, or effector function. Since they are relatively abundant compared to those from "foreign" pathogens, peptides derived from endogenous proteins are normally presented at sufficiently high levels to induce the deletion of any T cells that recognize them. However, if the combined effects of particular genetic variants decreases the stimulatory capacity of APC, this could preferentially diminish their ability to trigger the deletion, but not the positive selection or functional activation of autoreactive T cells. Indeed, the

Immunology of Type 1 Diabetes, Second Edition, edited by George S. Eisenbarth.

combined effects of genes both within and outside the MHC contribute to APC from NOD mice having a lower immunostimulatory capacity than those from control strains. This is likely to be an important factor in the development of autoreactive diabetogenic T cells in NOD mice.[2] MHC gene products likely contribute to decreased APC stimulatory capacity through their peptide binding and presentation functions. Non-MHC genes contribute to fewer APC reaching a full maturational state in NOD mice than in control strains, which could also lead to diabetogenic T cells not being activated to a deletional threshold. Defects in APC maturation have also been reported in human T1D patients.[7-9] While the presence of APC with a relatively low immunostimulatory capacity may have a dangerous component of impairing the deletion of autoreactive T cells, this situation could simultaneously confer a selective advantage of broadening the overall T cell repertoire allowing for resistance to a wider array of pathogens.

The advantage gained of broadening a pathogen responsive T cell repertoire by reducing the ability of APC to mediate deletional, but not effector mechanisms, would be negated if the autoreactive T cells generated as a byproduct of this process became functionally activated. Such a concern would be obviated if activation of pathogen reactive T cells eliminates or functionally suppresses the autoreactive effectors. Evidence that this may occur is provided by the fact that T1D development is strongly suppressed in NOD mice exposed to a wide range of microbial agents or other immunological stimuli.[10] Indeed, NOD mice survive infection with murine hepatitis virus, which is lethal to most strains, and at the same time are rendered T1D resistant.[11] The question is how this occurs? One possibility is that the array of cytokines induced in response to a pathogen infection, as well as Toll receptor stimulation, feeds back to APC in a way that now increases their stimulatory capacity to a level sufficient to trigger the deletion of autoreactive T cells, or perhaps also activate some other immunoregulatory mechanisms. As described above, since endogenous antigens would be relatively more abundant than those derived from pathogens, autoreactive rather than microbial reactive T cells would be the most likely to undergo deletion following an increase in APC stimulatory capacity. It should be noted that some infectious agents, in particular Coxsackie viruses, have been proposed to serve as a positive environmental trigger of T1D development in genetically susceptible individuals.[12,13] However, studies in NOD mice have indicated that a Coxsackie infection can only accelerate T1D onset if it occurs after a critical threshold level of autoimmune responses against beta cells have already developed.[14,15] Furthermore, a Coxsackie virus infection of NOD mice that have not yet developed significant levels of beta cell autoimmunity actually inhibits T1D development, perhaps through the mechanisms described above.[15]

The scenario described above for NOD mice could also partially explain why in humans there is a higher rate of T1D in sub-tropical rather than tropical populations, and that more advanced economic development is also associated with increased risk.[3,16,17] Of course, one contributing factor is that the frequency of certain MHC susceptibility variants, such as the DQ8 class II gene product, are enriched in high risk populations.[18] However, there is epidemiological evidence that when exposure to dangerous pathogens is high, what are now thought to be "T1D permissive" MHC variants also exert beneficial effects. This is illustrated by the fact that the Dutch colonists who survived the multiple tropical diseases of Surinam were preferentially those who inherited MHC alleles also associated with T1D susceptibility.[19] Such protection was not due to a founder effect. It was not determined whether these pathogen resistant Dutch settlers and their descendents developed T1D at a lower frequency than the relatively high rate characterizing MHC-matched individuals in their home country. However, given their subsequently high survival and reproductive rate, it is tempting to speculate that T1D was rare among the pathogen-exposed Dutch residents of Surinam that carried what are normally considered to be high risk MHC variants for this disease. If so, one contributing factor to the lower levels of T1D in tropical than sub-tropical environments might be that the former is more conducive to the outgrowth of pathogenic microbes, and this forces what would otherwise be T1D permissive MHC variants to exert an alternative beneficial function of warding off infection. Conversely, the more economically developed countries will usually possess advanced sanitation systems which might reduce exposures to microbial pathogens, and hence allow MHC variants that would normally protect against infections, to instead mediate deleterious autoimmune functions leading to T1D.

A further indication that T1D susceptibility genes do not represent inherently deleterious variants is increasing evidence they are actually common physiologically normal alleles which exert pathogenic functions only in certain combinatorial contexts. Jorn Nerup was the first to hypothesize this would be the case.[20] Supporting evidence has come from a study that was the first to determine the actual identity of a T1D susceptibility gene outside the MHC in NOD mice. This study found that two allelic variants of β2-microglobulin (β2m) differing by a single amino acid conferred susceptibility or resistance to T1D (21). However, neither variant represents an aberrantly functional mutant, as both dimerize with and induce normal expression of MHC class I molecules on the cell surface, but in a subtly different structural conformation which determines the extent to which they select diabetogenic T cells. Furthermore, while the pairing of particular β*f*2m and MHC class I allelic variants appears to be an important component of T1D development in NOD mice, this pairing is not inherently pathogenic since it also occurs in many other strains lacking autoimmune proclivity. Similarly, there is epidemiological evidence that when expressed in conjunction with particular class II alleles, the presence of common HLA-A2 MHC class I molecules (present in ~40% of Caucasians) can lead to a very rapid onset of beta cell autoimmunity and T1D.[22] This was supported by the finding that human HLA-A2.1 MHC class I molecules transgenically expressed in NOD mice can mediate autoreactive T cell responses that further accelerate T1D development.[23]

In conclusion, what we now think to be deleterious T1D susceptibility genes, may very well have been present among our earliest ancestors residing in the Olduvai Gorge. The fact that these genes were not subsequently removed by natural selection processes that would be expected to occur if their only function was to induce what was an early lethal disease throughout most of human evolution, indicates they actually have other primary functions. These alternative gene functions are hypothesized to entail a loosened selectional process that allows for development of a broadened repertoire of pathogen reactive T cells enabling the individuals carrying them to survive what would otherwise be lethal infections. The development of autoreactive T cells would be a byproduct of such a loosened selectional process. This would not be a problem if the autoreactive effectors were kept in check by mechanisms induced through recurrent activation of pathogen reactive T cells. However, the autoreactive T cells could elicit dangerous outcomes, including T1D development, if pathogen exposure was reduced through migration, improved sanitation, or other means. This scenario of conferring beneficial functions under environmental stresses that have greatly lessened in modern times, could explain why allelic variants currently defined as conferring susceptibility to T1D remain in the human gene pool, and now exert pathogenic rather than life saving activities.

References

1. Todd JA, Wicker LS. Genetic protection from the inflammatory disease type 1 diabetes in humans and animal models. Immunity 2001; 15: 387-395.
2. Serreze DV, Leiter EH. Genes and pathways underlying autoimmune diabetes in NOD mice. In: von Herrath MG, ed. Molecular Pathology of Insulin Dependent Diabetes Mellitus. New York: Karger Press, 2001:31-67.
3. Patterson CC, Dahlquist G, Soltesz G et al. Is childhood-onset type 1 diabetes a wealth related disease? An ecological analysis of European incidence rates. Diabetologia 2001; 44(1):9-16.
4. Jameson SC, Bevan MJ. T cell selection. Curr Opin Immunol 1998; 10:214-219.
5. Sebzda E, Mariathasan S, Ohteki T et al. Selection of the T cell repertoire. Annu Rev Immunol 1999; 17:829-874.
6. Abbas AK. Die and let live: Eliminating dangerous lymphocytes. Cell 1996; 84:655-657.
7. Jansen A, van Hagen M, Drexhage HA. Defective maturation and function of antigen-presenting cells in type 1 diabetes. Lancet 1995; 345:491-492.
8. Takahashi K, Honeyman MC, Harrison LC. Impaired yield, phenotype, and function of monocyte-derived dendritic cells in humans at risk for insulin dependent diabetes. J Immunol 1998; 161:2629-2635.

9. Litherland SA, Xie XT, Hutson AD et al. Aberrant prostaglandin synthase 2 expression defines an antigen-presenting cell defect for insulin-dependent diabetes mellitus. J Clin Invest 1999; 104:515-523.
10. Atkinson MA, Leiter EH. The NOD mouse model of type 1 diabetes: As good as it gets? Nature Med 1999; 5:601-604.
11. Wilberz S, Partke HJ, Dagnaes-Hansen F et al. Persistent MHV (mouse hepatitis virus) infection reduces the incidence of diabetes mellitus in non-obese diabetic mice. Diabetologia 1991; 34:2-5.
12. Rewers M, Atkinson M. The possible role of enteroviruses in diabetes mellitus. In: Rotbart HA, ed. Human enterovirus infections. Washington DC: Library of Congress, 1995:353-384.
13. Ramsingh AI, Chapman N, Tracy S. Coxsakieviruses and diabetes. Bioessays 1997; 19:793-800.
14. Horwitz MS, Bradley LM, Harbertson J et al. Diabetes induced by Coxsackie virus: Initiation by bystander damage and not molecular mimicry. Nature Medicine 1998; 4:781-785.
15. Serreze DV, Ottendorfer EW, Ellis TM et al. Acceleration of type I diabetes by a coxsackievirus infection requires a preexisting critical mass of autoreactive T cells in pancreatic islets. Diabetes 2000; 49:708-711.
16. Karoven M, Tuomilehto J, Libman I et al. A review of the recent epidemiological data on the worldwide incidence of type 1 (insulin-dependent) diabetes mellitus. Diabetologia 1993; 36:883-892.
17. Onkama P, Vaananen S, Karvonen M et al. Worldwide increase in incidence of type 1 diabetes-the analysis of data on published incidence trends. Diabetologia 1999; 42:1395-1403.
18. Thorsby E, Undlien D. The HLA associated predisposition to type 1 diabetes and other autoimmune diseases. J Ped Endocrin Metab 1996; 9:75-88.
19. van Rood JJ, Claas FHJ, de Vries RRP. HLA, plagues and people. In: de Vries RRP, Cohen IR, van Rood JJ, eds. The role of micro-organisms in non-infectious diseases. London: Springer-Verlag, 1990:1-5.
20. Nerup J, Mandrup-Poulsen T, Helqvist S et al. On the pathogenesis of IDDM. Diabetologia 1994; 37(2):82-89.
21. Hamilton-Williams EE, Serreze DV, Charlton B et al. Transgenic rescue implicates β2-Microglobulin as a diabetes susceptibility gene in NOD mice. Proc Natl Acad Sci USA 2001; 98:11533-11538.
22. Robles DT, Eisenbarth GS, Wang T et al. Identification of children with early onset and high incidence of anti-islet autoantibodies. Clin Immunol 2002; 102:217-224.
23. Marron MP, Graser RT, Chapman HD et al. Transgenic analyses in NOD mice provide direct evidence for the mediation of diabetogenic T cell responses by human HLA-A2.1 MHC class I molecules. Proc Natl Acad Sci USA 2002; 99:13753-13758.

Index

A

B

C

D

H

I

J

K

L

M

N

O

P

R

S

T

V

W

X